International Conference on Sustainable Systems and Structures (ICSSS 2019)

IOP Conference Series: Materials Science and Engineering
Volume 1025

Vizianagaram, India
15 - 16 November 2019

ISBN: 978-1-7138-3325-3
ISSN: 1757-8981

Printed from e-media with permission by:

Curran Associates, Inc.
57 Morehouse Lane
Red Hook, NY 12571

Some format issues inherent in the e-media version may also appear in this print version.

Institute of Physics
Dirac House, Temple Back
Bristol BS1 6BE UK

Phone: 44 1 17 929 7481
Fax: 44 1 17 920 0979

techtracking@iop.org

Additional copies of this publication are available from:

Curran Associates, Inc.
57 Morehouse Lane
Red Hook, NY 12571 USA
Phone: 845-758-0400
Fax: 845-758-2633
Email: curran@proceedings.com
Web: www.proceedings.com

TABLE OF CONTENTS

DURABILITY OF CONCRETE

RELIABILITY, OPTIMISATION AND COMPUTER APPLICATIONS

SUSTAINABLE ADDITIVES FOR SOIL STABILIZATION

SUSTAINABLE CONSTRUCTION MATERIALS

SUSTAINABLE INFRASTRUCTURE DEVELOPMENT

WATER RESOURCE MANAGEMENT

IOP Conf. Series: Materials Science and Engineering 1025 (2021) 011001 doi:10.1088/1757-899X/1025/1/011001

PREFACE ICSSS 2019

Proceedings of

International Conference on

SUSTAINABLE SYSTEMS AND STRUCTURES

15-16, November 2019

(Sponsored by AICTE, New Delhi)

INTRODUCTION

It gives me immense pleasure in writing the preface for the proceedings of The International Conference on Sustainable Systems and Structures (ICSSS'19) that was organised during November 15-16, 2019 by the Department of Civil Engineering, MVGR College of Engineering (Autonomous), Vizianagaram, 535005, Andhra Pradesh, INDIA. A proposal for GOC (Grant for Organising Conference) was submitted by myself Dr P Markandeya Raju, Professor, Department of Civil Engineering, MVGR in 2017. A matching grant of Rs. 5 Lakhs was sanctioned by AICTE, New Delhi ***(F. No. 67-20/RIFD/GOC/Policy-1/2017 dated 28-02-2019)***. The Conference was scheduled to be held in person and organised during November 15th - 16th 2019. The program included Five keynote lectures given by experts in various fields and 37 oral presentations given by renowned Professors, practising engineers and research scholars working in various areas of Sustainable Civil Engineering. All the five invited speakers delivered keynote lectures on topics of their expertise related to recent trends in sustainable Civil Engineering followed by brief interaction. Four technical sessions were organised during which authors made oral presentations followed by question and answer. One best paper was chosen from among the papers presented on each day based on the novelty, presentation and response to questions and the Best paper award were presented. The conference was held in the A/C Auditorium of MVGR. This preface presents an overview of this conference, details of Organising committee members, themes and sub-themes of the conference, review process adopted and schedule of the conference including the details of keynote lectures. An Annexure with details of the Management, College and the Department has been included at the end followed by Photo gallery of the conference.

ABOUT THE CONFERENCE

Civil Engineers are expected to develop safe, economical, serviceable, durable infrastructure and systems based on their knowledge and expertise. A lot of hard-earned tax payer's money is being invested by the government on sustainable structures and systems of public interest. Although the contractors and engineers ensure economy with quality and necessary specifications during construction, many hidden factors that affect the long-term performance and sustainability of structures. If the infrastructure has poor long-term performance, it will need frequent repairs and hence investment. The problem is more pertinent with structures in all the cities of India located in Coastal areas as they are exposed to the marine environment. Although many technologies and methods are available to ensure the durability of concrete structures, addressing this problem is twofold. Firstly, the engineers and contractors should be made aware of various methods of ensuring the durability and sustainability of proposed constructions. The other important point to be addressed is to identify and adopt suitable rehabilitation techniques to old constructions with durability related problems. Hence, Sustainability is the key component of future technology, innovations and development. Sustainable solutions for engineering and societal problems are the need of the hour as the resources available in the country are limited and the demand is rising. As Civil Engineers are expected to play a major role in developing the infrastructure of the country, they must look for sustainable solutions in the future. The conference (ICSSS'19) addresses the sustainable systems, solutions and their applications in the fields of Structural Engineering, Geotechnical engineering, Transportation engineering and Water resources engineering. It brings the young budding civil engineers, industry and the academic experts on one platform. The objective of the conference is to develop sustainable solutions for Civil

Engineering problems and initiate linkages for strengthening research, awareness and experience to Civil engineers.

ADVISORY COMMITTEE

CHIEF PATRONS
Sri. P Ashok Gajapathi Raju, Chairman, MANSAS,
Ms P Aditi Gajapathi Raju, Member, Trust Board, MANSAS

PATRON
Dr KVL Raju, Correspondent, MANSAS & Principal, MVGR

MEMBERS
Dr YMC Sekhar, Vice-Principal
Dr TVN Parthasarathi, Dean (Exams)
Dr Ch. Purnachandra Rao, Dean (A & E)
Prof. PS Sitharama Raju, Dean (IQAC)
Dr K Rajeswara Rao, Dean (Civil Infrastructure)
Prof. P. Ranga Raju, Dean (Admin)
Dr R Ramesh, Dean (Research & Development)
Dr M. Sunil Prakash, Dean (T & P)
Dr DR Prasada Raju, Dean (Faculty Develop)
Dr S. Mohan Kumar, Dean (Students)
Dr P. Ravindranadh, Dean (Strategic Planning)
Dr GVSSN Sannyasi Raju, Dean (Finance)
Dr S Chandramouli, Professor, Civil

TECHNICAL ADVISORY BOARD

INTERNATIONAL
Prof. Ali Kaveh
Editor-in-Chief for AJCE and IJOCE, Iran University of Science and Technology, Iran
Prof. Srinivas S Pulugurtha
Graduate Program Director, Civil and Environmental Engineering, UNCC, CHARLOTTE, USA
Prof. Ka-Veng Yuen, University of Macau, E11, Avenida da Universidade, Taipa, Macau, China
Prof. Bellie Sivakumar, The University of New South Wales, Australia
Prof. Srinivas Allena, Civil and Environmental Engineering, Washington State University, USA
NATIONAL
Prof. KVL Subramaniam, IIT Hyderabad
Prof. M. Amarnath Reddy, IIT Kharagpur
Prof. G. Appa Rao, IIT Madras
Prof. S. K. Barai, IIT Kharagpur
Prof. Debasis Roy, IIT Kharagpur
Prof. D. Ramaseshu, NIT Warangal
Prof. N V Umamahesh, NIT Warangal
Dr K. Rama Raju, Chief Scientist, CSIR-SERC
Prof. V Ravindra, JNTU Kakinada
Prof. D S R. Murthy, AU, Visakhapatnam
Prof. G Yesuratnam, JNTU Kakinada
Prof. KVGD Balaji, GITAM (Deemed to be) University
Mr Sandeep Mirmire, Principle ACS, Hyderabad

IOP Conf. Series: Materials Science and Engineering 1025 (2021) 011001 doi:10.1088/1757-899X/1025/1/011001

ORGANIZING COMMITTEE

CHAIRMAN
Dr P Markandeya Raju, Professor and Head, Department of Civil Engineering

ORGANIZING SECRETARY
Dr G Partheepan, Associate Professor, Department of Civil Engineering

COORDINATORS
Mr R P Singh; Mr V Vinay; Mr W Sai Deepak;

MEMBERS
Mr B Ramesh Raju; Dr R Maheswaran; Mr S Murali Sagar; Mr Ch V Ravi Sankar;
Mr A Varaprasad; Dr P Sudheer; Mr BV Joga Rao; Mr B Ramu; Mr TP Sreejani;
Mr S Purushotham Rao; Mr Kalyan AVS; Mr K Santosh Kumar; Mr SS Bhanu Sai Kumar;
Mrs T Jahnavi; Mrs D Praseeda; Mr B Jagadeesh; Mr G Rahul Reddy; Mr A Sai Kumar;
Mr BVSSR Bhaskar; Ms M Sai Priya;

CONFERENCE THEMES AND SUB-THEMES
STRUCTURAL ENGINEERING: •Design of Structures and Buildings for Sustainability • Characterization, Durability and Modelling behaviour of novel materials • Design of Energy Efficient Building • Sustainable use of construction materials • Recent Advances in sustainable construction practice • Repair, rehabilitation and retrofitting of structures using modern techniques and novel materials • Challenges in numerical modelling and design of structures • Service life assessment and Life cycle cost and durability of structures • Evaluation of in-situ properties of concrete structures • Non-destructive testing methods and its applications • Regulatory Framework and Government Initiatives for sustainable construction • Preservation of cultural heritage etc

SUSTAINABLE WATER RESOURCES SYSTEMS PLANNING & MANAGEMENT: •Integrated water resources management • Sustainable wastewater treatment methods • Sustainable groundwater development and management • Sustainable GIS applications for water resources management • Impact of climate change on water resources • sustainable solutions for food security

GEOTECHNICAL AND TRANSPORTATION ENGINEERING: • Handling of problematic soil • Sustainable materials and methods for soil stabilization • Transportation infrastructure and mobility • Sustainable Geo-environmental solutions using GIS • Usage of sustainable materials and systems for landfill and stabilization • Sustainable Urban transport solutions

REVIEW PROCESS
Over 100 papers were received by the organisers. The papers were received by mail (aicteicsss19@gmail.com), through post/courier and in person. 65 papers were accepted for presentation after review by experts. The review process adopted was as follows,

Step 1: The paper received was checked by Turnitin software for Similarity index. Papers with a similarity index less than 30% were sent for review.
Step 2: The review was conducted in two stages. The preliminary review was done by Dr P Markandeya Raju, The Chair of ICSSS'19. The papers were shortlisted based on relevance to the theme of the conference, the novelty of the work and presentation of content (quality of language, grammar, following formatting guidelines etc.).

IOP Conf. Series: Materials Science and Engineering 1025 (2021) 011001 doi:10.1088/1757-899X/1025/1/011001

Step 3: The shortlisted papers were sent for review (in blind - without details of authors) to one external reviewer depending on their expertise. The papers of authors who responded satisfactorily to the reviewer's comments were accepted for presentation.

The following reviewers have reviewed the papers.
1. Dr Bibhuti Bhusan Das, Associate Professor of Civil Engineering, NIT Surathkal, INDIA.
Theme: Sustainable Construction Materials, Structural Engineering
2. Dr Partheepan Ganesan, Associate Professor, MVGR College of Engineering (Autonomous) Vizianagaram, INDIA Theme: Sustainable Infrastructure Development, Structural Engineering
3. Dr S. Chandramouli, Professor, MVGR College of Engg. (Autonomous) Vizianagaram, INDIA
Themes: Water Resource Management, Environmental Sustainability, Reliability and Optimization.
4. Dr K. Rajeswara Rao, Professor, MVGR College of Engg. (Autonomous) Vizianagaram, INDIA
Themes: Sustainable Materials for Soil Stabilization, Geotechnical Engineering
5. Dr SSSV Gopala Raju, Professor and Director, IIT Nuzivid, INDIA
Themes: Durability of Concrete, Transportation Engineering and Highway Materials

In this context, it is appropriate to express my special gratitude to the reviewers for spending their valuable time despite their busy schedule for thoroughly reviewing the papers and for giving suitable comments that helped a lot in improving the quality of the papers selected for presentation at the conference and publication in the proceedings. Among the 65 papers presented in the conference, based on comments given by Chair and Co-chair of each session, some papers of high quality were shortlisted for publication in IOP Conference Proceedings: Materials Science and Engineering. The authors of these papers were allowed to write extended versions of these papers by improving the content reducing the similarity index (if possible) to less than 20%. Total 37 papers were submitted for inclusion in IOP Conference Proceedings: Materials Science and Engineering which included some invited papers from researchers on the theme of the conference received after the conduct of the conference. IOP Publishers accepted 35 papers for publication as a special issue in IOP Conference Series: Materials Science and Engineering, Volume 1025 with Dr. P Markandeya Raju as a Guest Editor. It is to benoted that this is a Scopus indexed journal and these papers were included in the proceedings after a thorough peer – review.

PROGRAM SCHEDULE

Over 100 delegates registered for the conference. Five keynote speakers delivered expert lectures and the accepted papers were presented in four technical sessions. The detailed program schedule was as follows.
About 56 papers were selected for presentation at the conference.

Time	Program Schedule
Day 01 – 15th November 2019	
8.45 AM – 9.25 AM Registration	
9.25 AM – 10.15 AM	**Inaugural function** Chief Guest: *Sri A Saibaba*, Chief Engineer –Indian Railways Guest of Honour: *Dr KVL Raju*, Correspondent MANSAS & Principal, MVGR
10.15 AM – 11.15 AM	Keynote Lecture on "Role of Civil Engineers in Developing Sustainable Infrastructure" by *Sri. A. Saibaba*, Chief Engineer-Indian Railways
11.15 AM – 11.30 AM High Tea	
11.30 AM – 12.30 AM	Keynote Lecture on "Precast system – Materials and Construction" by *Dr G. Appa Rao*, Professor of Civil Engineering IIT Madras.
12.30 AM – 1.30 PM Technical Session-1	
01.30 PM – 02.00 PM Lunch Break	
02.00 PM –	Keynote Lecture on "Developments in Road Infrastructure and Engineering

03.00 PM	in India" by *Dr M. Amarnath Reddy*, Associate Professor, IIT Kharagpur.
03.00 PM – 04.00 PM Technical Session-2	
Day 02 – 16th November 2019	
09.30 AM – 10.30 AM	Keynote Lecture on "Polavaram Project and Its impact on Socio-Economics Conditions" by *Sri. N. Jagan Mohan Rao*, Chief Engineer (Retired), North Coastal
10.30 AM – 10.45 AM Tea break	
10.45 AM – 12.30 PM Technical Session-3	
12.30 PM-01.30 PM	Keynote Lectures 1. Analytical and numerical methods for vibration control in Impact forging hammer system and surroundings. 2. Design and Distribution of Friction Dampers for Optimum Seismic Performance Enhancement of 20-Storey benchmark Steel Moment resisting frame. by *Dr K. Rama Raju*, Chief Scientist, Head-Vibration control group, CSIR-SERC, Chennai.
01.30 PM – 02.00 PM Lunch	
02.00 PM – 03.30 PM Technical Session-4	
03.30 PM-04.00 PM	**Valedictory**

The conference was sponsored by AICTE, New Delhi with a matching grant of 5 Lakhs sanctioned to Dr P Markandeya Raju, Professor and Head, Department of Civil Engineering, MVGR. I take this opportunity to sincerely thank AICTE for the grant. I express my sincere thanks to all the Organising committee members, Reviewers, participants, press and media for their support in successfully conducting this conference. Special thanks to publishers of IOP Conference Series: Materials Science and Engineering for accepting to publish the extended volume of selected papers submitted to ICSSS'19. I thank the almighty for letting me see this day.

ANNEXURE

ABOUT THE MANAGEMENT (MANSAS)
https://www.mvgrce.com/sites/default/files/2020-03/AboutMANSASTrust1.pdf

ABOUT THE COLLEGE (MVGRCE)
https://www.mvgrce.com/

ABOUT THE DEPARTMENT
https://www.mvgrce.com/department-civil-engineering

ABOUT THE CONFERENCE
https://sites.google.com/view/icsss-mvgr/home

Prof. Dr. P Markandeya Raju
Chair, ICSSS'19; Editor, Proceedings of ICSSS'19

CONTACT INFORMATION
Dr P Markandeya Raju,
Professor and Head of the Department,
Department of Civil Engineering,
MVGR College of Engineering (Autonomous)
VIZIANAGARAM – 535 005 Andhra Pradesh, INDIA
Office Phone: +91 08922 241199 Ext. 101,
Mobile: 9440528403 **WhatsApp:** 9177749097
Email: markandeyaraju@mvgrce.edu.in, hodcivil@mvgrce.edu.in

IOP Conf. Series: Materials Science and Engineering 1025 (2021) 011001 doi:10.1088/1757-899X/1025/1/011001

PHOTO GALLERY

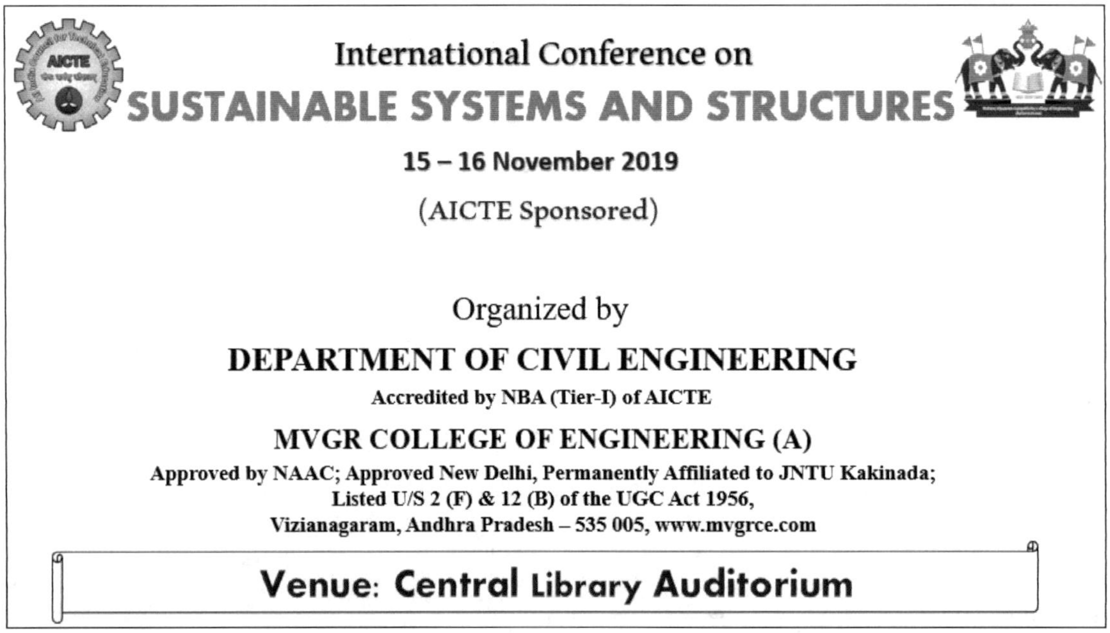

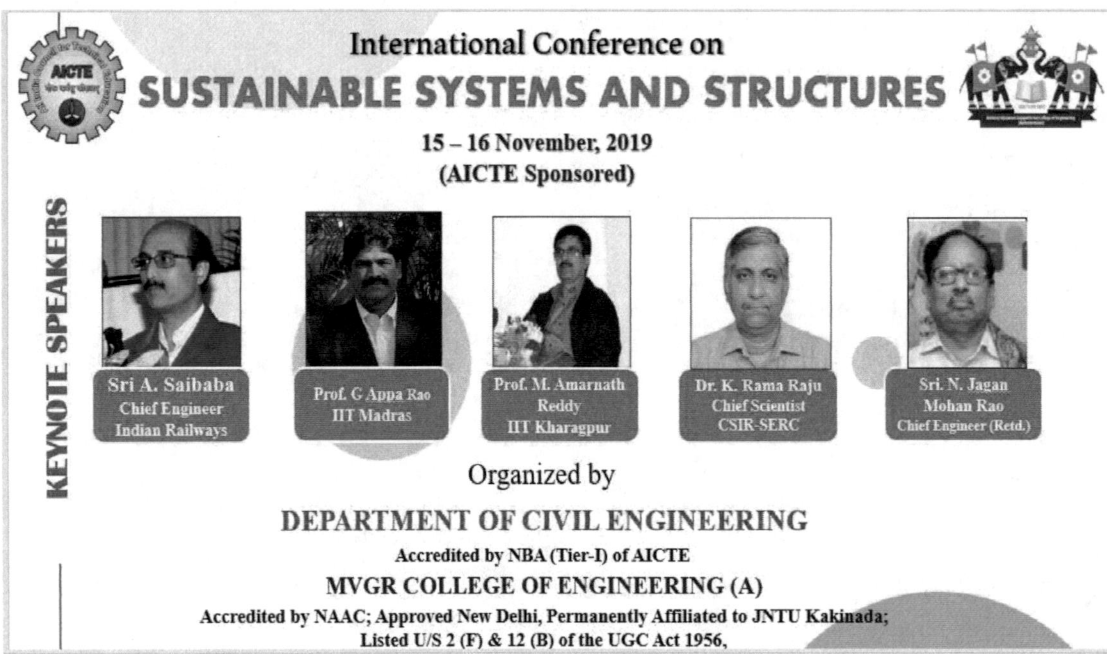

MVGR College of Engineering (Autonomous), Vizianagaram, INDIA

Department of Civil Engineering

College Library and A/c Auditorium (ICSSS'19 Venue)

Registration desk

Inaugural Function

SUSTAINABLE SYSTEMS AND STRUCTURES

SUSTAINABLE SYSTEMS AND STRUCTURES

IOP Conf. Series: Materials Science and Engineering 1025 (2021) 011001 doi:10.1088/1757-899X/1025/1/011001

Keynote speakers Day 1

Sri. Ankala Saibaba, Chief Engineer Indian Railways

Dr G Appa Rao. Professor IIT Madras

Dr Amarnath Reddy, Associate Professor IIT Kgp.

Some participants on Day 1 presenting their papers

IOP Conf. Series: Materials Science and Engineering 1025 (2021) 011001 doi:10.1088/1757-899X/1025/1/011001

Some of the participants on Day 1 receiving participation certificates

IOP Conf. Series: Materials Science and Engineering 1025 (2021) 011001 doi:10.1088/1757-899X/1025/1/011001

Other participants Day 1

Keynote speakers Day 2

Dr. K Rama Raju, Chief Scientist SERC Chennai

Sri. N Jagan Mohan Rao Chief Engineer (Retired), North Coastal Andhra Pradesh

Some participants on Day 2 presenting their papers

Some of the participants on Day 2 receiving participation certificates

Other participants Day 2

Best paper awards

Mr. P Sumanjali (Day 1) **(Day 2) Mr. B V Joga Rao**

IOP Conf. Series: Materials Science and Engineering 1025 (2021) 011001 doi:10.1088/1757-899X/1025/1/011001

PRESS COVERAGE

Experts and faculty members releasing a souvenir at a seminar on 'Sustainable Systems' at MVGR Engineering College in Vizianagaram on Friday

MVGR hosts seminar on Sustainable Systems

Vizianagaram: A two-day international conference on 'Sustainable Systems and Structures' was held at Department of Civil Engineering, MVGR College of Engineering (A). Indian Railways Chief engineer A Saibaba briefed the importance of these kind of conferences on sustainability and said the young students would be benefitted with this kind of meets as they transform knowledge into ideas. Principal Dr KVL Raju said that civil engineers would play a key role in development of sustainable systems as per the changing environmental requirements. Civil engineering HoD Dr P Markandeya Raju said the sustainability concept will address three ways namely, reuse of construction materials, minimum energy requirement for producing construction materials and increase the maximum life of structures. Dean Civil & Infrastructures Dr K Rajeswara Rao also shared the dais. Faculty members and students of department and over 50 participants from various colleges participated in the conference. About 100 research papers were submitted in the conference. Dr G Appa Rao, IIT Madras, Dr M Amarnath Reddy, IIT Kharagpur, N Jagan Mohan Rao, former Chief Engineer, APWRD and K Rama Raju, Chief Scientist, CSIR-SERC were also invited to deliver lectures on this two-day conference.

International Conference on Sustainable Systems and Structures

By Our Staff Reporter
Visakhapatnam, Nov 15

A two-day "International Conference on Sustainable Systems and Structures" has started at Department of Civil Engineering, MVGR College of Engineering A. Saibaba, Chief Engineer, Indian Railways was the chief guest of the inaugural function.

Addressing the gathering, Saibaba Ankala explained the importance of these kind of conferences on sustainability and with that younger students will be more beneficial by transferring the knowledge and ideas. He also added that it is an excellent opportunity from this conference to share about the sustainability to young minds.

Dr. KVL Raju, Principal, advised that civil engineers should play a key role to develop sustainable systems keeping in view of changes in environment due to population growth, technological advancements impact, so that they can create a better sustainable environment.

The Head of the Department, Dr. P. Markandeya Raju told that the sustainability concept is the need of an hour especially in civil engineering aspect and can address in 3 ways namely, 1. Reuse of construction materials, 2. Minimum energy requirement for producing construction materials and 3. Increase the maximum life of structures. Dean Civil & Infrastructures, Dr. K. Rajeswara Rao also shared the dias.

Dr. G Appa Rao, IIT Madras, Dr. M Amarnath Reddy, IIT Kharagpur, . N. Jagan Mohan Rao, Retd. Chief Engineer, APWRD and K Rama Raju, Chief Scientist, CSIR-SERC are also invited to deliver lectures on this two-day conference.All faculty members and students of the department, nearly 50 participants from various colleges all over India participated the conference. About 100 research papers were selected for the conference.

IOP Conf. Series: Materials Science and Engineering 1025 (2021) 011002 doi:10.1088/1757-899X/1025/1/011002

Peer review declaration

All papers published in this volume of *IOP Conference Series: Materials Science and Engineering* have been peer reviewed through processes administered by the Editors. Reviews were conducted by expert referees to the professional and scientific standards expected of a proceedings journal published by IOP Publishing.

Type of peer review: One Single blinded review by one of the appointed reviewers and other open review by the Conference Chair Dr P Markandeya Raju

Conference submission management system:

All papers were submitted to aicteicsss19@gmail.com

Number of submissions received: 65

Number of submissions sent for review: All submissions were sent for review

Number of submissions accepted: 37

Acceptance Rate: 56.92 %

(Number of Submissions Accepted / Number of Submissions Received X 100)

Average number of reviews per paper: 2

Total number of reviewers involved: 5

Any additional info on review process:

Refer PREFACE of the conference for detailed info on the review process.

Contact person for queries:

Dr P Markandeya Raju
Chair, ICSSS'19, Editor Proceedings of ICSSS'19
Professor and Head, Department of Civil Engineering
MVGR College of Engineering (Autonomous)
VIZIANGARAM 535001, AP, INDIA
Email: markandeyaraju@mvgrce.edu.in
Mobile: +91 9440528403

Crack width evaluation of Reinforced Concrete beam in flexure

Bimalendu Dash[1,2], P Markandeya Raju[3], Kapileswar Mishra[4]

[1]PhD Scholar, Centurion University of Technology and Management, Paralakhemundi, 761200 Odisha India
[2]Assistant Professor, GMR Institute of Technology, Rajam, 532127, Andhra Pradesh, India
[3]Professor and Head, Department of Civil Engineering, MVGR College of Engineering (Autonomous), Vizianagaram, 535005, India
[4]Professor, Centurion University of Technology and Management, Paralakhemundi, 761200 Odisha India
Email: bimalendudash.007@gmail.com, markandeyaraju@mvgrce.edu.in

Abstract. Crack width evaluation of flexural RC member is evaluated analytically by two significant formulas given by researchers and four standard code i.e. IS 456-2000, BSEN 1992-1-1-2004, ECP 203-2007, ACI 318-95 and compared. Parameters such as clear cover, the effect of diameter, variation of depth, number of bars with the same cross-section and increasing external moment are considered for computing the crack width. From the study of results of this study, it can be concluded that clear covers estimated based on IS 456 2000 are highest and those estimated based on EGYPTIAN CODE; 203-2007 are lowest as compared to other codes and mathematical models considered in this study. It can also be concluded that crack width increases with increase in clear cover or external moment while it decreases with increase in reinforcement diameter, depth of the beam or number of bars.
Keywords: Crack width, Crack Spacing, Flexural RC member, Steel Stress.

1. Introduction

Generally, crack width of cracks that develop during the loading of a member (flexural) should satisfy the serviceability criteria. The crack width of beams is evaluated when loading occurs at the tension or compression face of an RC beam. Due to low tensile strength, cracks occur in the RC beam. Control of cracking is important for obtaining long-term durability for concrete structures, especially for those that are subjected to aggressive environments. When the crack width is increasing, the service life of the structure will be decreasing by allowing more rapid penetration of chlorides to reach the reinforcement causing corrosion. Various factors such as high humidity, repeated loads, and gases with chemicals may cause corrosion of reinforcement and spalling of concrete. Cracking in reinforced concrete structures affects structural performance including stiffness, energy absorption, capacity etc.,

Consequently, there is an interest in the control of cracking by building codes and scientific organizations. Anyways, in some exceptional cases, we may check the crack width by some theoretical expressions. In some cases, shrinkage crack also may occur, we can reduce those cracks by adding extra reinforcement in reinforced concrete.

2. Objective of the work

The effect of a various parameter such as clear cover, the effect of diameter, variation of depth, number of bars with same cross-section and increasing external moment analytically determined crack width is studied for RC beam. The main objective of the paper is to compare the crack width of reinforced concrete beam as calculated by various codes of practices (i.e. IS: 456-2000[15], BSEN 1992-1-1-2004 [11], ECP 203-2007 [12], ACI 318-95 [13]) and two most commonly adopted formula proposed by literature.

3. An analytical method for crack width computation

3.1. Crack width prediction equations according to some building code provision

3.1.1. IS: 456-2000 [15] (See figure 1)

The crack width (W_{cr}) can be determined by the following formula

$$W_{cr}=\frac{3a_{cr}\varepsilon_m}{1+\frac{2(a_{cr}-c_{min})}{h-x}}$$

$$\varepsilon_m=\varepsilon_1-\frac{b(h-x)(a-x)}{3E_sA_s(d-x)}$$

$$\varepsilon_1=\frac{M\,x}{I_c(0.5E_c)}$$

$$I_C=\frac{bx^3}{3}+(m\,A_{st}(d-x)^2)$$

Where,

I_C= Moment of Inertia of the cracked section

W_{cr}= Crack width

a_{cr}=Distance considered from the point of the surface to the nearest longitudinal bar

ε_m= Average steel strain

C_{min}= Minimum cover of the longitudinal bar

$h(or)D$ = Overall depth of the member [2]

x= Depth of the neutral axis [2]

a=Distance from the compression face to point at which crack width is estimated [2]

E_S= Modulus of elasticity of steel [2]

$A_S(or)A_{St}$= Area of tension reinforcement

ε_1=Strain at the level considered, calculation by ignoring concrete in the tension zone

b = Breadth of beam

d=Effective depth

m = Modular ratio

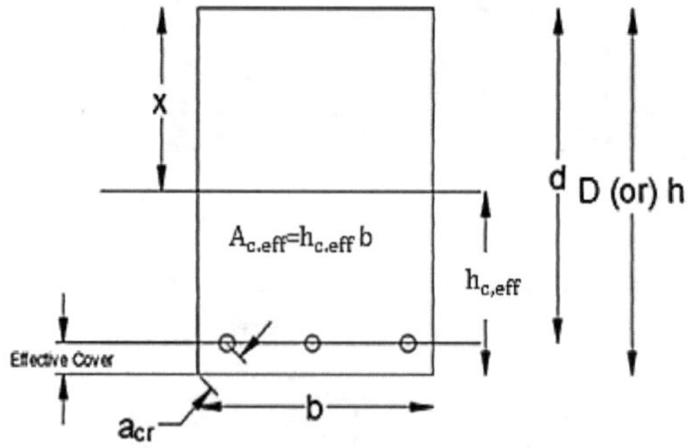

Figure 1. Cross-section of the beam.

3.1.2 BSEN 1992-1-1-2004 [11].

The equation for crack width as per EURO code

$$W_k=S_{r,max}(\varepsilon_{sm}-\varepsilon_{cm})$$

$$(\varepsilon_{sm}-\varepsilon_{cm})=\frac{(f_s-k_t(\frac{f_{ct,eff}(1+n\rho_{eff})}{\rho_{eff}})}{E_s}\geq 0.6\frac{f_s}{E_s}$$

IOP Conf. Series: Materials Science and Engineering 1025 (2021) 012001 doi:10.1088/1757-899X/1025/1/012001

$$n = \frac{E_s}{E_c}$$

$$\rho = \frac{A_s}{A_{c,eff}}$$

$$S_{r,max} = 3.4C + \frac{0.425\, k_1 k_2\, \emptyset}{\rho_{eff}}$$

Where,

W_k= Design crack width as per BS EN1992-1-1-2004 [11]

ε_{sm}=Mean steel strain mostly under the effect of tension stiffening or shrinkage 0

ε_{cm}= Mean strain in concrete

f_s= Stress in tension reinforcement

k_t= Factor that expresses the duration of loading

$f_{ct,eff}$= Mean value of tensile strength of the concrete

n=Modular ratio

$A_{c,eff}$=Effective tension area

k_1=Coefficient which considers the bond properties of bonded reinforcement

k_2= Coefficient strain distribution

Φ=Dia of bar

$S_{r,max}$=Average stabilized crack spacing

ρ_{eff}=Effective reinforcement ratio

E_s=Modulus of elasticity for steel

E_c= Modulus of elasticity for concrete

C=Cover provided at the longitudinal bar

A_s=Area of tension reinforcement

3.1.3 EGYPTIAN CODE; 203-2007 [12].

The formula for crack width as per Egyptian code is,

$$W_k = \beta \varepsilon_{sm} S_{rm}$$

$$\varepsilon_{sm} = \frac{f_s}{E_S}\left(1 - \beta_1 \beta_2 \left(\frac{f_{scr}}{f_s}\right)^2\right)$$

$$S_{rm} = 50 + 0.25 k_1 k_2 \Phi / \rho_{eff}$$

$$f_{scr} = \frac{mMd_c}{I_{cr}}$$

Where,

W_k=Design crack width as per EGYPTIAN CODE; 203-2007 [12]

E_S=Modulus of elasticity for steel

S_{rm}= Average stabilized crack spacing [3]

ε_{sm}=Mean steel strain mostly under the effect of tension stiffening or shrinkage 0

k_1= Coefficient which considers the bond properties of bonded reinforcement

k_2= Coefficient of strain distribution

β=*Coefficient which is related to average crack width*

β_1= Coefficient for bar bond characteristics

β_2= Coefficient which accounts load duration

f_{scr}= Stress in tension longitudinal reinforcement that causes under first crack stress in the tension reinforcement

f_s=Stress in tension reinforcement

Φ=Dia of bar

ρ_{eff}=Effective reinforcement ratio

d_c=Effective cover

M= Applied movement

m= Modular ratio

I_{cr}= Moment of inertia for cracked section

3.1.4 ACI 318-95 [13].

The equation for crack width as per ACI code is,

$$W_{max}=0.011\beta f_S \sqrt[3]{d_c A_o} 10^{-3} \text{ mm},$$
$$\beta=\frac{h-x}{d-x}$$

Where,

W_{max}=Design crack width as per ACI 318-95 [13]

h= Overall depth

x= Depth of neutral axis

β= Coefficient relating the average crack width to the design value [3]

d_c=Effective cover

A_o=Area of concrete surrounding each reinforcing bar, ($A_0 = \frac{2d_c b}{3}$)

f_s= Stress in tension reinforcement

d=Effective depth

3.2 Other important formulas are given by researchers for crack width

3.2.1 GERGELY AND LUTZ (1968) [9].

As per Gergely, the crack width

$$W_S=0.011\sqrt[3]{\left(c+\frac{\Phi}{2}\right)A_0}\,(f_S\text{-}34.45)\times 10^{-3}$$

Where,

W_S=Design crack width as per GERGELY AND LUTZ (1968) [9]

C_{min}= Minimum cover to the longitudinal bar[2]

f_s= Stress in tension reinforcement

Φ= Dia of bar

A_o=Area of concrete surrounding each reinforcing bar[2]

3.2.2 FROSCH (1999) [10]

As per Frosch, the equation of crack width,

$$W_{max}=2\frac{f_s}{E_s}d\beta$$

Where,

W_{max}=Design crack width as per FROSCH (1999) [10]

E_S=Modulus of elasticity for steel

f_S= Stress in tension reinforcement

d =Effective depth

β =Coefficient which is related to average crack width

4. Scope of the work

The limitation of crack width as per IS 456 2000 [15]for Normal and Aggressive environmental conditions are 0.1 mm and 0.3 mm respectively. The crack width as calculated based on expressions presented in section 3 for various clear cover, the effect of diameter, variation of depth, number of bars with the same cross-section and increasing external moment are presented in figure 2 to figure 6. Other details related to this parametric comparison are

Case 1: Effect of the clear cover

In this case, the clear cover is varied from 25 to 50mm with an incremental order of 5mm. However, the width is (b) 250mm, depth is (D) 400mm, area of reinforcement is (A_{st}) 3 numbers of 28mm diameter and moment is (M) 100kNm as considered for the comparative study based on different code provisions as well as models from literature.

Case 2: Effect of reinforcement diameter

In this case the diameter of reinforcement considered in increasing order as 20mm, 22mm, 25mm, 32mm, 36mm and 40mm with the geometrical properties like width (b) 250mm, depth (D) 400mm, clear cover of 25mm and also moment (M) 100kNm taken for the comparative study based on different code provisions as well as models from literature [9][13].

Case 3: Effect of depth variations for RC beam

In this case, the variation of depth is considered in ascending order as 400mm, 450mm 500mm, 550mm, 600mm, 650mm with width (b) 250mm, the clear cover of 25mm, area of Reinforcement (A_{st}) 3 numbers of 28mm diameter and moment (M) 100kNm are taken for the comparative study based on different code provisions as well as models from literature.

Case 4: Effect of increasing the number of bars for constant width

In this case the variation of the number of reinforcement bars considered in ascending order as 3 to 8 with an incremental order of 1. However, width (b) 250mm, the clear cover of 25mm, the diameter of reinforcement 28mm diameter and moment(M) 100kNm are taken for the comparative study based on different code provision as well as models from literature.

Case 5: Effect of an externally applied moment concerning the constant geometrical properties

In this case the variation of moment considered in ascending order as 100 to 600kNm with an incremental order of 100kNm. However, width (b) 250mm, clear cover of 25mm, area of reinforcement (A_{st}) 3 numbers of 28mm diameter are taken for the comparative study based on different code provisions as well as models from literature.

5. Results and Discussion

From the results, for the above 5 cases (figure) the following observation can be made [15].

1. The cases that are up to 0.1mm acceptable in Aggressive environment as per IS: 456-2000.
2. The cases that in between 0.1mm and 0.3mm are acceptable in Normal environment as per IS: 4562000.

Case 1: Effect of the clear cover

The effect of clear cover considering the crack width evaluation by various codes and models from literature in figure 2. From the figure 2, it is observed that the EGYPTIAN CODE; 203-2007 [12] gives the lowest value 0.1 mm, whereas the other codes give crack width in between 0.1 - 0.2 mm and are normal according to IS 456 2000 [15]. From figure 2 it is found that all the clear covers considered, IS 456-2000[15] shows higher crack width and EGYPTIAN CODE; 203-2007 [12] shows lowest as compared to other codes and formula consider in this study. Further, it is also observed from figure 2 that, when the clear cover increases the crack width also increases [4].

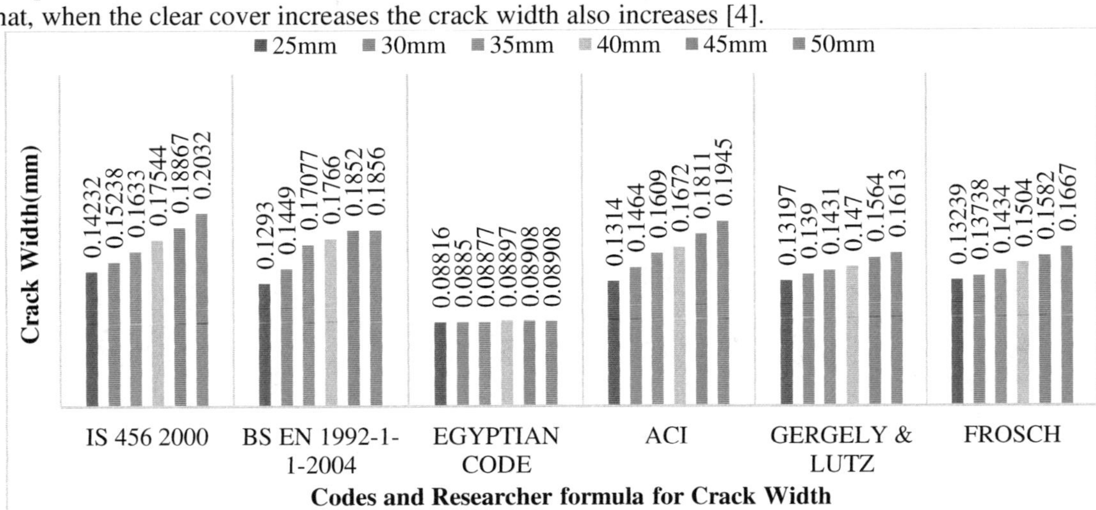

Figure 2. Crack Width Values for Effect of the clear cover

Case 2: Effect of reinforcement diameter

The effect of diameter of the reinforcement on the crack width evaluation analysis is presented in figure 3 [11]. From figure 3, it is observed that the crack width as per EGYPTIAN CODE; 203-2007 [12]

offered for 20mm diameter applicable for the normal and aggressive environment. Whereas for 22mm diameter the value of crack width for Egyptian code is less when compared to remaining codes and researchers. For 25mm, 32mm diameter the values of crack width for BS EN 1992-1-1-2004 [11] [14], Egyptian Code; ECP 203-2007 [12] are less when compared to remaining codes and researchers. For 40mcm diameter, the values of crack width for Egyptian code is giving less value among the remaining codes and researchers. However, it is also observed from figure 3 that, when the reinforcement diameter increases the crack width reduced.

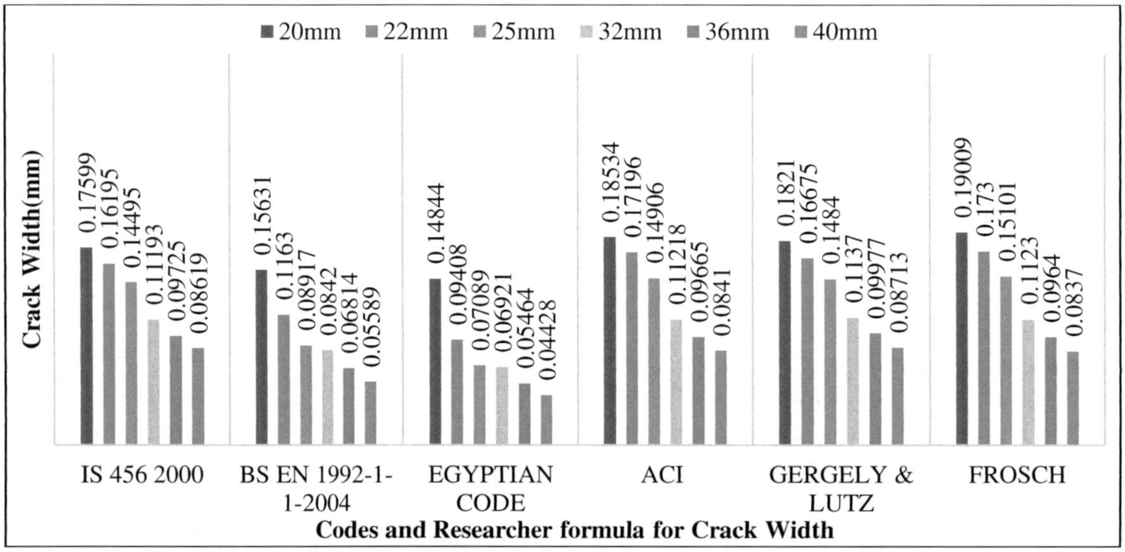

Figure 3. Crack Width Values for Effect of reinforcement diameter

Case 3: Effect of depth variations

The effect of depth variations consider for the crack width evaluation analysis is presented in figure 4. From the figure 4, it is observed that the EGYPTIAN CODE; 203-2007 [12] offered as aggressive in nature, whereas the other considered code prevising are given normal according to IS 456-2000 [15]. From the graph it is found that all the depth variation considered, Frosch (1999) [10] shown higher crack width and ACI 318-95 [13] shown lowest at a maximum depth of 650 mm as compared to other code consider in this study. However, it is also observed from figure 4 that, when the depth of the beam increases the crack width decreased.

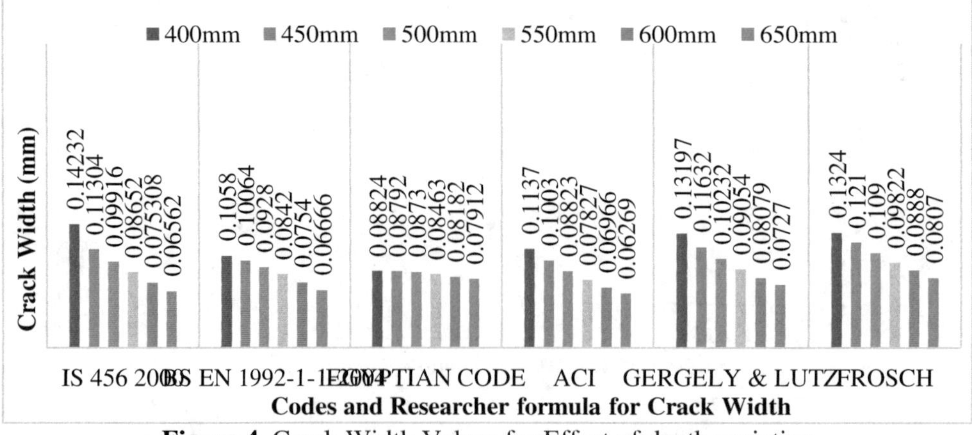

Figure 4. Crack Width Values for Effect of depth variation

Case 4: Effect of increasing the no. of bars to the constant width

The effect of increasing the no. of bars to constant width consider for the crack width evaluation analysis is presented in figure 5. From the figure 5, it is observed that the EGYPTIAN CODE; 203-2007 [12] offered as aggressive in nature, whereas the other considered code prevising are given normal according to IS 456 2000 [15]. From the graph it is found that all the increase in the no. of bars considered, IS 456 2000 [15] shown higher crack width and EGYPTIAN CODE; 203-2007 [12] shown lowest as compared to other code consider in this study. However, it is also observed from figure 5 that, when the number of bars increases the crack width decreased.

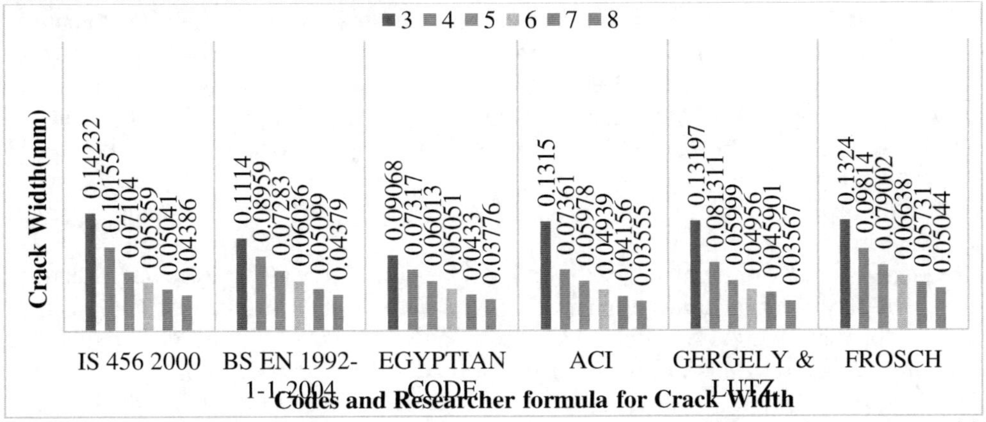

Figure 5. Crack Width Values for Effect of increasing the no. of bars to the constant width

Case 5: Effect of the externally applied moment to the constant geometrical properties

The effect of moment consider for the crack width evaluation analysis is presented in figure 6. From the figure 6, it is observed that the EGYPTIAN CODE; 203-2007 [12] offered as normal in nature, whereas the other considered code prevising are given normal according to IS 456 2000 [15]. Form graph it is found that all the moments considered, all the codes and researchers shown higher values. However, it is also observed from figure 6 that, when the moment increases the crack width also increased.

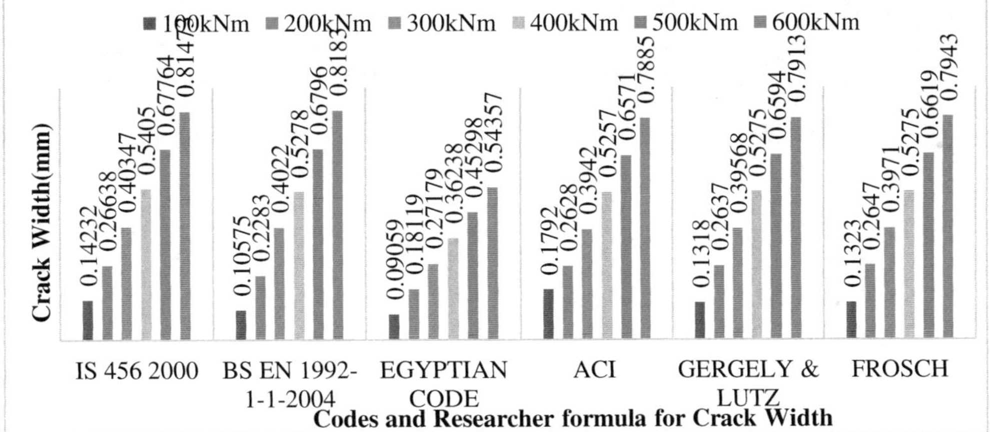

Figure 6. Crack Width Values Effect of the externally applied moment to the constant geometrical properties

6. Conclusion

The crack width evaluation of RC beam based on equations given by two researchers and standard codes of different countries for different cases like the effect of clear cover, reinforcement diameter, depth variation, increasing the number of bars and externally applied moment for the same cross-section are presented in this paper. From the study of results of this study, it can be concluded that clear covers estimated based on IS 456 2000 [15] are highest and those estimated based on EGYPTIAN CODE; 203-2007 [12] are lowest as compared to other codes and mathematical models considered in this study. It can also be concluded that crack width increases with increase in clear cover or external moment while it decreases with increase in reinforcement diameter, depth of the beam or number of bars.

References

[1] Said M. Allam Mohie S Shoukry Gehad E Rashad, Amal S Hassan 2012 Crack width evaluation for flexural RC members *Alexandria Engineering Journal* **51 (3)** p 211–20

[2] Šmejkal, Jiří and Jaroslav Procházka 2015 Calculation of the Minimum Reinforcement for Limitation of Crack Width According to Change of National Annex for ČSN EN 1992-1-1 *Advanced Materials Research* **1106** p 217-20.

[3] Afefy Hamdy Mohy El-Din, Mohamed Hussein Mahmoud 2014 Structural Performance of RC Slabs Provided by Pre-Cast ECC Strips intension cover zone *Construction and Building Materials* **65** p 103-13

[4] Ziari A 2009 Investigation of flexural cracking and leakage in RC liquid containing structures *Engineering Structures* **31 (05)** p 1056-67

[5] Chris G Karayannis, Constantin E Chalioris 2013 Design of partially prestressed concrete beams based on the cracking control provisions *Engineering Structures* **48** p 402-16

[6] Chau K, Albermani F 2003 A coupled knowledge-based expert system for design of liquid retaining Structures *Automation in Construction* **12 (5)** p 589-602.

[7] Huanzi Wang Abdeldjelil Belarbi 2011 Ductility characteristics of fiber reinforced concrete concrete beams reinforced with FRP rebars *Construction and Building Materials* **25 (5)** p 2391-401

[8] Said M Allam Mohie S Shoukry Gehad E Rashad Amal S. Hassan 2013 Evaluation of tension stiffening effect on the crack width calculation of flexural RC members *Alexandria Engineering Journal* **52 (2)** p 163-73

[9] Gergely Peter, Leroy A Lutz 1968 Maximum Crack Width in Reinforced Concrete Flexural Members *ACI Symposium Publication* **20** p 87–117

[10] Frosch R J 1999 Another Look at Cracking and Crack Control in Reinforced Concrete *ACI Structural Journal*, **96 (3)** p 437–42

[11] Euro Code 2: Design of Concrete Structures – Part 1: General Rules and Rules for Buildings 1992-1, *European Committee for Standardization*, October 2001, and Belgium.

[12] The EGYPTIAN CODE; 203-2007 for design and construction of Reinforced Concrete structures ECP 203-2007 *Ministry of Housing* Egypt

[13] ACI 318-95R-95 1995 Commentary on Building Code Requirements for Reinforced Concrete *American Concrete Institute* Detroit

[14] BS 8110: Part 2 1997 Structural Use of Concrete: Code of Practice for Special Circumstances *British Standard Institution* London

[15] IS 456 2000 Indian Standard code of practice - Plain and Reinforced Concrete (Fourth Revision) *Bureau of Indian Standards* New Delhi

Evaluation of corrosion resistance of corrosion inhibitors in concrete structures by Impressed voltage test

V L Satish[1], V Ravindra[2]

[1]Research Scholar JNTU College of Engg. Kakinada, KAKINADA, Andhra Pradesh, INDIA
[2] Professor of Civil Engineering JNTU College of Engg. Kakinada Andhra Pradesh, India
Email: isrosatish@gmail.com

Abstract. There are several commercially marketed corrosions inhibiting admixtures for concrete, presently not tested adequately and lack long term performance results. In this study laboratory studies were conducted on two commercially available Corrosion inhibitors to determine the effectiveness and their corrosion resistance parameters of concrete with OPC and PPC. This study examined the durability parameters in concretes mixed with Anodic Inhibitor (Calcium Nitrite based) and Bipolar Corrosion inhibitor (amino alcohol-based). The Accelerated corrosion testing (Impressed voltage test) was conducted to determine the efficacy of inhibitors. For PPC mixes, the Calcium Nitrite Inhibitor has shown better results in reducing the rate of corrosion of the embedded steel reinforcement. For OPC mixes both the inhibitors have given better results compared to control mix in reducing the corrosion rate of embedded steel.
Keywords: Durability, Chloride induced corrosion, Anodic Inhibitor, Bipolar Corrosion inhibitor, accelerated corrosion testing, Impressed voltage test.

1. Introduction

Numerous options are available and new technologies continue to make an impact in the field of concrete protection. Environmental loading plays an important role in corrosion of reinforced concrete structures induced by the presence of chlorides. Chloride induced corrosion is predominant where the environmental loading exceeds the resistance of the structure against corrosion. For the protection of concrete structures, the environmental loading can be reduced or increase the resistance of the concrete or do a combination of both.

The primary deterioration mechanism in chloride-induced corrosion affects the reinforcement and damages the protective layer around the reinforcement. The physical applications over the concrete surfaces like barrier coatings prevent or delay the entry of chlorides, oxygen, and dampness through the concrete cover to the rebars. Admixtures, sealing compounds, overlays and protective coatings on steel reinforcing bars are being used in practice now. Membrane coatings made with materials such as resins and epoxies etc. like anti-carbonation coatings are used to prevent the ingress of harmful agents into the concrete.

In recent times, the utilization of corrosion inhibitors has become a lucrative solution for preventing or delaying the corrosion associated damage in reinforced concrete structures.
There are several commercially marketed corrosion inhibiting admixtures for concrete available at present which have not undergone adequate testing and lack long term performance results to justify their use in important structures expected to serve for a longer period [1].

It has been proven that some admixtures, inhibit corrosion of the reinforcement in the chloride laden environment [2]. Calcium nitrate is one of the widely used corrosion inhibitors for concrete that is believed to be giving promising results [3]. Earlier studies focussed on various corrosion inhibition admixtures with the primary attention on Potassium chromate, Sodium nitrite, Sodium benzoate and stannous chloride [4]. Corrosion inhibitors based on their action are categorized as anodic inhibitors, catholic inhibitors and Ambiodic inhibitors.

Generally, the mechanism of an inhibitor can be explained in is one or more of the below-mentioned processes:

- The inhibitor is chemically adsorbed on the surface of the reinforcement and forms a protective layer with inhibitor effect
- The inhibitor forms a protective oxide layer on of the base metal;
- The inhibitor reacts with the corrosive component present in aqueous media and thereby forming a complex product.

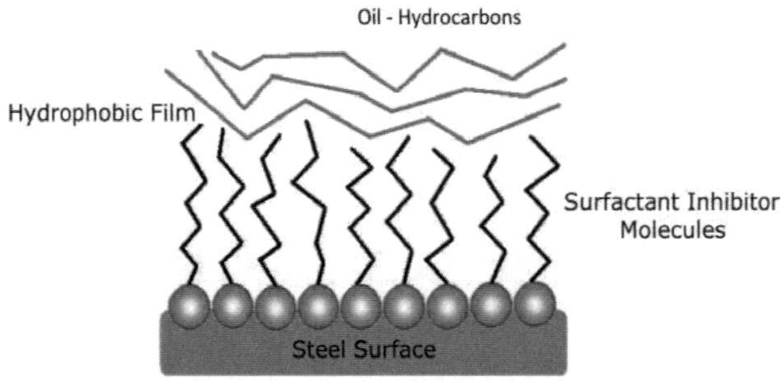

Figure 1. Mechanism of a conventional corrosion inhibitor

2. Materials and Methods

Two commercially available corrosion inhibitors, Anodic inhibitor (Calcium Nitrite based) and Bipolar Inhibitor mixed (Amino alcohol-based) in concrete made from Portland pozzolana cement (PPC) and Ordinary Portland cement (OPC) are studied in terms of modification of the workability and compressive strength of concrete.

2.1 Materials

M 35 grade concrete using OPC and PPC was used to assess the effect of corrosion inhibitors in terms of corrosion resistance. The properties of each of these materials used in the concrete are tested and used for mix design of concrete. The materials are tested in conformation with the IS code.

The materials used for this experimental work are:

- Cement: OPC & PPC
- Sand: Zone II
- Coarse aggregate: HBG 20 mm graded
- Water: Potable water
- Commercially available Anodic (Calcium nitrate based) corrosion inhibiting admixtures (CNIA)
- Commercially available Bipolar corrosion inhibiting admixtures (BIA)

2.1.1. Ordinary Portland Cement (OPC): 53 grade Ordinary Portland Cement conforming to IS 8112 is used in this study.

2.1.2 Portland Pozzolana Cement (PPC): PPC conforming to IS 1489 is used in this study

2.1.3 Anodic Corrosion inhibitors (Calcium Nitrite based): A commercially available corrosion inhibitor which works by anodic action Calcium Nitrite based inhibitor) was used in this study. This is a chloride-free aqueous solution of calcium nitrate. Calcium nitrate is a component that chemically reacts to stop the corrosion caused by chlorides. Nitrites (Calcium or sodium salt) comes under anodic inhibitors, as they form ferric oxide film at the anode by attacking the chloride ions [5]. A major study examining the effects of calcium nitrite revealed that calcium nitrite significantly increased the resistance to pitting corrosion which is the quality of an anodic inhibitor [6].

The test parameters of the corrosion inhibitors are given in table 1.

Table 1. Test parameters of the corrosion inhibitors

Sl. No	Test Conducted	Calcium Nitrite Inhibitor	Bi-Polar Inhibitor
1	Dry Material content (by % weight)	31.78	16.15
2	pH	6.03	8.17
3	Ash content (by % weight)	15.55	17.59
4	Relative density	1.18	1.1
5	Chloride (%)	Nil	Nil

2.2 Mix proportions

M35 grade concrete using OPC was used to assess the effect of corrosion inhibitors. Based on the properties of the ingredients, the mix was designed for M35 grade of concrete as per IS 10262:2009 [9]. The concrete specimens were cast to study the modification of the strength and durability parameters of concrete. The various mix proportions used in this study are given in table 2. The Calcium Nitrate Inhibitor is designated as CN and Bi-polar inhibitor is designated as BI.

Table 2. Details of Concrete mix proportions

Sl. No.	Specimen Combination	Cement	Coarse Agg		Fine Aggregate	Water	Inhibitor	Remarks
		kg	20mm kg	12mm kg	kg	Litre	Litre	
1	M35 Grade Concrete with (OPC)	360	690	460	710	167	0.00	OPC Control Mix
2	M35 OPC + CN2%	360	690	460	710	159.80	7.20	2% Inhibitor
3	M35 OPC + CN5%	360	690	460	710	149.00	18.00	5% Inhibitor
4	M35 OPC + BI2%	360	690	460	710	159.80	7.20	2% Inhibitor
5	M35 OPC + BI5%	360	690	460	710	149.00	18.00	5% Inhibitor
6	M35 Grade Concrete with (PPC)	390	676	482	740	172	0.00	PPC Control Mix
7	M35 PPC + CN2%	390	676	482	740	164.20	7.80	2% Inhibitor
8	M35 PPC + CN5%	390	676	482	740	152.50	19.50	5% Inhibitor
9	M35 PPC + BI2%	390	676	482	740	164.20	7.80	2% Inhibitor
10	M35 PPC + BI5%	390	676	482	740	152.50	19.50	5% Inhibitor

2.3 Workability and Strength tests:

The slump cone test was conducted on all the mixes for determining the workability. Concrete cubes of size 150 × 150 × 150 mm were cast for M35 grade concrete using OPC cement. Both the inhibitors viz the anodic inhibitor (Calcium Nitrite based) and the Bipolar inhibitor in dosages of 2% and 5% by weight of cement are mixed in the concrete mixes and cubes were cast. These dosages are adopted based on the literature review and the manufacturer's dosage range. The cubes are cured underwater and are tested for determining the compressive strength at 3 days, 7 days and 28 days. The Testing of specimens for compressive strength was carried out according to IS: 516-1959 [10].

2.4 Accelerated Corrosion Testing (Impressed Voltage Test (ASTM-C876)).

Impressed voltage technique is used for accelerating the corrosion process of reinforcing steel in the laboratory. The impressed current technique is in vogue for laboratory studies for monitoring the effect of reinforcement corrosion on the deterioration of concrete.

2.4.1 Apparatus and procedure for Impressed Voltage Test (ASTM-C876) [11]: The Impressed voltage test is based on the principle of electrochemical polarization. To evaluate the corrosion resistance under accelerated test conditions, concrete cylinders of size 75 mm diameter and 150 mm length with a steel reinforcement bar of 16mm diameter is embedded centrally into it (refer schematic in figure 2). After curing, the dried cylinder specimens are immersed in a 3% Na Cl solution. The steel bar projecting out of the cylinder at the top is connected to the positive terminal of the power pack (anode) and the stainless steel plate is connected to the negative terminal (cathode). A constant voltage of 6 volts from the D.C power pack is supplied to the circuit. The experimental setup is presented in figure 3. The applied voltage is maintained at a constant value and the variation of the current to time is monitored.

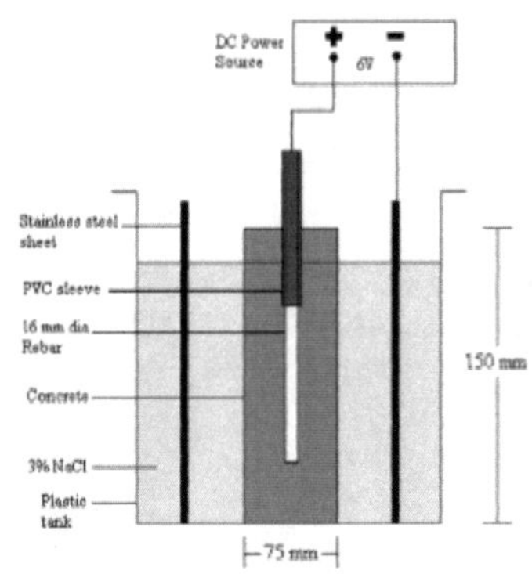

Figure 2. Schematic of Impressed voltage technique

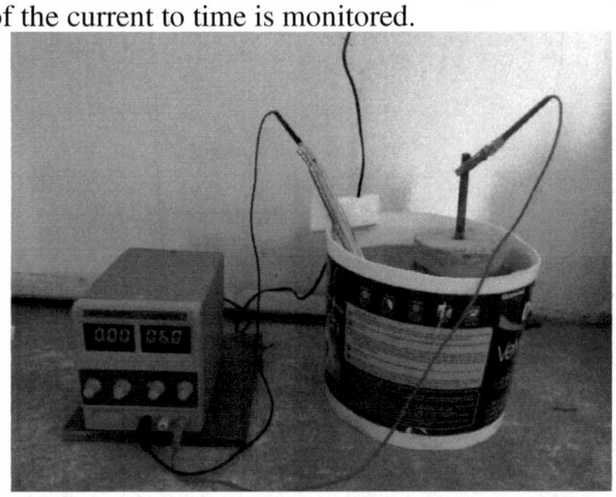

Figure 3. Experimental set up for impressed voltage test

3. Results and Discussion

The results of workability, compressive strength and accelerated corrosion test are presented in subsequent sections

3.1 Strength Parameters:

There is no improvement in workability by the addition of corrosion inhibitors in concrete. A decrease in workability is observed by the addition of both Calcium Nitrite inhibitor and Bipolar inhibitor compared to control mix concrete. The results of the compressive strength are shown in figure 4. A rheological study is required to find out reasons for the decrease in workability.

For OPC mixes, by addition of 2% Calcium Nitrite inhibitor, there is an increase in compressive strength by 10% at 28 days. However, by addition of 2% Bipolar inhibitor, there is a very marginal increase in compressive strength by 2.5% at 28 days. Addition of Bipolar inhibitor at a dosage of 5% resulted in the reduction of compressive strength by about 33% to 36%. The reduction in compressive strength may be due to the hydrophobic nature of inhibitors which forms microbubbles making slightly porous and reducing the strength. For PPC Mixes by addition of 2% Calcium Nitrite inhibitor, there is a marginal increase in compressive strength by 3.4% at 28 days. Addition of Bipolar inhibitor at a dosage of 5% resulted in the reduction of compressive strength by about 3.5% at 28 days.

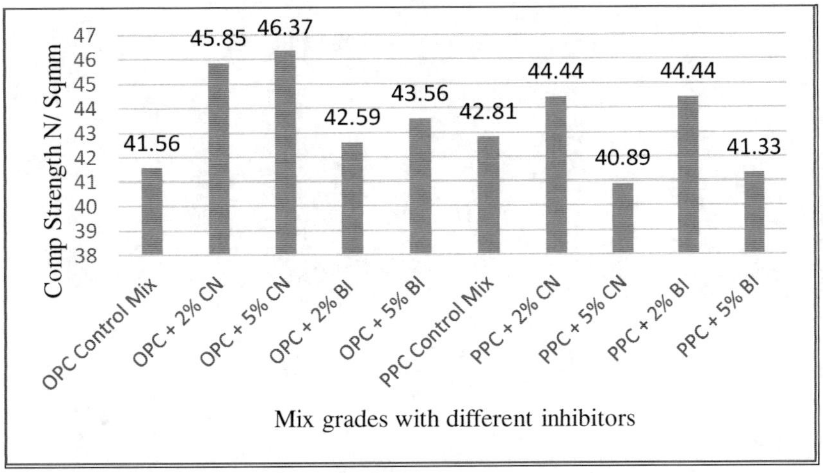

Figure 4: Graph showing the Compressive strengths of various mixes

3.2 Accelerated corrosion testing results

The corrosion of reinforcing steel was accelerated using the impressed voltage technique. The concrete cylinders of control mix and with admixed inhibitors were cast steel bar of 16mm diameter embedded centrally into it. The specimens were cured for 28 days and are dried for 24 hours before carrying out the impressed voltage test. The tested specimens showing the corrosion stains are shown in figure 5.

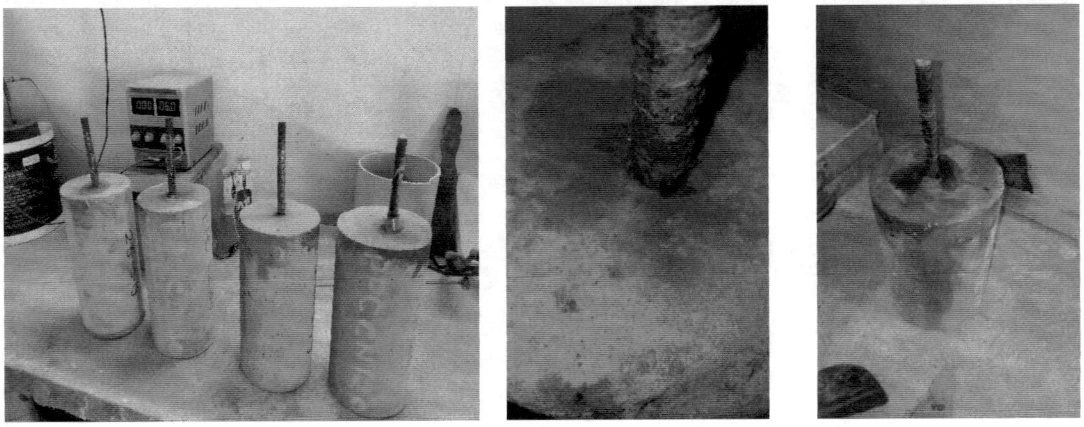

Figure 5. Test Specimens under the Impressed voltage test

3.2.1 Impressed voltage test results for PPC mixes

Observations from figure 6:

- The results indicate that the maximum anodic currents are recorded to be 52, 35 and 50 mA for Control mix, 2% CN inhibitor and 2% BI inhibitor respectively (refer figure 6).
- 2% of CN has shown reduced anodic current measurements. Hence the time taken for the initial crack was prolonged.
- It was observed that the time for initiation of crack was found to be 132, 204 and 180 hours for Control mix, 2% CN inhibitor and 2% BI inhibitor respectively.

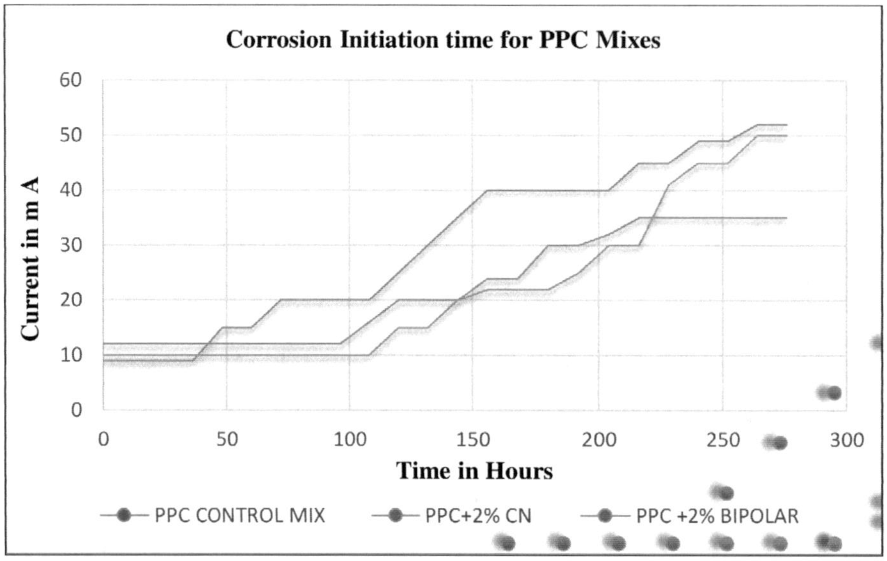

Figure 6: Graph showing the Corrosion Initiation Time and Change in Current values

3.2.2 Impressed voltage test results for OPC mixes

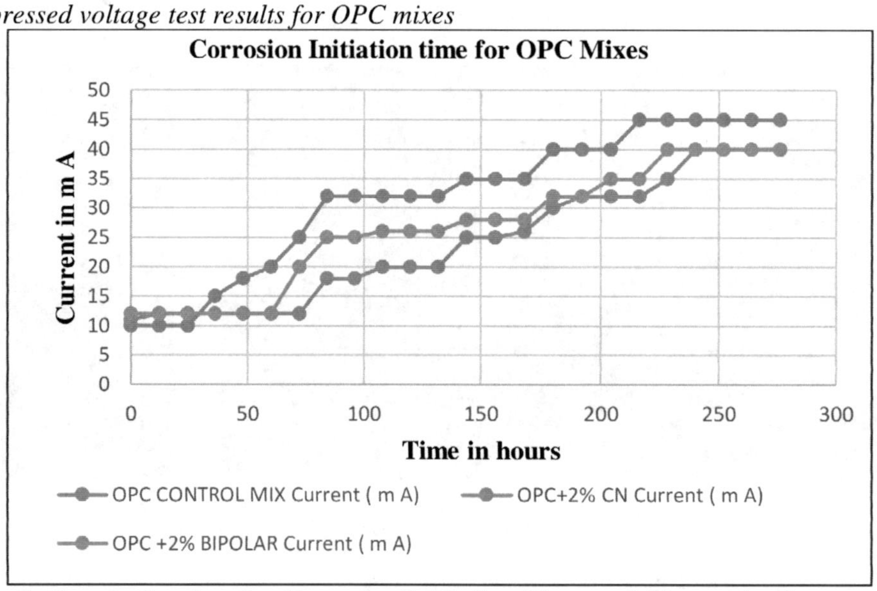

Figure 7: Graph showing the Corrosion Initiation Time and Change in Current values

Observations from figure 7:

- The results indicate that the maximum anodic currents are recorded to be 45, 40 and 40 mA for Control mix, 2% CN inhibitor and 2% BI inhibitor respectively (refer figure 7).
- 2% CN and 2% BI have has shown reduced anodic current measurements. Hence the time taken for the initial crack was prolonged.
- It was observed that the time for initiation of crack was found to be 132, 204 and 180 hours for Control mix, 2% CN inhibitor and 2% BI inhibitor respectively.

4. Conclusions

1. There is no improvement in the workability of OPC concrete mixes or PPC concrete mixes by addition of both the Corrosion inhibitors. Lower dosage of Corrosion inhibitors i.e., @ 2% yielded better workability compared to higher dosages. The dosages recommended by the manufacturers also matches with the above result.
2. There is a very marginal increase in compressive strength by addition of both the inhibitors at 2% dosage, whereas there is a decrease in compressive strength by addition of inhibitors at 5% dosage.
3. From the Impressed voltage test results, the maximum anodic current measured was found to be 35 and 40 mA for the Calcium Nitrite Inhibitor. On the other hand, Bipolar inhibitor has shown the maximum anodic current as 45 and 40 mA.
4. For PPC mixes, the Calcium Nitrite Inhibitor has shown better results in reducing the effect of corrosion. For OPC mixes both the inhibitors exhibited better performance compared to control mix.

References

[1] Taylor, S. Ray 1996 An Investigation of New Inhibitors to Mitigate Rebar Corrosion in Concrete, Virginia *Transportation Research Council (VTRC).*
[2] Srinivasan P Corrosion of reinforcement in concrete structures- Assessment & control *Advanced course on structural health monitoring, SERC Chennai* Page L13.1.
[3] Alonso C, Andrade C 1990 Effect of nitrite as a corrosion inhibitor in contaminated and chloride-free carbonated mortars *ACI Materials Journal* **87 (2)** p. 130–7
[4] Saraswathy V, Muralidharan S, Kalyanasundaram, Thangavel K, Srinivasan S 2001 Evaluation of composite corrosion –inhibiting admixture and its performance in concrete *Cement and Concrete Research* **31** p 789- 94
[5] Dhouibi L, Triki E, Salta M, Rodrigues P, Raharinaivo A 2003 Studies on corrosion inhibition of steel reinforcement by phosphate and nitrate. *Materials and Structures.***36 (8)** 530-40
[6] Berke N. S 1986 The Use of Anodic Polarization to Determine the Effectiveness of Calcium Nitrite is an Anodic Inhibitor in Corrosion Effect of Stray Currents and the Techniques for Evaluating Corrosion of Rebars in Concrete (V. Chaker, ed.), ASTM STP 906. *American Society for Testing and Materials* Philadelphia, Pa. p 78-91
[7] Jamil H E, Shriri A, Boulif R, Bastos C, Montimor M F, Ferreira M G S 2004 Electrochemical the behaviour of amino alcohol-based inhibitors used to control corrosion of reinforcing steel. *Electro chemica Acta.* **49 (17-18)** p 2753-60
[8] Soylev T A, Mc Nally C, Richardson M 2007 Effectiveness of amino alcohol-based surface-applied corrosion inhibitors in chloride-contaminated concrete *Cement and Concrete Research* **37 (6)** 972-7
[9] IS 10262 2009 Guidelines for concrete mix design proportioning *BIS Standards* New Delhi India
[10] IS 516 1959 Reprint 2006 Method of Tests for Strength of Concrete" *BIS Standards* New Delhi India
[11] ASTM C876-15 Standard Test Method for Corrosion Potentials of Uncoated Reinforcing Steel in Concrete *ASTM International*

Effect of alkali curing environment on M30 grade Sugarcane Bagasse Ash (SCBA) Concrete

P V Rambabu[1], G V Ramarao[2]

[1]Assistant Professor, Civil Engineering Department, SRKR Engineering College, Bhimavaram, Andhra Pradesh, India
[2]Professor, Civil Engineering Department, Andhra University College of Engineering (Autonomous), Visakhapatnam, Andhra Pradesh, India
Email: pvr.civil@gmail.com, ramaraogv_2000@yahoo.com

Abstract. The present paper deals with the replacement of Ordinary Portland Cement with SCBA at different percentages of 0%, 5%, 10%, 15% and 20% by weight of cement in M30 grade concrete, the mechanical properties of SCBA blended concrete were studied when subjected to alkali environment. The concrete specimens were cured in normal water, sodium sulphate solution (1%, 3% and 5%) and Magnesium Sulphate solution (1%, 3% and 5%) at ages of 28, 60 and 90 days and tested. From the experimental observation, the SCBA blended concrete provides resistance against sulphate attack.

1. Introduction

Ordinary Portland cement is the main constituent in the production of concrete. Replacement of cement with agricultural and industrial wastes such as sugar cane Bagasse ash, Rice husk ash, fly ash, blast furnace slag etc was the present trend in concrete technology [1-4]. A few studies were conducted on the ashes obtained directly from the Argo and other industries to study the pozzolanic activity and their suitability as binders, partially replacing cement [5-9]. Sulphate attack on marine Structures was major Concern in the durability aspect of structures [10-15]. Leaching of the calcium compounds was a major problem due to sulphate attack. Sulphate attack leads to loss of calcium silicate hydrate gel and results in deterioration of the cement constituents [16]. The sulphate attack depends on several factors [17]. Use of pozzolanic materials as a partial replacement of ordinary cement has been found to resist sulphate attack [18-20]. The sugar cane Bagasse ash as a partial replacement of cement was taken to study the performance of concrete. The main objective of this research was to study the durability Characteristics of Sugar cane Bagasse ash blended concrete Exposed to Alkali Environment (Sodium And magnesium Sulphates)

2. Materials

The materials utilised in this work were Ordinary Portland Cement of grade 53 (with setting time of 121minutes (initial) and setting of 242 minutes (final) and consistency of 31%), Fine aggregate (of Zone II, with water absorption 1% Specific gravity 2.6 and fineness modulus 2.99), Coarse aggregate (with water absorption0.5%, fineness modulus7.26and Specific gravity 2.8) and Water (suitable according to Indian standards for use in concrete: potable water). Sugar Cane Bagasse Ash (SCBA) was obtained from Sugar factory at Vuyurru (KCP Factory), Andhra Pradesh. Each 1000 kg of sugarcane generates approximately 0.60 % of residual Bagasse Ash. The major component of the residue was silicon dioxide. The specific gravity of SCBA was found to be 2.49. The physical properties of SCBA are presented in Table 1.

Table 1: Physical properties of Sugar Cane Bagasse Ash

No	Property	Test Results
1	Specific gravity	2.49
2	Particle shape	Spherical
3	Density	375 kg/m³
4	Specific surface area	420 m²/kg

3. Experimental Investigation

Mixing: the cementations materials we remixed properly and aggregates (fine and coarse) were added and then mixed with water till consistency was achieved. The casting of specimen: Before placing concrete into moulds, the moulds were cleansed. The moulds were placed on the level ground and Compacted by vibrator. A total number of 126 cubes,126 cylinders and 126 beams were caste Curing: After Casting the moulds were left for 24 hours and de moulded, de moulded specimens cured in water and sulphate solutions forth desired period. Durability: Here, the durability tests were conducted on SCBA concrete against salts such as Sodium and Magnesium Sulphate. The response of $Na_2SO_4$and $MgSO_4$ attack on sugar cane Bagasse ash concrete for various percentages was studied by observations like the loss in mechanical properties.

4. Experimental Results

The graphs below show the various strength results of concrete specimens with 0%, 5%, 10%, 15% and 20% weight replacement of cement with SCBA cured in water and various percentages of Na_2SO_4 and $MgSO_4$ solutions for 28, 60 and 90 days.

4.1 Compressive strength of concrete cured in water:

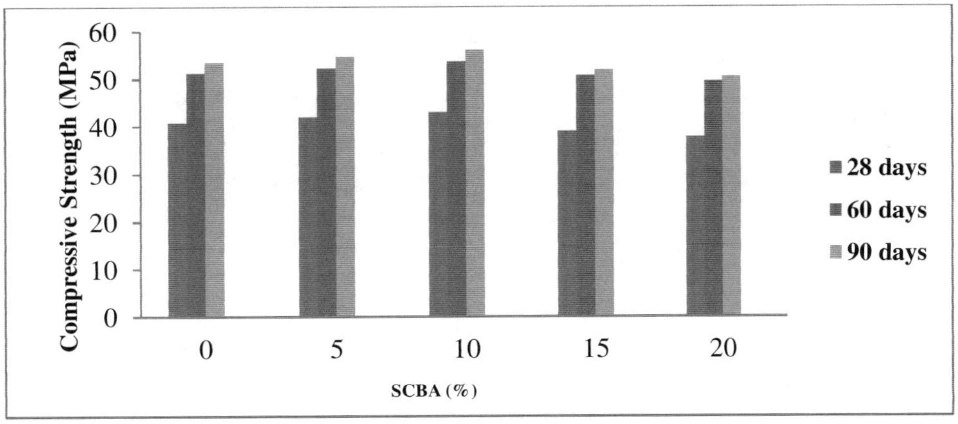

Figure 1. Compressive strength of concrete cured in water

From figure 1, it was noticed that, strength increase up to 2.99%, 1.4%, 2.3% at 28, 60, 90 days for 0% to 5%, the strength increase up to 2.3%, 2.8%, 2.7% at 28, 60, 90 days for 5% to 10%, strength decreases to 10.2%, 5.8%, 8.1% at 28, 60, 90 days for 10% to 15% and strength decreases to 3.04%, 2.4%, 2.9% at 28, 60, 90 days for 15% to 20%. The strength results of SCBA replaced concrete cubes cured in water was shown below graphically. The decline in compressive strength was noticed at fifteen and twenty percentage replacements and a surge in compressive strength was noticed at ten percentage replacements.

4.2 Compressive strength of concrete cured in 1% Na_2SO_4 solution:

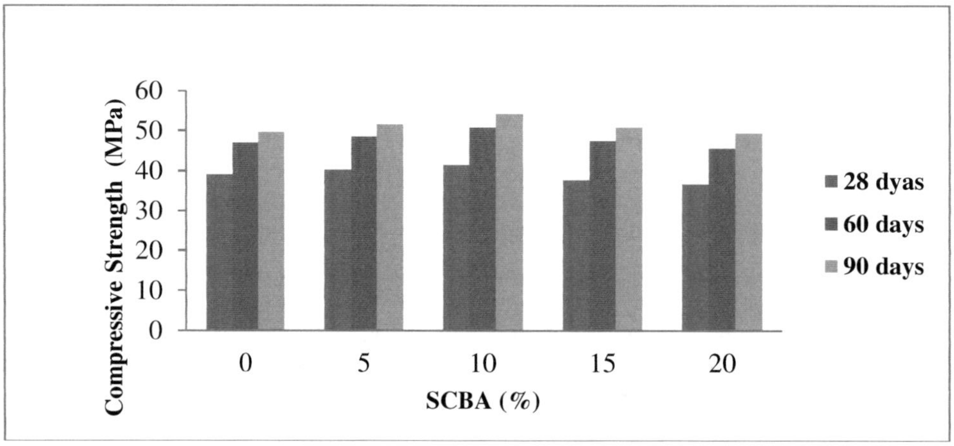

Figure 2. Compressive strength of concrete cured in 1% Na_2SO_4 solution

From figure 2, it was noticed that, for 0% to 5% strength increase up to 3.21%, 2.95%, 3.49% at 28, 60, 90 days, for 5% to 10% the strength increase up to 3.03%, 4.76%, 5.28% at 28, 60, 90 days, for 10% to 15% strength decreases to 9.84%, 6.75%, 6.34% at 28, 60, 90 days and for 15% to 20% strength decreases to 3.23%, 3.63%, 2.4% at 28, 60, 90 days. The graph shows the results for SCBA cubes cured 1% Na_2SO_4 solution, for all curing periods. Decline in compressive strength was noticed at fifteen and twenty percentage replacements and surge in compressive strength was noticed at ten percentage replacements.

4.3 Compressive strength cured in 3% Na_2SO_4 solution:

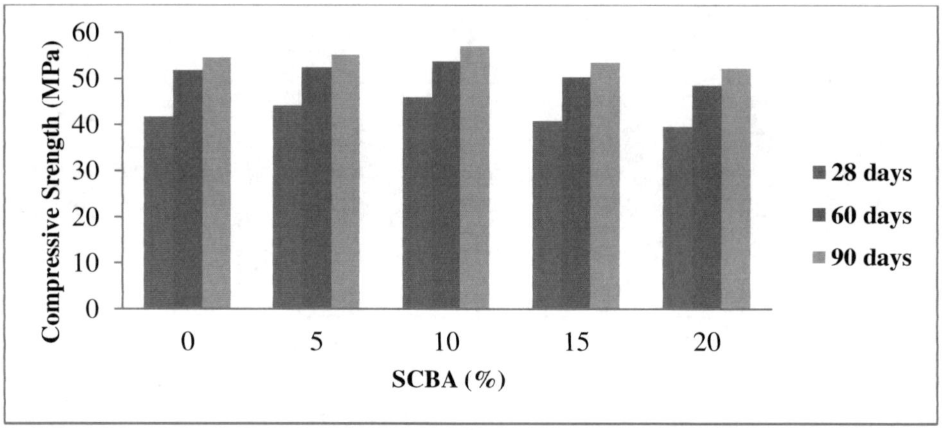

Figure 3. Compressive strength of concrete cured in 3% Na_2SO_4 solution

From figure 3, it was noticed that, for 0% to 5% strength increase up to 5.91%, 1.04%, 1.33% at 28, 60, 90 days, for 5% to 10% the strength increase up to 3.91%, 2.71%, 3.11% at 28, 60, 90 days, for 10% to 15% strength decreases to 11.29%, 6.38%, 6.02% at 28, 60, 90 days and for 15% to 20% strength decreases to 3.04%, 3.41%, 2.37% at 28, 60, 90 days. The graph shows the results for SCBA cubes cured 3% Na_2SO_4solution, for all curing periods. Decline in compressive strength was noticed at fifteen and twenty percentage replacements and surge in compressive strength was noticed at ten percentage replacements.

4.4 Compressive strength of concrete cured in 5% Na_2SO_4solution:

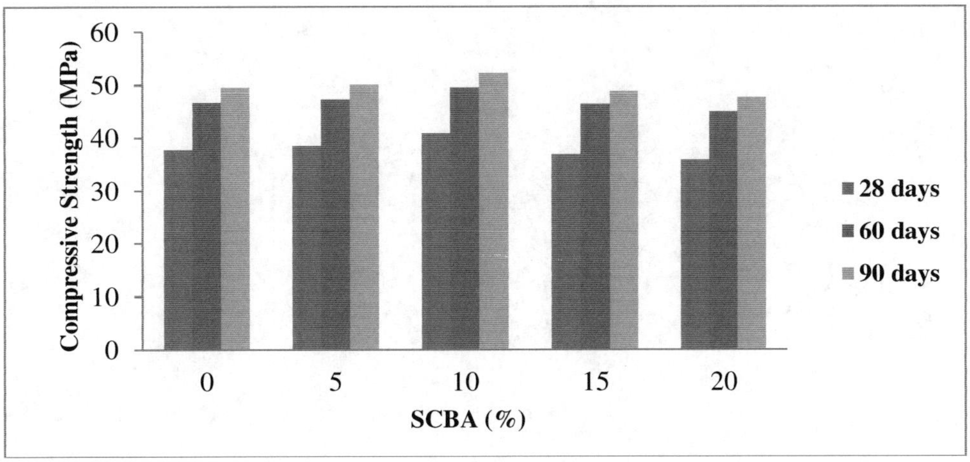

Figure 4. Compressive strength of concrete cured in 5% Na_2SO_4 solution

From figure 4, it was noticed that, for 0% to 5% strength increase up to 1.9%, 1.58%, 0.98% at 28, 60, 90 days, for 5% to 10% the strength increase up to 6%, 4.67%, 4.43% at 28, 60, 90 days, for 10% to 15% strength decreases to 9.51%, 6.4%, 6.54% at 28, 60, 90 days and for 15% to 20% strength decreases to 3%, 3.17%, 2.5% at 28, 60, 90 days. The graph shows the results for SCBA cubes cured 5% Na_2SO_4solution, for all curing periods. Decline in compressive strength was noticed at fifteen and twenty percentage replacements and surge in compressive strength was noticed at ten percentage replacements.

4.5 Compressive strength of concrete cured in 1% $MgSO_4$ solution:

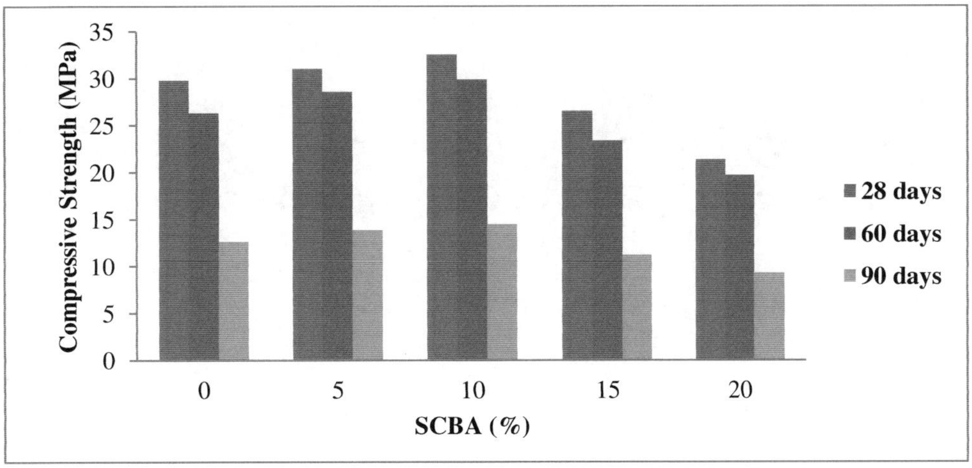

Figure 5. Compressive strength of concrete cured in 1% $MgSO_4$ solution

From figure 5, it was noticed that, for 0% to 5% strength increase up to 1.22%, 2.9%, 1.44% at 28,60, 90 days, for 5% to 10% the strength increase up to 3.04%, 1%, 2.8% at 28,60, 90 days, for 10% to 15% strength decreases to 7.61%, 6.5%, 7.3% at 28,60, 90 days and for 15% to 20% strength decreases to 2.5%, 2.6%, 1.36% at 28,60, 90 days. The graph shows the results for SCBA cubes cured 1% $MgSO_4$ Solution, for all curing periods. Decline in compressive strength was noticed at fifteen and twenty percentage replacements and surge in compressive strength was noticed at ten percentage replacements.

4.6 Compressive strength of concrete cured in 3% $MgSO_4$ solution:

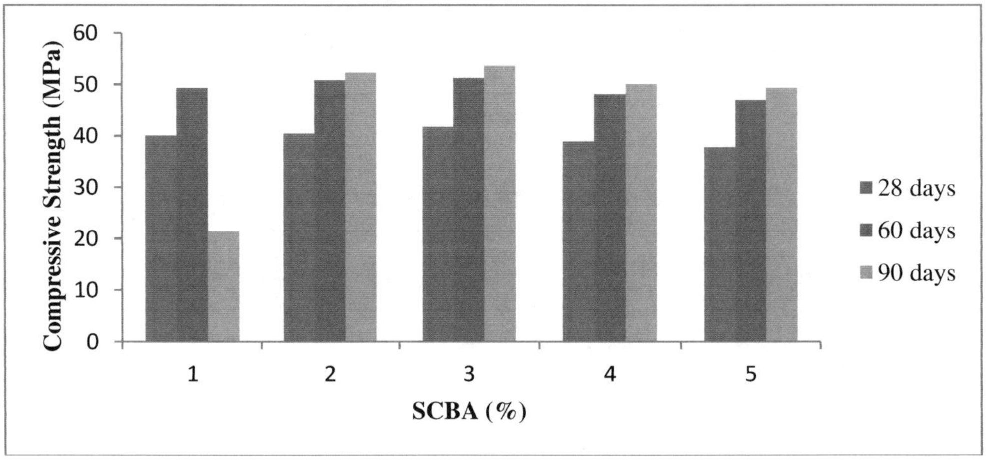

Figure 6. Compressive strength of concrete cured in 3% $MgSO_4$ solution

From figure 6, it was noticed that, for 0% to 5% strength increase up to 3.5%, 2.8%, 2.26% at 28,60, 90 days, for 5% to 10% the strength increase up to 1.62%, 1.12%, 2.56% at 28,60, 90 days, for 10% to 15% strength decreases to 6.4%, 5.4%, 5.8% at 28,60, 90 days and for 15% to 20% strength decreases to 4.3%, 2.9%, 2.7% at 28,60, 90 days. The graph shows the results for SCBA cubes cured 3% $MgSO_4$ solution, for all curing periods. Decline in compressive strength was noticed at fifteen and twenty percentage replacements and surge in compressive strength was noticed at ten percentage replacements.

4.7 Compressive strength of concrete cured in 5% $MgSO_4$ solution:

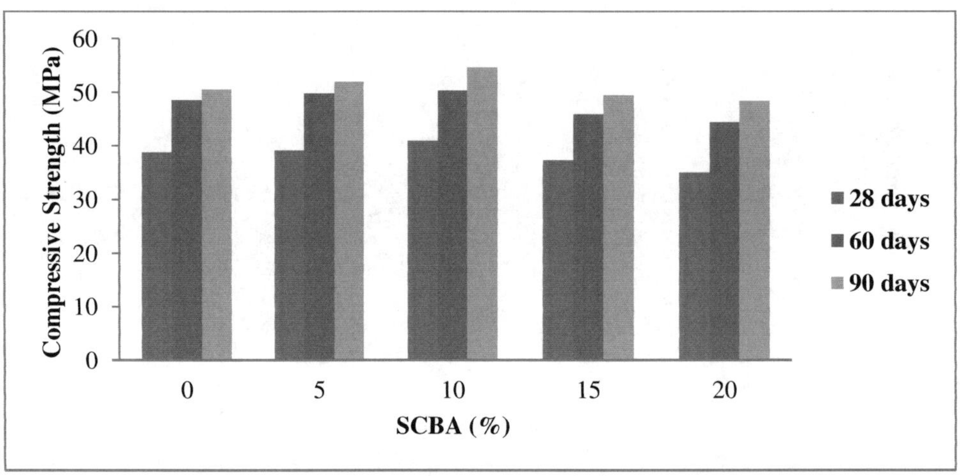

Figure 7. Compressive strength of concrete cured in 1% $MgSO_4$ solution

From figure 7, it was noticed that, for 0% to 5% strength increase up to 0.98%, 2.5%, 2.95% at 28,60,90 days, for 5% to 10% the strength increase up to 4.7%, 1.08%, 5.23% at 28,60, 90 days, for 10% to 15% strength decreases to 9.8%, 9.4%, 10.4% at 28,60, 90 days and for 15% to 20% strength decreases to 6.4%, 3.7%, 2.5% at 28,60, 90 days. The graph shows the results for SCBA cubes cured in 5% $MgSO_4$ Solution, for all curing periods. Decline in compressive strength was noticed at fifteen and twenty percentage replacements and surge in compressive strength was noticed at ten percentage replacements.

4.8 Split tensile strength of concrete cured in water:

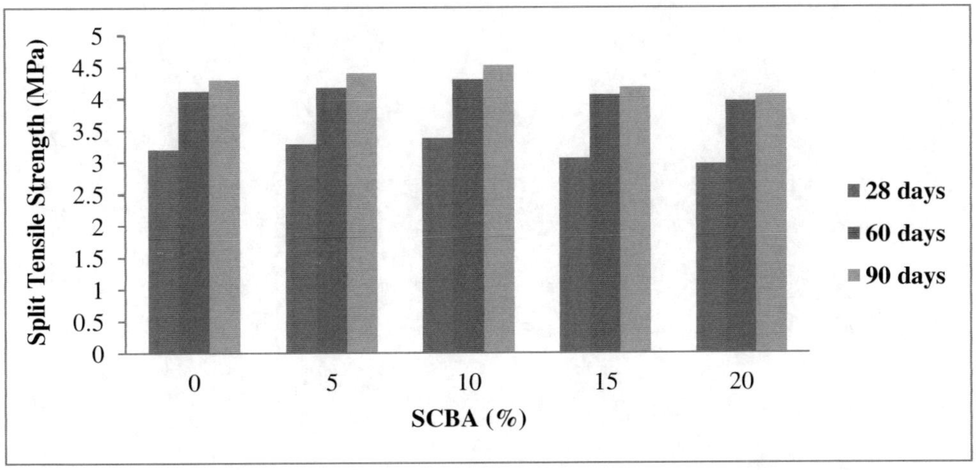

Figure 8. Split tensile strength of concrete cured in water

From figure 8, it was noticed that, for 0% to 5% strength increase up to 2.8%, 1.4%, 2.3% at 28, 60, 90 days, for 5% to 10% the strength increase up to 2.43%, 2.9%, 2.7% at 28,60, 90 days, for 10% to 15% strength decreases to 10.1%, 5.9%, 8.13% at 28,60, 90 days and for 15% to 20% strength decreases to 3.03%, 2.5%, 2.9% at 28,60, 90 days. The graph shows the results for SCBA cylinders cured in water, for all curing periods. The decline in compressive strength was noticed at fifteen and twenty percentage replacements and a surge in compressive strength was noticed at ten percentage replacements.

4.9 Split tensile strength of concrete cured in 1% Na_2SO_4 solution

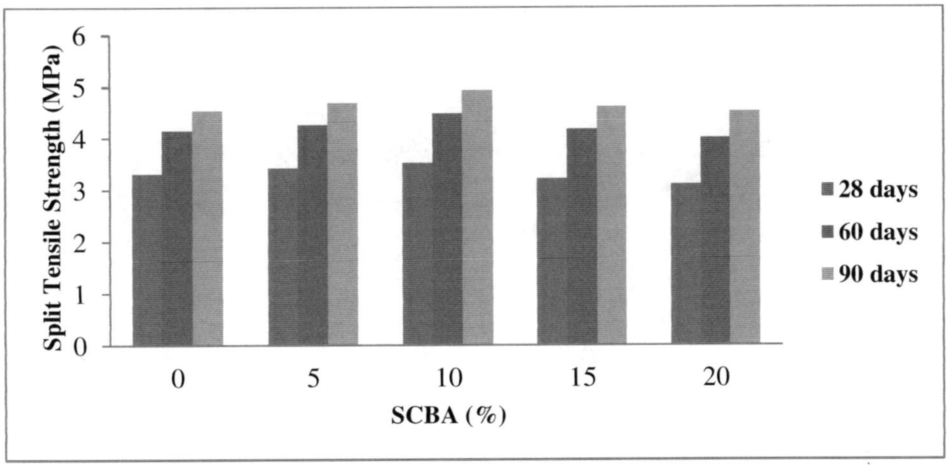

Figure 9. Split tensile strength of concrete cured in 1% Na_2SO_4 solution

From figure 9, it was noticed that, for 0% to 5% strength increase up to 2.32%, 2.89%, 3.31% at 28,60, 90 days, for 5% to 10% the strength increase up to 2.92%, 4.92%, 5.34% at 28,60, 90 days, for 10% to 15% strength decreases to 8.8%, 6.7%, 6.28%at 28,60, 90 days and for 15% to 20% strength decreases to 3.11%, 3.83%, 2.38% at 28,60, 90 days. The graph shows the results for SCBA cylinders cured in1% Na_2SO_4solution, for all curing periods. Decline in compressive strength was noticed at fifteen and twenty percentage replacements and surge in compressive strength was noticed at ten percentage replacements

4.10 Split tensile strength of concrete cured in 3% Na_2SO_4 solution:

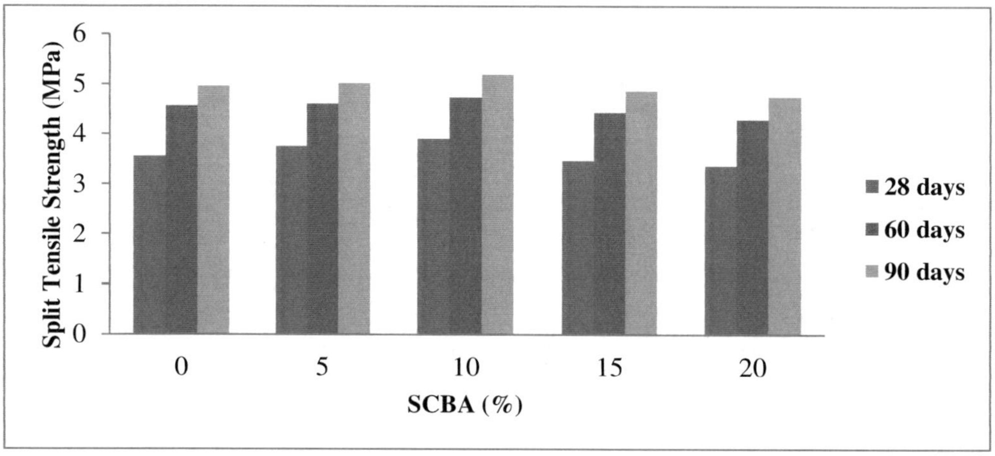

Figure 10. Split tensile strength of concrete cured in 3% Na_2SO_4 solution

From figure 10, it was noticed that, for 0% to 5% strength surges up to 5.91%, 0.87%, 1.2% at 28,60, 90 days, for 5% to 10% the strength increase up to 3.72%, 2.82%, 3.18% at 28,60, 90 days, for 10% to 15% strength decreases to 11.28%, 6.4%, 5.98% at 28,60,90 days and for 15% to 20% strength decreases to 2.89%, 3.38%, 2.46% at 28,60, 90 days. The graph shows the results for SCBA cylinders cured in 3% Na_2SO_4solution, for all curing periods Decline in compressive strength was noticed at fifteen and twenty percentage replacements and surge in compressive strength was noticed at ten percentage replacements.

4.11 Split tensile strength of concrete cured in 5% Na_2SO_4 solution:

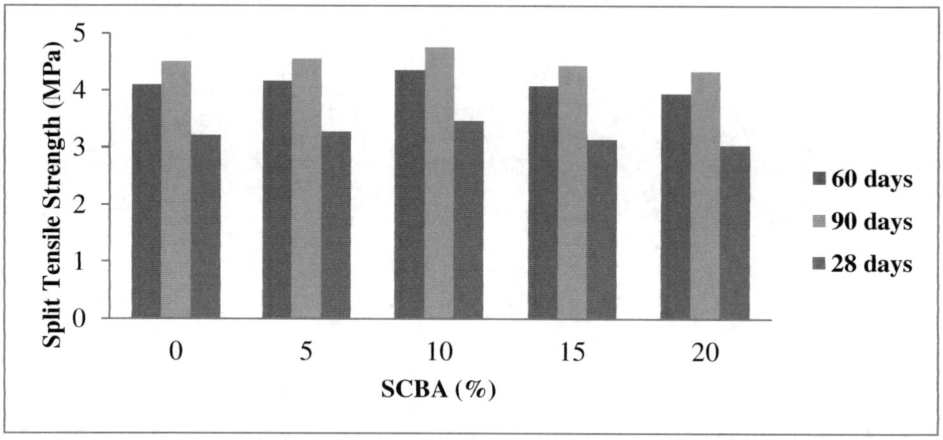

Figure 11: Split tensile strength of concrete cured in 5% Na_2SO_4 solution

From figure 11, it was noticed that, for 0% to 5% strength increase up to 1.86%, 1.46%, 0.88% at 28,60, 90 days, for 5% to 10% the strength increase up to 6.11%, 4.8%, 4.44 % at 28,60, 90 days, for 10% to 15% strength decreases to 9.51%, 6.42%, 6.52% at 28,60, 90 days and for 15% to 20% strength decreases to 3.18%, 3.2%, 2.47% at 28,60, 90 days. The graph shows the results for SCBA cylinders cured in 5% Na_2SO_4solution, for all curing periods. Decline in compressive strength was noticed at fifteen and twenty percentage replacements and surge in compressive strength was noticed at ten percentage replacement

4.12 Split tensile strength of concrete cured in 1% $MgSO_4$ solution:

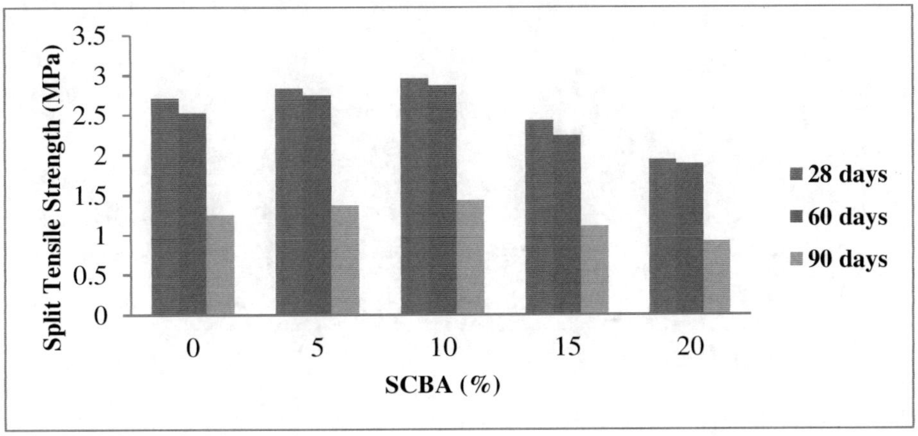

Figure 12. Split tensile strength of concrete cured in 1% $MgSO_4$ solution

From figure 12, it was noticed that, for 0% to 5% strength increase up to 1.16%, 2.9%, 1.5% at 28,60, 90 days, for 5% to 10% the strength increase up to 3.16%, 0.89%, 2.76% at 28,60, 90 days, for 10% to 15% strength decreases to 7.8%, 6.6%, 7.3% at 28,60, 90 days and for 15% to 20% strength decreases to 2.46%, 2.6%, 1.35% at 28,60, 90 days. The graph shows the results for SCBA cylinders cured in 1% $MgSO_4$ solution, for all curing periods. Decline in compressive strength was noticed at fifteen and twenty percentage replacements and surge in compressive strength was noticed at ten percentage replacements.

4.13 Split tensile strength of concrete cured in 3% $MgSO_4$ solution:

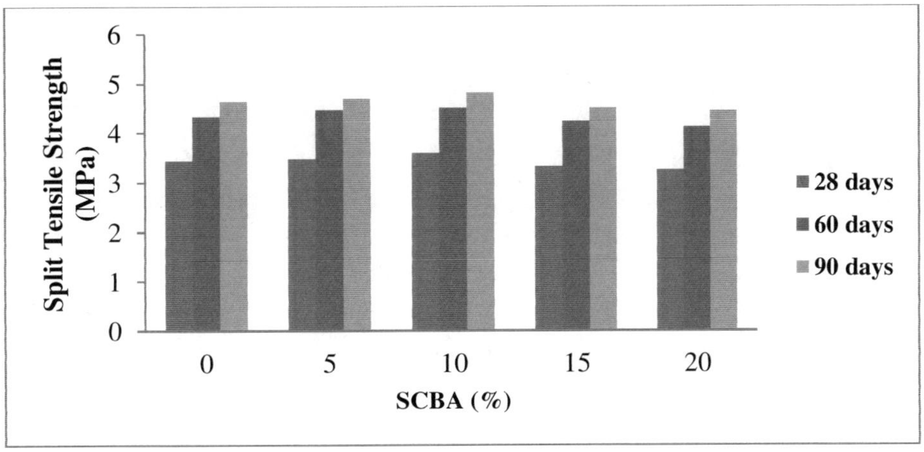

Figure 13. Split tensile strength of concrete cured in 3% $MgSO_4$ solution

From figure 13, it was noticed that, for 0% to 5% strength increase up to 3.58%, 2.8%, 2.2% at 28,60, 90 days, for 5% to 10% the strength increase up to 1.59%, 1.06%, 2.5% at 28,60, 90 days, for 10% to 15% strength decreases to 6.4%, 5.5%, 5.8% at 28,60, 90 days and for 15% to 20% strength decreases to 4.34%, 2.9%, 2.7% at 28,60, 90 days. The graph shows the results for SCBA cylinders cured in 3% $MgSO_4$ solution, for all curing periods. Decline in compressive strength was noticed at fifteen and twenty percentage replacements and surge in compressive strength was noticed at ten percentage replacements.

4.14 Split tensile strength of concrete cured in 5% $MgSO_4$ solution:

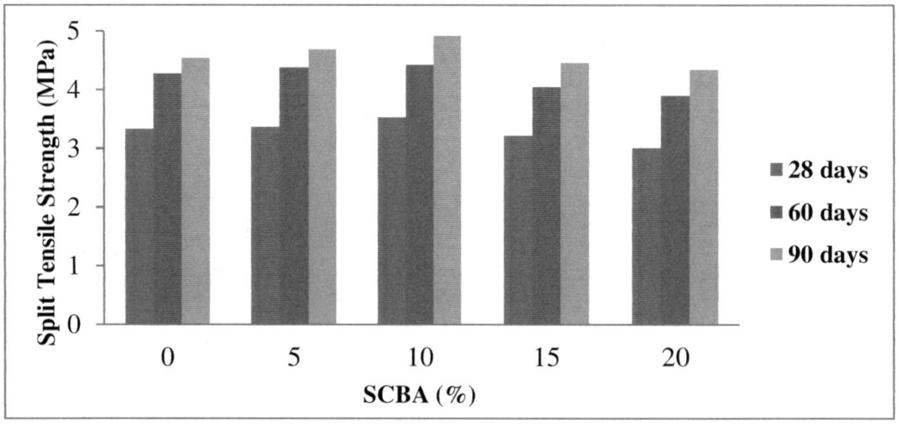

Figure 14. Split tensile strength of concrete cured in 5% $MgSO_4$ solution

From figure 14, it was noticed that, for 0% to 5% strength increase up to 1.79%, 1.61%, 0.99% at 28,60, 90 days, for 5% to 10% the strength increase up to 5.94%, 4.57%, 4.42% at 28, 60, 90 days, for 10% to 15% strength decreases to 9.35%, 6.39%, 6.43% at 28,60, 90 days, 10.3% and for 15% to 20% strength decreases to 3.21%, 3.05%, 2.52% at 28,60, 90 days. The graph shows the results for SCBA cylinders cured in 5% $MgSO_4$ solution, for all curing periods. Decline in compressive strength was noticed at fifteen and twenty percentage replacements and surge in compressive strength was noticed at ten percentage replacements.

4.15. Flexural strength of concrete cured in water:

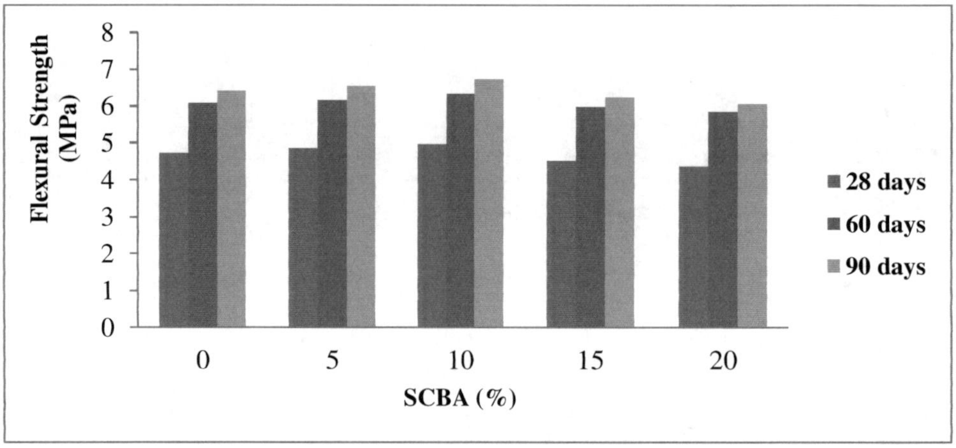

Figure 15. Flexural strength of concrete cured in water

From figure 15, it was noticed that, for 0% to 5% strength increase up to 2.9%, 1.3%, 2.18% at 28,60, 90 days, for 5% to 10% the strength increase up to 2.5%, 2.9%, 3.1% at 28,60, 90 days, for 10% to 15% strength decreases to 9.65%, 9.4%, 10.3% at 28,60, 90 days and for 15% to 20% strength decreases to 6.6%, 3.8%, 2.5% at 28,60, 90 days. The graph shows the results for SCBA beams cured in water, for all curing periods. Decline in compressive strength was noticed at fifteen and twenty percentage replacements and surge in compressive strength was noticed at ten percentage replacements.

4.16 Flexural strength of concrete cured in 1% Na_2SO_4 solution:

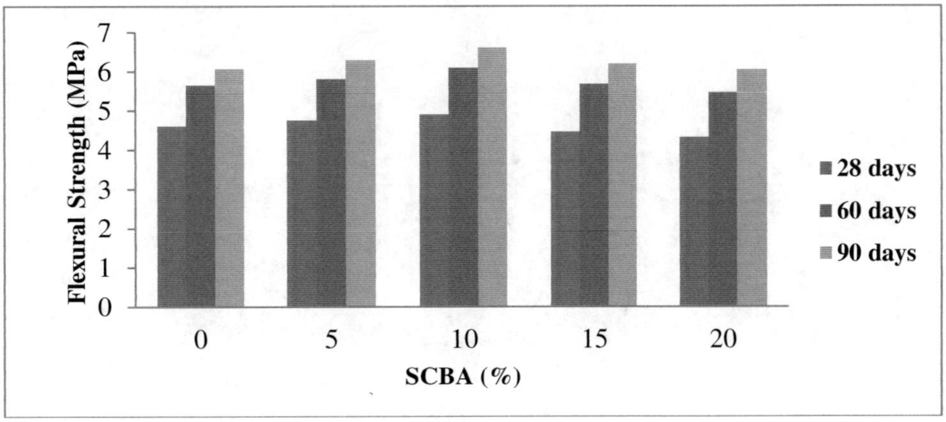

Figure 16. Flexural strength of concrete cured in 1% Na_2SO_4 solution

From figure 16, it was noticed that, for 0% to 5% strength increase up to 3.26%, 2.83%, 3.45% at 28,60, 90 days, for 5% to 10% the strength increase up to 2.94%, 4.81%, 5.25% at 28,60, 90 days, for 10% to 15% strength decreases to 8.79%, 6.73%, 6.35% at 28,60, 90 days and for 15% to 20% strength decreases to3.36%, 3.7%, 2.42% at 28,60, 90 days. The graph shows the results for SCBA beams cured in1% Na_2SO_4 solution, for all curing periods. Decline in compressive strength was noticed at fifteen and twenty percentage replacements and surge in compressive strength was noticed at ten percentage replacements.

4.17 Flexural strength of concrete cured in 3% Na_2SO_4solution:

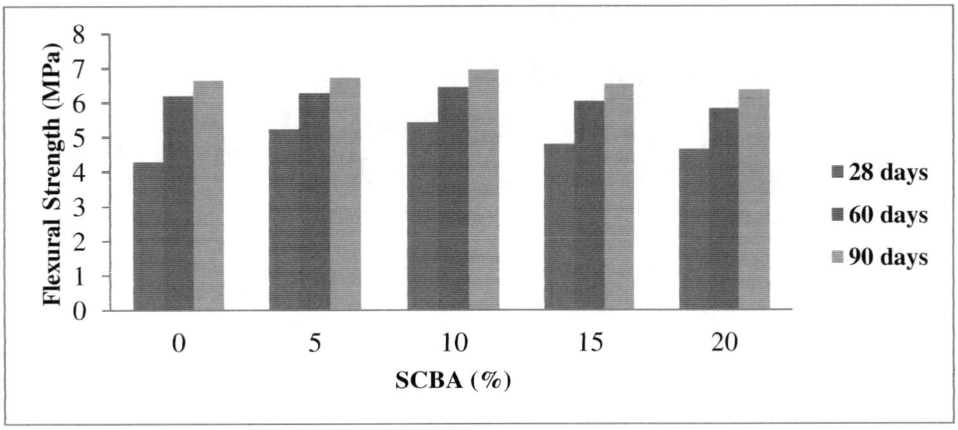

Figure 17/ Flexural strength of concrete cured in 3% Na_2SO_4 solution

From figure 17, it was noticed that, for 0% to 5% strength increase up to 6.09%, 1.12%, 1.35% at 28,60,90 days, for 5% to 10% the strength increase up to 3.83%, 2.7%, 3.11% at 28,60, 90 days, for 10% to 15% strength decreases to 11.25%, 6.35%, 6.04% at 28,60, 90 days and for 15% to 20% strength decreases to3.11%, 3.47%, 2.45% at 28,60, 90 days. The graph shows the results for SCBA beams cured in 3% Na_2SO_4 solution, for all curing periods. Decline in compressive strength was noticed at fifteen and twenty percentage replacements and surge in compressive strength was noticed at ten percentage replacements.

4.18 Flexural strength of concrete cured in 5% Na_2SO_4 solution:

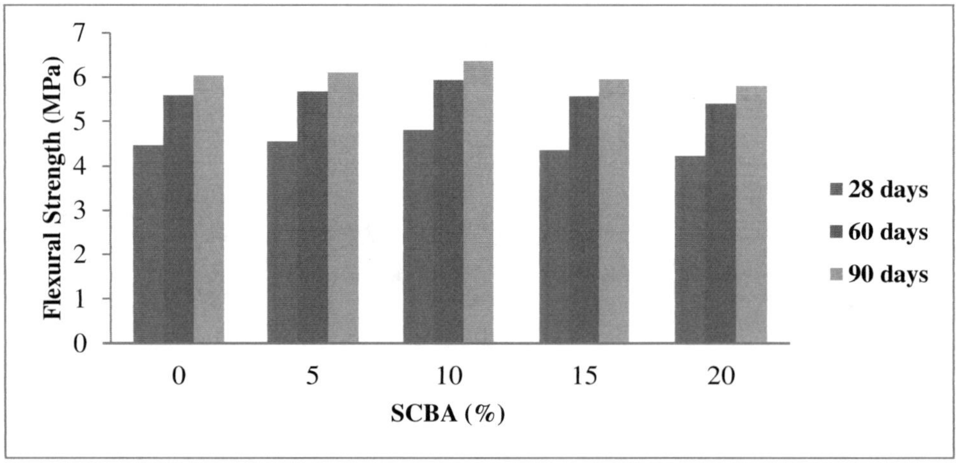

Figure 18. Flexural strength of concrete cured in 5% Na_2SO_4 solution

From figure 18, it was noticed that, for 0% to 5% strength increase up to 4.2%, 8.5%, 9.3% at 28,60, 90 days, for 5% to 10% the strength increase up to 4.5%, 4.6%, 4.7% at 28, 60, 90 days, for 10% to 15% strength decreases to 22.6%, 28.3%, 29.2% at 28,60, 90 days and for 15% to 20% strength decreases to 24.6%, 18.6%, 20.6% at 28,60, 90 days. The graph shows the results for SCBA beams cured in 5% Na_2SO_4solution, for all curing periods. Decline in compressive strength was noticed at fifteen and twenty percentage replacements and surge in compressive strength was noticed at ten percentage replacements.
4.19 Flexural strength of concrete cured in 1% $MgSO_4$ solution:

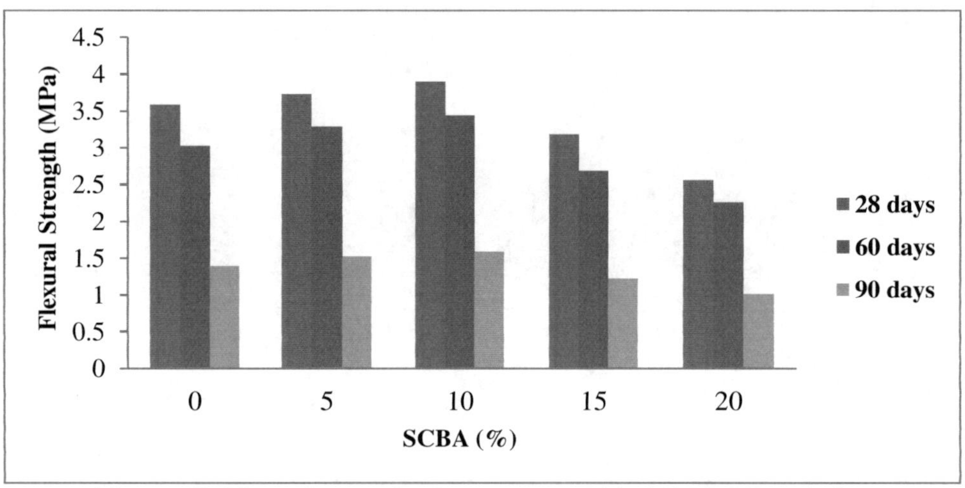

Figure 19. Flexural strength of concrete cured in 1% $MgSO_4$ solution

From figure 19, it was noticed that, for 0% to 5% strength increase up to 1.25%, 2.9%, 1.57% at 28,60, 90 days, for 5% to 10% the strength increase up to 2.9%, 0.97%, 2.8% at 28,60, 90 days, for 10% to 15% strength decreases to 7.5%, 6.6%, 7.4% at 28,60, 90 days and for 15% to 20% strength decreases to 2.6%, 2.6%, 1.31% at 28,60, 90 days. The graph shows the results for SCBA beams cured in 1% $MgSO_4$ solution, for all curing periods. Decline in compressive strength was noticed at fifteen and twenty percentage replacements and surge in compressive strength was noticed at ten percentage replacements.

4.20 Flexural strength of concrete cured in 3% $MgSO_4$ solution:

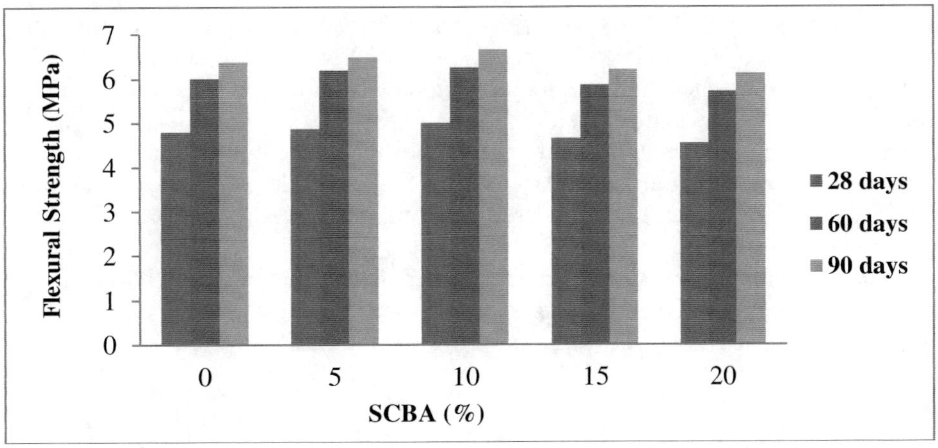

Figure 20. Flexural strength of concrete cured in 3% $MgSO_4$ solution

From figure 20, it was noticed that, for 0% to 5% strength increase up to 3.5%, 2.8%, 2.6% at 28,60, 90 days, for 5% to 10% the strength increase up to 1.5%, 1.1%, 2.6% at 28,60, 90 days, for 10% to 15% strength decreases to 6.4%, 5.4%, 5.4% at 28,60, 90 days and for 15% to 20% strength decreases to 4.4%, 2.9%, 2.9% at 28,60, 90 days. The graph shows the results for SCBA beams cured in 3% $MgSO_4$ solution, for all curing periods. Decline in compressive strength was noticed at fifteen and twenty percentage replacements and surge in compressive strength was noticed at ten percentage replacements.

4.21 Flexural strength of concrete cured in 5% $MgSO_4$ solution:

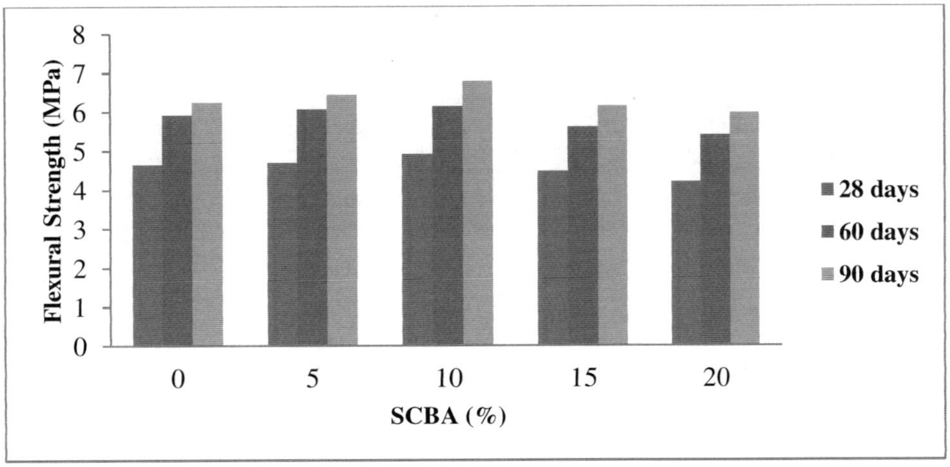

Figure 21. Flexural strength of concrete cured in 5% $MgSO_4$ solution

From figure 21, it was noticed that, for 0% to 5% strength increase up to 0.86%, 2.5%, 2.9% at 28,60, 90 days, for 5% to 10% the strength increase up to 4.9%, 1.15%, 5.3% at 28,60, 90 days, for 10% to 15% strength decreases to 10.1%, 9.4%, 10.4% at 28,60, 90 days and for 15% to 20% strength decreases to 6.4%, 3.6%, 2.5% at 28,60, 90 days. The graph shows the results for SCBA beams cured in 5% $MgSO_4$ solution, for all curing periods. Decline in compressive strength was noticed at fifteen and twenty percentage replacements and surge in compressive strength was noticed at ten percentage replacements.

5. Conclusions

The following are the conclusions from this study.

1. The mechanical properties of SCBA (sugarcane Bagasse ash) concrete specimens were improved when compared to concrete specimens without SCBA when cured in water.
2. SCBA blended concrete when cured in Sodium sulphate solution and Magnesium sulphate solution shows better results than normal concrete.
3. The increment in compressive strength of sugarcane Bagasse ash blended concrete was noticed up to replacement level of 10%. Further replacement of SCBA results in the decrement of strength.
4. Improvement in split tensile and flexural strength strengths of SCBA blended concrete was noticed at 10% replacement of SCBA.
5. SCBA concrete specimens provide sulphate resistance than specimens without SCBA.
   Hence Ten percentage replacement of sugarcane Bagasse ash with cement in concrete is an optimum percentage and can be used in places vulnerable to sulphate environment.

References

[1] Papadakis V G, Tsimas S 2002 *Cement and Concrete Research* **32** 1525 – 1532

[2] Nicole P H, Monteiro P J M, Carasek H 2000 *Materials Journal* **97** 486-492

[3] Alireza Naji Givi, Suraya Abdul Rashid, Farah Nora A. Aziz, Mohamad Amran Mohd Salleh 2010 *Journal of American Science* **06** 157-165

[4] Nehdi M, Duquette J, El Damatty A 2003 *Cement and Concrete Research* **33** 1203–1210

[5] Aigbodion, V S, Hassan S B, Ause T, Nyior G B 2010 *J. Miner. Mater. Charact. Eng.,* **9** 67-77

[6] Chusilp N, Jaturapitakkul C, Kiattikomol K 2009 *Constr. Build. Mater.,* **23 (12)** 3523-3531

[7] Fairbairn EM R, Americano B B, Cordeiro G C, Paula T P, Filho R D T, Silvoso M M, 2010 *J. Environ. Manage.,* **91 (9)** 1864-1871

[8] Frías M, Villar E, Savastano H 2011 *Cement Concrete Compos.,* **33 (4)** 490-496

[9] Srinivasan R, Sathiya K 2010 *Int. J. Serv. Learn. Eng.,* **5 (2)** 60-66

[10] Ganesan K, Rajagopal K, Thangavel K 2007 *Anti-Corr. Methods Mater* **54 (4)** 230-236

[11] Chatveera B, Lertwattanaruk P 2011 *Journal of Environmental Management* **92** 59-66

[12] Mehta P K 1983 *Cement and Concrete Research* **13** 401–406

[13] Wee T H , Suryavanshi A K, Wong S F, Anisur Rahman A KM 2000 *ACI Materials Journal* **97** 536–549

[14] Brown P W 1981 *Cement and Concrete Research* **11** 719–727

[15] Ouyang C, Nanni A, Chang W F 1988 *Cement and Concrete Research* **18** 699–709

[16] Tumidajski P J , Chan G W, Philipose K E 1995 *Cement and Concrete Research* **25** 1159–1163

[17] Hooton R D 1993 *ACI Materials Journal* **90** 143–151

[18] Tikalski P J, Carrasquillo R L1992 *ACI Materials Journal* **89** 69–75

[19] Al-Amoudi O S, Maslehuddin M, Saadi M M 1995 *ACI Materials Journal* **92** 15–24

[20] Mangat P S, Khatib J M 1995 *ACI Materials Journal* **92** 542–552

Compressive strength and durability of Metakaolin blended concrete exposed to acid and sulphate attack

D Viswanadha Varma[1], G V Rama Rao[2], P Markandeya Raju[3], M Pavan Kumar[4]

[1]Research Scholar, Department of Civil Engineering, College of Engineering (Autonomous), Andhra University, Visakhapatnam, India.
[2]Professor, Department of Civil Engineering, College of Engineering (Autonomous), Andhra University, Visakhapatnam, India.
[3]Professor and Head, Department of Civil Engineering, MVGR College of Engineering (Autonomous), Vizianagaram 535 005, India.
[4]Assistant Professor, Department of Civil Engineering, SVP Engineering College, Visakhapatnam India
E-mail: markandeyaraju@mvgrce.edu.in

Abstract. Sustainable development of the cement and concrete industry requires the utilization of industrial and agricultural waste components. At present, for a variety of reasons, the concrete construction industry is not sustainable. In general, high strength concrete made with ordinary cement is not durable and does not have good resistance to acids except for some weak acids if the exposure is occasional. Metakaolin is a supplementary cementitious material used as an admixture to produce high strength and high-performance concrete. The objective of this study is to arrive at Optimal quality of metakaolin that can partially replace the cement for M20, M30, M40, M50, M60, M70, M80 grades of concrete to get better strength and durability. The mix proportions of OPC concrete are obtained for M20, M30, M40, M50, M60, M70, M80 as per IS 10262 2009 [1]. Mix proportions with partial replacements of cement by 0%, 10%, 15%, 20%, 25%, 30% of Metakaolin with concrete were determined. Concrete specimens for compressive strength (150 mm×150 mm×150 mm cubes) are cast and Durability studies are conducted for all grades with partial replacement of cement by Metakaolin for 1% and 0.5% concentrations of HCl and H_2SO_4. From the results, it was concluded that the compressive strength of concrete is maximum at 15% replacement of cement by Metakaolin. The compressive strength of concrete showed better result at 15% replacement of Metakaolin for 0.5% and 1% HCl and H_2SO_4 at the age of 28days of strength. The effect of HCl on strength of the Metakaolin concrete is lower than the effect of H_2SO_4.
Keywords: Metakaolin; compressive strength; acid attack; sulphate attack;

1. Introduction
Sustainable development of the cement and concrete industry requires the utilization of industrial and agricultural waste components. At present, for various reason, the concrete construction industry is not sustainable. Firstly, it consumes huge quantities of virgin materials which need to remain for the next generations. Secondly, the principal binder in concrete is Portland cement, the production of which is a major contributor to greenhouse gas emissions that is implicated to global warming and climate change. Thirdly, many concrete structures suffer from a lack of durability and thus wasting precious natural resources. So, finding a solution to substitute a part of the cement for improved durability seems to be desirable for sustainable development. Metakaolin ($Al_2Si_2O_7$) exhibits pozzolanic properties and hence can be used as a concrete constituent, replacing part of the cement content. Metakaolin differs from

other supplementary cementitious materials like fly ash, slag or silica fume as it is not a by-product of an industrial process. It is a fine, natural white clay manufactured by the calcination of kaolinitic clay at a temperature ranging between 650° C and 850° C to particle structure making it a highly reactive, amorphous pozzolans for a specific purpose under controlled conditions.

Al-Akhras 2006 [2] studied the durability of metakaolin to sulfate attack. Bai 2002 [3] investigated the temperature change and heat evolution of mortar incorporating PFA and metakaolin. Brooks and Johari 2001 [4] studied the effect of metakaolin on creep and shrinkage of concrete. Courard et al 2003 [5] researched on the durability of mortars modified with metakaolin. Frías et al. 2000 [6] experimented on the effect of pozzolanic reaction of metakaolin on the heat evolution in metakaolin-cement mortars. Gruber et al. 2001 [7] observed that there is an increase in concrete durability with high-reactivity metakaolin. Khatib and Clay 2004 [8] experimented on the absorption characteristics of metakaolin concrete. Lee et al. 2005 [9] observed the effect of solution concentrations and replacement levels of metakaolin on the resistance of mortars exposed to magnesium sulfate solutions. Poon et al. 2001 [10] experimented on the rate of pozzolanic reaction of metakaolin in high-performance cement pastes. Poon et al. 2003 [11] studied the performance of metakaolin concrete at elevated temperatures. Qian and Li 2001 [12] studied the relationships between stress and strain for high-performance concrete with metakaolin. Ramlochan et al. 2000 [13] studied the effect of metakaolin on the alkali-silica reaction in concrete.

Many others investigated the basic mechanical and microscopic properties of cement mortar and concrete produced with partial replacement of cement with metakaolin. The reason for these improved properties of concrete was attributed to high siliceous content in metakaolin also called as High Reactivity Metakaolin (HRM). During the cement hydration process, water reacts with Portland cement and forms calcium – silicate - hydrate (C-S-H). The by-product of this reaction is the formation of calcium hydroxide (lime). This lime forms a weak link in concrete, and hence reduces the effect of the C-S-H. When Metakaolin is added in the hydration process, it reacts with the free lime to form additional C-S-H material, thereby making the concrete denser, stronger. In addition to strength, concrete has to be durable for the structures to be sustainable. The problem is more so in structures exposed to sulphate attack (seawater, sewage, industrial waste, salts in groundwater, and delayed release of clinker) and acid attack (acid rains).

Many researchers also attempted to study the effect of acid exposure on various mechanical properties of High-performance concrete made with Metakaolin blended cement concrete. They observed that partial replacement of cement by Metakaolin reduces the thickness of the interfacial zone, thus improving the adhesive bond between the hardened cement paste and particles of sand or aggregate. Metakaolin is obtained from the 20 Microns Limited Company at Vadodara in Gujarat. The specific gravity of Metakaolin is about 2.65.

Sulphate attack is caused by the chemical reaction between sulphate ions and hydration products, leading to ettringite and gypsum formation. Mono-sulphate, CH, and water combine to form ettringite. The expansive force generates tensile stresses in concrete. This leads to severe damage and cracking. Sulphuric acid (H_2SO_4) is particularly aggressive because, in addition to the sulphate attack of the aluminates phase, acid attack on $Ca(OH)_2$ and C-S-H takes place. Acid rains, which consists mainly of sulphuric acid and nitric acid and has a pH value between 4.0 and 4.5, may cause surface weathering of exposed concrete.

1.1. *Corrosion of embedded steel in concrete*

Deterioration of concrete containing embedded metals, such as conduits, pipes, and reinforcing and prestressing steel, is attributable to the combined effect of more than one cause. Corrosion of steel in concrete is an electrochemical process. The electrochemical potentials to form the corrosion cells may be generated in the following ways. Composition cells may be formed when two dissimilar metals are embedded in concrete, such as steel rebar's and aluminium conduit pipes, or when significant variations exist in surface characteristics of the steel. Concentration cells may be formed due to differences in concentration of dissolved ions in the vicinity of steel, such as alkalies, chlorides and oxygen.

1.2. Concluding remarks on literature
It was observed from the literature survey that the experiments are not carried out on all grades. The behaviour cannot be predicted with a limited number of grades and with a little number of concrete cubes. Very little quantity of papers identified on metakaolin concrete subjected to acid attack. So, the work is framed to conduct experiments on a greater number of concrete cubes by considering arrange of grades, various percentages of metakaolin and acids for curing to model a greater number of samples and to determine the acceptable results on metakaolin concrete.

2. Objective, Scope and Methodology
This section deals with the objective, scope of the study and planning of entire work in a stepwise manner along with properties of the materials.

2.1. The objective of the study
The objective of this work is to evaluate the optimum percentage of Metakaolin replacement in cement for maximum compressive strength and durability of concrete exposed to acid and sulphate enhancement of its strength properties.

2.2. Scope of the study
The results of this experimental study are applicable, only to the effect on compressive strength of concrete, caused by the replacement of cement by Metakaolin to the extends of 0, 10, 15, 20, 25 and 30 percentage in respect of OPC. Fine aggregate passing through 4.75 mm I.S. Sieve and retained on 300 microns I.S. Sieve was used in this study. Coarse aggregate passing through 12.5 mm sieve and retained on 4.75mm sieve, was adopted. Ordinary potable water was used for mixing and curing. Immersion curing for a period of 28 days was adopted. The effect on compressive strength, caused by the replacement of cement by Metakaolin to the extends of 0%, 10%, 15%, 20%, 25%, 30% in respect of OPC and acids (for H_2SO_4 – 0.5% and 1% concentration and for HCl – 0.5% and 1% concentration) was studied.

2.3. Need for the present study
The behaviour of high strength concrete in strength and durability has been known earlier for normal and other concretes whereas, for Metakaolin blended concrete, they have not been known. Hence this study is an emphasis on mechanical properties of M20, M30, M40, M50, M70, M80 grades of concrete with partially replaced cement.

2.4. Need for the present study
Based on the preliminary investigations carried out, the experimental investigation is planned as under.

1. Find the properties of the materials such as cement, sand, coarse aggregate and Metakaolin.
2. Obtain the mix proportions of OPC concrete for M20, M30, M40 and M50 by IS method (IS 10262 2009 [1]). Entropy and Shacklock's empirical method was adopted for mix proportioning of M60, M70 and M80.
3. Calculate the mix proportions for weight batchingwith partial replacements such as 0%, 10%, 15%, 20%, 25%, 30% of Metakaolin with concrete.
4. Preparation of Testing Specimens such as concrete specimens of size 150 mm × 150 mm × 150 mm for durability studies in laboratory with 0%, 10%, 15%, 20% 25% 30% replacement of OPC with Metakaolin for M20, M30, M40, M50, M60, M70 and M80 grade concrete.
5. To evaluate the 28 days compressive strength and durability (RCPT) of M20, M30, M40, M50, M60, M70 and M80 grade concrete with Metakaolin replacement exposed to 1% and 0.5% concentrations of HCl and H_2SO_4.

3. Experimental program
3.1. Material Properties
The materials used in the experimental work namely cement, Metakaolin, fine aggregate and coarse aggregate (20mm passing and 10mm retained) have been tested in the laboratory for use in the mix design. Ordinary Portland cement of 43 grade confirming to IS 8112: 1989 [14] was used in this

research. Fine Aggregate. Aggregates smaller than 4.75mm and up to 0.075mm are considered as fine aggregate. The specific gravity of fine aggregate is 2.60.

Sieve analysis to determine the particle size distribution of the coarse and fine aggregates is done by sieving the aggregates as per IS 2386 (Part 1) 1963 [15]. A set of IS Sieves of sizes- 80mm, 40mm, 20mm, 16mm, 10mm, 4.75mm, 2.36mm, 1.18mm, 600µm, 300µm and 150 µm were used for finding fineness modulus of fine and coarse aggregate. IS Sieves of sizes from 80mm to 4.75mm were used for coarse aggregate analysis and from 4.75mm to 150 µm IS sieves were used for the analysis of fine aggregates.

Fine aggregate

Table 1. Sieve Analysis of Fine Aggregate

IS Sieve size (mm)	Weight retained (kgs)	Cumulative weight retained (kgs)	Cumulative % weight retained (w1)	% Passing
10	0.005	0.005	0.5	99.5
4.75	0.005	0.010	1	99
2.36	0.01	0.02	2	98
1.18	0.075	0.095	9.5	90.5
0.6	0.16	0.255	25.5	74.5
0.3	0.43	0.685	68.5	31.5
0.15	0.260	0.945	94.5	5.5

Table 2. Properties of Fine Aggregate

Properties	Test values
Specific gravity	2.60
Bulk density (gm/cc)	1.68
Fineness modulus	2.66

Fineness modulus= $\frac{W_1}{100} = \frac{201{\cdot}5}{100}$=2.015;

Hence Fine aggregate (sand) belongs to **ZONE – III.**

Coarse aggregate.

Aggregates greater than 4.75mm are considered as coarse aggregate.

Table 3. Sieve Analysis of Coarse Aggregate

IS Sieve size (mm)	Weight retained (kgs)	Cumulative weight retained (kgs)	Cumulative % weight retained(w2)	Cumulative % passing
80	0	0	0	100
40	0	0	0	100
20	1.37	1.37	27.4	72.6
10	3.545	4.915	98.3	1.7
4.75	0.085	5.0	100	0
2.36	-	-	100	0
1.18	-	-	100	0
0.6	-	-	100	0
0.3	-	-	100	0
0.15	-	-	100	0

Table 4. Properties of Coarse Aggregate

Properties	Test values
Specific gravity	2.64
Bulk density (loose) (gm/cc)	1.38
Bulk density (compacted)(gm/cc)	1.53

Sieve Analysis
Coarse aggregate: 20mm passed
Fine aggregate: Confirming to Zone III of Table 4 of IS 383 1970 [16]
Test data for materials.

Cement	OPC 43 grade
Specific Gravity of Cement	3.15
Specific Gravity of Coarse Aggregate	2.71
Specific Gravity of Fine Aggregate	2.52
Specific Gravity of Water	1
Water Absorption of 20 mm Aggregate	0.5%
Water Absorption of Sand	1.0%
Free (Surface) Moisture of 20 mm Aggregate/Sand	Nil

3.2. Manufacturing of concrete and its workability
Concrete is mixed, placed, compacted, cured, de-moulded and tested.
3.3 Tests on workability
Slump cone test. Slump cone test is a very common test for determination of workability of concrete. This test was carried out for M80, before casting the specimens.
Compaction Factor Test. This test is more accurate than slump cone test and this test is used to determine the workability of low water-cement ratio concrete more accurately. This test is conducted as per IS 1199 1959 [16].

Table 5. Compaction Factor Values

Description	M20	M30	M40	M50	M60	M70	M80
Plain Concrete	0.94	0.92	0.89	0.88	0.87	0.86	0.84
Metakaolin 10%	0.89	0.88	0.87	0.86	0.84	0.83	0.82
Metakaolin 15%	0.88	0.87	0.86	0.84	0.83	0.81	0.80
Metakaolin 20%	0.873	0.86	0.84	0.82	0.80	0.79	0.78
Metakaolin 25%	0.86	0.85	0.83	0.80	0.79	0.76	0.75
Metakaolin 30%	0.85	0.84	0.81	0.79	0.76	0.74	0.72

Table 6. Slump Cone Values

Description	M20	M30	M40	M50	M60	M70	M80
Plain Concrete	40	38	34	30	26	22	20
Metakaolin 10%	35	33	30	26	24	20	18
Metakaolin 15%	33	28	26	23	19	17	15
Metakaolin 20%	30	26	20	19	17	14	13
Metakaolin 25%	28	25	23	19	17	14	11
Metakaolin 30%	26	21	19	17	14	12	10

IOP Conf. Series: Materials Science and Engineering 1025 (2021) 012004 doi:10.1088/1757-899X/1025/1/012004

4. Results and Discussions

This section deals with the analysis of the results after conduction of compression test concrete cube specimens. Discussions are made by comparing the results between with and without Metakaolin variation in acids by keeping a control specimen as reference.

4.1. Compressive strength

Determination of Compressive Strength for Different Grades with variation in % of Metakaolin for 28 and 90 days.

4.1.1. Compressive strength for 28 days.

For M20 grade concrete % increase of compressive strength for replacement of cement with Metakaolin by 10% is 6.81%, 15% is 12.58%, 20% is 4.86%, 25% is 3.03% with respect to control specimen (0% Metakaolin at 27°c curing). The % decrease in compressive strength for replacement of cement with Metakaolin by 30% is 0.39% for control specimen (0% Metakaolin at 27°c curing).

For M30 grade concrete % increase in compressive strength for replacement of cement with Metakaolin by 10% is 4.42%, 15% is 12.10%, 20% is 9.94%, 25% is 5.6%, 30% is 1.98% with respect to control specimen (0% Metakaolin at 27°c curing).

For M40 grade concrete % in compressive strength for replacement of cement with Metakaolin by 10% is 3.34%, 15% is 11.66%, 20% is 9.20%, 25% is 4.64%, and 30% is 0.74% with respect to control specimen (0% Metakaolin at 27°c curing).

For M50 grade concrete % increase in compressive strength for replacementof cement with Metakaolin by 10% is 1.69%, 15% is 3.38%, 20% is 3.38%, 25% is 6.25%, 30% is 7.30% with respect to control specimen (0% Metakaolin at 27°c curing).

For M60 grade concrete % increase in compressive strength for replacementof cement with Metakaolin by 10% is 2.61%, 15% is 9.02%, 20% is 8.36%, 25% is 4.29%, 30% is 0.89% with respect to control specimen (0% Metakaolin at 27°c curing).

For M70 grade concrete % increase in compressive strength for replacement of cement with Metakaolin by 10% is 7.26% and 15% is 9.50% to control specimen (0% Metakaolin at 27°c curing). The % decrease in compressive strength for replacement of cement with Metakaolin by 20% is 3.34%, 25% is 9.49% and 30% is 13.96% to control specimen (0% Metakaolin at 27°c curing).

For M80 grade concrete % increase in compressive strength for replacementof cement with Metakaolin by 10% is 9.13% and 15% is 11.68%, 20% is 6.60% to control specimen (0% Metakaolin at 27°c curing). The % decrease in compressive strength for replacementof cement with Metakaolin by 25% is 1.00%, 30% is 5.07% to control specimen (0% Metakaolin at 27°c curing).

For all grades, the compressive strength increases with the replacement of cement by Metakaolin up to 15% and then decreases. Maximum compressive strength of 11.68% is obtained for M80 grade concrete with replacement of cement by 15% Metakaolinfor 28 days curing to control specimen (0% Metakaolin at 27°C curing).

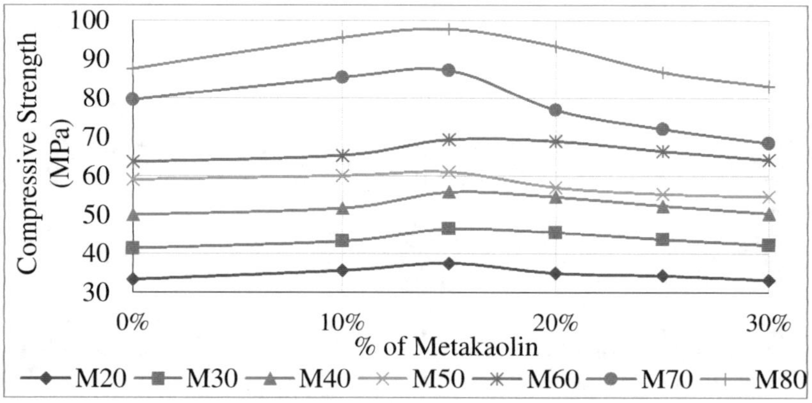

Figure 1. Compressive Strength for Different Grades with Variation in % of Metakaolin for 28 days

4.1.2. Compressive strength for 90 days.

For M20 grade concrete % increase in compressive strength for replacementof cement with Metakaolin by 10% is 8.64%, 15% is 17.44%, 20% is 14.82%, 25% is 12.47%, 30% is 5.59% with respect to control specimen (0% Metakaolin at 27°c curing) and respectively.

For M30 grade concrete % increase in compressive strength for replacementof cement with Metakaolin by 10% is 6.74%, 15% is 19.0%, 20% is 15.73%, 25% is 8.35%, 30% is 1.93% with respect to control specimen (0% Metakaolin at 27°c curing).

For M40 grade concrete % increase in compressive strength for replacementof cement with Metakaolin by 10% is 5.10%, 15% is 11.25%, 20% is 7.17%, 25% is 0.16%, and 30% is 0.40% with respect to control specimen (0% Metakaolin at 27°c curing).

For M50 grade concrete % increase in compressive strength for replacementof cement with Metakaolin by 10% is 1.63% 15% is 3.27%, 20% is 4.91%, 25% is 2.57%, 30% is 6.36% with respect to control specimen (0% Metakaolin at 27°c curing).

For M60 grade concrete % increase in compressive strength for replacementof cement with Metakaolin by 10% is 6.99%, 15% is 11.20%, 20% is 4.41%, 25% is 2.65%, 30% is 3.66% with respect to control specimen (0% Metakaolin at 27°c curing).

For M70 grade concrete % increase in compressive strength for replacementof cement with metakaolin by 10% is 3.06%, 15% is 23%, 20% is 5.37%, 25% is 2.96% and 30% is 1.42% with respect to control specimen (0% Metakaolin at 27°c curing).

For M80 grade concrete % increase in compressive strength for replacementof cement with Metakaolin by 10% is 3.00%, 15% is 10%, 20% is 7.56%, 25% is 2.24% and slightly decreased for 30% is 0.50% with respect to control specimen (0% Metakaolin at 27°c curing).

For each grade, the compressive strength increases with the replacement of cement by Metakaolin up to 15% and then decreases. Maximum compressive strength of 10.00% is obtained for M80 grade concrete with replacement of cement by 15% Metakaolin for 90 days curing to control specimen (0% Metakaolin at 27°c curing).

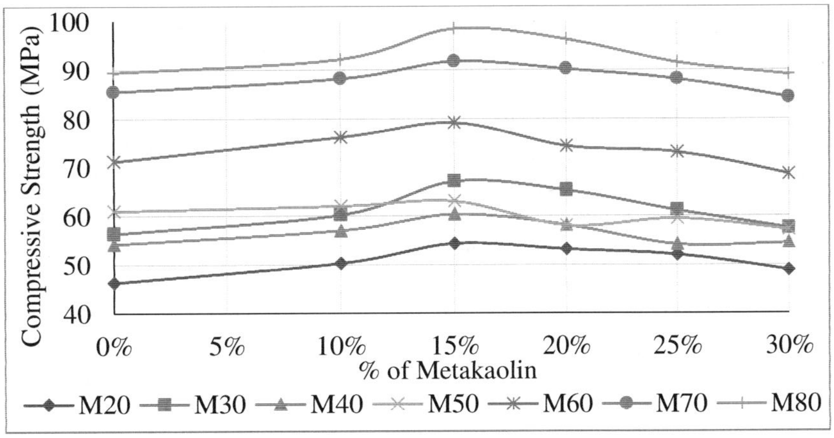

Figure 2. Compressive Strength for Different Grades with Variation in % of Metakaolin for 90 days

4.1.3. Summary of effect of Metakaolin on Compressive Strength of Concrete.

It is observed that at 15% replacement of cement with Metakaolin, concrete attains its maximum compressive strength to all grades (M20 - M80) for both 28 days and 90 days curing. When the replacement exceeds 15% with Metakaolin, the compressive strength is found to be decreasing slightly. The reduction in compressive strength for above 15% replacement of Metakaolin, when compared to Metakaolin with 10%, is expecting because of the result of a clinker dilution effect. The dilution effect is a result of replacing a part of cement by the equivalent quantity of Metakaolin. In Metakaolin concrete, the filler effect, the pozzolanic reaction of Metakaolin with calcium hydroxide and compounding effect (synergetic effect of mineral admixture) react opposite of the dilution effects Metakaolin improves the pozzolanic reaction by filler effect and accelerate the hydration of cement.

4.1.4. Effect of compressive strength with variation in acid of 0.5% HCl (Grade vs. % of Metakaolin).

The % decrease of compressive strength with the addition of 0.5% HCl for M20 grade concrete, replacement of cement with Metakaolin by 0% is 6.6% respectively. The % increase for 10% is 0.8% and 15% is 5.8% respectively. The % decrease for 20% is 1.14%, 25% is 2.31%.

The % decrease of compressive strength with the addition of 0.5% HCl for M30 grade concrete, replacement of cement with Metakaolin by 0% is 5.37%. The % decrease for 10% is 0.41%. The % increase for 15% is 6.70%, 20% is 5.10%, 25% is 1.30% when compared with control specimen (normal water curing). The % decrease for 30% is 2.85% respectively when compared with control specimen (normal water curing).

The % decrease of compressive strength with the addition of 0.5% HCl for M40 grade for concrete, replacement of cement with Metakaolin by 0% is 4.44%. The % decrease for 10% is 0.66%. The % increase for 15% is 7.20%, 20% is 5.20%, 25% is 1.08%. The % decrease for 30% is 3.26% respectively when compared with control specimen (normal water curing).

The % decrease of compressive strength with the addition of 0.5% HCl for M50 grade concrete, replacement of cement with Metakaolin by 0% is 3.76%. The % decrease for 10% is 1.35%. The % increase for 15% is 1.67%. The % decrease for 20% is 6.52%, 25% is 9.27%, 30% is 10.69% respectively when compared with control specimen (normal water curing).

There is no change in compressive strength with the addition of 0.5% HCl for M60 grade concrete, replacement of cement with Metakaolin by 0%. The % increase of compressive strength for replacement of cement with Metakaolin by 10% is 2.61%, 15% is 9.02%, 20% is 8.36%, 25% is 4.29%, 30% is 0.89% respectively when compared with control specimen (normal water curing).

The % loss of compressive strength with the addition of 0.5% HCl for M70 grade concrete, replacement of cement with Metakaolin by 0% is 2.79%. The % increase for 10% is 4.75%, 15% is 6.70%. The % decrease for 20% is 5.85%, 25% is 11.72%, 30% is 16.48% respectively when compared with control specimen (normal water curing).

The % loss of compressive strength with the addition of 0.5% HCl for M80 grade concrete for replacement of cement with Metakaolin by 0% is 6.85%. The % increase for 10% is 9.13%, 15% is 4.31%. The % decrease for 20% is 3.03%, 25% is 7.35%, 30% is 22.06% respectively when compared with control specimen (normal water curing).

For each grade, the compressive strength increases with increase in % of Metakaolin up-to 15% and then decreases. The maximum compressive strength for 0.5% HCl curing is obtained for M80 grade concrete as 5.8% with the replacement of cement by 15% Metakaolin when compared with control specimen (normal water curing).

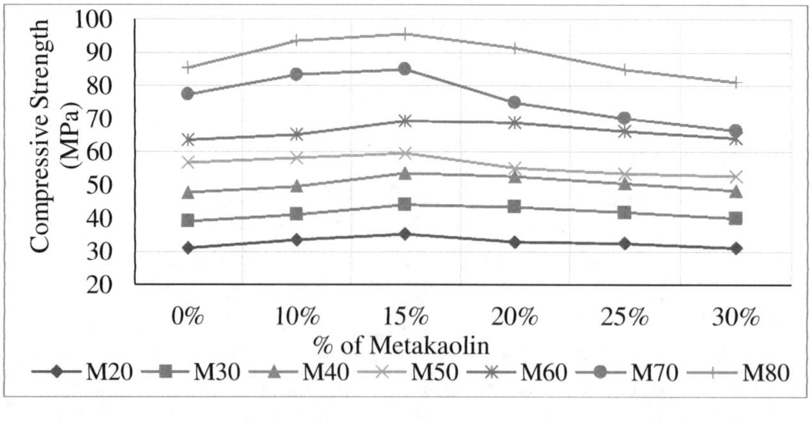

Figure 3. Effect of Compressive Strength with Variation in % of Metakaolin for 0.5% of HCl (% of Metakaolin vs. grade)

4.1.5. Effect of compressive strength with variation in acids 1% HCL (Grade vs % of Metakaolin)
The % loss of compressive strength with addition of 1% HCl for M20 grade concrete, replacement of cement with Metakaolin by 0% is 13.33%, 10% is 6.51% ,15% is 0.78% ,20% is 8.46%, 25% is 8.97%, 30% is 13.06% respectively when compared with control specimen (normal water curing).
The % loss of compared with compressive strength with the addition of 1% HCl for M30 grade concrete, replacement of cement with Metakaolin by 0% is 10.74%. The % decrease of compressive strength for replacement of cement with Metakaolin by 10% is 6.31%. The % increase of compressive strength for replacement of cement with Metakaolin by 15% is 1.33%. The % decrease of compressive strength for replacement of cement with Metakaolin by 20% is 0.79%, 25% is 4.06%, 30% is 8.22% respectively when compared with control specimen (normal water curing).
The % loss of compressive strength with addition of 1% HCl for M40 grade concrete, replacement of cement with Metakaolin by 0% is 8.0%. The % increase for 10% is 3.34%, 15% is 11.66%, 20% is 9.20%, 25% is 4.64% 30% is 0.74% respectively when compared with control specimen (normal water curing).
The % decrease of compressive strength with addition of 1% HCl for M50 grade concrete, for replacement of cement with Metakaolin by 0% is 7.52%. The % decrease for 10% is 5.49%, 15% is 2.76%, 20% is 10.66%, 25% is 13.03%, 30% is 14.45% respectively when compared with control specimen (normal water curing).
The % decrease of compressive strength with the addition of 1% HCl for M60 grade concrete, replacement of cement with Metakaolin by 0% is 6.98%. The % decrease by 10% is 4.37%. The % increase of compressive strength for replacement of Metakaolin in cement by 15% is 2.02%, 20% is 1.38%, % decrease for 25% is 1.99%, 30% is 5.74% respectively when compared with control specimen (normal water curing).
The % decrease of compressive strength with the addition of 1% HCl for M70 grade concrete, replacement of cement with Metakaolin by 0% is 5.02%. The % increase of compressive strength for replacement of cement with Metakaolin 10% is 0.98%, 15% is 3.90%. The % decrease for 20% is 8.92%, 25% is 14.51%, 30% is 19.27% respectively when compared with control specimen (normal water curing).
The % loss of compressive strength with addition of 1% HCl for M80 grade concrete, replacement of cement with Metakaolin by 0% is 5.07%, The % increase for 10% is 4.06%, 15% is 6.60%, 20% is 1.53%. The % decrease for 25% is 5.57%, 30% is 9.89% respectively when compared with control specimen (normal water curing).
For each grade, the compressive strength increases with increase in % of Metakaolin up-to 15% and then decreases. The maximum compressive strength for 1% HCl curing is obtained for M80 grade concrete as 6.60% with the replacement of cement by 15% Metakaolin when compared with control specimen (normal water curing).

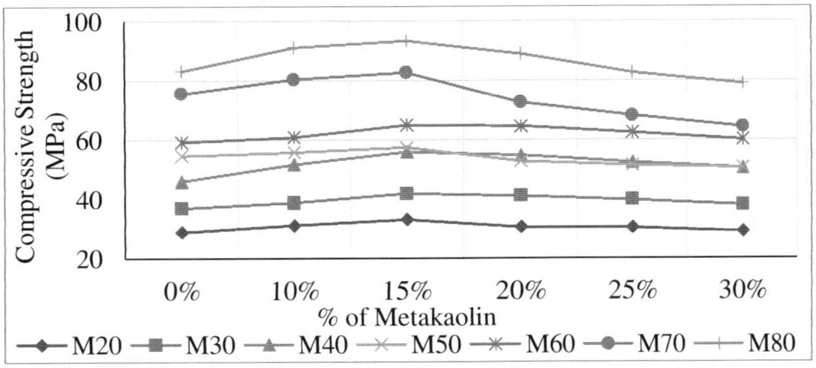

Figure 4. Effect of Compressive Strength with Variation in % of Metakaolin for 1% of HCl (% of Metakaolin vs. grade)

4.1.6. Effect of compressive strength with variation in acids 0.5% H_2SO_4 (Grade vs % of Metakaolin).
The % decrease of compressive strength with addition of 0.5% H_2SO_4 for M20 grade concrete, replacement of cement with Metakaolin by 0% is 19.33% 10% is 11.86%, 15% is 8.10% ,20% is 15.13%, 25% is 15.01%, 30% is 19.06% respectively when compared with control specimen (normal water curing).
The % decrease of compressive strength with addition of 0.5% H_2SO_4 for M30 grade concrete, replacement of cement with Metakaolin by 0% is 15.58%, 10% is 10.62%, 15% is 4.57%, 20% is 6.17%, 25% is 8.93%, 30% is 13.06% respectively when compared with control specimen (normal water curing).
The % decrease of compressive strength with addition of 0.5% H_2SO_4 for M40 grade concrete, replacement of cement with Metakaolin by 0% is 12.88%, 10% is 9.10% ,15% is 2.12%, 20% is 4.12%, 25% is 7.38% ,30% is 11.70% respectively when compared with control specimen (normal water curing).
The % decrease of compressive strength with addition of 0.5% H_2SO_4 for M50 grade concrete, replacement of cement with Metakaolin by 0% is 10.91%, 10% is 8.50%, 15% is 6.89%, 20% is 14.42%, 25% is 16.44%, 30% is 17.84% respectively when compared with control specimen (normal water curing).
The % decrease of compressive strength with addition of 0.5% H_2SO_4 for M60 grade concrete, replacement of cement with Metakaolin by 0% is 10.13%, 10% is 7.17%, 15% is 1.80%, 20% is 2.10%, 25% is 5.15%, 30% is 8.88% respectively when compared with control specimen (normal water curing).
The % decrease of compressive strength with addition of 0.5% H_2SO_4 for M70 grade concrete, replacement of cement with Metakaolin by 0% is 8.09%, 10% is 0.55%, % increase for 15% is 0.84%, % decrease for 20% is 11.71%, 25% is 9.49%, 30% is 13.96% respectively when compared with control specimen (normal water curing).
The % decrease of compressive strength with addition of 0.5% H_2SO_4 for M80 grade concrete, replacement of cement with Metakaolin by 0% is 7.35%. The % increase for 10% is 2.03%, 15% is 3.81%. The % decrease for 20% is 1.00%, 25% is 7.86%, 30% is 12.17% respectively when compared with control specimen (normal water curing).
For each grade, the compressive strength increases with increase in % of Metakaolin up-to 15% and then decreases. The maximum compressive strength for 0.5% H_2so4 curing is obtained for M80 grade concrete as 3.81%with replacement of cement by 15% Metakaolin when compared with control specimen (normal water curing).

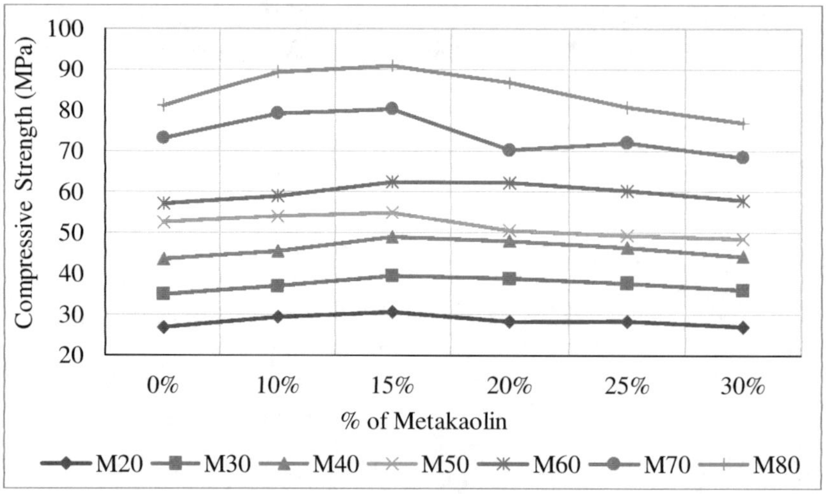

Figure 5. Effect of Compressive Strength with Variation in % of Metakaolin for 0.5% of H_2SO_4 (% of Metakaolin vs. grade)

4.1.7. Effect of compressive strength with variation in acids 1% H_2SO_4 (Grade vs % of Metakaolin)
The % loss of compressive strength with addition of 1% H_2SO_4 for M20 grade concrete, replacement of cement with Metakaolin by 0% is 35.34%, 10% is 29.21%, 15% is 24.80%, 20% is 30.51%, 25% is 31.68%, 30% is 33.09% respectively when compared with control specimen (normal water curing).
The % loss of compressive strength with addition of 1% H_2SO_4 for M30 grade concrete, replacement of cement with Metakaolin by 0% is 28.48%, 10% is 10.09%, 15% is 18.03%, 20% is 18.56%, 25% is 22.36% , 30% is 24.37% respectively when compared with the % loss of compressive strength with addition of 1% H_2So4 for M40 grade concrete, replacement of cement with Metakaolin by 0% is 23.54%, 10% is 20.66%, 15% is 13.24%, 20% is 14.36%, 25% is 18.48%, 30% is 21.04% respectively when compared with control specimen (normal water curing).
The % loss of compressive strength with addition of 1% H_2SO_4 for M50 grade concrete, for replacement of cement with Metakaolin by 0% is 19.94%, 10% is 18.30%, 15% is 16.32%, 20% is 23.10%, 25% is 25.84%, 30% is 25.76% respectively when compared with control specimen (normal water curing). The % loss of compressive strength with addition of 1% H_2So4 for M60 grade concrete, replacement of cement with Metakaolin by 0% is 18.51%, 10% is 16.26%, 15% is 10.55%, 20% is 10.16%, 25% is 13.89%, 30% is 16.23% respectively when compared with control specimen (normal water curing).
The % loss of compressive strength with addition of 1% H_2SO_4 for M70 grade concrete, replacement of cement with Metakaolin by 0% is 14.79%, 10% is 7.81%, 15% is 6.14%, 20% is 18.15%, 25% is 24.02%, 30% is 27.79% respectively when compared with control specimen (normal water curing).The % loss of compressive strength with addition of 1% H_2So4 for M80 grade concrete, replacement of cement with Metakaolin by 0% is 7.35%. The % increase for 10% is 2.03%, 15% is 3.81%. The % decrease for 20% is 1.00%, 25% is 7.86%, 30% is 12.17% respectively when compared with control specimen (normal water curing).
For each grade, the compressive strength increases with increase in % of Metakaolin up-to 15% and then decreases. The maximum compressive strength for 1% H_2so4 curing is obtained for M 80 grade concrete as 3.81% with the replacement of cement by 15% Metakaolin when compared with control specimen (normal water curing).

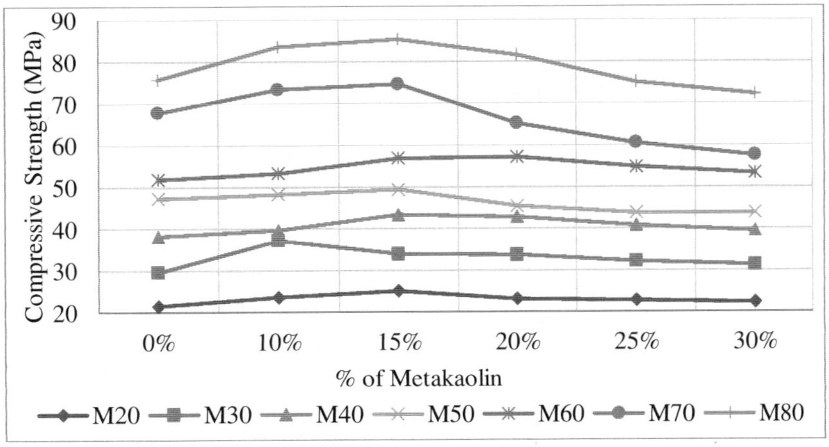

Figure 6. Effect of Compressive Strength with Variation in % of Metakaolin for 1% of H_2so4 (% of Metakaolin vs. grade)

4.1.8. Summary on Effect of Acids on Compressive Strength of Concrete. It was observed that ordinary Portland Cement (OPC) is highly alkaline in nature and having the pH values above 12. So, whenever the concrete or matrix paste comes into contact with the acids. The reaction between the concrete and acid will start and finally leads to the disintegration of its components If pH decreases to values lower than stability limits of cement hydrates, then the corresponding hydrate loses calcium and decomposes to the amorphous hydrogel. The final reaction products of acid attack are the corresponding calcium salts of the acid besides hydrogels of siliceous, aluminium, and ferric oxides. Concrete is susceptible to acid because of its alkaline nature. The components of the cement paste break during contact with acids.

5. Conclusions

Based on the analysis of experimental results and discussions the following conclusions are made.

1. Maximum compressive strength of 11.68% and 10.00% is obtained for M80 grade concrete with replacement of cement by 15% Metakaolin for 28 days curing and 90 days curing to control specimen (0% Metakaolin at 27°c curing).
2. The maximum compressive strength for 0.5% HCl, 1% HCl curing is obtained for M80 grade concrete as 5.8%, 6.60% with the replacement of cement by 15% Metakaolin when compared with control specimen (normal water curing). The maximum compressive strength for 0.5% H_2SO_4, 1% H_2SO_4 curing is obtained for M80 grade concrete as 3.81%, 3.81% with the replacement of cement by 15% Metakaolin when compared with control specimen (normal water curing).

Summary

The compressive strength of concrete is maximum at 15% replacement of cement by Metakaolin. The compressive strength of concrete showed better result at 15% replacement of Metakaolin for 0.5% and 1% HCl and H_2SO_4 at the age of 28 days of strength. The effect of HCl on strength of the Metakaolin concrete is lower than the effect of H_2SO_4. ·

The structures that exposed to the severe acidic environment should be given special attention while designing the structure, especially while selecting the concrete compositions and a higher safety factor should be adopted. If possible, special types of cement should be allowed reducing the deterioration effect due to the harsh acidic environment. Acidic curing environment hurts the compressive strength and density of the concrete. The rate of reduction of compressive strength for a lower concentration of acid (5%) which decreases as the concentration is raised from 5% to10%. The compressive strength is lower at a higher concentration of acid irrespective of the type of acid and method of attack. In case of concrete structures which have to perform in an acidic environment, particular attention should be given to the design, a higher factor of safety almost double for strength should be used and if feasible, special admixtures to mitigate the effect of the acidic environment should be recommended.

6. References

[1] IS 10262 2009 Indian Standard Concrete Mix Proportioning - Guidelines (First Revision) *Bureau of Indian Standards* New Delhi
[2] Al-Akhras NM 2006 Durability of metakaolin to sulfate attack *Cement and Concrete Research* **36 (9)** p 1727–34
[3] Bai J, Wild S 2002 Investigation of the temperature change and heat evolution of mortar incorporating PFA and metakaolin *Cement & Concrete Composites* **24** p 201–9
[4] Brooks JJ, Johari MMA 2001 Effect of metakaolin on creep and shrinkage of concrete *Cement & Concrete Composites* **23** p 495–502
[5] Courard L, Darimont A, Schouterden M, Ferauche F, Willem X, Degeimbre R 2003 Durability of mortars modified with metakaolin *Cement and Concrete Research* **33** p 1473–9
[6] Frías M, Sánchez de Rojas M I, Cabrera J 2000 The effect that the pozzolanic reaction of metakaolin has on the heat evolution in metakaolin-cement mortars *Cement and Concrete Research* **30** p 209–16
[7] Gruber K A, Ramlochan T, Boddy A, Hooton R D, Thomas M D A 2001 Increasing concrete durability with high-reactivity metakaolin *Cement and Concrete Research* **23 (6)** p 479–84
[8] Khatib J M, Clay R M 2004 Absorption characteristics of metakaolin concrete *Cement and Concrete Research* **34 (1)** p 19–29
[9] Lee S T, Moon H Y, Hooton R D, Kim J P 2005 Effect of solution concentrations and replacement levels of metakaolin on the resistance of mortars exposed to magnesium sulfate solutions *Cement and Concrete Research* **35** p 1314–23

[10] Poon C S, Lam L, Kou S C, Wong Y L, Wong R 2001 Rate of pozzolanic reaction of metakaolin in high-performance cement pastes *Cement and Concrete Research* **31** p 1301–6
[11] Poon C S, Azhar S, Anson M, Wong Y L 2003. Performance of metakaolin concrete at elevated temperatures *Cement & Concrete Composites* **25** p 83–9
[12] Qian X, Li Z 2001 The relationships between stress and strain for high-performance concrete with metakaolin *Cement and Concrete Research* **31** p 1607–11
[13] Ramlochan, T, Thomas M, Gruber K A 2000 The effect of metakaolin on alkali-silica reaction in concrete *Cement and Concrete Research* **30** p 339–44
[14] IS 8112 1989 Indian Standard Ordinary Portland Cement 43 Grade — Specification (Second Revision) *Bureau of Indian Standards* New Delhi
[15] IS 2386 (Part 1) 1963 Indian Standard Methods of Test for Aggregates for Concrete Part I Particle Size and Shape *Bureau of Indian Standards* New Delhi
[16] IS 383 1970 Specification for Coarse and Fine Aggregates from natural sources for concrete *Bureau of Indian Standards* New Delhi
[17] IS 1199 1959 Indian Standard Methods of Sampling and analysis of concrete *Bureau of Indian Standards* New Delhi

Cost optimization of a rectangular singly reinforced concrete beam by Generalized Reduced Gradient method

P Markandeya Raju[1], A Manasa[2], G Rohini[2]

[1]Professor and Head, Civil Engineering, MVGR College of Engineering (A), Vizianagaram 535005 Andhra Pradesh, India
[2]Under graduate students of Civil Engineering, Dr Lankapalli Bullayya College of Engineering (Women), New Resapuvanipalem, Visakhapatnam 530013, Andhra Pradesh, India
Email: markandeyaraju@mvgrce.edu.in

Abstract. Cost Optimization of a singly reinforced rectangular beam using GRG method is presented in this paper. The objective function is chosen as the overall cost of the beam given by the sum of the products of unit rates of steel and concrete in rupees per unit volume (m^3) with their volumes. 28 combinations of simply supported beams of different spans subjected to different point loads and uniformly distributed loads are analyzed using EXCEL Solver. The constraints are adopted to satisfy various guidelines as per IS 456:2000 code of practice.
Keywords: Singly reinforced beam, Optimization, Excel solver, Cost-effective, Structural Design

1. Introduction

The word optimization denotes a process of making a design or a decision as effective as possible. It is generally done by minimizing or maximizing an objective function that satisfies various constraints. It plays a very important role in structural engineering by helping the engineers in designing a good, efficient, safe and economical structure. Various Optimization techniques developed by mathematicians have the potential to reduce the overall cost of engineering construction through the effective and relevant mathematical methodology. The main purpose of structural optimization is to find the best design out of several alternate designs without compromising on safety, serviceability and durability by effective utilization of available resources.

In this paper, Cost Optimized design of a singly reinforced rectangular beam using GRG (Generalized Reduced Gradient) method is presented. The objective function is chosen as the overall cost of the beam given by the sum of the products of unit rates of steel and concrete in rupees per unit volume (m^3) with their volumes. 480 combinations of simply supported beams of different spans subjected to different point loads and uniformly distributed loads are analyzed using EXCEL Solver [1] where the data, Objective function, other formulae for calculation and constraints are written in order. The different combinations are solved and finally, the most optimum combination of design structure is taken. The optimization is done for the singly reinforced beam. The constraints are adopted to satisfy various clauses of IS 456:2000 [2]. The rates of concrete and steel are taken in Indian currency i.e. Rupees per unit volume as per SSR

2. Literature review

Many researchers have been working in this area of cost optimization of RC beams. Some of the works and corresponding findings are presented as follows.

Vishakha Tushiram Holambe et al [4] used to define a constrained optimization function for a per meter cost of singly reinforced rectangular concrete (SRRC) beam. 21 combinations of the span of beam and loading were analyzed using the SWARM Optimization approach. The unit cost of materials and unit cost for side and bottom formwork was considered to be a major cost, the result which was obtained were compared with the depth of the beam, area of steel required and cost of beam per meter for balanced sections. Stency Mariam Thomas et al [5] presented an optimal design of reinforced concrete beams, its main objective is to minimize the total cost of the beam. The optimization was done for different grades of concrete and steel to know the better grade of concrete and steel the design was done through MATLAB software. The problem was formulated as a non-linear constrained minimization it was solved through fmincon SQP algorithm. A model of optimal design of rectangular reinforced concrete section is presented by Barros et al [6] considered the stress-strain diagram which was described in EC2-2001 and MC90, the expressions like economic bending moment, optimal area of steel and optimal steel ratio between upper and lower steel was developed. He had concluded that in non-dimensional form the equations were nearly coincident for both singly and doubly reinforcement. Cost optimization is implemented in code was compared with other optimum models based on the ultimate design of ACI. Sonia et al [7] presented a particle swarm optimization (POS) to achieve optimal design of Reinforced Concrete (RC) beams. They have developed an algorithm to search for a minimum cost solution that has to satisfy Indian code requirements for R.C beams. The objective function consisted of the cost of concrete and rebars as prevalent at the place of construction. Many examples were presented to show the effectiveness of the formulation for achieving optimal design. Cost optimization of concrete structures is presented by Kamal C Sarma and Hojjat Adeli [8] they presented a series of paper on cost optimization of concrete structures which was published in archival journals. The concrete structures must involve at least 3 different materials that are concrete, steel and formwork. The structures which are included are beams, slabs, columns, frame structures, bridges, water tanks, folded plates, shear walls, pipes and a tensile member. A review of reliability-based cost optimization is also included in this paper. They have concluded that there is a need to perform research on cost optimization of realistic 3-dimensional structures, especially large structures with hundreds of members where optimization can result in a substantial saving. Kaveh and Ahangaran [9] presented a paper on discrete cost optimization of composite floor system using social harmony research model. They have included the costs of concrete, steel beam and shear studs in the cost function. The design is based on AISC load and resistance factor design specification and plastic design concept. They have considered 6 decision variables for the objective function. They have presented a model which results in significant cost saving. A parametric study is performed which is used to investigate the effects of beam spans and loading. Optimal seismic design of the reinforced concrete shear wall-frame structure is presented by Ali Kaveh and Pooya Zakian [10] they have performed the seismic design of reinforced concrete (RC) dual system as an optimization problem for which the charged system search algorithm is utilized as an optimize. The databases were created on ACI seismic criteria for beams, columns and shear walls. Ordinary design constraints and effective seismic design constraints were taken. According to the result, the proposed methodology can be considered as a suitable practical approach for optimal seismic design of reinforced concrete shear wall-frame structures. Shahzad Umar et al [11] presented a paper on cost optimization of RC godown. The main aim of this paper is to achieve optimal design of reinforced concrete structures, optimal sizing and reinforcing for beam and column member in multi-bay and multi-storey structures which results in cost saving over typical practice design. Siddhant Lakkad and Panchal [12] presented a paper on cost optimization of reinforced concrete structure elements, the main aim of this paper is to achieve the optimal and ideal design of the reinforced concrete structure, reducing size and reinforcement for beam member and column member in multi-bay and multi-storey structures. Which is used in several software likes STAAD PRO, ETABS, STRUDS. Programming of the design of structural elements

IOP Conf. Series: Materials Science and Engineering 1025 (2021) 012005 doi:10.1088/1757-899X/1025/1/012005

beam and column has been done using the MATLAB program. This optimization task reduces 24.76% of the total approximate cost in beams and 13.79% in columns.

From the above review literature, it can be observed that each author has considered the objective function differently. Some have considered material quantity and some have considered the overall cost. Further, most of the authors missed some of the key constraints that define the performance of the RC beam. Hence there is a need to comprehensively study the cost optimization of RC beam considering all possible constraints.

3. Methodology

The objective of this paper is to optimize the cost of a singly reinforced rectangular beam. The following step-by-step methodology was followed.

Step 1: 480 combinations of simply supported beams (SSB) of different spans subjected to different point loads were considered for the study. The combinations are obtained by varying the following parameters.

Span (L) : 1 m, 2 m,4 m and 8 m
Grade of Concrete : M20, M25, M30, M35 and M40.
Grade of Steel : Fe 250, Fe 415 and Fe 500.
Central Point Load (kN) : 10, 20, 40, 80
Uniformly distributed load (kN/m) : 3, 7, 10, 20

Step 2: An EXCEL [1] sheet to analyze and determine the optimized design of the beam for distributed and point loads are developed. GRG Method was adopted. The cost of different grades of steel and concrete are taken from SSR (2018-19) [03]. By changing the grade of concrete, the grade of steel (thereby their costs), Ultimate moment of resistance (Mu), Ultimate shear force (Vu), different objective functional values are obtained for different spans.

Step 3: From this value, the overall cost and percentage of steel are calculated for various point loads and uniformly distributed loads. While the effective span 'L' is changeable, the optimum breadth 'B' and 'depth' d are given by the EXCEL Solver.

Step 4: Graphs are drawn for different spans of the beam based on overall cost and percentage of steel combinations and analyzed.

The objective function which we have used for this singly reinforced beam is the total cost of the beam that includes the total cost of steel and concrete including fabrication and labour cost.

The expression is given by:

$f(x) = Cr\times (L\times B\times D- Vs) + Cs\times Vs\times \rho$

f(x) = (Rate of concrete in rupees per unit volume (Rs. $/m^3$)) × (Volume of concrete (m^3)) + (Rate of steel (Rs. /kg.) × (Volume of steel in m^3 × Density of steel in kg. $/m^3$)

Where,

Vs = Volume of Steel in m^3

Cr= Cost of concrete (Rs. $/m^3$)

Details required

Cs= Cost of steel (Rs. /kg.)

Cost of steel (Cs): 36,42,41.6 (Rs/kg) for different grades of steel

L= Length of the beam in m

B= Breadth of the beam in m

D= Overall depth of the beam in m

ρ= Density of steel = 7700 kg/m^3

Other notation/data used for derivation of objective function and determination of constraints are as follows:

Mu = Ultimate (factored) bending moment N/mm (due to dead and live load)

Vu = Ultimate (factored) vertical shear force in N (due to dead and live load)

fck = Compressive Strength of concrete = 20 MPa, 25 MPa, 30 MPa, 35 MPa, 40 MPa

For grades of Concrete: M20, M25, M30, M35, M40 respectively

fy = Grade of Steel = 250 N/mm^2, 415 N/mm^2, 500 N/mm^2
For grades of steel: Fe250, Fe 415, Fe 500 respectively.
C = Force of compression = 0.36×fck×B×Xu
T = Force of tension = 0.87×fy×Ast
Here,
Ast = Area of steel in tension zone
Moment of Resistance = MR = 0.87×fy×Ast×(d - 0.42×Xu)$(Clause\ 38.1\ of\ IS\ 456\ 2000)$
Mu, lim = Ultimate moment of resistance of the mid-span = k×fck×b×d^2
Where,
k = Moment of resistance constant depending on the grade of steel)
k = 0.148 for Fe 250, 0.138 for Fe 415 and 0.133 for Fe 500

$$Xu = Depth\ of\ neutral\ axis\ of\ the\ section = \frac{0.87 \times fy \times Ast}{0.36 \times fck \times b}$$

$(Clause\ 38.1\ of\ IS\ 456\ 2000)$

$$\frac{Xu, lim}{d} = 0.53\ for\ Fe\ 250, 0.48\ for\ Fe\ 415\ and\ 0.46\ for\ Fe\ 500$$

$(Clause\ 38.1\ of\ IS\ 456\ 2000)$

Initial values of cross-section and longitudinal dimensions
L= 2000mm
B= 200mm (changeable)
D= 341mm (changeable)
d'= Clear cover= 25mm (minimum clear cover if bar diameter less than or equal to 25mm)
$(Clause\ 26.4\ of\ IS\ 456\ 2000)$

ds = Diameter of stirrups (8 mm, 10 mm, 12 mm)
dts = Diameter of largest bar of tensile steel
d = effective depth = D – (effective cover) = D – (d'+ ds + (dts/2))
Area of Longitudinal steel
2 numbers of 12 mm diameter bars are provided as anchor bars in compression zone.
Asc= Area of steel in compression=$(2 \times \frac{\pi}{4} \times 12^2)$= 226.19 mm^2
Ast = Area of steel tension (limited to range within 226.28 mm^2 and 1473.21 mm^2)

The area of steel tension (limited to range within 226.28 mm^2 and 1473.21 mm^2) for different bar combinations adopted in this study is presented in Table 1.

Table 1. *Area of tension steel (mm^2) for various bar combinations*

Area of tension steel (mm^2)	Bar diameter (mm)	No. of bars	Bar diameter (mm)	No. of bars
0226.28	12	2	00	0
0339.42	12	3	00	0
0402.28	16	2	00	0
0515.42	16	2	12	1
0603.42	16	3	00	0
0628.57	20	2	00	0
0741.71	20	2	12	1
0829.71	20	2	16	1
0942.85	20	3	00	0
0982.14	25	2	00	0
1095.28	25	2	12	1
1183.28	25	2	16	1
1296.42	25	2	20	1
1473.21	25	3	00	0

Area of lateral reinforcement

Sv= Spacing of 2 legged vertical stirrups of diameter '*ds*' in mm (without bent up bars)

$$S_v = \frac{(0.87 \times fy \times Asv \times d)}{Vus} \quad (Clause\ 40.4\ of\ IS\ 456\ 2000)$$

Where

$$Asv = Area\ of\ 'ds\ mm' diametre\ 2\ legged\ vertical\ stirrups = \frac{2 \times \pi \times ds^2}{4}$$

Strength of the shear reinforcement $= Vus = Vu - \mathrm{Tc} \times B \times D$

Nc = Values are taken from Table 19 of IS456:2000 based grade of concrete and Ast provided

$$\mathrm{n = Number\ of\ Stirrups} = \frac{\text{Length over which stirrups are provided}}{\text{spacing of stirrups}} + 1$$

Length over which stirrups are provided = L – 2 × d'

Hence,

$$n = \frac{\mathrm{L} - 2\mathrm{d'}}{S_v} + 1$$

Length of stirrup = perimeter of stirrup + length of hooks

Perimeter of stirrup = 2× $\{(b - 2d') + (D - 2d')\}$

Length of hooks = 2× (24×ds)

$(Clause\ 26.2.2.1\ of\ IS\ 456\ 2000)$

Length of stirrup = 2× $\{(b - 2d') + (D - 2d') + (24 \times ds)\}$

$$Asts = Area\ of\ stirrups = \frac{\pi \times ds^2}{4}$$

$$\mathrm{Tv} = \frac{Vu}{B \times D} \quad (Clause\ 40.4\ of\ IS\ 456\ 2000)$$

Cost of different grades of concrete

Table 2 presents the Nominal concrete mix ratios (by volume and weight) that are considered for the estimation of the cost of different grades of concrete. The rate of Cement, Fine aggregate, Coarse aggregate, fabrication and labour cost were adopted as per SSR [2].

Table 2. *Nominal mix ratios for different grades of concrete (by volume and weight)*

M5	1	5	10	Parts
	146	751	1502	kg
M7.5	1	4	8	Parts
	180	740	1480	kg
M10	1	3	6	Parts
	235	721	1443	kg
M15	1	2	4	Parts
	335	688	1376	kg
M20	1	1.5	3	Parts
	427	657	1315	kg
M25	1	1	2	Parts
	589	603	1207	kg
M30	1	0.75	1.5	Parts
	726	557	1115	kg
M35	1	0.5	1	Parts
	946	484	969	kg
M40	1	0.25	0.5	Parts
	1357	347	695	kg

Table 3 presents the sample calculation for gross cost of M30 grade concrete and Fe 415 steel including centering cost, contractors profit and labour cost.

Note: Supply and placing of the Design Mix Concrete corresponding to IS 456 using Weigh Batcher / Mixer with 20mm size graded machine crushed hard granite metal (coarse aggregate) from approved quarry including cost and conveyance of all materials like cement, fine aggregate (sand) coarse aggregate, water etc., to site and including Seigniorage charges, sales & other taxes on all materials including all operational, incidental and labour charges such as weigh batching, machine mixing, laying concrete, curing etc., complete but excluding cost of steel and its fabrication charges for finished item of work (APSS No. 402)with minimum cement content as per IS code from standard suppliers approved by the department including pumping, centering, shuttering, laying concrete, vibrating, curing etc. complete but excluding cost of steel and its fabrication charges for finished item of work.

Table 3. *Sample calculation for cost of M30 grade concrete and Fe 415 grade steel*

FF Beams plain (not Hilly) areas			
PLAIN CONCRETE BASIC COST per 1 cum (M30 grade concrete) up to 150m			5037.67
Centering charges for structural members (Beams) as per SSR 2018-19 (in Rs. per cubic meters)	cum	1.000	2868.00
Total			7905.67
Contractor profit @13.615%			1076.36
MA @ 40% on labor (labour compound on centering @1473/-)			1117.66
Grand Total			10099.8
Fe415 grade steel	MT	1.000	42500

4. Derivation for Objective function

The Volume of steel = Volume of Longitudinal steel + Volume of Lateral steel

The volume of Longitudinal steel = product of Sum of the areas of steel in tension and compression zone and length of the bars

Here, Length of bars = Overall span – 2 times clear cover = L – 2× d'

Hence, the Volume of Longitudinal steel = $(Ast + Asc) \times (L - 2d')$

The volume of Lateral steel (stirrups) = number of stirrups × area of stirrup × Length of the stirrup

$$V_s = \left[\frac{L - 2d'}{S_v} + 1\right] \times \left[\frac{\pi \times ds^2}{4}\right] \times [2 \times \{(b - 2d') + (D - 2d') + (24 \times ds)\}]$$

The volume of concrete = Total Volume of Beam – Volume of Steel

Vc = V – Vs = L×B×D- Vs

5. Constraints

1) Neutral axis constraint: $Xu \leq Xu,max$
2) Moment constraint: $Mu \leq Mu,lim$
3) Deflection constraint:

$$d \leq \sqrt{\frac{Mu}{k \times fck \times b}} \; (Clause\ 38.1\ of\ IS\ 456\ 2000)$$

4) For spans up to 10m constraint:

$$\frac{l}{d} \leq 20 \;\; (Clause\ 23.2.1\ of\ IS\ 456\ 2000)$$

5) For spans > 10m constraint:

$$\frac{l^2}{d} \leq 200 \; if \; l < 10m, else \; \frac{l}{d} \leq 20 \; (Clause \; 23.2.1 \; of \; IS \; 456 \; 2000)$$

6) The maximum area of steel constraint:

$$\frac{Ast}{B \times d} \leq 0.04 \; (Clause \; 26.5.1.1(b) \; of \; IS \; 456 \; 2000)$$

7) The minimum area of steel constraint:

$$\frac{Ast}{B \times d} \geq \frac{0.85}{fy} \; (Clause \; 26.5.1.1(a) \; of \; IS \; 456 \; 2000)$$

8) Constraint to restrict the beam to be under-reinforced (to be designed as singly reinforced):

$$\frac{\frac{Xu}{d}}{\frac{Xu,lim}{d}} - 1 < 0 \; (Clause \; 38.1 \; of \; IS \; 456 \; 2000)$$

9) Recommended minimum beam depth to width ratio:

$$b < \frac{d}{1.5}$$

10) Recommended maximum beam depth to width ratio:

$$b > \frac{d}{3}$$

11) Minimum width constraint (assumed): B ≥ 200
12) Maximum width constraint (assumed): B ≤ 600
13) Minimum depth constraint (assumed): d ≥ 200
14) Maximum width constraint (assumed): d ≤ 600
15) Minimum Ast constraint (rounded off to nearest multiple of 100): Ast ≥ 200
16) Maximum Ast constraint (rounded off to nearest multiple of 100): Ast < 1500
17) The constraint for ensuring positive value for Volume of steel: The volume of steel > 0
18) The constraint for ensuring that Moment of resistance of the mid-span section is greater than the Ultimate (factored) bending moment: MR > Mu

The objective function and the constraints are programmed into the EXCEL Solver. Table 4 presents the EXCEL result sheet on running the EXCEL Solver for the following sample data.

Grade of Concrete	: M40	Ultimate Moment	: 40 kN_m
Grade of steel	: Fe415	Ultimate Shear force	: 80 kN
Effective span	: 2000 mm		

Table 4. Excel sheet calculations

Objective function	**2551.9**		d	300	mm
CR (Cost of Concrete)	10099.68	Rs./m³	Vus *(Cl. 40.4)*	1	
Grade of concrete	M40		C	143319.25	N
Grade of steel	Fe415		T	143319.25	N
k	0.138		Xu *(Cl. 38.1)*	49.76	mm
fy	415	N/mm²	Mu *(Cl. 38.1)*	40000300.02	N_mm
Cs (Cost of Steel)	42	Rs./kg	Mu,lim *(Cl. 38.1)*	99360000	N_mm
L (Effective Span)	2000	mm	Xu,lim./d *(Cl. 38.1)*	0.48	
Mu (Ultimate Moment)	40	kN_m	Min.Sv	225	mm
Vu (Ultimate Shear force)	80	kN	(L^2/d)-200	13133.33	
fck (Char. Comp. Strength of Concrete)	40	N/mm²	(L/d)-20	-13.33	

B	200	mm		C1	-94.24	<=0
D	341	mm		C2	-59359699.98	<=0
Ast	396.95	mm^2		C3	299.99	>=0
Bar Combination.				C4	-14.13	<=0
d'	**25**	mm		C5	-13.33	<=0
w (Density of Steel)	7700	kg/m^3		C6	-0.034	<=0
Asc	226.19	mm^2		C7	0.005	>=0
Vol. of Steel	3748390.22	mm^3		C8	-0.654	<0
100Ast/BD	0.498	%		C9	0	<0
Tv *(Cl. 40.4)*	1.17	N/mm^2		C10	100	>0
Tc (Cl. 40.1)	0.568	N/mm^2		C11	0	>=0
Sv *(Cl. 40.4)*	225	N/mm^2		C12	-400	<=0
ds	8	mm		C13	100	>=0
Vc	132651609.8	N		C14	-300	<=0
dc	25	mm		C15	196.95	>=0
Vsc	45909.75	N		C16	-1103.05	<0
N	10			C17	3748390.22	>0
dts	16	mm		C18	300.02	>0
Xu,lim	144	mm				

6. Results and discussion

The results presented graphically in 4 parts. Although the calculations are performed for all grades of concrete ranging from M20 to M40 (in intervals of 5, the results are presented here only for M20 and M40 to minimize the length of the manuscript.

Part 1 presents the variation of the percentage of steel with span for different point loads. Each graph presents results for a given grade of steel and concrete.

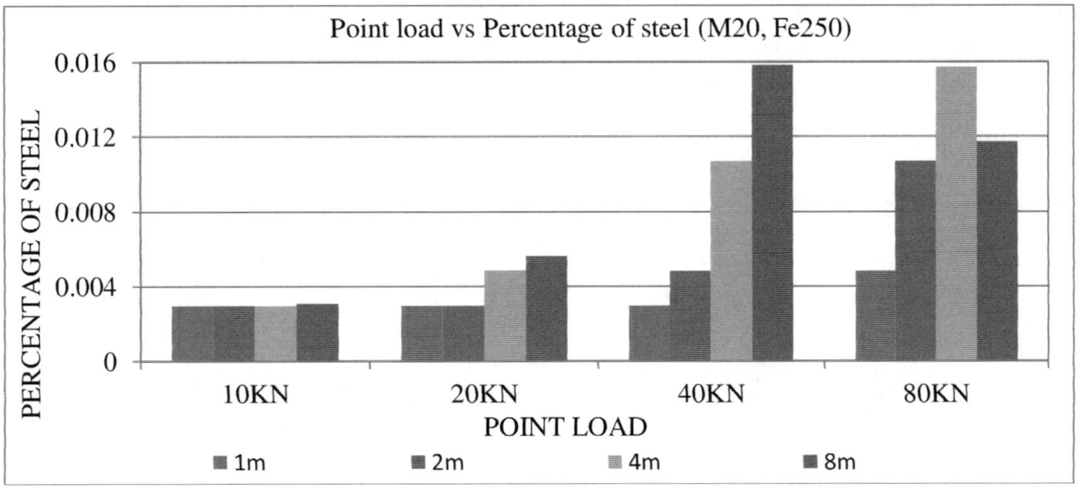

Figure 1. Variation of the percentage of steel with span for different loads (for Fe250, M20)

From figure. 1, it can be observed that for a span 1 m, at 10 KN, the percentage of steel is 0.002969. It slightly decreases to 0.002965 at 20kN and remains constant till 40 kN. Then increases to 0.004829 at 80 kN. For a span 2 m, at 10 kN, the percentage of steel is 0.002965 and remains constant till 20 kN. Then it increases to 0.004829 at 40kN and gradually increases to 0.010699 at 80 kN. For a span 4 m, at 10 kN, the percentage of steel is 0.002982. It increases to 0.004847 at 20 kN. Then it increases to 0.010699 at 40kN and gradually increases to 0.015736 at 80kN. For a span 8 m, at 10 kN, the percentage of steel is 0.003085. Then increases to 0.00564 at 20kN and it increases to 0.015852 at 40kN, later it decreases to 0.011728 at 80 kN.

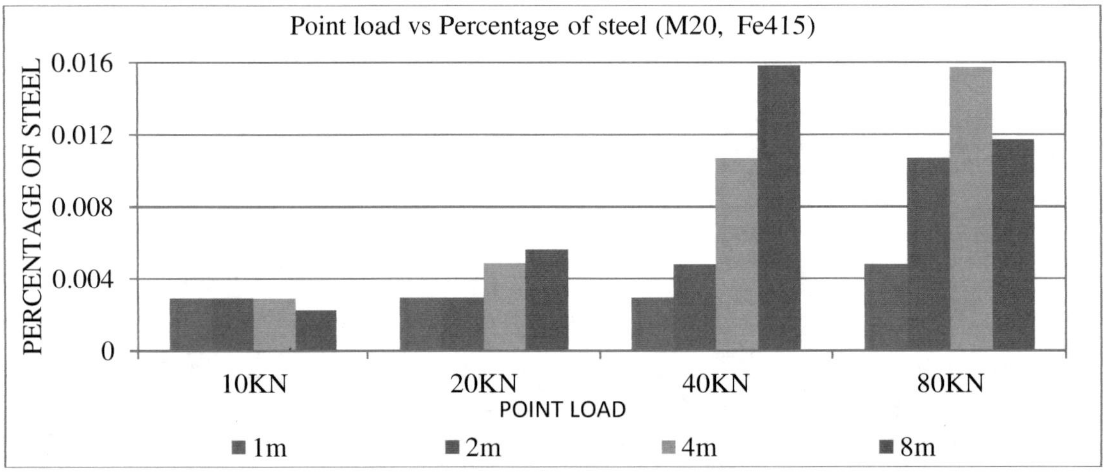

Figure 2. Variation of the percentage of steel with span for different loads (for Fe415, M20)

From figure 2, it can be observed that for a span 1 m, at 10 kN, the percentage of steel is 0.002923. It slightly increases to 0.002964 at 20kN and remains constant till 40 kN. Then it increases to 0.004829 at 80kN. For a span 2 m, at 10 kN, the percentage of steel is 0. 002923. It slightly increases to 0.002964 at 20 kN. Then it increases to 0.004828 at 40kN and 0.010699 at 80kN. For a span 4 m, at 10 kN, the percentage of steel is 0.002932. It increases to 0.004846 at 20 kN. Then it increases to 0.010699 at 40kN and it increases to 0.015735 at 80kN. For a span 8 m, at 10 kN, the percentage of steel is 0.002247. Then it increases to 0.00563 at 20kN and increases to 0.015851 at 40 kN. Then gradually decreases to 0.011727 at 80 kN.

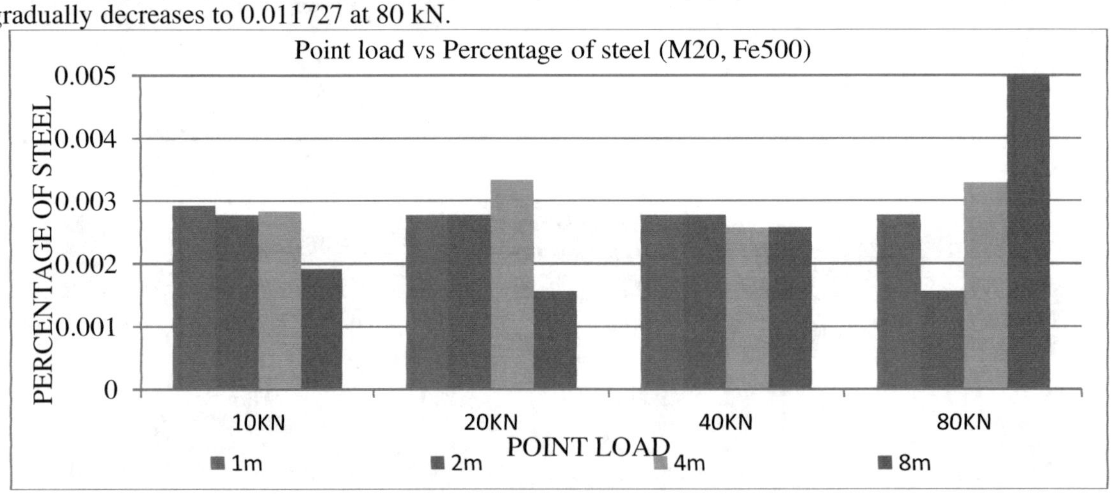

Figure 3. Variation of the percentage of steel with span for different loads (for Fe500, M20)

IOP Conf. Series: Materials Science and Engineering 1025 (2021) 012005 doi:10.1088/1757-899X/1025/1/012005

From figure 3, it can be observed that for a span 1m, at 10 kN, the percentage of steel is 0.002923. It slightly decreases to 0.002778 at 20kN and remains constant till 80 kN. For a span 2m, at 10 kN, the percentage of steel is 0.002777 and remains constant till 40kN. Then it decreases to 0.00156 at 80 kN. For a span 4m, at 10 kN, the percentage of steel is 0.002834. Then it increases to 0.003336 at 20 kN. Later it decreases to 0.002568 at 40 kN. Then it gradually increases to 0.003294 at 80kN. For a span 8m, at 10 kN, the percentage of steel is 0.001923. Then decreases to 0.00156 at 20kN, increases to 0.002576 at 40kN and further increases to 0.005021 at 80kN.

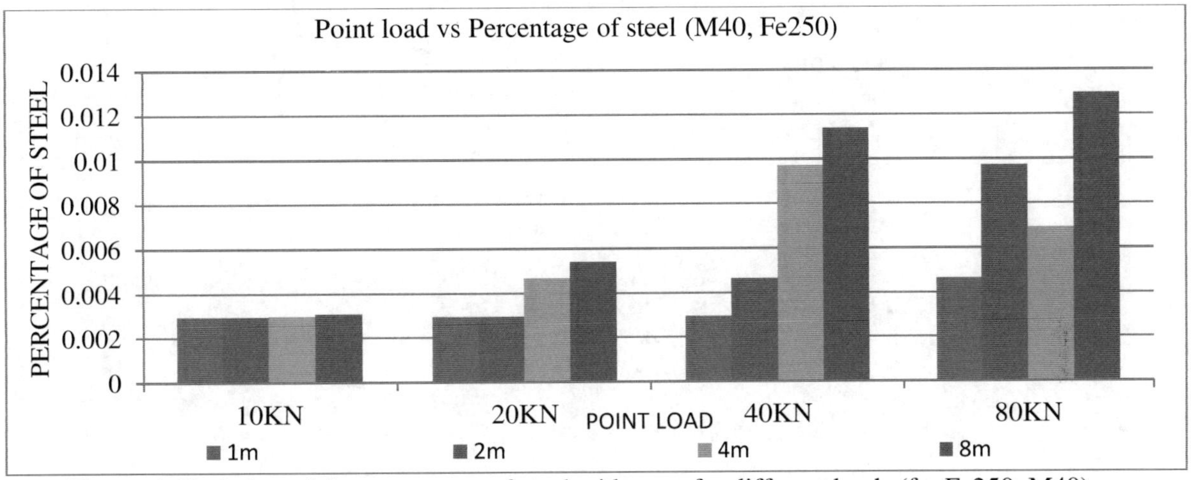

Figure 4. Variation of the percentage of steel with span for different loads (for Fe250, M40)

From figure 4, it can be observed that for a span 1m, at 10 kN, the percentage of steel is 0.002964 and remains constant till 40 kN. Then increases to 0.004644 at 80 kN. For a span 2m, at 10 kN, the percentage of steel is 0.002964 and remains constant till 20kN. Then it increases to 0.004643 at 40kN and 0.009731 at 80 kN. For a span 4 m, at 10 kN, the percentage of steel is 0. 002982. It increases to 0.004674 at 20 kN. Then it increases to 0.009731 at 40 kN and decreases to 0.0069227 at 80kN. For a span 8 m, at 10 kN, the percentage of steels 0.003085. Then increases to 0.005378 at 20 kN and increases to 0.0011388 at 40 kN. Then increases to 0.012953 at 80 kN.

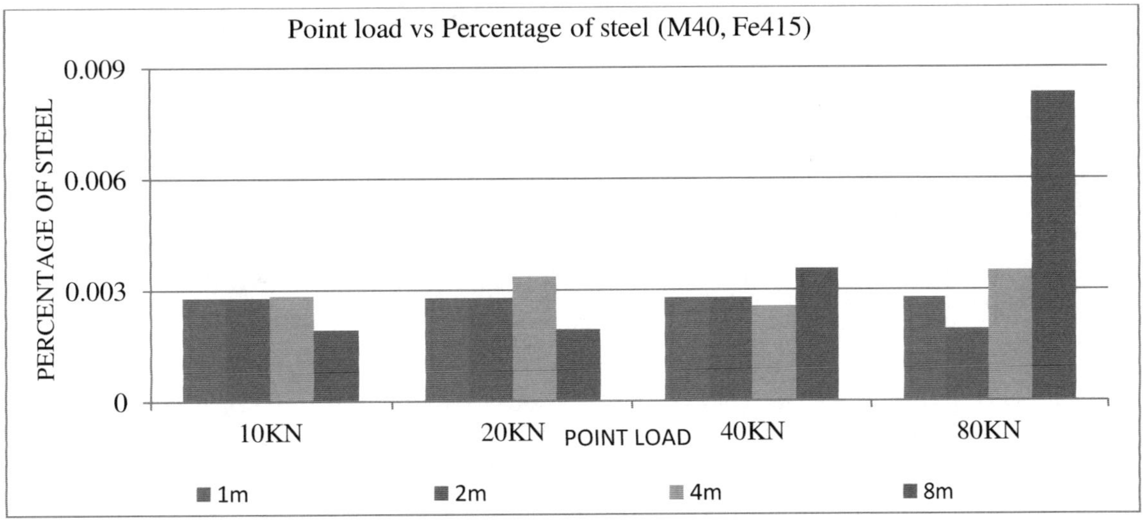

Figure 5. Variation of the percentage of steel with span for different loads (for Fe415, M40)

IOP Conf. Series: Materials Science and Engineering 1025 (2021) 012005 doi:10.1088/1757-899X/1025/1/012005

From figure 5, it can be observed that for a span 1 m, at 10 kN, the percentage of steel is 0.002777 and remains constant till 80 kN. For a span 2 m, at 10 kN, the percentage of steel is 0.002777 and remains constant till 40kN. Then it decreases to 0.001940 at 80kN. For a span 4 m, at 10 kN, the percentage of steel is 0.002834. It increases to 0.003340 at 20 kN. Then it decreases to 0.002554 at 40 kN and increases to 0.003505 at 80kN. For a span 8 m, at 10 kN, the percentage of steels 0.001923. Then increases to 0.001940at 20 kN and increases to 0.003542 at 40 kN. Then increases to 0.008306 at 80 kN.

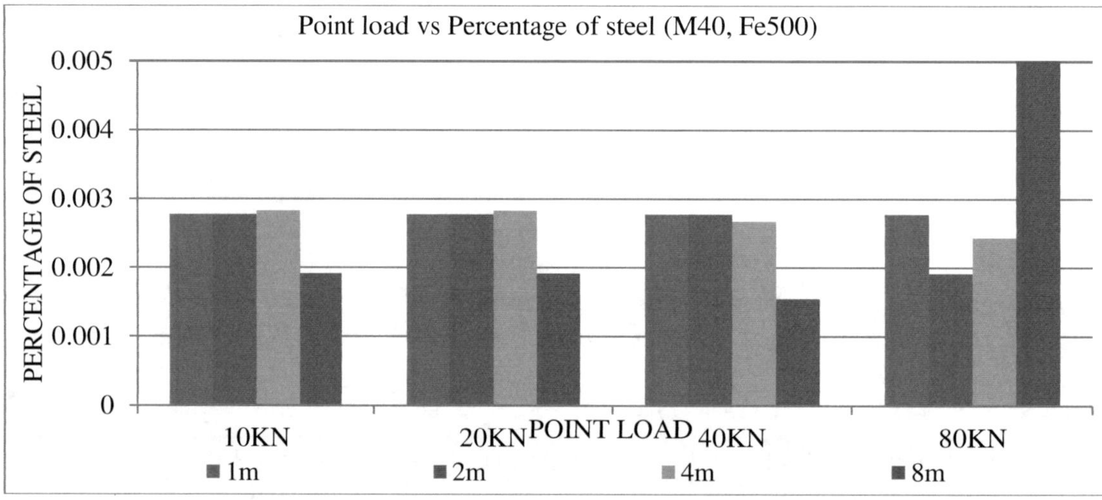

Figure 6. Variation of the percentage of steel with span for different loads (for Fe500, M40)

From figure 6, it can be observed that for a span 1 m, at 10 kN, the percentage of steel is 0.002777 and remains constant till 80kN. For a span 2 m, at 10 kN, the percentage of steel is 0.002777 and remains constant till 40 kN. Then it decreases to 0.001923 at 80kN. For a span 4 m, at 10 kN, the percentage of steel is 0.002834. It increases to 0.002835 at 20 kN. Then it decreases to 0.002674 at 40 kN and decreases 0.002432 at 80kN. For a span 8 m, at 10 kN, the percentage of steels 0.001923 and remains constant till 20kN and decreases to 0.001549 at 40 kN. Then increases to 0.005021 at 80 kN.

Part 1 present's variation of the optimum percentage of steel for a given span and point load. From the results, it can be concluded that for lower grades of steel, the magnitude of the optimum percentage of steel is almost same for all spans at lesser loads while it increases with span for higher loads. However, for higher grades of steel, the variation of the percentage of steel with span and point load is not very significant.

Part 2 presents the variation of the percentage of steel with span for different uniformly distributed loads. Each graph presents results for a given grade of steel and concrete.

From figure 7, it can be observed that for a span 1 m, at 3 kN/m, the percentage of steel is 0.002965 and remains constant till 20kN/for a span 2 m, at 3 kN/m, the percentage of steel is 0.002965 and remains constant till 20kN/m. For a span 4 m, at 3 kN/m, the percentage of steel is 0.002982. It increases to 0.003328 at 7 kN/m. Then it increases to 0.004847 at 10 kN/m and 0.010699 at 20kN/for a span 8 m, at 3 kN/m, the percentage of steel is 0. 003232. Then increases to 0.008186 at 7 kN/m and increases to 0.012681 at 10 kN/m. Then decreases to 0.011802 at 20 kN/m.

IOP Conf. Series: Materials Science and Engineering 1025 (2021) 012005 doi:10.1088/1757-899X/1025/1/012005

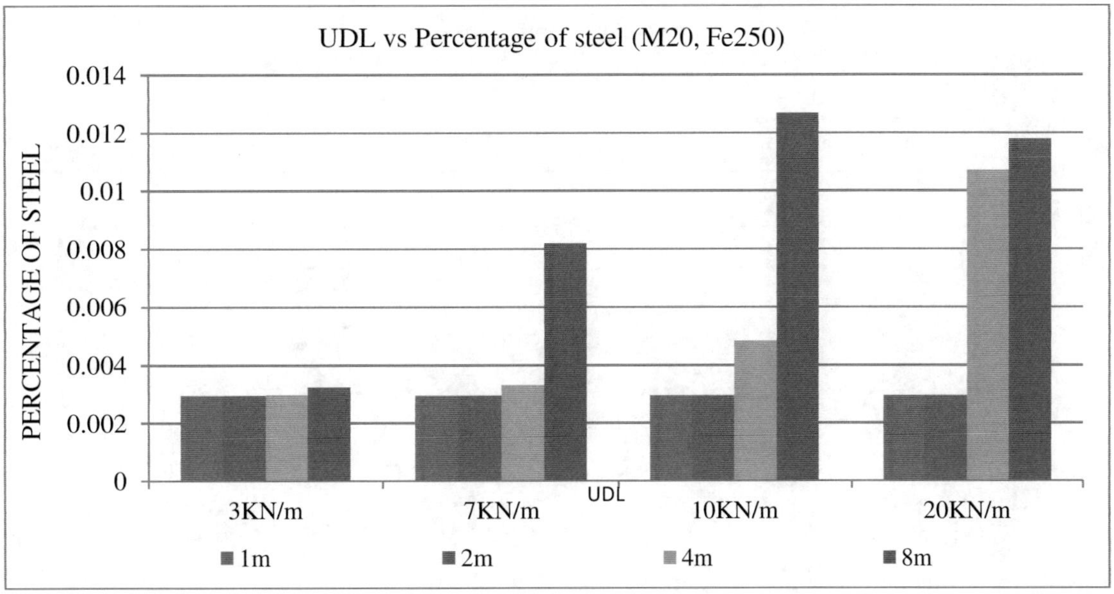

Figure 7. Variation of the percentage of steel with span for different loads (for Fe250, M20)

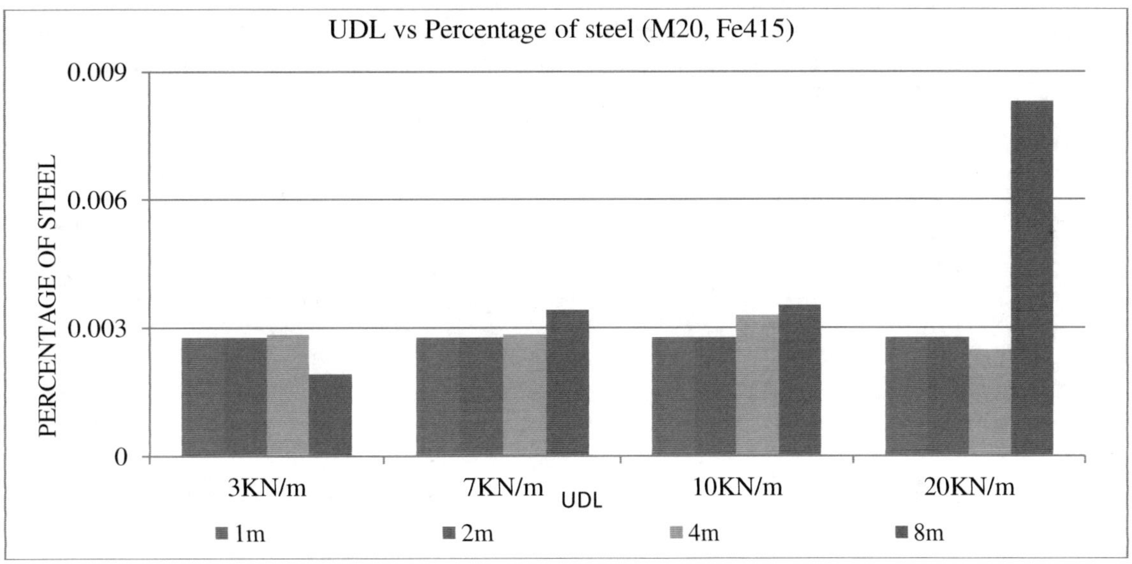

Figure 8. Variation of the percentage of steel with span for different loads (for Fe415, M20)

From figure 8, it can be observed that for a span 1 m, at 3 kN/m, the percentage of steel is 0.002777 and remains constant till 20 kN/for a span 2 m, at 3 kN/m, the percentage of steel is 0.002777 and remains constant till 20 kN/m. For a span 4 m, at 3 kN/m, the percentage of steel is 0.002834 and remains constant till 7kN/m. Then it increases to 0.003277 at 10 kN/m and decreases to 0.002482 at 20kN/for a span 8 m, at 3 kN/m, the percentage of steel is 0. 001923. Then increases to 0.003425at 7 kN/m and increases to 0.003542 at 10 kN/m. Then increases to 0.008306 at 20 kN/m.

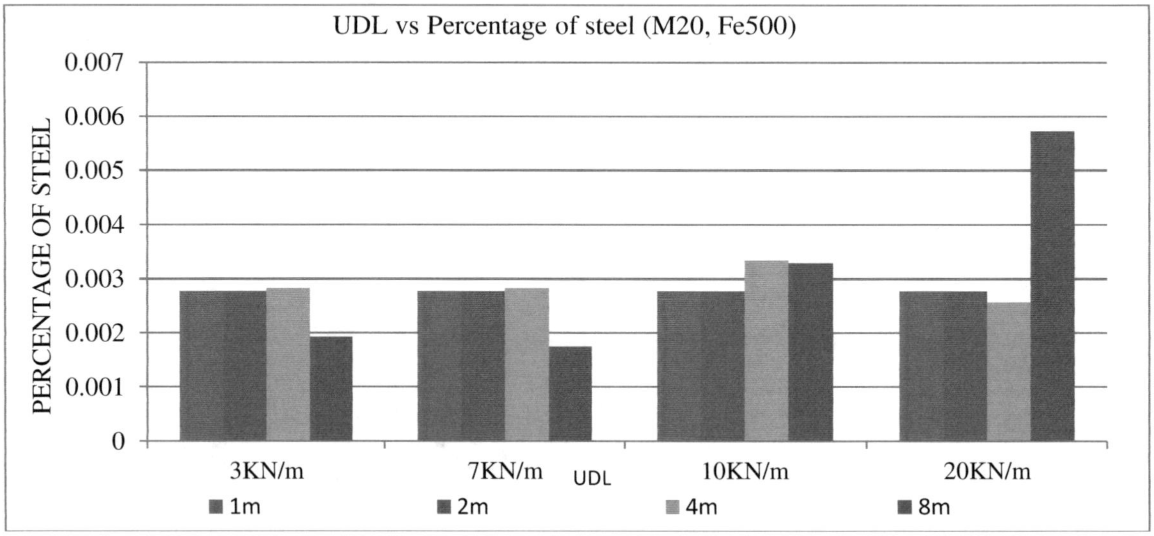

Figure 9. Variation of the percentage of steel with span for different loads (for Fe500, M20)

From figure 9, it can be observed that for a span 1 m, at 3 kN/m, the percentage of steel is 0.002778 and remains constant till 20 kN/for a span 2 m, at 3 kN/m, it is 0.002778 and remains constant till 20kN/m. For a span 4 m, at 3 kN/m, it is 0.002835 and remains constant till 7 kN/m. Then it increases to 0.003336 at 10 kN/m and decreases to 0.002568 at 20kN/for a span 8 m, at 3 kN/m, the percentage of steel is 0. 001923. Then decreases to 0.001749 at 7kN/m and increases to 0.003302 at 10 kN/m and then increases to 0.005731 at 20 kN/m.

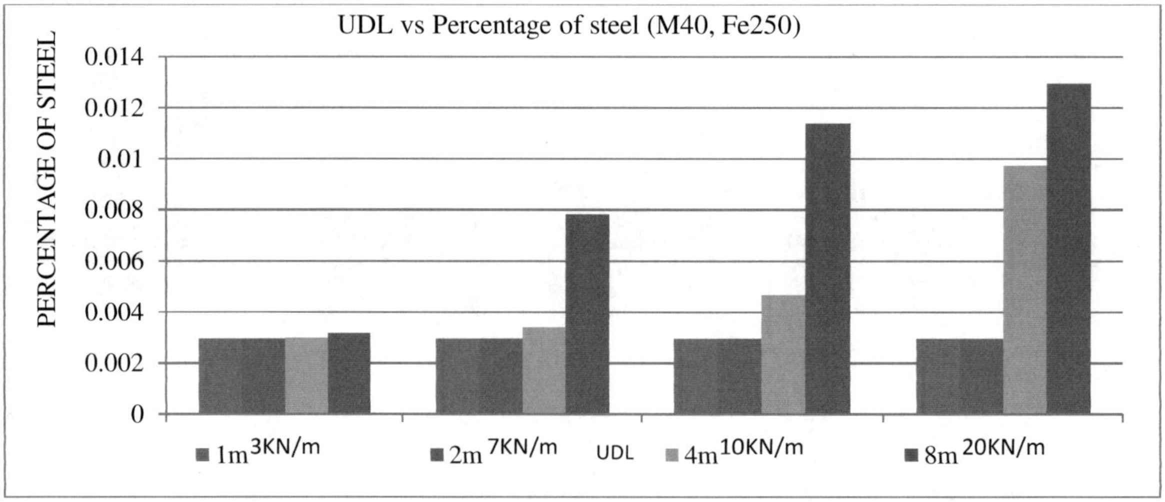

Figure 10. Variation of the percentage of steel with span for different loads (for Fe250, M40)

From figure 10, it can be observed that for a span 1 m, at 3 kN/m, the percentage of steel is 0.00296 and remains constant till20 kN/for a span 2 m, at 3 kN/m, the percentage of steel is 0.00296 and remains constant till 20kN/m. For a span 4 m, at 3 kN/m, the percentage of steel is 0.00298. It increases to 0.00338 at 7 kN/m. Then it increases to 0.00467 at 10 kN/m and 0.00973 at 20kN/for a span 8 m, at 3 kN/m, the percentage of steel is 0. 00317. Then increases to 0.00782 at7 kN/m and increases to 0.01139 at 10 kN/m. Then increases to 0.01295 at 20 kN/m.

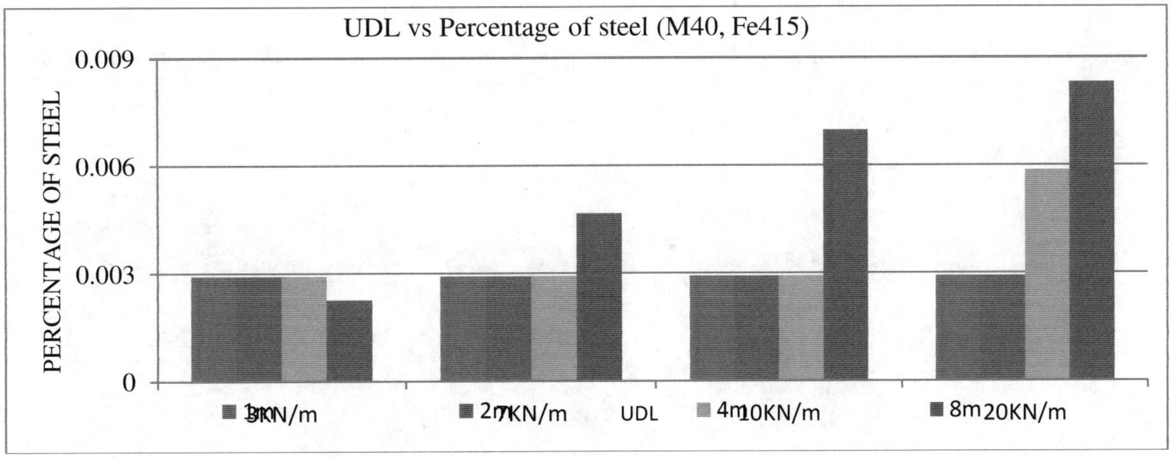

Figure 11. Variation of the percentage of steel with span for different loads (for Fe415, M40)

From figure 11, it can be observed that for a span 1 m, at 3 kN/m, the percentage of steel is 0.00292 and remains constant till 10 kN/m. Then it increases to 0.002923 at 20kN/for a span 2 m, at 3 kN/m, the percentage of steel is 0.00292 and remains constant till 10 kN/m. Then it increases to 0.002923 at 20kN/for a span 4 m, at 3 kN/m, the percentage of steel is 0.00293 and remains constant till10 kN/m. Then it increases to 0.005847 at 20kN/for a span 8 m, at 3 kN/m, the percentage of steel is 0. 00225. Then increases to 0.00467 at7 kN/m and increases to 0.00699 at 10 kN/m. Then increases to 0.008307 at 20 kN/m.

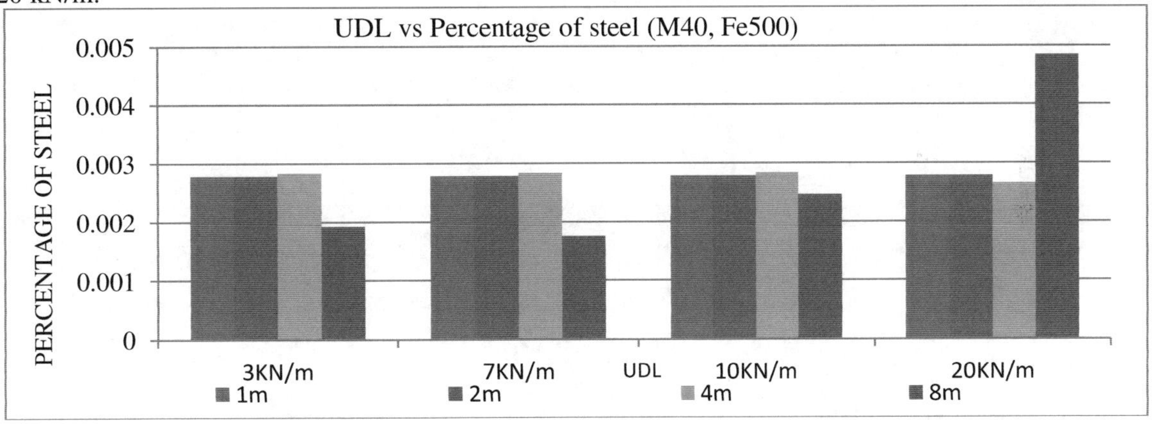

Figure 12. Variation of the percentage of steel with span for different loads (for Fe500, M40)

From figure 12, it can be observed that for a span 1 m, at 3 kN/m, the percentage of steel is 0.002777 and remains constant till 20 kN/for a span 2 m, at 3 kN/m, the percentage of steel is 0.002777 and remains constant till 20kN/m. For a span 4 m, at 3 kN/m, the percentage of steel is 0.002834 and remains constant till 7 kN/m. Then it increases to 0.002835 at 10 kN/m and 0.002659 at 20kN/for a span 8 m, at 3 kN/m, the percentage of steel is 0. 001923. Then decreases to 0.001754 at7 kN/m and increases to 0.002468 at 10 kN/m. Then increases to 0.004849 at 20 kN/m.

Part 2 presents the variation of the optimum percentage of steel for a given span and uniformly distributed load. From the results, it can be concluded that for lower grades of steel, the magnitude of the optimum percentage of steel is almost same for all spans at lesser loads while it increases with span for higher loads. However, for higher grades of steel, the variation of the percentage of steel with span and point load is not very significant.

Part 3 presents the variation of cost/m^3 in Rs. with span for different loads. Each graph presents results for a given grade of steel and concrete.

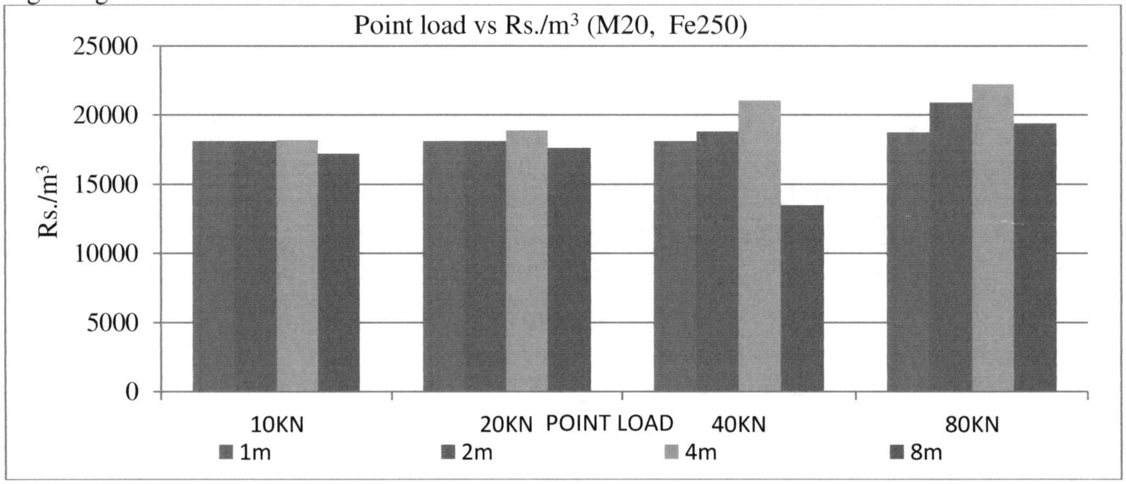

Figure 13. Variation of cost/m^3 (in Rs.) with span for different loads (for Fe250, M20)

From figure 13, it can be observed that for a span 1 m, at 10 kN, the Cost /m^3 is Rs. 18069.35. It slightly decreases to 18067.76at 20 kN. Then it increases to 18069.35 at 40 kN and 18725.89 at 80 kN. For a span 2 m, at 10 kN, it is 18123.9 and remains constant till 20 kN. Then it increases to 18799.68 at 40 kN and 20984.46 at 80 kN. For a span 4 m, at 10 kN, it is 18144.98. It increases to 18843.45 at 20 kN. Then it increases to21049.18 at 40 kN and 22180.54 at 80 kN. For a span 8 m, at 10 kN, it is 17161.72 and slightly decreases to 17161.71 at 20 kN and decreases to 13442.79 at 40 kN. Then increases to 19386.2 at 80 kN.

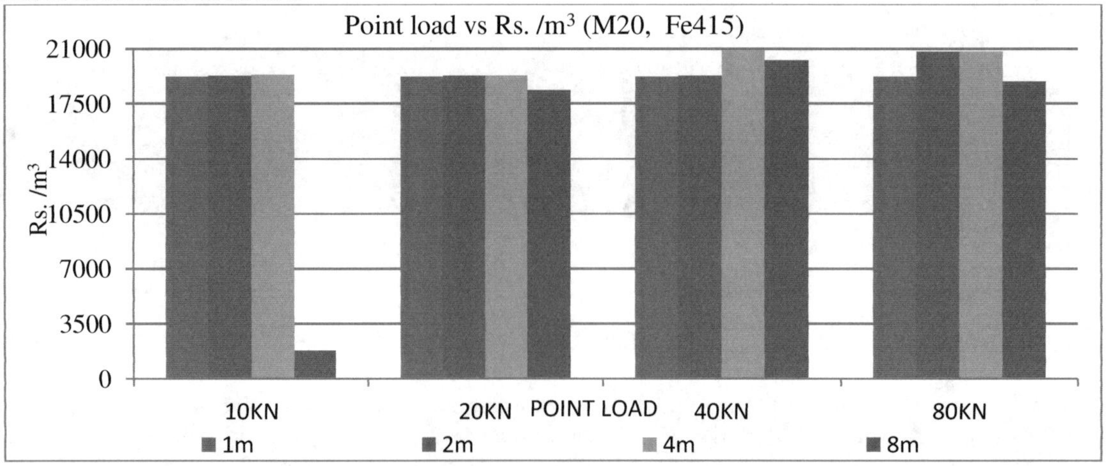

Figure 14. Variation of cost/m^3 (in Rs.) with span for different loads (for Fe415, M20)

From figure 14, it can be observed that for a span 1 m, at 10 kN, the cost/m^3 is 19230.23 and remains constant till 80 kN. For a span 2 m, at 10 kN, the cost/m^3 is 19295.34 and remains constant till 40 kN and increases to 20827.05 at 80 kN. For a span 4 m, at 10 kN, the cost/m^3 is 19347.33. It decreases to 19332.12 at 20 kN. Then it increases to 20886.37 at 40 kN and decreases to 20825.93 at 80 kN. For a span 8 m, at 10 kN, the cost/m^3 is 1783.41. It increases to 18368.30 at 20 kN and increases to 20274.71 at 40 kN. Then decreases to 18948.83 at 80 kN.

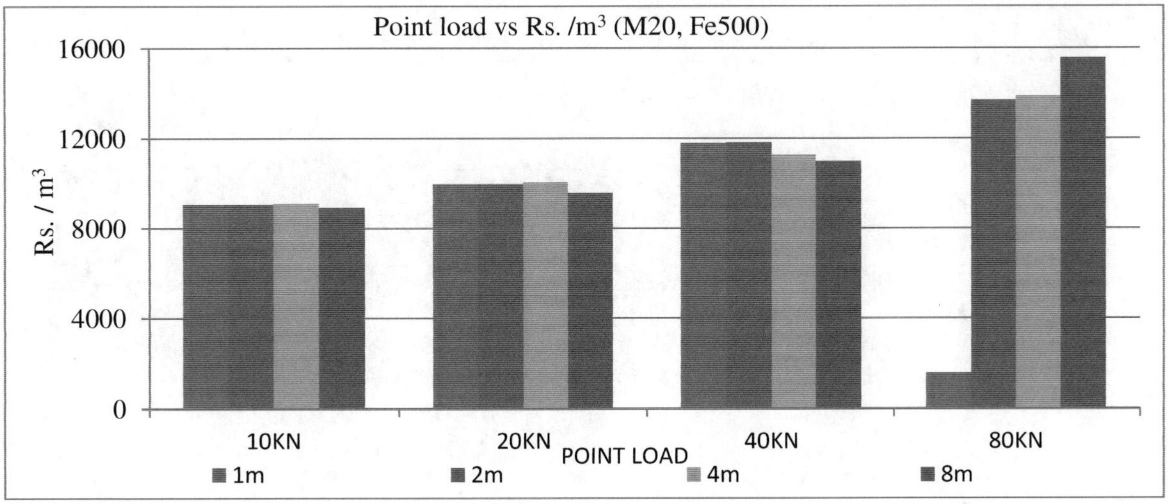

Figure 15. Variation of cost/m^3 in Rs. with span for different loads (for Fe500, M20)

From figure 15, it can be observed that for a span 1 m, at 10 kN, the cost/m^3 is Rs. 9069.305. It increases to 9983.527 at 20 kN. Then it increases to 11812.48 at 40 kN and decreases to 1581.486 at 80 kN. For a span 2 m, at 10 kN, it is 9073.25. It increases to 9993.076 at 20 kN. Then it increases to 11832.72 at 40 kN and 13687.92 at 80 kN. For a span 4 m, at 10 kN, the cost/m^3 is 9083.983. It increases to 10051.48 at 20 kN. Then it increases to11290.52 at 40 kN and 13884.09 at 80 kN. For a span 8 m, at 10 kN, it is 8946.585. It increases to 9579.107 at 20 kN and increases to 10984.52 at 40 kN. Then increases to 15550.20 at 80 kN.

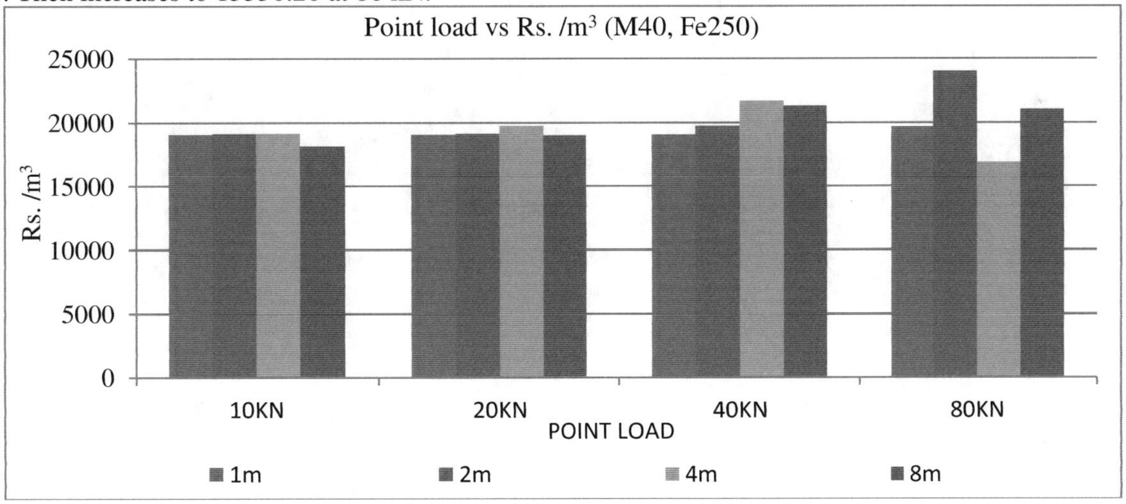

Figure 16. Variation of cost/m^3 in Rs. with span for different loads (for Fe250, M40)

From figure 16, it can be observed that for a span 1 m, at 10 kN, the cost/m^3 is Rs. 19049. It increases to 19049.03 at 20 kN and remains constant till 40 kN. Then it increases to 19639.05 at 80 kN. For a span 2 m, at 10 kN, it is 19105.02 and remains constant till 20 kN. Then it increases to 19710.78 at 40 kN and 23989.38 at 80 kN. For a span 4 m, at 10 kN, the cost/m^3 is 19126.04. It increases to 19755.38 at 20 kN. Then it increases to21660.96 at 40 kN and decreases to 16895.34 at 80 kN. For a span 8 m, at 10 kN, it is 18145.39. It increases to 19007.53 at 20 kN and increases to 21291.12 at 40 kN. Then decreases to 21044.44 at 80 kN.

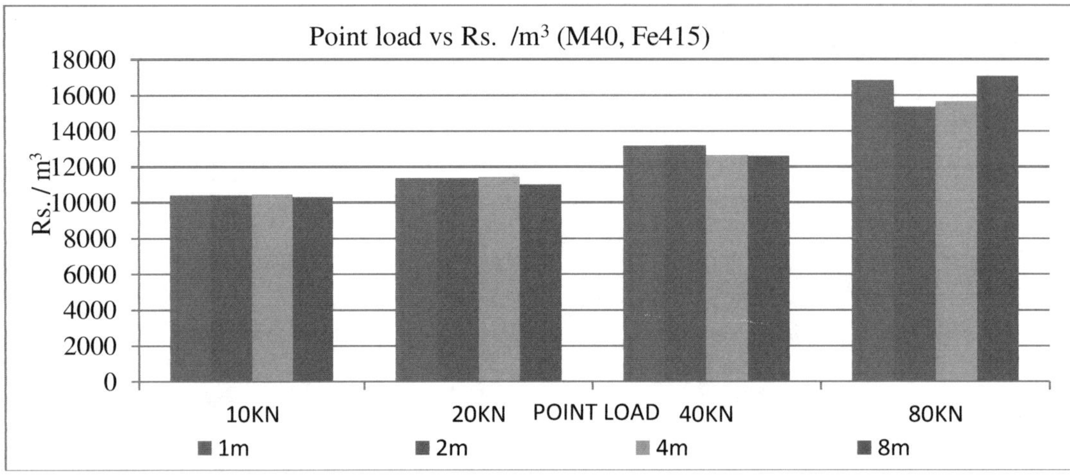

Figure 17. Variation of cost/m^3 in Rs. with span for different loads (for Fe415, M40)

From Figure 17, it can be observed that for a span 1 m, at 10 kN, the cost/m^3 is Rs. 10431.4. It increases to 11345.87 at 20 kN. Then it increases to 13174.81 at 40 kN and 16832.72 at 80 kN. For a span 2 m, at 10 kN, it is 10435.4. It increases to 11355.22 at 20 kN. Then it increases to 13194.86 at 40 kN and 15361.41 at 80 kN. For a span 4 m, at 10 kN, it is 10445.63. It increases to 11412.76 at 20 kN. Then it increases to12655.97 at 40 kN and 15642.82 at 80 kN. For a span 8 m, at 10 kN, it is 10314.59. It increases to 11018.64 at 20 kN and increases to 12606.85 at 40 kN. Then increases to 17068.59 at 80 kN.

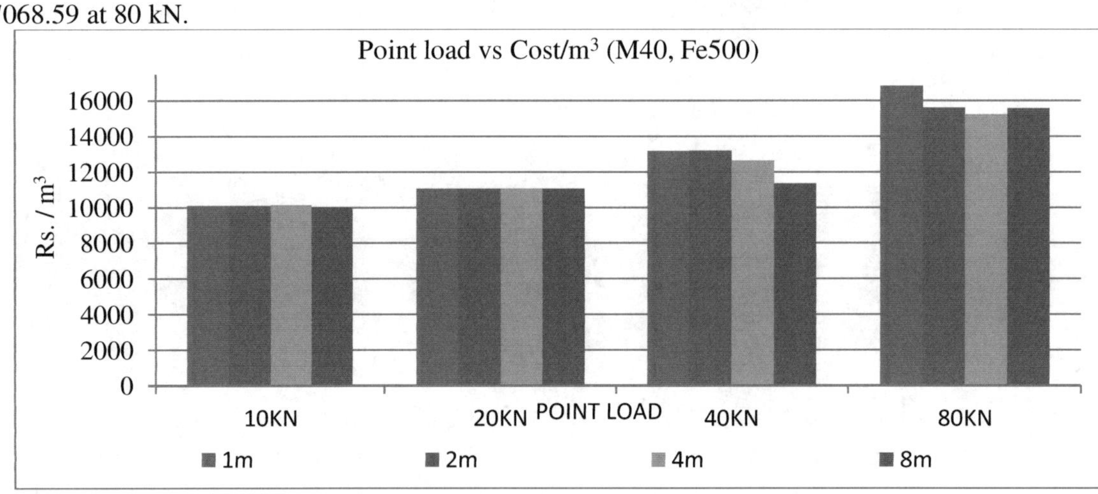

Figure 18. Variation of cost/m^3 in Rs. with span for different loads (for Fe500, M40)

From figure 18, it can be observed that for a span 1 m, at 10 kN, the cost/m^3 is Rs. 10132.19. It increases to 11046.66 at 20 kN. Then it increases to 13174.34 at 40 kN and 16832.23 at 80 kN. For a span 2 m, at 10 kN, it is 10136.23. It increases to 11056.05 at 20 kN. Then it increases to 13194.38 at 40 kN and 15599.32 at 80 kN. For a span 4 m, at 10 kN, the cost/m^3 is 10146.57. It increases to 11080.14 at 20 kN. Then it increases to12672.54 at 40 kN and 15256.35 at 80 kN. For a span 8 m, at 10 kN, it is10014.14. It increases to 11072.17 at 20 kN and increases to 11357.81 at 40 kN. Then increases to 15561.16 at 80 kN.

Part 3 presents the variation of cost /m^3 in Rs. for a given span and central point load. From the results, it can be concluded that for lower grades of steel and concrete, the cost/ m^3 in Rs. is almost the same for all spans and loads. At higher grades of concrete, it is constant for a given load and independent of the span. However, the cost /m³ increase with span for higher grades of concrete.

Part 4 presents the variation of cost/m³ in Rs. with span for different uniformly distributed loads. Each graph presents results for a given grade of steel and concrete.

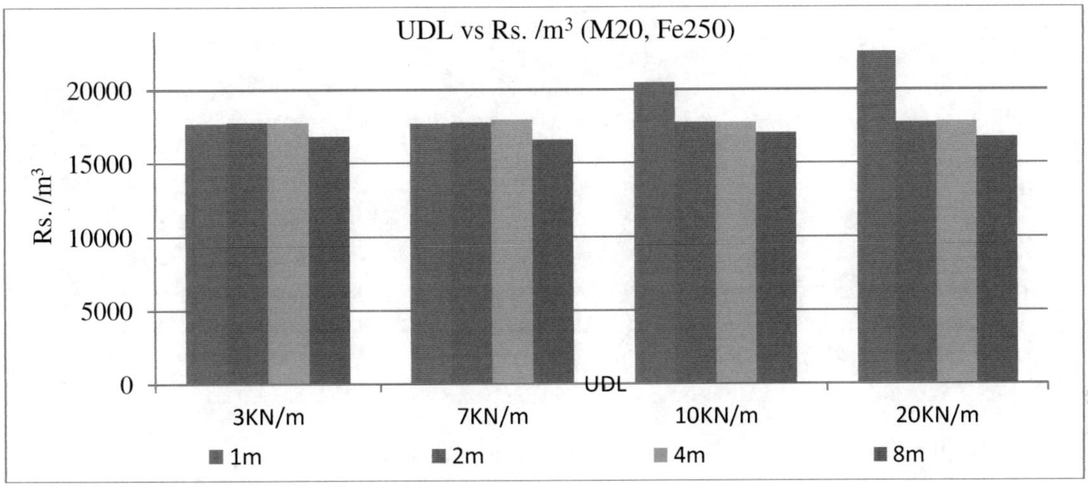

Figure 19. Variation of cost/m³ in Rs. with span for different loads (for Fe250, M20)

From figure 19, it can be observed that for a span 1 m, at 3 kN/m, the cost/m³ is Rs. 17688.61 and remains constant till 7 kN/m. Then it increases to 20449.46 at 10 kN/m and 22553.06 at 20 kN/m. For a span 2 m, at 3 kN/m, it is17744.81 and remains constant till 20kN/m. For a span 4 m, at 3 kN/m, it is 17769.62. It increases to 17910.08 at7 kN/m. Then it decreases to17756.03 at 10 kN/m and 17772.91 at 20kN/m. For a span 8 m, at 3 kN/m, it is 16832.19. Then decreases to 16601.21at 7kN/m and increases to 17047.64 at 10 kN/m. Then decreases to 16767.77 at 20 kN/m.

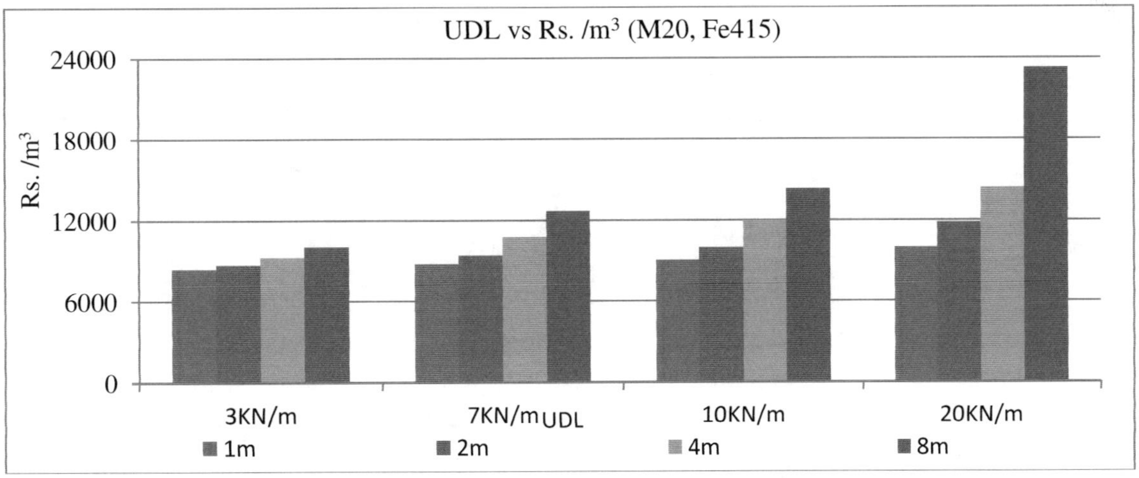

Figure 20. Variation of cost/m³ in Rs. with span for different loads (for Fe415, M20)

From figure 20, it can be observed that for a span 1 m, at 3 kN/m, the cost/m³ is Rs. 8429.20. It increases to 8795 at 7 kN/m. Then it increases to 9069.34 at 10 kN/m and 9983.81 at 20 kN/m. For a span 2 m, at 3 kN/m, it is 8705.60. It increases to 9441.46 at 7 kN/m. Then it increases to 9993.36 at 10 kN/m and 11833 at 20 kN/m. For a span 4 m, at 3 kN/m, it is 9270.96. It increases to 10764.70 at 7 kN/m. Then it increases to11948.50 at 10 kN/m and 14371.59 at 20kN/m. For a span 8 m, at 3 kN/m, the cost/m³ is 10008.13. Then increases to 12683.64 at7 kN/m and increases to 14290.88 at 10 kN/m. Then increases to 23270.42 at 20 kN/m.

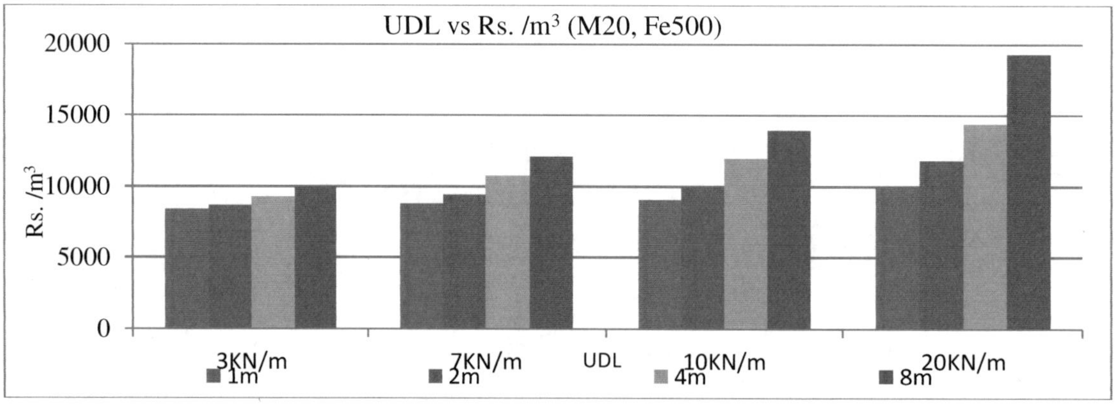

Figure 21. Variation of cost/m³ in Rs. with span for different loads (for Fe500, M20)

From figure 21, it can be observed that for a span 1 m, at 3 kN/m, the cost/m³ is Rs. 8429.22. It increases to 8795.009 at 7 kN/m. Then it increases to 9069.351 at 10 kN/m and 9983.824 at 20 kN/m. For a span 2 m, at 3 kN/m, it is 8705.611. It increases to 9441.468 at 7 kN/m. Then it increases to 9993.362 at 10 kN/m and 11833.01 at 20 kN/m. For a span 4 m, at 3 kN/m, it is 9270.963. It increases to 10764.76 at 7 kN/m. Then it increases to 11956.91 at 10 kN/m and 14396.79 at 20kN/m. For a span 8 m, at 3 kN/m, it is 10008.14. Then increases to 12077.82 at 7 kN/m and increases to 13926.43 at 10 kN/m. Then increases to 19291.98 at 20 kN/m.

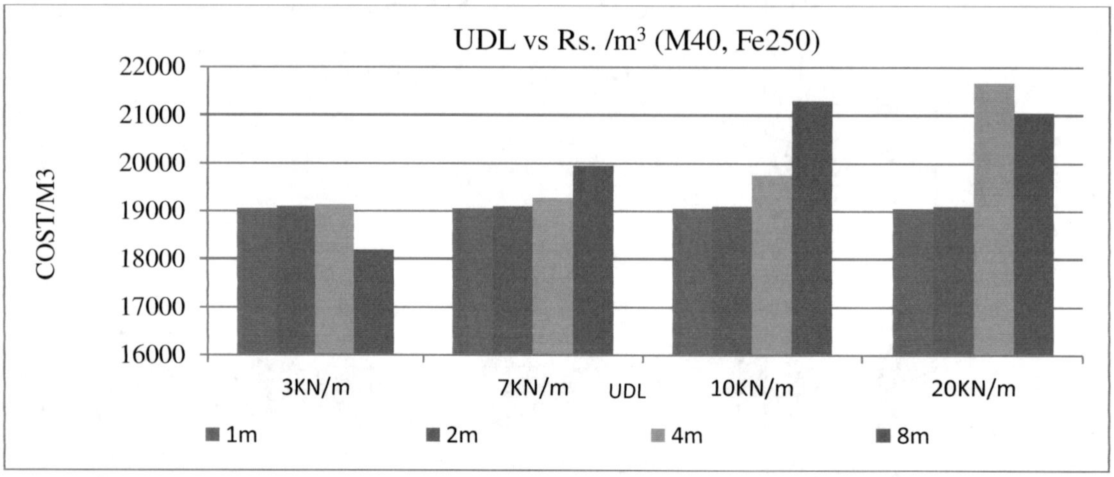

Figure 22. Variation of cost/m³ with span for different loads (for Fe250, M40)

From figure 22, it can be observed that for a span 1 m, at 3 kN/m, the cost/m³ is Rs. 19049.5 and remains constant till 20 kN/m. For a span 2 m, at 3 kN/m, the cost/m³ is Rs. 19105.5 and remains constant till20 kN/m. For a span 4 m, at 3 kN/m, the cost/m³ is 19130.2. It increases to 19275.1 at7 kN/m. Then it increases to19755.9 at 10 kN/m and 21661.4 at 20kN/m. For a span 8 m, at 3 kN/m, the cost/m³ is 18177. Then increases to 19950.6 at7 kN/m and increases to 21291.6 at 10 kN/m. Then decreases to 21044.9 at 20 kN/m.

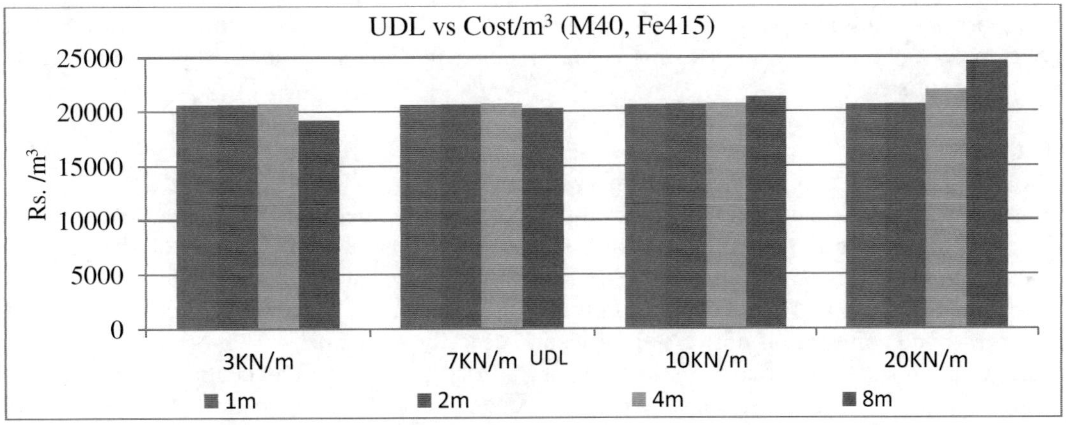

Figure 23. Variation of cost/m³ Rs. with span for different loads (for Fe415, M40)

From figure 23, it can be observed that for a span 1 m, at 3 kN/m, the cost/m³ is Rs. 20591.6 and remains constant till 10 kN/m. Then it decreases to 20591.56 at 20 kN/m. For a span 2 m, at 3 kN/m, the cost/m³ is 20656.5 and remains constant till 10 kN/m. Then it decreases to 20656.47 at 20kN/m. For a span 4 m, at 3 kN/m, the cost/m³ is 20708 and remains constant till 7 kN/m. Then it decreases to20694.7 at 10 kN/m and increases to 21983.3 at 20kN/m. For a span 8 m, at 3 kN/m, the cost/m³ is19195.5. Then increases to 20302.8 at7 kN/m and increases to 21353.8 at 10 kN/m. Then increases to 24632.21 at 20 kN/m.

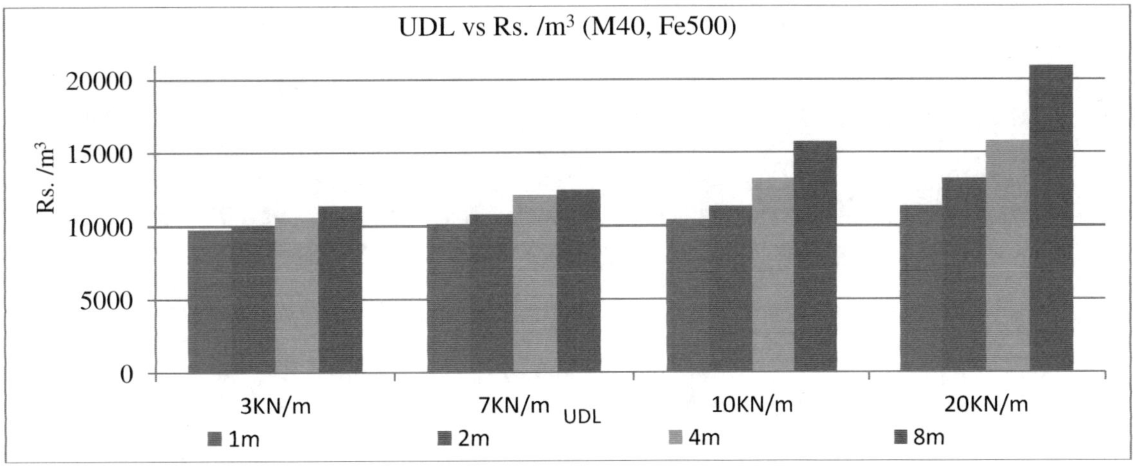

Figure 24. Variation of cost/m³ in Rs. with span for different loads (for Fe500, M40)

From figure 24, it can be observed that for a span 1 m, at 3 kN/m, the cost/m³ is 9791.277. It increases to 10157.06 at7 kN/m. Then it increases to 10431.41 at 10 kN/m and 11345.88 at 20 kN/m. For a span 2 m, at 3 kN/m, the cost/m³ is 10067.47. It increases to 10803.33 at 7 kN/m. Then it increases to 11355.22 at 10 kN/m and 13194.87 at 20 kN/m. For a span 4 m, at 3 kN/m, the cost/m³ is 10632.33. It increases to 12126.23 at 7 kN/m. Then it increases to13246.3 at 10 kN/m and 15790.61 at 20kN/m. For a span 8 m, at 3 kN/m, the cost/m³ is 11375.87. Then increases to 12467.96 at 7 kN/m and increases to 15737.48 at 10 kN/m. Then increases to 20885.83 at 20 kN/m.

Part 4 presents the variation of cost /m^3 in Rs. for a given span and uniformly distributed load. From the results, it can be concluded that for lower grades of steel and concrete, the cost/ m^3 in Rs. is almost the same for all spans and loads. At higher grades of concrete, it increases with an increase in span and load. However, the increase in cost /m^3 is significant when higher grades of steel are used than at lower grades of steel.

CONCLUSIONS

The following general conclusions can be drawn from this study.

1. Based on the study of variation of optimum percentage of steel for a given span and central point load/uniformly distributed load, it can be concluded that for lower grades of steel, the magnitude of the optimum percentage of steel is almost same for all spans at lesser loads while it increases with span for higher loads. However, for higher grades of steel, the variation of the percentage of steel with span and point load is not very significant.
2. Based on the study of variation of cost /m^3 in Rs. for a given span and central point load, it can be concluded that for lower grades of steel and concrete, the cost/ m^3 in Rs. is almost same for all spans and loads. At higher grades of concrete, it is constant for a given load and independent of the span. However, the cost /m^3 increases with span for higher grades of concrete.
3. Based on the study of variation of cost /m^3 in Rs. for a given span and uniformly distributed load, it can be concluded that for lower grades of steel and concrete, the cost/ m^3 in Rs. is almost same for all spans and loads. At higher grades of concrete, it increases with an increase in span and load. However, the increase in cost /m^3 is significant when higher grades of steel are used than at lower grades of steel.

References

[1] Joseph Billo E 2007 EXCEL for Scientists and Engineers Numerical Methods *WILEY INTERSCIENCE* A John Wiley & Sons, Inc., Publication
[2] IS 456 2000 Code of practice for Indian Plain and Reinforced Concrete (Fourth revision) *Bureau of Indian Standards* New Delhi
[3] SSR 2018 19 Schedule of Rates for Building Works *Government of Andhra Pradesh*
[4] Vishakha Tulshiram Holambe, Vasudev Raghunath Upadhye, Ajay Gulabrao Dahake 2016 Optimum Design of singly reinforced beams using particle swarm optimization *Advances in Civil and Structural Engineering* Mantech Publications **1** **(1)** p 01-10
[5] Stency Mariam Thomas, Prince Arulraj G 2017 Optimization of Singly Reinforced RC Beams *International Journal of Research GRANTHAALAYAH* **5** **(2)** February 2017 p 199-207
[6] Helena Barros, Martins R A F, Barros AFM 2005 Cost optimization of singly and doubly reinforced concrete beams with EC2 2001 *Structural and Multidisciplinary Optimization* **30** **(3)** September 2005 p 236-42
[7] Sonia Chutani, Jagbir Singh 2017 Design Optimization of Reinforced Concrete Beams *Journal of Institution of Engineers (India) Series A* **98** **(4)** December 2017 p 429-35
[8] Kamal C Sarma, Hojjat Adeli 1998 Cost Optimization of Concrete Structures *Journal of Structural Engineering* **124** **(5)** p 570-8
[9] Ali Kaveh, Ahangaran M 2012 Discrete cost optimization of composite floor system using social harmony search model *Applied Soft Computing* **12** **(1)** January 2012 p 372-81
[10] Ali Kaveh, Pooya Zakian 2014 Optimal seismic design of Reinforced Concrete shear wall-frame structures *KSCE Journal of Civil Engineering* **18** p 2181-90
[11] Shahzad Umar, Anwar Ahmad, Syed Aqeel Ahmad and Samarul Huda 2017 Cost Optimization of RC Godown *International Journal of Civil Engineering and Technology* **8** **(3)** 244-51
[12] Lakkad Siddhant, Panchal V R 2019 Cost Optimization of Reinforced concrete structure element *i-Managers Journal on Structural Engineering* **8** **(1)** p 18-28

Reliability analysis of doubly Reinforced Concrete Beams retrofitted by plate bonding

P Markandeya Raju[1], Sk. Lal Mohiddin[2]

[1]Professor and Head, Department of Civil Engineering, MVGR College of Engineering (Autonomous), Vizianagaram 535005
[2]Assistant Professor, Department of Civil Engineering, Vignan's Institute of Information Technology (Autonomous), Visakhapatnam 530046
Email: markandeyaraju@yahoo.com, mohiddin2804@gmail.com

Abstract: This paper concerns the safety of the simply supported reinforced concrete beams subjected to only bending loads: so that the reliability index and probability of failure are to be calculated. Reliability index was carried out by using Hasofer-Lind method for the doubly reinforced beam with and without retrofitted beams having externally bonded plates (porous and solid) with variable thickness for the limit state in bending. Reliability index can be calculated by using MS-EXCEL. The reliability index for control beam is 0.4022 and the FR-A4 beam is 3.9791. So, the reliability index is high for retrofitted beam only. The Probability of failure (3.5E-05) for FR-C4 beam is very low under the combination of porous plate and 3mm plate thickness.
Keywords: Reinforced concrete beams, Retrofitting, Plate bonding, Bending, Reliability analysis, Limit state function, AFOSM, Reliability Index, Probability of failure.

1. Introduction

1.1 General

Reliability-based techniques have been used broadly in structural engineering. This approach allows designers to realistically assess the possibility of structural failure. It involves the use of probability and definition of a safety index to achieve a balance between safety and cost. Evaluation of structural safety related to the design procedure for reinforced concrete beams in pure bending. The so-called reliability index method was also used to design reinforced concrete beams and structural steel members.

The performance of a structure is evaluated by its safety, serviceability, and economy. The information about input variables is never certain, precise, and complete. The sources of uncertainties may be (i) inherent uncertainty, i.e. physical uncertainty, (ii) limited data, i.e. statistical uncertainty, (iii) imperfect knowledge, i.e., model uncertainty, and (iv) gross errors. In the presence of uncertainties, the absolute safety of a structure is difficult due to (i) the unpredictability of (a) loads on a structure during its life, (b) in-place material strengths, and (c) human mistakes, (ii) structural inventions in forming the mathematical model of the structure to calculate its response or behaviour, and (iii) the limitations in numerical methods. Therefore, some risk of unacceptable performance must be tolerated. For the risk of life, structural safety is important. In the conventional deterministic analysis and design methods, it is assumed that all parameters (loads, strengths of materials, etc.) are not subjected to probabilistic variations. The safety factors provided in the existing codes and standards, primarily based on practice, decision, and understanding, may not be satisfactory and economical.

The concept of reliability has been applied to many fields and has been understood in many ways the most common meaning, and accepted by all, of reliability is that reliability is the probability of an item performing its intended function over a given period under the operating conditions encountered. It is important to note that the above definition stresses · -' four · significant -elements, viz. (i) probability, (ii) intended function, (iii) time, and (iv) operating conditions. Because of the uncertainties, the reliability is a probability which is the first element in the definition. The second point, intended function, signifies that the reliability is a performance characteristic. For a structure to be reliable, it must perform a certain function or functions reasonably for which it has been designed, i.e. safety against shear or flexure or torsion, etc. The reliability is always related to time. In the case of structure, it is related to the lifetime of the structure. During this specified life of the structure, it must perform the assigned function satisfactorily. The last point is the operating conditions. This establishes the actions or stresses that will be imposed on the structure. These may be loads, temperature, shock, vibrations, corrosive atmosphere, etc. Reliability also changes for quality control, workmanship, production procedure, inspection, etc.

In structural reliability analyses, a performance function or failure function, G, can be defined as the resistance minus the loading (Thoft-Christensen & Baker, 1982). The probability of failure refers to the probability that an undesired performance occurs. Besides, both the resistance (R) and the loading (S) are regarded as continuous random variables. Therefore, each has a probability density function (PDF) as shown in Figure 1. Furthermore, the failure function, G = R - S, can also be observed as a random variable with its own probability density function. The probability of failure corresponds to the shaded area which is shown in Figure 1 and Figure 2. In this study, the resistance was characterised by the moment carrying capacity and the loading was signified by the ultimate dead and live load moment that could be maintained by the beam. When the failure action, G, was greater than zero, a safe structure state existed. Besides, when the failure function G was less than or equal to zero, a failure state existed as shown in Figure 2 (Steinberg, 1997).

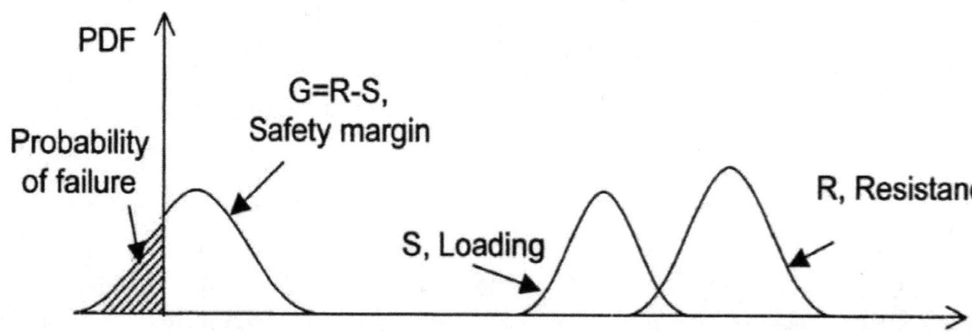

Figure 1. PDFs of Load Effect and Resistance and Failure Function

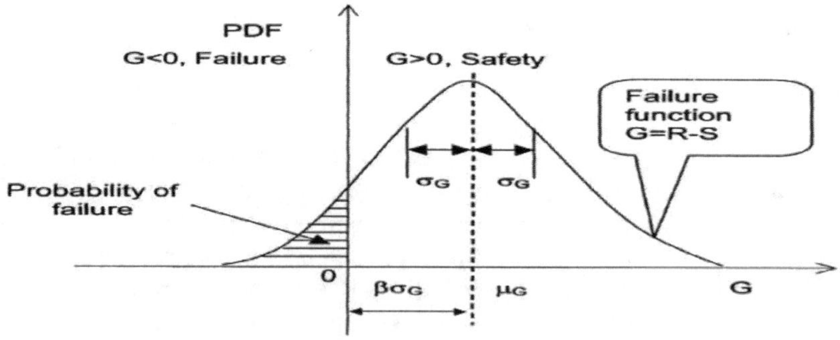

Figure 2. Distribution of Failure Function G = R – S

1.2 Objective and Scope of work

1) The objective of this study is to compare the reliability index and probability of failure for RC beams and those strengthened by externally bonded plates of varying thickness.

1.3 Scope

1) The thickness of Plates considered: 1mm and 3mm.
2) Types of plates considered - Solid plate and perforated plate.
3) Size of beam considered is doubly reinforced beams for 120×210 mm^2 size beam.

2. Literature review

Nikolaos et al. (1995) [1] opined that the Strengthening of concrete structures in flexure with superficially epoxy-bonded carbon-fibre-reinforced plastic (CFRP) laminates is becoming a progressively popular retrofit technique among scholars and engineers worldwide. Strength reduction factors are derived, to attain a uniform reliability index of about three over a broad range of design situations. A general strength reduction factor Ø = 0.80 is offered. Carol Shield (2009) [2] says that the structural reliability of concrete flexural members reinforced with fibre reinforced polymer (FRP) reinforcement is examined. Flexural designs using either ACI 440.1R-03 or ACI 440.1R-06 provide satisfactory reliability, with reliability indices between 3.5 and 4.8, with the older versions of ACI 440.1R yielding higher reliability. Hany and Antonio (2013) [3] explained the concept of reliability index, an interim index of "comparative reliability" is proposed that bypasses the loading variables and weighs the resistances of two structural fundamentals with the same ultimate limit state directly against each other. This concept is put into practice by calculating flexural and shear-strength reduction factors for fiber-reinforced polymer (FRP) reinforced concrete (RC) members by comparison with conventional steel-reinforced concrete beams possessing the same ultimate capacity. Salma et al. [4] (2017) observed that the prevailing approach of designing rectangular bolstered concrete beam is based on limit state plan philosophy which makes use of partial safety essentials for fabric electricity and cargo. A strengthened concrete body is modelled in ETABS software program, the moment and vicinity of anxiety reinforcement values are extracted from the equal software program for statistical analysis. Probability of failure is acquired by Monte Carlo Simulation method. The complete reliability analysis turned into applied thru developing software in MATLAB software program.

3. Experiment work

3.1 Details of beam:

A total of 5 beams, with 120 mm × 210 mm rectangular cross-section with an overall length of 1500 mm and a clear span of 1440 mm were cast. A clear cover of 20 mm is adopted. The beams are designed and cast as flexure deficit beams with M20 grade of concrete, Fe500 grade of steel. The beams are tested under a two-point loading condition. Out of 5 beams, 1 beam is taken as control beam and the remaining 4 beams were retrofitted with steel plates having different thickness and the plates with perforations. To describe the specimens conducted in the study, the following notation is adopted:
CB: Control beam; FR: Flexure retrofitted beam

Table 1 Details of beam

Name of specimen	Shear reinforcement	Asc	Thickness of plate (mm)	Type of plate	Ast	Thickness of epoxy layer (mm)
CB	8mm Dia. 2 legged. vertical stirrups with spacing 120 mm	10 mm, Dia. 2 bars	-	-	12 mm Dia. 2 bars	-
FR-1			1	Solid		1
FR-2			1	porous		1
FR-3			3	Solid		1
FR-4			3	Porous		1

3.2 Beam reinforcement details

The beam reinforcement details and details of plates bonded to the beams are presented in table 2 and figure 3, 4, 5 and 6.

Table 2 Beam reinforcement details

Beam designation	Width of beam (mm)	Depth of beam (mm)	Longitudinal Reinforcement (mm)		Thickness of plate (mm)	Depth of beam after retrofitting with a 1mm thickness of an epoxy layer (mm)
			Top (2 No.'s)	Bottom (2 No.'s)		
CB	120	210	10	12		210
FR-1	120	210	10	12	1	212
FR-2	120	210	10	12	1	212
FR-3	120	210	10	12	3	214
FR-4	120	210	10	12	3	214

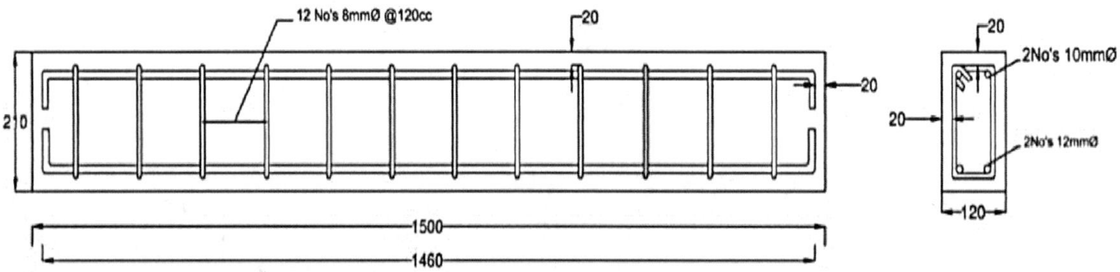

Figure 3. Detailed drawing of Beam reinforcement

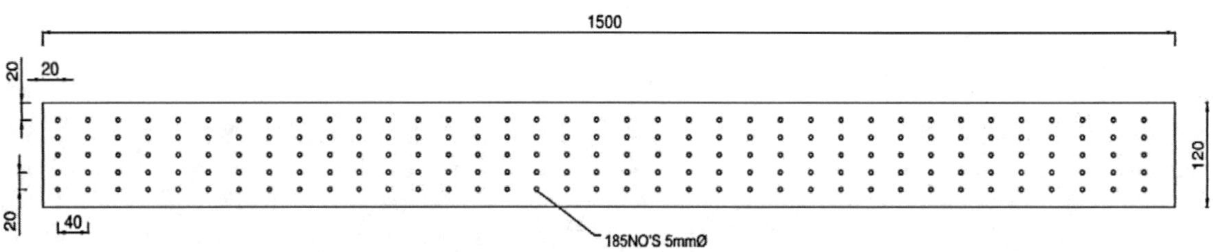

Figure 4. Detailed drawing of a solid plate

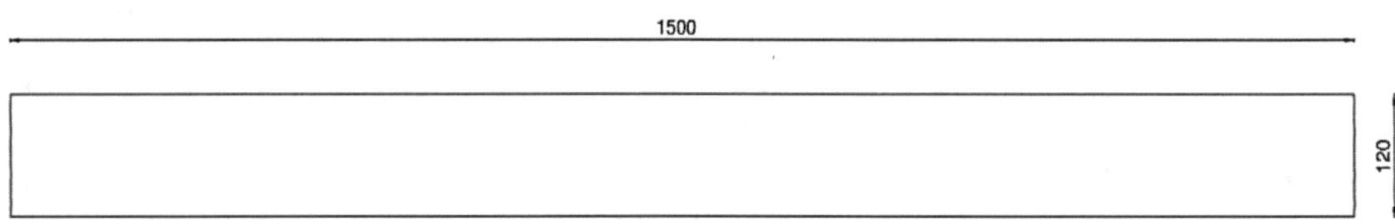

Figure 5. Detailed drawing of the porous plate

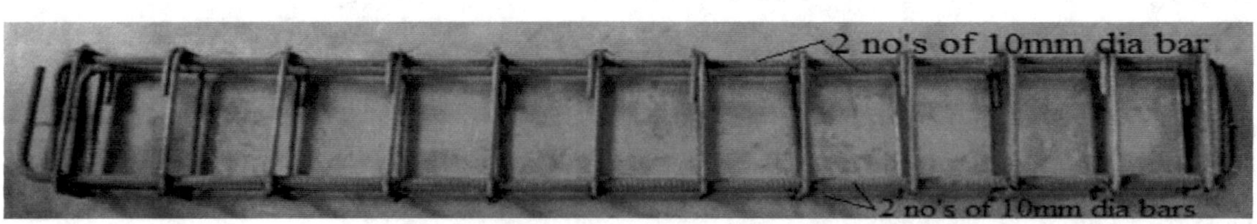

Figure 6. Longitudinal sectional details for all beams

Table 3 Practical values of strengths & deflections and ultimate load value:

Beam designation	**fck (N/mm²)**	**1st flexure crack Load (kN)**	**Deflection at First Flexure crack (mm)**	**1st shear crack Load (kN)**	**Deflection at 1st shear crack (mm)**	**Ultimate failure load (kN)**	**Deflection at Ultimate load (mm)**
CB	36.46	91.096	8.638	89.098	7.739	94.72	11.709
FR-1	32.70	94.719	4.33	104.58	4.891	117.248	6.523
FR-2	37.81	84.868	8.678	-	-	90.209	18.267
FR-3	29.40	108.99	17.826	108.98	19.44	110.658	22.089
FR-4	35.50	73.926	3.45	-	-	84.96	4.588

4. Reliability analysis of reinforced concrete control beam

The step by step procedure for Reliability analysis is presented as follows.

1. Limit state equation in terms of basic variables is

$$g(x) = 0.133 f_{ck} b d^2 + 0.87 f_y A_{st2}(d-d^1) - \frac{WL}{6} \quad (1)$$

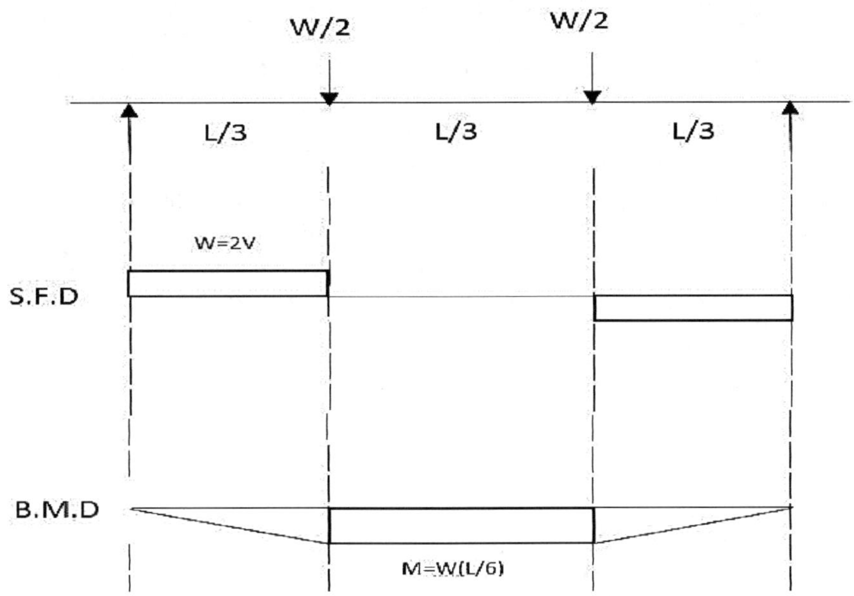

Figure 7. Bending moment diagram of a simply supported beam

2. The basic variables are normalized as follows.

$$Z_1 = \frac{fck - \mu fck}{\sigma fck};\ Z_2 = \frac{b - \mu b}{\sigma b}; Z_3 = \frac{d - \mu d}{\sigma d}$$

$$Z_4 = \frac{fy - \mu fy}{\sigma fy}; Z_5 = \frac{Ast2 - \mu Ast2}{\sigma Ast2}; Z_6 = \frac{w - \mu w}{\sigma w}; Z_7 = \frac{l - \mu l}{\sigma l}$$

IOP Conf. Series: Materials Science and Engineering 1025 (2021) 012006 doi:10.1088/1757-899X/1025/1/012006

3. The mean and standard deviations are calculated by using the below formula

Table 4 Mean and Standard deviations for random variables

Random variable	Mean of random variable $\mu_{fy} = \frac{x_{fy}}{1 - 1.645cov_{fy}}$	Standard deviation of random variable $\sigma_{fy} = \mu_{fy} \times cov_{fy}$
fy	598.444	59.844
$Ast2$	69.073	2.762
D	181.987	3.639
fck	26.5516	3.9823
	124.082	2.481
W	71.2761	14.2552
l	1488.987	29.779

4. Substituting mean and standard deviations values in normal distributions. Basic variables in terms of normal distributions are calculated.

$$f_{ck} = 3.9823Z_1 + 26.551$$
$$b = 2.481Z_2 + 124.082$$
$$d = 3.639Z_3 + 181.987$$
$$f_y = 59.844Z_4 + 598.444$$
$$A_{st2} = 2.762Z_5 + 69.073$$
$$w = 14.2552 \times 10^3 Z_6 + 71.2761 \times 10^3$$
$$l = 29.779Z_7 + 1488.987$$

5. Substituting the above values in equation (1) we obtain

g(x) = 0.133{(3.982Z_1 + 26.551)(2.481Z_2 + 124.082)(3.639Z_3 + 181.987)²} + 0.87{(59.844Z_4 + 598.444)(2.762Z_5 + 69.073)(3.639Z_3 + 181.987)} − {17.4(59.844Z_4 + 598.444)(2.762Z_5 + 69.073)} − {(70.7499 × $10^3 Z_7 Z_6$ + 3.537585 × $10^6 Z_6$ + 353.7546 × $10^3 Z_7$ + 17.6881729 × 10^6)}

Hence the limit state equation in terms of the normalized coordinate system is g(x) = 0

6. At the design point adopting $Z_i = \alpha_i \beta_i$, g (z) in terms of α_i, β is re-written by using the Initial iteration-1: β=3, α_1, α_2, α_3, α_4, α_5 = −0.377

$$\alpha_6 \alpha_7 = 0.377$$

Then the equation becomes

$$\beta_{new} = \frac{-3.326959918 \times 10^6}{\begin{gathered}17.3996 \times (\beta^3)\alpha_1\alpha_2\alpha_3^2 + 43.5168 \times (10^3)\beta\alpha_1\alpha_2 + 1.7403 \times (10^3)\beta^2\alpha_1\alpha_2\alpha_3 + 870.2120\beta^2\alpha_1\alpha_3^2 \\ +2.176413937 \times 10^6\alpha_1 + 87.0389 \times 10^3\beta^1\alpha_1\alpha_3 + 116.0172 \times \beta^2\alpha_2 + 290.1609 \times 10^3\alpha_2 + \\ +11.6040 \times 10^3\beta^1\alpha_2\alpha_3 + 5.8023 \times 10^3\beta^1\alpha_3^2 + 580.3541 \times 10^3\alpha_3 + 523.2920\beta^2\alpha_3\alpha_4\alpha_5 + 26.1699 \\ \times 10^3\beta^1\alpha_4\alpha_5 + 13.0866 \times 10^3\beta^1\alpha_3\alpha_4 654.4626 \times 10^3\alpha_4 + 5.2329 \times 10^3\beta^1\alpha_3\alpha_5 + 261.7017 \times 10^3\alpha_5 \\ +130.8678 \times 10^3\alpha_3 - 2.8760 \times 10^3\beta^1\alpha_4\alpha_5 - 71.9247 \times 10^3\alpha_4 + 1.6529 \times 10^3\alpha_5 - 70.7499 \times 10^3\beta\alpha_7\alpha_6 \\ 7.5375849 \times 10^6\alpha_6 - 353.7546 \times 10^3\alpha_7\end{gathered}}$$

$$\beta_{new} = 0.747523$$

7. A derivative of g(x) with respect to Z1, Z2, Z3, Z4, Z5, Z6 and Z7 will result in

$$\frac{\partial g(x)}{\partial z_1} = 17.3996Z_2Z_3^2 + 43.5168 \times 10^3 Z_2 + 1.7403 \times 10^3 Z_2 Z_3 + 870.2120 \times Z_3^2$$
$$+ 2.176413937 \times 10^6 + 87.0389 \times 10^3 Z_3$$

Substituting $Z_1 = \alpha_1\beta$, $Z_2 = \alpha_2\beta$, $Z_3 = \alpha_3\beta$ then $\frac{\partial g(x)}{\partial z_1} = 2032070$

Like this similarly, the derivatives of g(x) with respect to Z_2, Z_3, Z_4, Z_5, Z_6 *and* Z_7 and then substitute $Z_2 = \alpha_2\beta$, $Z_3 = \alpha_3\beta$, $Z_4 = \alpha_4\beta$, $Z_5 = \alpha_5\beta$, $Z_6 = \alpha_6\beta$, $Z_7 = \alpha_7\beta$.

IOP Conf. Series: Materials Science and Engineering 1025 (2021) 012006 doi:10.1088/1757-899X/1025/1/012006

Then we get the $\frac{\partial g(x)}{\partial z_2} = 230168.6, \frac{\partial g(x)}{\partial z_3} = 581048.2, \frac{\partial g(x)}{\partial z_4} = 542060.9, \frac{\partial g(x)}{\partial z_5} = 228454.4,$ $\frac{\partial g(x)}{\partial z_6} = -7617603, \frac{\partial g(x)}{\partial z_7} = -433773$

8. Calculating the k value:

$$k = \sum_{i=1}^{n} (\delta g1/\delta g2)^2 = 794241$$

The new values of α_i, are determined as follows.

Table 5 Design points for different random variables

Design point α_i	$\alpha_i = \frac{1}{k}\left(\frac{\partial g_1}{\partial g_2}\right)$ $i = 1, 2, 3 \ldots \ldots n$
α1	-0.25585
α2	-0.02898
α3	-0.07316
α4	-0.06825
α5	-0.02876
α6	0.95910
α7	0.054615

	A	B	C	D	E	F	G	H	I	J	K	L
1		Iteration1			Iteration2			Iteration3			Iteration4	
2		B	3		B	0.747523		B	0.420353		B	0.420228
3		a1	-0.377		a1	-0.25585		a1	-0.27414		a1	-0.27442
4		a2	-0.377		a2	-0.02898		a2	-0.03552		a2	-0.03596
5		a3	-0.377		a3	-0.07316		a3	-0.08748		a3	-0.08836
6		a4	-0.377		a4	-0.06825		a4	-0.07342		a4	-0.07348
7		a5	-0.377		a5	-0.02876		a5	-0.03266		a5	-0.03273
8		a6	0.377		a6	0.959104		a6	0.952295		a6	0.952258
9		a7	0.377		a7	0.054615		a7	0.051082		a7	0.04826
10												
11		Bnew	0.747523		Bnew	0.420353		Bnew	0.420228		Bnew	0.420226
12		g1	2032070		g1	2170716		g1	2172566		g1	2172527
13		g2	230168.6		g2	281222.1		g2	284727		g2	284719.3
14		g3	581048.2		g3	692710		g3	699570		g3	699555.7
15		g4	542060.9		g4	581322		g4	581737.1		g4	581731.9
16		g5	228454.4		g5	258575.7		g5	259138.1		g5	259135.8
17		g6	-7617603		g6	-7540473		g6	-7539104		g6	-7539020
18		g7	-433773		g7	-404479		g7	-382076		g7	-382066
19												
20		K	7942415		K	7918212		K	7917080		K	7916986
21		a11	-0.25585		a11	-0.27414		a11	-0.27442		a11	-0.27441
22		a22	-0.02898		a22	-0.03552		a22	-0.03596		a22	-0.03596
23		a33	-0.07316		a33	-0.08748		a33	-0.08836		a33	-0.08836
24		a44	-0.06825		a44	-0.07342		a44	-0.07348		a44	-0.07348
25		a55	-0.02876		a55	-0.03266		a55	-0.03273		a55	-0.03273
26		a66	0.959104		a66	0.952295		a66	0.952258		a66	0.952259
27		a77	0.054615		a77	0.051082		a77	0.04826		a77	0.048259
28												
29		Knew	1		Knew	1		Knew	1		Knew	1

Figure 8: Iterations for control beam (CB)

Table 6 Results

Sl. No	Beam designation	Reliability Index (β)	Probability of failure (Pf)
1	CB	0.4022	3.4E-01
2	FR-1	1.6029	5.4E-02
3	FR-2	3.0045	1.3E-03
4	FR-3	2.5332	5.7E-03
5	FR-4	3.9791	3.5E-05

5. Results discussion

5.1 *Comparison of reliability index for control beam and flexure retrofitted beam having different thickness of the plate*

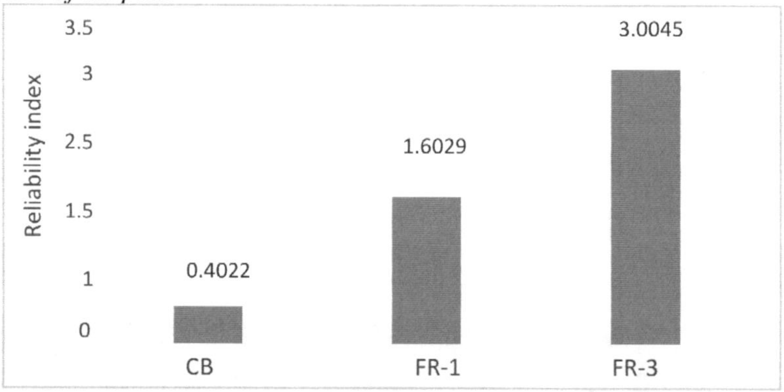

Figure 9. Reliability index for control beam and flexure retrofitted beam

From figure 9, it can be observed that the reliability index is more for flexure retrofitted beam when compared to the control beam. Further, if the thickness of the plate is increased then the reliability index also increases and the probability of failure is decreased.

5.2 Comparison of reliability index for control beam and flexure retrofitted beam having different thickness of plate with different types of plates.

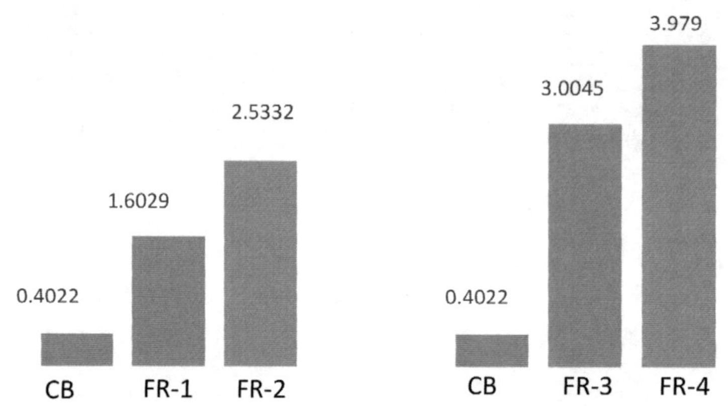

Figure 10. Reliability index for control beam and flexure retrofitted beam having different thickness with different types of plates

From figure 10, it can be observed that the reliability index is relatively high for 3mm porous plate thickness when compared to the solid plate having a 1mm and 3mm plate thickness. The Lowest probability of failure corresponds to PPT3 beam, Pf = 3.5×10^{-5} (β=3.9791) i.e. which is having more thickness of the plate and the porous plate is retrofitted to beam. The highest probability of failure corresponds to CB beam is Pf = 3.4×10^{-1} (β=0.4022).

6. Conclusion

Based on the results obtained from the analytical investigation of the present study on Reliability Analysis of Retrofitted Reinforced Concrete Beams, the following specific conclusions can be made:

1. The reliability index for control beam and flexure retrofitted beam is different. For the failure mode (flexure) the reliability index against CBA is lower than the other types of beams in this research.
2. It is observed that the reliability index varies from 0.4022 to 3.9791 which corresponds to a probability of failure of 3.4E-01 to 3.5E-05 for the size of beam (120 x120) mm having 2 No's 12 mm dia of bars in the bottom for the control and flexure retrofitted beam for two different types of plates (solid and porous) with varying thickness of the plate. It can be seen that as the thickness of plate increases and the type of plate change from solid to porous there is an increment in reliability index which leads to lesser value in the probability of failure.
3. The Probability of failure (3.5E-05) for FR-C4 beam is very low under the combination of porous plate and 3mm plate thickness.

Acknowledgements

The authors would like to thank the management of MVGR for providing excellent e-learning resources and encouraging learning environment that has made the successful completion of this study possible.

Appendix

b = width of cross section;
d = effective depth;
R = resistance;
S = loading;
G = limit state function;
CB = control beam;
FR = flexure retrofitted;
β = reliability index;
Pf = probability of failure;
α = design points;
μ = mean of random variable;
σ = standard deviation of random variable;
fck = concrete compressive strength;
Fy = steel tensile strength;
Ast = area of steel in tension zone;
Asc= area of steel in compression;
COV = coefficient of variation;
W = ultimate load;
PDF = probability density function;
L= span of the beam: FR-A1= retrofitted beam with solid plate having 1mm plate thickness;
FR-A2= retrofitted beam with porous plate having 1mm plate thickness;
FR-A3= retrofitted beam with solid plate having 3mm plate thickness;
FR-A4= retrofitted beam with porous plate having 3mm plate thickness.

References

[1] Plevris N, Triantafillou T C, Veneziano D 1995 Reliability of RC beams strengthened with CFRP laminates *ASCE Journal of Structural Engineering* **121 (7)** p 1037–44
[2] Carol Shield 2009 On the Reliability of FRP Reinforced Concrete *Life Cycle Assessment of Sustainable Infrastructure Material* University of Minnesota USA
[3] Hany Jawaheri Zadeh, Antonio Nanni 2013 Reliability analysis of concrete beams internally reinforced with fibre-reinforced polymer bars *ACI Structural* **110 (6) p** 1023-31
[4] Salma Taj, Karthik B M, Mamatha N R, Rakesh J, Goutham D 2017 Prediction of Reliability Index and Probability of Failure for Reinforced Concrete Beam Subjected to Flexure *International Journal of Innovative Research in Science, Engineering and Technology* **6 (5)** p 8616-22
[5] Ranganathan R 1999 Structural reliability analysis and design *Jaico publishing house*
[6] Devaraj Ravindra 2017 Reliability-based analysis and design *IK International Publishing House Pvt, Ltd.*

Influence of non-biodegradable wastes on mechanical properties of concrete – A Neural Network approach

Sridhar J[1], Vivek D[2]

[1]Associate Professor, Department of Civil Engineering, GMR Institute of Technology, Rajam, Andhra Pradesh, India
[2]Assistant Professor, Department of Civil Engineering, KPR Institute of Engineering and Technology, Coimbatore, Tamil Nadu, India
Email: sridharjayaprakash@gmail.com

Abstract. The environmental pollution due to different non-biodegradable waste is not only making the surroundings vulnerable but also makes a serious defect on the humans. Because of the non-biodegradable property of plastics, decomposition is not possible which makes them remain in the environment for a longer period creating ecological problems. These non-biodegradable materials will be a significant source of plastic waste management if these materials can be substituted as a construction material by using in concrete. An experimental investigation is made on the usage of Non-Biodegradable plastic ravage materials as fine aggregates in M20 concrete with substitution from 0%, to 50% by weight of fine aggregate. Subsequently, the ravage glass powder is included in concrete as a substitution for cement with a percentage of replacement of 0% and 5% by weight of cement. Specimen for concrete in compression, tension and flexure are tested for 28 days curing. The experimental outcome shows that concrete with 10% replacement of plastic waste by weight of fine aggregate with 5% replacement of glass powder by weight of cement had an elevated influence on the mechanical properties. Further Artificial Neural Network Analysis is performed to validate the experimental results. The classification of Neural networks (NN), such as -Leven Berg-Marquardt (NN-LM) and Gradient Descent (NN-GD) is used to perform the analysis.
Keywords: Concrete, Non-biodegradable, Plastic waste, Glass powder, compressive strength, split tensile strength, flexural strength and Neural Network

1. Introduction

Plastics are taken into account as a significant technical breakthrough in recent years, such that they influenced new inventions substituted alternative materials in the existing product. They are lightweight, robust, adaptable and immune to humidness, chemical product, and degradation. Plastics utilization has reduced the problem of solid urban waste accumulation and in turn, promoted plastic containers [1]. The waste plastic may be a key concern for the property of the surroundings because it pollutes rivers, land, and oceans. However, its light-weight, flexibility, moisture-resistant, and price effectiveness will build it a substitute for several existing composite materials like concrete [2]. The current trend in concrete research revolves around sustainability [3].

Large quantities of plastic waste can be used for concreting if a compromise is made on their strength [4]. Plastics are utilized in concrete as fine aggregates and plastic fibres. Between the two, plastic aggregates are cheap and involve less variety of process steps for usage compared to fibres. [5] PET may be a quite polyesters manufactured from the antifreeze and terephthalic acid's composition and its chemical name is synthetic resin Terephthalate or "PET" [6]. Analysis relating to the employment of plastic waste as an associate combination for surface microstructure study shows some progress worth considering. [7]. PET fibres have been used as a fractional and complete substitute for sand in concrete with the proportion varying from 2%–100% [8]. Concrete with PET 15% established lower modulus of the physical property and splitting tensile strength with relation to standard concrete and an increasing trend for compressive strength and flexural strength with 5% of PET replacement for fine aggregate [6].

The compressive strength of concrete reduces with the increase of waste plastic ratios at different ages of curing. Compressive strength improves by the addition of retarders to the mix. Flexural strength of waste plastic concrete reduces with the rise in the proportion of plastic ravage [4].

Higher compressive strength and Young's modulus will be obtained for PET particle sizes of 2.5% volume which is cured for twenty-eight days [9]. No effects will be found on the density of concrete with hydrogen peroxide solution (H_2O_2) and calcium hypochlorite treated plastic waste whereas the slump due to roughened face of treated plastic aggregates reduces [5]. The addition of Nano-SiO_2 effectively improves the mechanical properties, however, magnifies its autogenic shrinkage when compared to standard mortar [10]. When PET used as aggregate in concrete it reduces the compressive strength [11]. Plastic waste substitution for sand in concrete increases the compressive strength up to 5% to 10 % replacement and reduces for 15% and 20% substitution.

The flexural and split tensile strength can be improved up to 100 % addition of the sand with PET [12]. Concrete in compression and tension decreases with the rise in recycled plastic substance [13]. In concrete containing waste plastic mechanical properties reduces with rising the percentages of waste plastic bottle and waste plastic bags [14]

Compressive strength improves for concrete with glass powder as a substitution for binding material when compared to concrete without glass powder [15]. Concrete with the waste glass powder as a replacement for fine aggregate in various percentages of 5% to 50% improves for 25% glass powder replacement with fine aggregate for M20 concrete [16]. Lightweight concrete with 10% replacement of glass powder for cement shows higher compressive strength [17]

Based on the above research results, an attempt has been made to investigate the influence of plastic ravage as part of fine aggregates varied from 0% to 50% by weight with glass powder as substitution of 5% for cement. Mechanical properties such as compressive strength, tensile strength and flexural strength will be studied for different combinations

2. Material Properties

2.1 Cement.

OPC of 43 grade with specific gravity 3.1 and initial setting time 32 min was used in this investigation and its physical properties comply the requirements as per Indian code provision of IS 8112-2013 [18]

2.2 Aggregates.

Coarse aggregate of size 20mm and sand conforming to zone II were utilized as per IS 10262-2019 [19]. The specific gravity of fine aggregate and coarse aggregate is 2.69 and 2.81 respectively.

2.3 Plastic waste.

Plastic waste particles obtained from PET bottles and plastic bags ground to a maximum size of 4.75mm as shown in figure 1 are used as fine aggregates in concrete. The specific gravity was found to be 1.11 g/cm^3. It has an adsorption property and so it increases the water requirement, as plastic is elastic in nature it will boost the tensile strength of the concrete for only some percentage of replacement.

2.4 Glass powder.

Glass powder shown in figure 2 as fine as cement can be used as a replacement of cement in concrete. Glass powder in concrete increases the mechanical properties efficiently when compared with the conventional concrete. The specific gravity is 2.5.

2.5 Concrete mix design. The concrete mixture has been designed as per IS 10262 2019 [19] for M20 grade of concrete. The proportion of the concrete mixture is 1:1.5:2.5 with w/c ratio 0.44. To review the mechanical properties of concrete the mix was made ready in specified partial substitution of fine aggregate with plastic wastes starting from 0% to 50% by weight and cement was partially replaced from 0% and 5% by weight by glass powder. Coarse aggregate of 1113 kg/m^3, the sand of 637 kg/m^3, the cement of 204 kg/m³ and water of 81 litres/m³ were adopted for all proportions studied. The particulars of the other ingredients of the concrete mix proportions are given in Table 1.

Figure 1. Plastic powder.

Figure 2. Glass powder.

Table 1. Concrete Mix Proportion

Mix designation	Plastic waste (%)	Glass powder (%)	Plastic (kg/m³)	Glass powder (kg/m³)
CPG0	0	0	0	0
CPG1	0	5	0	10.2
CPG2	10	0	63.7	0
CPG3	10	5	63.7	10.2
CPG4	20	0	127.4	0
CPG5	20	5	127.4	10.2
CPG6	30	0	191.1	0
CPG7	30	5	191.1	10.2
CPG8	40	0	254.8	0
CPG9	40	5	254.8	10.2
CPG10	50	0	318.5	0
CPG11	50	5	318.5	10.2

2.6 Experimental investigation. For each of the above mix proportions mentioned in table 1 concrete cubes of size, 15 cm × 15 cm × 15 cm, cylinders of size 30 cm × 15 cm and prisms of size 10 cm × 10 cm x 50 cm were cast. The cube specimens were tested for compression, a cylindrical specimen for split tensile strength and prism were assessed for flexural strength for 28 days of curing as per IS 516 1959 [20]. The strength of concrete under compression and Split tension and flexure was found with Universal testing machine with capacity 1000kN.

3. Artificial Neural Networks

Artificial neural networks (ANNs) are unit calculations recreating human neurons. They are an assortment of Artificial Intelligence, that puts forth an attempt to make the organizations of the neuron of the life central system framework named 'nervosum' a man-made nerve cell, conjointly known as a unit or a hub, which takes many key affiliations (dendrites inside the existence neuron) that are unit or node. The unit at that point decides the all out of the weighted information sources and applies for an opening work (closely resembling body inside the natural neuron). The result of the unit is then passed mistreatment the output association (axon function).In the ongoing years, such smart skilled frameworks are with success applied in a few fields of engineering, among that, they are broadly utilized for foreseeing the mechanical properties of homogenized cement-based concrete. It was proven that

ANN-based strength prediction might be successful to calculate the strength of concrete for different combinations of ingredients for different curing period [7]. In the present work, Artificial Neural Network has been set up utilizing MATLAB code, Neural Network chest.

3.1 Network Architecture. Artificial neural organization might be significant information demonstrating instrument that can keep and represent troublesome input and output affiliations. The design of the ANN model needs to recognize the necessity (i.e., an assortment of input neurons, output neurons, hidden layers, and neurons in each concealed layers) and the organization settings (activation function). As shown in Figure.3 three neural networks carry with at least 3 layers, i.e., a partner in the nursing input layer, one or many hidden layers, and partner in nursing an output layer. The adopted requirement consists of the following:

(i) Six neurons (Ni =6) inside the input layer, that imply the factors of cement content (CC: kg), fine aggregate content (FA; metric weight unit), coarse aggregate content (CA; kg), water content (W; litres), Plastic waste substance (PW; kg), and Glass powder content (GP; kg)

(ii) One vegetative cell inside the output layer, that represents the value of the corresponding compressive strength (MPa), splitting tensile strength (MPa), and flexural strength (MPa).

(iii) One hidden layer with ten neurons.

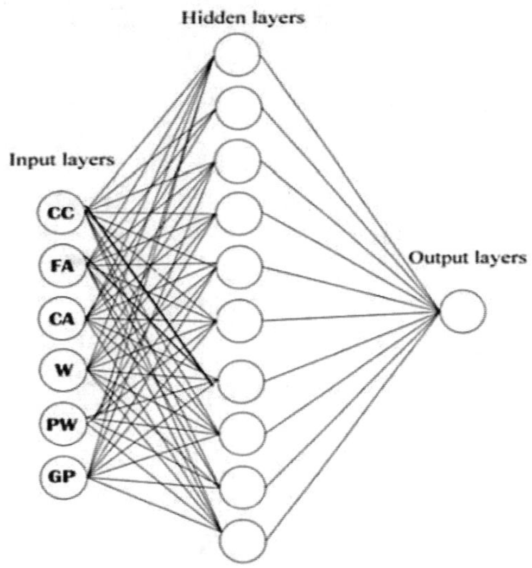

Figure 3. Architecture of ANN

The classification of Neural Network training functions, such as Leven Berg-Marquardt (NN-LM) and Gradient Descent (NN-GD) is for the analysis. Feed-forward backpropagation neural networks have been used in this study.

4. Results and discussions

4.1 Compression strength. To investigate the concrete under compression specimens with varied plastic wastes with glass powder were factory-made. Compressive strength is determined with samples of size 15 cm × 15 cm × 15 cm. As shown in figure 4 and table 2, initially the compressive strength increases with an increase in the percentage of substitution of sand by plastic waste, however, it decreases after a while. The maximum percentage increase in compressive strength of 16.61% was obtained for concrete specimen CPG3 i.e. 10% substitute of sand volume with plastic waste with 5% of glass powder replacement for cement. Also, the specimen CPG1, CPG2 and CPG3 have an increase in compressive strength of 2.56%, 6.77% and 6.07% respectively. Moreover, it observed that when the plastic waste substitution is increased beyond 10% for fine aggregate with 5% substitution of glass powder by weight of cement, compressive strength decreases.

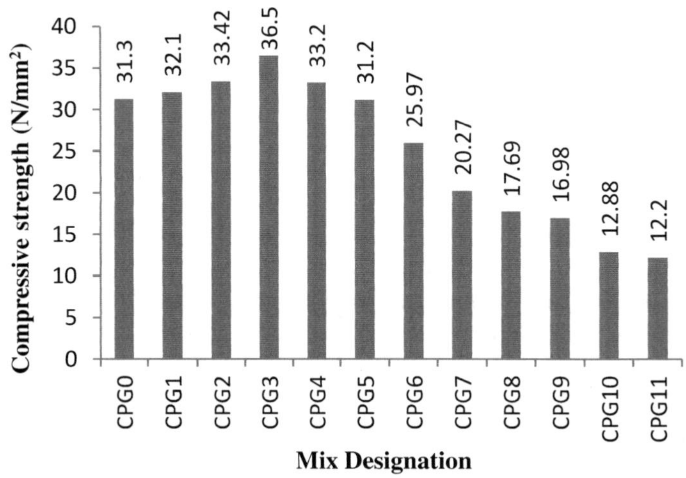

Figure 4. Compressive strength of concrete

Table 2. Mechanical Properties of concrete

Mix	Plastic waste (%)	Glass powder (%)	Comp. strength (MPa)	% increase in Comp. strength	Tensile strength (MPa)	% increase in Split tensile strength	Flexural strength (MPa)	% increase in Flexural strength
CPG0	0	0	31.3	0.00	2.13	0.00	4.03	0.00
CPG1	0	5	32.1	2.56	2.52	18.31	4.23	4.96
CPG2	10	0	33.42	6.77	2.73	28.17	4.83	19.85
CPG3	10	5	36.5	16.61	3.22	51.17	6.01	49.13
CPG4	20	0	33.2	6.07	2.82	32.39	5.23	29.78
CPG5	20	5	31.2	-0.32	2.52	18.31	4.76	18.11
CPG6	30	0	25.97	-17.03	2.23	4.69	4.23	4.96
CPG7	30	5	20.27	-35.24	2.18	2.35	4.15	2.98
CPG8	40	0	17.69	-43.48	2.10	-1.41	4.01	-0.50
CPG9	40	5	16.98	-45.75	2.09	-1.88	3.98	-1.24
CPG10	50	0	12.88	-58.85	1.98	-7.04	3.82	-5.21
CPG11	50	5	12.2	-61.02	1.85	-13.15	3.72	-7.69

4.2 Split tensile strength. Split tensile strength is determined with cylindrical samples of size 30 cm × 15 cm. In figure 5 and table 2, the effect of substituting plastic waste with various percentages of substitutes by weight of fine aggregate proportion with glass powder for cement is presented. The outcome shows that the highest percentage increase in split tensile strength of 51.17% was obtained for CPG3 i.e. concrete with 10% plastic waste by weight of fine aggregate with 5% glass powder by weight of cement. Whereas for CPG1, GPG2 percentage increases in split tensile strength were 18.31%, 28.17% respectively. For CPG4, CPG5, CPG6, CPG7 the percentage increase is 32.39%, 18.31%, 4.69% and 2.35% respectively. Moreover, the split tensile strength decreases by 1.41%, 1.88%, 7.04%, 13.15% for CPG8, CPG9, CPG10 and CPG11 respectively. In general, the trend of splitting tensile strength is escalating up to 10% substitution of fine aggregate by plastic wastes with 5% glass powder by weight of cement and after which the split tensile strength reduces. This may be because of the pessimistic result of an even face texture on the bond strength of plastic wastes [6] and contribution of amorphous structure of finely grounded glass powder which offers excellent pozzolanic property [17]

4.3 Flexural strength. Table.2 and Figure.6 shows the flexural strength of Prism samples of size 10 cm × 10 cm × 50 cm. From the experimental results, it is obvious that maximum proportion of raise in flexural strength was obtained for CPG3i.e concrete with 10% substitution of fine aggregate by PET waste with 5% glass powder for cement which is 49.13% superior to the control specimen. The percentage increase in flexural strength of specimen CPG1, CPG2, CPG4, CPG5, CPG6 and CPG7 is 4.96%, 19.85%, 29.78%, 18.11%, 4.96% and 2.98% respectively. Whereas the flexural strength decreases for specimen CPG8, CPG9, CPG10, CPG11 by 0.50%, 1.24%, 5.21% and 7.69 respectively. The flexural strength increases up to 10% substitution of fine aggregate by plastic ravage with 5% replacement of glass powder with cement, beyond which it decreases due to the increase of the waste plastic ratio which reduces bonding between plastic waste and cement paste [4].

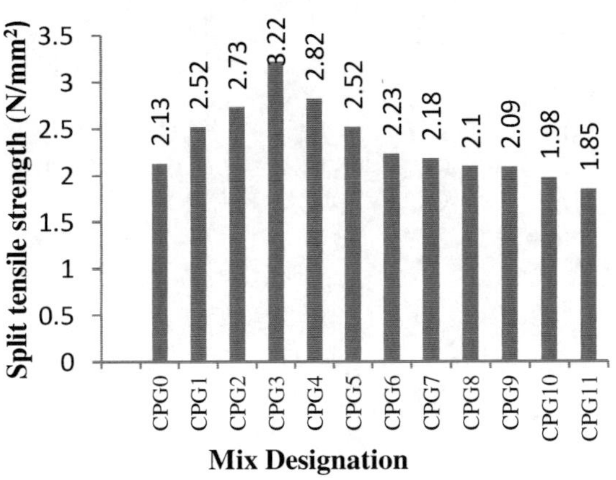

Figure 5. Split tensile strength of concrete

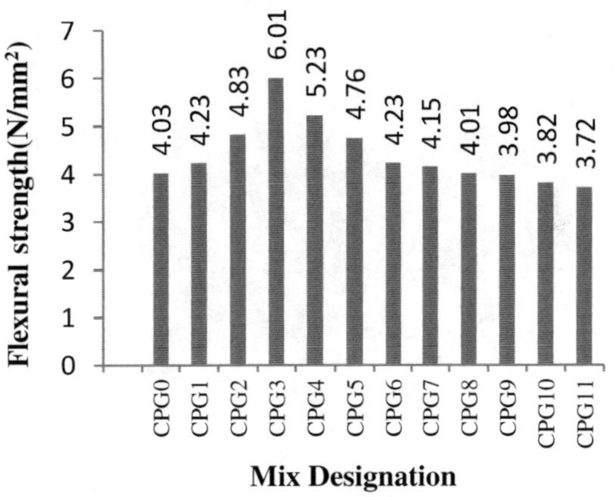

Figure 6. Flexural strength of concrete

4.4 Prediction using NN. This segment determines the fundamental options of the joint result of plastic wastes with glass powder in concrete for mechanical properties. Here, the comparison uses the plastic waste of various proportions by weight fine aggregate and glass powder by weight of cement. The particular values of the combinations of plastic waste and glass powder combinations were already mentioned. Consequently, this segment compares the mechanical properties with a distinct grouping of plastic waste with 0% and 5% glass powder using Neural Network-Gradient Descent and Neural Network–Levenberg–Marquard. Initially, figure 7 and table 3 shows the evaluation of the experimental and predicted values of concrete. Here, 12 different combinations of the weight of plastic wastes with 0% and 5% glass powder are determined and further compared with the experimental values. The maximum percentage increase in compressive strength predicted by NN-LM was found maximum in case of CPG11 which is 11.39 % higher than experimental values and maximum percentage in a decrease in compressive strength was obtained for CPG9 which 10.54% lesser than experimental values. Similarly, the percentage increase in the predicted compressive strength by NN-GD ranges from 2.01% to 11.02%. The percentage decrease in the predicted compressive strength by NN-GD ranges from 1.70% to 7.62%.

Figure 8 and Table 3 Illustrates the split tensile strength comparison of experimental and predicted values. The predicted values in NN-LM shows that the maximum percentage increase was obtained for CPG1 which is 11.51% higher than the experimental values and percentage decrease is maximum for BCPG3 which is 12.42% lesser than the experimental values. Similarly, the predicted valued in NN-GD the maximum percentage increase was obtained for CPG8 which is 8.10% higher than the actual values and percentage decrease is maximum for CPG4 which is 3.90% lesser than the predicted values. Figure.9 and table 3 shows the comparison of experimental and predicted values for flexural strength.

Table 3. Comparison of experimental and predicted values of concrete

Mix	Comp. strength (MPa)			Split Tensile strength (MPa)			Flexural strength (MPa)		
	Experi-mental	Predicted		Experi-mental	Predicted		Experi-mental	Predicted	
		NN-LM	NN-GD		NN-LM	NN-GD		NN-LM	NN-GD
CPG0	31.30	31.307	31.93	2.13	2.16	2.28	4.03	4.03	4.10
CPG1	32.10	32.75	33.00	2.52	2.81	2.52	4.23	4.29	4.60
CPG2	33.42	33.41	32.44	2.73	2.72	2.73	4.83	4.96	4.67
CPG3	36.50	35.84	33.72	3.22	2.82	3.20	6.01	5.90	5.83
CPG4	33.20	33.10	31.79	2.82	2.81	2.71	5.23	5.22	5.16
CPG5	31.20	30.19	30.50	2.52	2.57	2.52	4.76	4.89	5.02
CPG6	25.97	26.03	24.88	2.23	2.25	2.23	4.23	4.65	4.44
CPG7	20.27	21.02	20.32	2.18	2.24	2.18	4.15	4.23	4.48
CPG8	17.69	16.68	17.39	2.10	2.08	2.27	4.01	4.00	3.99
CPG9	16.98	15.19	15.77	2.09	1.92	1.93	3.98	3.97	3.90
CPG10	12.88	13.80	14.30	1.98	1.98	1.98	3.82	3.82	3.87
CPG11	12.20	13.59	12.83	1.85	1.86	1.86	3.72	3.75	3.80

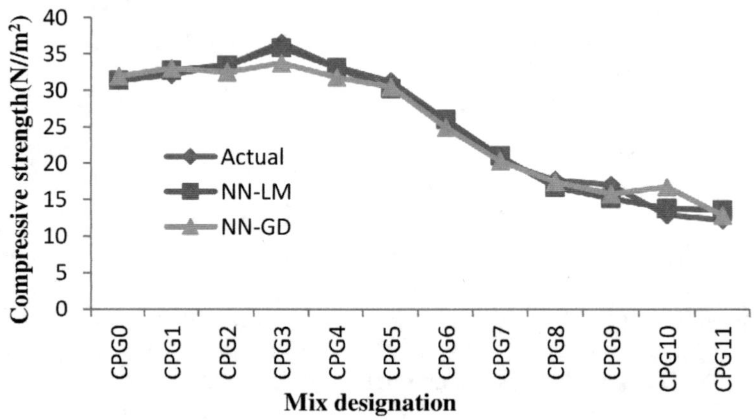

Figure. 7 Graphical depictions of a neural network in terms of Compression strength

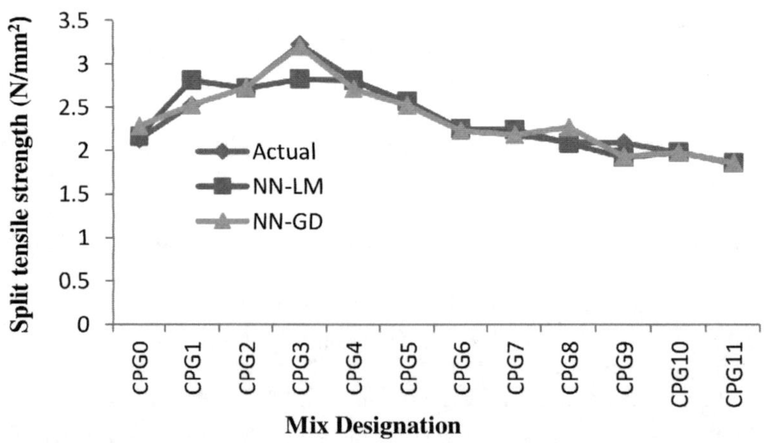

Figure 8. Graphical depiction of neural network in terms of Split tensile strength

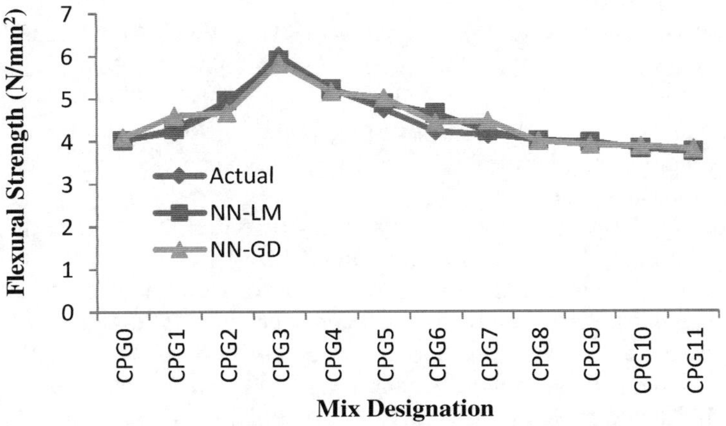

Figure 9. Graphical depiction of neural network in terms of Flexural strength

The predicted values in NN-LM shows that the maximum percentage increase was obtained for CPG6 which is 9.93% higher than the experimental values and percentage decrease is maximum for CPG3 which is 1.83% lesser than the experimental values. The same performance is attained in the case of NN-GD based approach.

5. Conclusion

In this study, the mechanical properties of concrete containing plastic wastes as a substitution for fine aggregate and glass powder as a partial substitution for cement were investigated. Furthermore, the prediction of mechanical properties of concrete with plastic waste and glass powder were made using Network-Gradient Descent and Neural Network–Levenberg–Marquard methods. The conclusions achieved from the experimental and ANN outcomes are listed as follows:

1. The concrete specimens consisting of 10% substitution of fine aggregates with plastic wastes with 5% glass powder obtained the optimal compressive strength. In fact, for 10% of plastic waste content with 5% glass powder has a maximum percentage increase of compressive strength of 16.61%. Further increase in plastic waste contents beyond 10%, the compressive strengths decrease.
2. The split tensile strength strength was maximum for concrete with 10% plastic waste and 5% glass powder which is 51.17% superior to the control specimen.
3. The highest percentage increase in flexural strength was obtained for 10% plastic waste and 5% glass powder which is 49.13% higher than the control specimen
4. Eventually, it can be said that usage of plastic wastes as fine aggregates in concrete enhances the mechanical properties of can be an eco-friendly solution for plastics.
5. The predicted mechanical properties using Neural Network–Levenberg–Marquard and Neural Network-Gradient Descent have attained its maximum performance.

References

[1] Ashwini Manjunath B T 2016 Partial replacement of E-plastic Waste as Coarse-aggregate in Concrete *Procedia Environmental Sciences* **35** p 731 – 9

[2] Adewumi John Babafemi, Branko Savija, Suvash Chandra Paul, Vivi Anggraini 2018 Engineering properties of Concrete with Waste Recycled Plastic A Review *Sustainability* **10** p 1-26

[3] Naga Rajesh K, Rath M K, Markandeya Raju P 2019 A research on sustainable micro-concrete *International Journal of Recent Technology and Engineering* **8(2 Special Issue 3)** p 1137-9

[4] Baboo Rai, Tabin Rushad S, Bhavesh Kr, Duggal S K 2012 Study of Waste Plastic Mix concrete with Plasticizer *ISRN Civil Engineering* **2012** p 1-5

[5] Zoe Harmonie Lee, Suvash Chandra Paul, Sih Ying Kong, Susilawati Susilawati, Xu Yang 2019 Modification of Waste Aggregate PET for Improving the Concrete Properties *Advances in*

Civil Engineering **2019** p 1-10
[6] Rahmani A E, Dehestani M, Beygi M H A, Allahyari H, Nikbin I M 2013 On the Mechanical properties of concrete containing waste PET particles *Construction and Building Materials* **47** p 1302-8
[7] Aref M. al-Swaidani, Waed T Khwies 2018 Applicability of Artificial Neural Networks to Predict Mechanical and Permeability Properties of Volcanic Scoria-Based Concrete *Advances in Civil Engineering* **2019** p 1-16
[8] Francisco Casanova-del-Angel, Jorge Luis Vazquez-Ruiz 2012 Manufacturing Light Concrete with PET Aggregate *ISRN Civil Engineering* **2012** p 1- 10
[9] Liliana Ávila Córdoba, Gonzalo Martínez-Barrera, Carlos Barrera Díaz, Fernando Ureña Nuñez, Alejandro Loza Yanez 2013 Effects on Mechanical Properties of Recycled PET in Cement-Based Composites *International Journal of Polymer Science* **2013** p 1-6
[10] Haiming Chen, Yangchen Xu, Donglei Zhang, Lingxia Huang, Yuntao Zhu, Le Huang 2019 The Influence of Nano-SiO_2 and Recycled Polypropylene Plastic Content on Physical, Mechanical, and Shrinkage Properties of Mortar *Advances in Civil Engineering* **2019** p 1-12
[11] Nabajyoti Saikiaa B, Jorge de Britoa 2013 Waste Polyethylene Terephthalate as an Aggregate in Concrete *Materials Research* 16 p 341-50
[12] Altamashuddinkhan Nadimalla, Siti Aliyyah Binti Masjuki, Asmahani Binti Saad, Kamsiah Binti Mohd Ismail, Maisarah Bt Ali 2019 Polyethylene Terephthalate (PET) Bottles Waste as Fine Aggregate in Concrete *International Journal of Innovative Technology and Exploring Engineering* **8** (**11**) p 77 – 80
[13] Rafat Siddique, Jamal Khatib, Inderpreet Kaur 2008 Use of recycled plastic in concrete: A review *Waste Management* **28** p 1835–52
[14] Manhal A Jibrael, Farah Peter 2016 Strength and Behavior of Concrete Contains Waste Plastic *Journal of Ecosystem & Ecography* **6** p 1-4
[15] Afif Rahma, Nabil El Naber, Sherzad Issa Ismail 2017 Effect of glass powder on the compression strength and the workability of concrete *Cogent Engineering* **4 (1)** p 1-12
[16] Bharat Nagar V, Bhargava P 2016 Effect of Glass Powder on Various Properties of Concrete *International Journal of Science, Engineering and Technology* 4 p 567-73
[17] Lim Sheau Hooi, Phang Jia Min 2017 Potential of substituting waste glass in aerated light weight concrete Proceedings of sustainable Civil Engineering Structures and Construction material SCESCM 2019 *Procedea Engineering* **171** p 633-9
[18] IS 8112 2013 Ordinary Portland Cement 53 Grade Specification *Bureau of Indian Standards* New Delhi India
[19] IS 10262 2009 Concrete Mix Proportioning Guidelines *Bureau of Indian Standards* New Delhi India
[20] IS 516 1959 Hardened Concrete Methods of Test *Bureau of Indian Standards* New Delhi India

Mathematical Model for Natural Frequency of Prestressed Concrete beam using STAAD.Pro.

P Markandeya Raju[1], M Pavan Kumar[2], S Adiseshu[3], V V Sai Sravan Kumar[4]

[1]Professor of Civil Engineering, MVGR College of Engineering (A), Vizianagaram – 535005 Andhra Pradesh, India.
[2]Assistant Professor, Department of Civil Engineering, SVP Engineering College, Visakhapatnam - 530041
[3]Professor of Civil Engineering, Andhra University, VISAKHAPATNAM. AP, India.
[4]Former PG Student of Structures, College of Engineering (A), Andhra University, VISAKHA-PATNAM – 530003 (AP), India.
Email: markandeyaraju@mvgrce.edu.in

Abstract. The problem of assessing the variation in natural frequency of a prestressed and un-prestressed beam has been of considerable interest to Structural Engineers. The objective of the present work is to find the effect of pre-stressing on Natural frequency of a Concrete beam and thereby establishing a relationship between the natural frequency of 1st mode of vibration and other modes. Conventional methods were used to calculate the natural frequency of 1 to 5 modes of vibration for un-prestressed beams while STAAD.Pro was used for assessing the same for prestressed beams. A 20 m span beam with different live loads and cross-sections were considered for the study. From the results, it was observed that as the cross-section of the beam increases, the natural frequency decreases. Comparing the natural frequency of 1st mode of vibration for beams with and without prestressing, it can be observed that within the scope of this work, the natural frequency of the un-prestressed beam with live load 15 kN/m is more than that for the prestressed beam. For the other two load-carrying capacities, it is vice-versa. It was also observed that the % difference of Natural frequency of a particular mode of vibration w.r.t mode 1 natural frequency is almost constant. In other words, the effect of live load and cross-section is negligible on the % difference of Mode 'i' to Mode 1 natural frequency. Based on this characteristic, a mathematical model has been developed.
Keywords: Prestressed concrete; reinforced concrete; simply supported; natural frequency; modes of vibration; STAAD.Pro.;

1. Introduction

The application of the prestressed tendons made from the solid bars has an essential influence on the architectural expression for large-span structures. Historically, pre-stressing was developed simultaneously for concrete and steel structures **(Trotsky, 1989)**. Prestressed members are widely being used presently in the construction industry. Because of the need for earthquake resistance of all constructions, determination of their natural frequency is necessary. Many empirical formulae are available but are based on Assumptions as per Literature. This paper demonstrates the application STAAD.Pro for finding the natural frequency of prestressed concrete beams. On comparing the natural frequency of prestressed and un-prestressed concrete beams, it is observed that just like in un-prestressed. Beams, there is a relationship between the natural frequency of the 1st mode of vibration and other modes. This phenomenon is applied to develop a mathematical model for finding the natural frequency of any mode of vibration if the natural frequency of 1st mode is known.

2. Literature Review

In the available literature concerning pre-stressed structures, it was found that the literature on the natural frequency of pre-stressed structures is less. However, a review of past studies on the analysis of the natural frequency of pre-stressed steel beams is presented. **Kato and Shimada (1986)** have found a decrease in natural frequency with damage in a bridge girder. They have attributed this to the loss of flexural rigidity due to opening up the cracks as the load is increased. One of the oldest papers on this topic has been written by **Tang and Leu (1989).** They studied a three-span countryside prestressed concrete bridge and concluded that the change of dynamic characteristics can be used as an indicator for damage detection of the bridge. It is therefore clear that prestressing force does not have the compression softening effect but influences the natural frequency of the prestressed beams differently. Prestressing force closes the cracks in concrete (microcracks as well as the structural cracks, if any). Cracks reduce the beam stiffness (natural frequency) and closing of cracks should increase the stiffness (and hence the natural frequency). However, there are serious difficulties in evaluating accurate values of modulus of elasticity and moment of inertia of concrete sections and hence the natural frequencies of vibration. Similar observations about the decrease in flexural stiffness due to damage, resulting in a reduction in natural frequency, have been reported by **Mirza et al. (1990)** and **Ambrosini et al. (1999). Singh et al. (1991)** treated the prestressed beam as a beam-column and took its theoretical basis for establishing the variation in the natural frequency with prestressing force. He carried out an experimental study on several prestressed girders to study the variation of natural frequency with loss of prestress and the presence of cracks. The natural frequency, obtained by ambient vibration measurements in the laboratory, increases with increase in prestress force up to a point; beyond it, the increasing rate is either very low or there is a decrease in frequency. The development of cracks in the beam results in a decrease of natural frequency. **Saiidi et al. (1994)** recalled the well-known “compression softening” effect and reported the discrepancy between theoretical and experimental results. **Saiidi et al. (1994)** carried out an experimental study of a prototype post-tensioned bridge with 155ft (47.2 m) span and 45 ft. width and a laboratory specimen. They found that the natural frequency of the bridge decreased with loss of prestress. The testing of the laboratory specimen also supported this result. The results available from the literature are different from the dynamic characteristics of an isotropic beam subjected to an axially compressive force which was put forward by **Clough et al. (1995)**. The reason is that the beams were considered as isotropic, homogeneous, linear elastic material in the traditional analysis method. However, more accurate results are required in the analysis of frequency of PSC beam. The constitutive model of PSC member is analyzed based on the microstructure of concrete in this paper. The orthotropic linear elastic model is used to analyze the relationship between dynamic frequency and prestressing force of the concrete beam, at the same time the equivalent stiffness of prestressed tendon relating to the prestressing force is added to the bending deformation stiffness of the beam. The analytical value agrees well with the test result, indicating that the current analysis method in this paper is feasible to the full-prestressed concrete beam. **Grace and Ross (1996)** conducted an analytical study aimed at examining the influence on girders’ natural frequencies and mode shapes of vibration of many parameters like prestress force level and the shape of the bottom prestressing strands. **Jain et al. (1996)** conducted an experimental study to determine the effect of cracking and prestress loss of the concrete beam on their fundamental frequency has been carried out. It has been argued that the change in the natural frequency of the prestressed concrete beam is due to change in the stiffness of the beam caused by closing or opening of cracks as the prestressing force is applied or removed, not due to compression softening effect discussed by some researchers. The study aims to contribute towards a non-destructive test method to assess the soundness of the prestressed concrete (PSC) bridge girders. **Jain and Goel (1996)** have pointed out that prestressing force is an internal force of the prestressed girder system and therefore prestressing force does not cause compression softening effect. **Dall’Asta and Leoni (1999)** studied the free vibrations of a class of concrete beams prestressed using unbonded internal cables to test the influence of the elastic characteristic of the materials, as well as the influence of the path and stress of the prestressing cables on the beam free vibrations. **Miyamoto et al. (2000)** studied the effect of the external tendons on the frequency of vibration of prestressed concrete beams. **Miyamoto et al. (2000)** studied the pre-stressing technique using external tendon to analytically investigate the dynamic behaviour of prestressed composite girder bridges. **Hamed and Frostig (2006)** presented a study dealing with the effect of the magnitude of the prestress force on the Natural Frequencies of prestressed beams with bonded and unbonded tendons. **Sunkyu et al. (2010)** studied the flexural behaviour and strengthening effect of a bridge using a steel I-beam that had been externally prestressed with unbonded tendons. **Zhang and Ruige (2007)** conducted dynamic tests on three bonded and two un-bonded fully prestressed concrete beams. **Marco Breeolotti et al. (2009)** explored the effects of prestress forces on modal parameters of concrete beams. **Vadims et al. (2012)** presented the determination of natural frequencies (as mode shapes, of the model depending on a prestressing level) of a physical model of a prestressed suspension bridge.

3. Objective and Scope

The objective of the present work is

1. To find the effect of pre-stressing on Natural frequency of a Concrete beam using STAAD.Pro
2. To establish a relationship between the natural frequency of the 1st mode of vibration and other modes.
3. Compare the results with those of un-prestressed beams obtained by conventional methods.

The scope of the work is limited to the following.

Grade of concrete: M35

Modulus of Elasticity: $E = 5000 \sqrt{f_{ck}}$ Here, $f_{ck} = 35$ N/mm^2 = 35000 kN/m^2

Therefore $E = 5000\sqrt{35} = 2.958 \times 10^7$ kN/m^2

Span of the beam: L = 20 m; Live load: w = 5 kN/m, 10 kN/m, 15 kN/m

m = mass per unit length (including live load)

C/s of the beam: B x D = 300 mm x 600 mm, 300 mm x 750 mm, 300 mm x 900 mm

Pre-stressing force: P is calculated assuming that the stress at soffit is nullified at working loads after all the losses. % loss of prestressing is assumed as 25.

Eccentricity: $e = (\frac{D}{2} - 50)$ mm (maximum possible)

Mode of vibration: 1, 2, 3, 4 and 5.

The nomenclature for Un-prestressed / Reinforced concrete beam and Pre-stressed concrete beam that is presented in Table 1 and Table 2 is adopted for the study.

Table 1. Reinforced Concrete Beams

Sl. No	Notation	Size of member (m) B	Size of member (m) D	Load kN/m
1	C1R	0.3	0.6	5
2	C2R	0.3	0.6	10
3	C3R	0.3	0.6	15
4	C4R	0.3	0.75	5
5	C5R	0.3	0.75	10
6	C6R	0.3	0.75	15
7	C7R	0.3	0.9	5
8	C8R	0.3	0.9	10
9	C9R	0.3	0.9	15

Table 2. Prestressed Concrete Beams

Sl. No	Nomenclature	B (m)	D (m)	P (kN)	e (m)	Load (kN/m)
1	C1P	0.3	0.6	1357	0.25	5
2	C2P	0.3	0.6	2071	0.25	10
3	C3P	0.3	0.6	2786	0.25	15
4	C4P	0.3	0.75	1181	0.325	5
5	C5P	0.3	0.75	1736	0.325	10
6	C6P	0.3	0.75	2292	0.325	15
7	C7P	0.3	0.9	1068	0.4	5
8	C8P	0.3	0.9	1523	0.4	10
9	C9P	0.3	0.9	1977	0.4	15

4. Methodology and Results

To find Natural frequency of a prestressed beam, STAAD.Pro is used and for Concrete beams, the classical methods available are used.

4.1 The natural frequency of un-prestressed beams

The following Standard formula is considered for determination of the natural frequency of an RC beam.

$$n = \frac{A}{l^8} \times \sqrt{\frac{EI}{m}}$$

Where

E - Young's modulus
I - area moment of inertia
L - Length of beam
m - mass per unit length of the beam
A = $(n\,\pi)^2$ for Mode n, where n = 1, 2, 3, 4 and 5.

The 1st 3 modes of vibration for the simply supported beam are shown in figure 1.

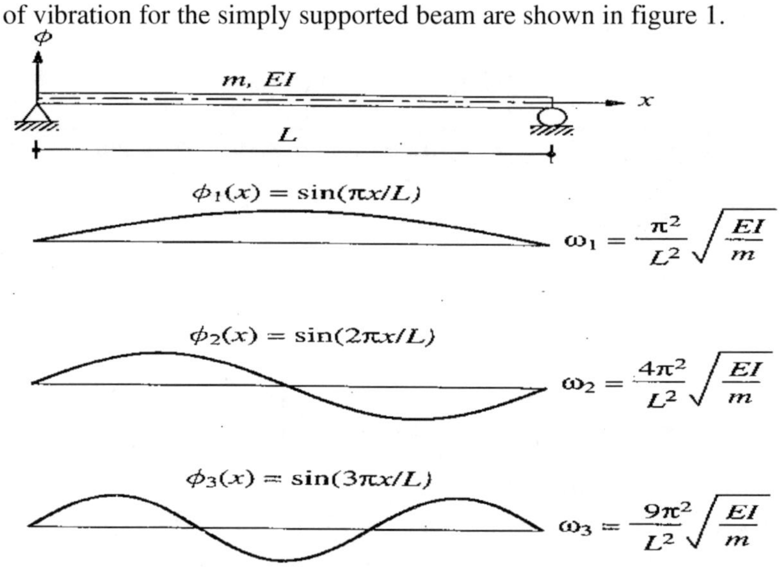

Figure 1. Natural vibration modes and frequencies of prismatic simply supported Un-prestressed beams

The natural frequencies and modes of vibration of a prismatic beam simply supported at both ends with properties defined in Table 1 are determined in cycles/sec. and presented in Table 3.

Table 3. Natural Frequencies of un-prestressed Concrete Beams

Types of Beams	**Natural Frequencies of un-prestressed Concrete Beams (Cycles/sec)**								
Case	C1R	C4R	C7R	C2R	C5R	C8R	C3R	C6R	C9R
MODE 1	1.59	2.10	2.63	1.29	1.73	2.20	1.11	1.51	1.93
MODE 2	6.38	8.43	10.54	5.16	6.95	8.83	4.45	6.05	7.75
MODE 3	14.44	18.98	23.73	11.62	15.65	19.87	10.02	13.62	17.44
MODE 4	25.53	33.74	44.18	20.67	27.82	35.33	17.82	24.22	31.00
MODE 5	39.90	52.73	65.91	32.29	43.48	55.19	27.85	37.84	48.45

4.2 Model Calculation for natural frequency of un-prestressed beams (for C1 R Second Frequency)

$$n = \frac{A}{l^8} \times \sqrt{\frac{EI}{m}}$$

$$I = \frac{300\ X\ 600^3}{12}$$

$E = 5000\sqrt{35} = 29580.39\ N/mm^2$

$EI = 1.59\ X\ 10^{14}\ N/mm^2$

$l = 20{,}000\ mm$

For 2nd mode of vibration

$A\pi^2 = (2\pi)^2 = 4\pi^2$

$m = (5 + 0.3\ X\ 0.6\ X\ 25)\ kN/m$

$$= \frac{(5 + 0.3\ X\ 0.6\ X\ 25)}{9.81\ X\ 10^3}$$

$= 9.683\ X\ 10^{-4}\ kg/mm$

$$f_2 = 4\pi^2\ X \sqrt{\frac{1.59\ X\ 10^{14}}{9.683\ X\ 10^{-4}\ X\ (20{,}000)^4}}$$

$f_2 = 40.12$ Radian/sec

$f_2 = \frac{1}{2\pi} X\ 40.12 = 6.385$ cycles/sec

$f_2 = 6.385$ cycles/sec

4.3 Natural Frequency of Prestressed Beams

The natural frequencies for the first few modes of vibration for prestressed beams are computed for the flexural motion of a prestressed simply supported beam using STAAD.Pro. The simply supported beam is divided into spanwise beam elements. For example, if it is divided into 20 elements, STAAD.Pro can give the natural frequencies of the first 20 modes of vibration. At nodes 1 and 21, all degrees of freedom except the rotation about the Z-axis is restrained. For the remaining nodes, only the translation along Y and the rotation about Z are permitted. Both shear deformation and rotary inertia have been excluded from the model. The mass matrix is diagonal. The cross-section, material and loads are assigned as per requirements to all the 20 elements. While assigning loads, the prestressing force with corresponding tendon profile and eccentricity are also assigned to all the 20 elements. And lastly, in loads command, the following step is included for finding natural frequency as shown in figure 2.

Page controls → Load and Definition → Load case details Load case 1 → New → Load item → Frequency → Modal calculation

Figure 2. The final step in Loads and Definitions for finding Natural Frequency of PSC beam

It is important to note that all the loads including self-weight, live load, prestress along with the command for modal calculations have to under only one load case. The STAAD.Pro analysis for all 9 cases is performed and the results are tabulated in Table 4.

Table 4. Natural Frequency of Prestressed Concrete Beams

Type of Beams	Natural Frequency of Prestressed Concrete Beams (Cycles/sec)								
Case	**C1P**	**C4P**	**C7P**	**C2P**	**C5P**	**C8P**	**C3P**	**C6P**	**C9P**
MODE 1	1.58	2.24	2.58	1.28	1.85	2.35	1.18	1.61	2.06
MODE 2	5.10	8.94	8.09	4.16	7.37	9.33	4.73	6.41	8.19
MODE 3	8.74	19.96	13.64	7.14	16.45	20.76	10.59	14.32	18.22
MODE 4	13.58	35.08	21.46	11.06	28.93	36.33	18.69	25.18	31.88
MODE 5	22.06	54.01	35.21	17.90	44.54	55.61	28.93	38.76	48.80

Natural Frequency of Prestressed Concrete Beams with Different cross-sectional areas for different Modes of Vibration given in Table 4 is presented in figure 3.

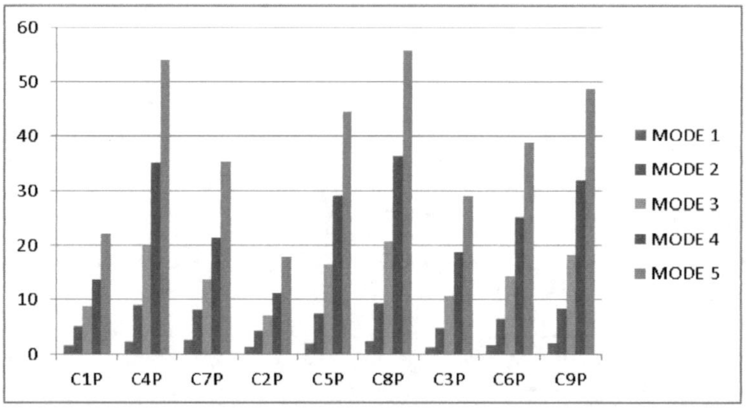

Figure 3. Natural Frequency of Prestressed Concrete Beams

Natural Frequency of un-prestressed Concrete Beams with Different cross-sectional areas for different Modes of Vibration given in Table 3 is presented in figure 4.

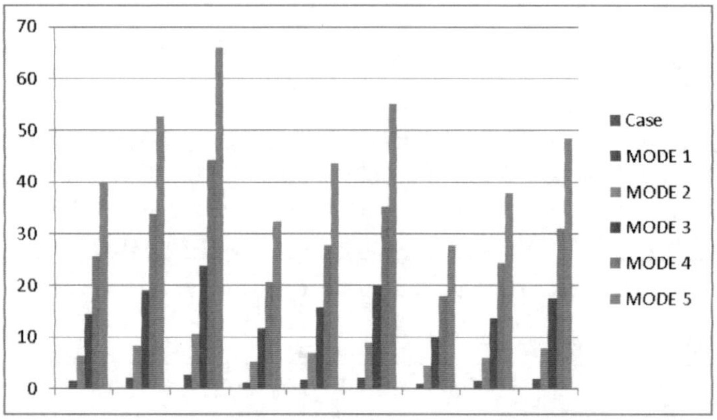

Figure 4. Natural Frequencies of un-prestressed Concrete Beams

5. Discussions on Results

For both Prestressed Concrete and un-prestressed Concrete beams, the following observations can be made from the results. For the same cross-section, the increase in load decreases the natural frequency of the pre-stressed beam. However, as the cross-section of the beam increases, the natural frequency increases. Obviously, the natural frequency also increases with an increase in the mode of vibration. Comparing the natural frequency of the 1st mode of vibration for Prestressed Concrete and Reinforced Concrete, the following observations can be made.

Table 5. Comparison of Natural Frequencies of Prestressed Concrete and un-prestressed Concrete beam for 1st Mode of Vibrations

Types of Beams	Natural Frequencies (Cycle/Sec)								
Case	C1	C4	C7	C2	C5	C8	C3	C6	C9
PSC	1.58	2.24	2.58	1.28	1.85	2.35	1.18	1.61	2.06
Un-prestressed	1.59	2.10	2.63	1.29	1.73	2.20	1.11	1.51	1.93

For the case study under consideration, the natural frequency of the un-prestressed Concrete beam with live load 5 kN/m is more than that for the prestressed beam. For the other two load-carrying capacities, it is vice-versa. The percentage difference of a mode of frequency to 1st mode is calculated for both prestressed and un-prestressed Concrete beams and presented in Table 6 and Table 7.
For example,
% diff. of Mode n to Mode 1, NF = (Mode n NF – Mode 1 NF) x 100/(Mode n NF).

Table 6. Percentage Difference of Natural Frequency of Mode i to Mode 1 for Prestressed Concrete Beams

Percentage difference of Natural Frequency of Prestressed Concrete Beams									
Beam→	C1P	C4P	C7P	C2P	C5P	C8P	C3P	C6P	C9P
% diff. of Mode 5 w.r.t Mode 1 NF	95.82	95.89	95.75	95.82	95.89	95.75	95.82	95.89	95.75
% diff. of Mode 4 w.r.t Mode 1 NF	93.58	93.64	93.51	93.58	93.64	93.51	93.58	93.64	93.51
% diff. of Mode 3 w.r.t Mode 1 NF	88.73	88.79	88.66	88.73	88.79	88.66	88.73	88.79	88.66
% diff. of Mode 2 w.r.t Mode 1 NF	74.87	74.91	74.81	74.87	74.92	74.81	74.87	74.92	74.80

Table 7. Percentage Difference of Natural Frequency of Mode i to Mode 1 for un-prestressed Concrete Beams

Percentage difference of Natural Frequency of un-prestressed Concrete Beams									
Beam→	C1R	C4R	C7R	C2R	C5R	C8R	C3R	C6R	C9R
% diff. of Mode 5 w.r.t Mode 1 NF	96.00	96.00	96.00	96.00	96.00	96.00	96.01	96.01	96.00
% diff. of Mode 4 w.r.t Mode 1 NF	93.75	93.75	93.74	93.75	93.74	93.75	93.76	93.58	93.77
% diff. of Mode 3 w.r.t Mode 1 NF	88.89	88.88	88.88	88.89	89.05	88.88	88.91	88.91	88.90
% diff. of Mode 2 w.r.t Mode 1 NF	75.00	74.99	74.98	75.00	74.98	74.99	75.03	75.04	74.37

From the Tables 6 and 7, it can be observed that although the natural frequencies for Prestressed and un-prestressed Concrete beams are different, the % difference of Natural frequency of a particular mode of vibration to mode 1 natural frequency is almost constant. That is the effect of live load and cross-section is negligible on the % difference of Mode n to Mode 1 natural frequency.
The approximate values are observed as
% diff. of Mode 5 to Mode 1 NF = 96
% diff. of Mode 4 to Mode 1 NF = 93
% diff. of Mode 3 to Mode 1 NF = 89
% diff. of Mode 2 to Mode 1 NF = 75
Let
n_i = natural frequency of the ith mode for a given prestressed beam
n_1 = natural frequency of the 1st mode for given prestressed beam
k_i = (% difference of Mode i to Mode 1 NF for rectangular beam)/100
For Example
k_2 = % difference of Mode 2 with respect to Mode 1 NF of rectangular beam = 75/100 = 0.75.
This is calculated as
% difference of Mode i to Mode 1 NF

$$= \frac{(n_i - n_1)}{n_i} \times 100$$

k_i = (% difference of Mode i with respect to Mode 1 NF) / 100

$$= (\frac{(n_i - n_1)}{n_i} \times 100)/100$$

$$= \frac{(n_i - n_1)}{n_i} \times 100$$

$$k_i = 1 - \frac{n_1}{n_i}$$

$$1 - k_i = \frac{n_1}{n_i}$$

$$(1 - k_i)^{-1} = \frac{n_i}{n_1}$$

$$n_1(1 - k_i)^{-1} = n_i$$

$$\boldsymbol{n_i = n_1(1 - k_i)^{-1}}$$

6. Conclusion

For both types of Beams Prestressed Concrete and Reinforced Concrete, the following conclusions can be made.

1. For the same cross-section, the increase in load decreases the natural frequency of the pre-stressed beam.
2. As the cross-section of the beam increases, the natural frequency increases.
3. Comparing the natural frequency of 1st mode of vibration for beams Prestressed Concrete and un-prestressed Concrete, it can be observed that the natural frequency of the un-prestressed Concrete beam with live load 5 kN/m is more than that for the prestressed beam. For the other two load-carrying capacities, 5 kN/m and 10 kN/m, the natural frequency of Prestressed Concrete Beams is more than un-prestressed Concrete Beams.

   Although the natural frequencies for Prestressed and un-prestressed Concrete beams are different, the % difference of Natural frequency of a particular mode of vibration w.r.t mode 1 natural frequency is almost constant. In other words, the effect of live load and cross-section is negligible on the % difference of Mode n to Mode 1 natural frequency.
   The approximate values are observed as
   % difference of Mode 5 to Mode 1 NF = 96
   % difference of Mode 4 to Mode 1 NF = 93

IOP Conf. Series: Materials Science and Engineering 1025 (2021) 012008 doi:10.1088/1757-899X/1025/1/012008

% difference of Mode 3 to Mode 1 NF = 89
% difference of Mode 2 to Mode 1 NF = 75

4. Hence, if the natural frequency of the first mode of vibration of a prestressed beam is known along with the natural frequencies of first and ith mode of vibration of the same cross-section when un-prestressed, then, the ith mode of vibration of the prestressed beam can be calculated from

$$\boldsymbol{n_i = n_1(1 - k_i)^{-1}}$$

Where
n_i = natural frequency of the ith mode of a prestressed concrete beam
n_1 = natural frequency of the 1st mode of a prestressed concrete beam
k_i = (% difference of mode i to Mode 1 NF for an un-prestressed beam)/100

References

[1] Sunkyu Park, Taewan Kim, Kwangsoo Kim and Sung-Nam Hong 2010 Flexural Behavior of steel I-beam pre-stressed with externally unbonded tendons Journal of Construction Steel Research 66 (1) p 125–32
[2] Miyamoto A, Katsuji T, Nakamura H, Bull J W 2000 Behaviour of pre-stressed beam strengthened with external tendons ASCE Journal of Structural Engineering 126 (9) p 1033-44
[3] Singh S N 1991 Effect of Prestress on Natural Frequency of PSC Beams M.Tech Thesis, Department of Civil Engineering, Indian Institute of Technology, Kanpur.
[4] Saiidi M, Douglas B 1994 Prestress force effect on vibration frequency of concrete bridges ASCE Journal of Structural Engineering 120 (7) p 2233-41
[5] Grace N F, Ross B 1996 Dynamic characteristic of post-tensioned girders with web openings ASCE Journal of Structural Engineering 122 (6) p.643-50
[6] Saiidi M, Douglas B, Feng S 1994 Prestress force effect on vibration frequency of concrete bridges ASCE Journal of Structural Engineering 120 (7) p 2233-41
[7] Jain S K, Goel S C 1996 Discussion on Prestress Force Effect on Vibration Frequency of Concrete Bridges ASCE Journal of Structural Engineering 122 (4) p 459-60
[8] Kato M, Shimada S 1986 Vibration of PC Bridge during Failure Process ASCE Journal of Structural Engineering 112 (7) p 1692-703
[9] Ambrosini D, Luccioni B, Danesi R 1999 Theoretical-Experimental damage Determination in Pre-stressed Concrete Beams International Symposium on NDT Contribution to the Infrastructure Safety Systems Santa Maria RS Brazil
[10] Mirza M S, Ferdjani O, Hadj-Arab A, Joucdar K, Khaled A 1990 An Experimental Study of Static and Dynamic Responses of Prestressed Concrete Box Girder Bridges *Canadian Journal of Civil Engineering* **17** p 481-93
[11] Tang J P, Leu K M 1989 Vibration Tests and Damage Detection of P/C Bridges *Proceedings of lCOSSAR '89 5 International Conference on Structural Safety and Reliability* San Francisco p 2263-6
[12] Dall'Asta A, Leoni G 1999 Vibration of beams prestressed by internal frictionless cables *Journal of Sound and Vibration* **222 (1)** p 1-18
[13] Miyamoto A, Tei K, Nakamura H, Bull J W 2000 Behavior of prestressed beam strengthened with external tendons *ASCE Journal Structural Engineering* **126 (9)** p 1033-44
[14] Hamed E, Frostig Y 2006 Natural frequencies of bonded and unbonded prestressed beams-prestress force effects *Journal of Sound and Vibration* **295** p 28-39
[15] Marco Brecolotti M, Ubertini F, Venanzi I 2009 Natural frequencies of prestressed concrete beams: theoretical prediction and numerical validation *Proceeding of the XIX Aimeta Conference* Ancona Italy p 14-7
[16] Vadimsgoremikins K, Rocens D, Serdjuks 2012 Cable Truss Analyses for Prestressed Suspension Bridge *Proceedings of the 8th International Conference of DAAAM Baltic Industrial Engineering* Tallinn Estonia April 19-21 2012 p 45-50
[17] Znang Taoting, CaiLijian, Zhang Changpao 2005 A summary of research status on the detection

method of prestressed concrete beams *Journal of Industrial Construction* **135 (II)** p 84-7
[18] Clough R W, Penzien J 1995 Dynamics of Structures (M) 2nd Edition *Computers and Structures* Inc. Berkeley California USA
[19] Jain S K, Goel S C 1996 Discussion on Prestress force effect on vibration frequency of concrete bridges *Structural Engineering* **121** p 459-60
Codes of Practice
[20] IS 456 1978 Indian Standard Code of Practice for Plain and Reinforced Concrete *Bureau of Indian Standards* New Delhi
[21] IS 456 2000 Indian Standard Code of Practice for Plain and Reinforced Concrete *Bureau of Indian Standards* New Delhi
[22] IS 516 1959, Indian Standard Method of Test for Strength of Concrete *Bureau of Indian Standards* New Delhi
[23] IS 1343 1980 Indian Standard Code of Practice for Prestressed Concrete *Bureau of Indian Standards* New Delhi
[24] IS 1343-2012 Indian Standard Code of Practice for Prestressed Concrete *Bureau of Indian Standards* New Delhi
[25] IS 1785 (Part 1 and 2) 1983 Indian Standard Specification for Plain and Hard Steel Wire for Prestressed Concrete *Bureau of Indian Standards* New Delhi

A comparative study of slope stability analysis of Manganese slag stabilized soil to an ordinary soil

Sumanjali Parimi[1], P V V Satyanarayana[2]

[1]Assistant Professor, Department of Civil Engineering, Raghu institute of technology, Visakhapatnam, India
[2]Professor, Department of Civil Engineering, College of Engineering (Autonomous), Andhra University, Visakhapatnam, India
Email: sumanjali.parimi676@gmail.com

Abstract. Slopes are required in the construction of Highways, Railways, Earthen dams, canal banks, levees and at many other locations. The construction of embankment over the ground is not possible in a vertical way. The Soil can be piled up with slope only. The cost of earthwork is minimum if the slopes are steepest but the Steeper slopes are not safe as per stability conditions because it may fail in shear. The only force acting on the piled earth is gravity force. This gravitational force produces shear in the soil and causes slippage. The failure lead to loss of property and life. There is a lesser chance of slippage of soil if we provide more bottom width compared to its top width. But it requires more area and more earth to fill. So, it is uneconomical.

In the present study, the soil was collected from the highway extension project, from the geotechnical investigation it was identified that the soil shear strength was quite low. To improve shear characteristics and to decrease the dry density of soil, it is to be stabilized with a lightweight material, possessing good frictional characteristics. A waste by-product was identified from the industry of nearest village called Manganese (Mn) slag. The soil was stabilized with various percentages of Mn slag, from the geotechnical characterization, an optimum % of slag is selected as 10 %. An embankment slope is assumed for further slope stability analysis. The analysis is carried out on two slopes, one is made with ordinary soil and the other is soil stabilized with optimum Mn slag %. Both stabilized soil embankment and ordinary soil embankment are analysed by using the Method of slices and the final results were compared.
Keywords: Red soil, Manganese slag, Embankments, shear characteristics, Slope stability analysis;

1. Introduction

An earth slope is an unsupported, inclined surface of a soil mass. The cost of earthwork would be minimum if the slopes are made steepest. However, very steep slopes are may not be stable. A compromise has to be made between economy and safety, and the slopes provided are neither too steep nor too flat. In other words, the steepest slopes which are stable and safe should be provided.
In this regard, to provide safe slopes, the soil shear characteristics has to be improved. The addition of frictional material will increase the angle of internal friction of stabilized material.

In this study the analysis of embankment slopes is carried out for two types of soils, one is with un-stabilized soil which is directly collected from the site and the other is soil stabilized with a waste by-product called Manganese (Mn) slag. In these two cases, the Factor of safety is calculated by analysis method ‘Method of slices. The factor of safety is the ratio of Resisting moment (M_R) to Overturning moment (M_O). These two moments are depending on shear parameters of the material – Cohesion (c) and Angle of internal Friction (ϕ). By increasing the angle of internal friction of soil, the resisting moment has to be increased and also by decreasing specific gravity of stabilized material the Overturning moment will be decreased. Therefore, if the soil is stabilized with a frictional material which is light in weight, the factor of safety could be increased. So that the chances of failure of the slope are minimized.

In this regard, a Waste material is identified in a nearby village called Mn slag, which is a by-product of Iron ore melting, observed at "FERRO ALLOYS CORPORATION" in Garividi village near Vizianagaram.

For the analysis, the soil used was collected at the site area of highway extension project of Vizianagaram. The single lane of NH 43 is extending as a two-lane road. The project includes several embankments at various heights.

Many researchers studied the effect of the addition of slag material to different soils, and many studies are described various methods of slope stability analysis. **Ibrahim M Asi et. al. (2017)** studied the effect of the addition of steel slag in the stabilization of expansive clay. In their conclusions, they stated that the addition slag will decrease plasticity and in increase dry density and frictional characteristics of clayey soils.

1.1 Objectives of the present study

The objectives of the present study are:

- To increase the stability of slopes economically.
- To make use of Industry waste material which is hugely available.
- To reduce the cost of construction by decreasing the bottom width and by increase side slopes.

2. Materials

2.1 Red soil

The soil was collected from highway extension project of NH 43 at a depth 0.5 m from G.L. The soil itself using as an embankment material. The soil was primarily identified as Red soil, possessing both cohesion and friction and particle size is a nearly sand-sized particle. Further Geotechnical investigation is carried out.

2.2 Manganese slag

Slag is the glass-like by-product left over after a desired metal has been separated from its raw ore. Slag is usually a mixture of metal oxides and silicon dioxide. The main product of the smelting process is a carbon-saturated ferroalloy consisting of 76 to 80% manganese, 12 to 15% of iron, 7.5% of carbon, and 1.2% of silicon. 70 to 80% of the manganese is recovered in the melt, and a slag consisting of 30 to 42% manganese is also achieved.

From the primary observation, the particle size of slag is medium sand and it is completely cohesionless material possessing high frictional characteristics. The material is also light in weight. Further geotechnical classification is done by tests.

3. Methodology

3.1 Material Investigation

The soil which was collected at the site was further investigated by using various Geotechnical tests such as Specific gravity test, Consistency limits, Dry sieve analysis, IS compaction test, Direct shear test. The same Geotechnical characterization was also done to the other material Mn slag from the tests. Now, the soil was mixed with different % of Mn slag such as 5, 10 and 15. The material proportion is mixed based on dry mix proportion. The same Geotechnical characterization is done at three % of Mn slag.

3.2 Analysis procedure

The slope stability analysis of an embankment slope is done by using the Method of Analysis. As many methods are available for the analysis of slopes, Method of slices was selected for its accuracy of analysis of finite slopes.

4. Results and Discussions

4.1 Geotechnical characterization of Materials

The Red soil and Manganese slag materials are examined by conducting various Geotechnical tests to know specific gravity, Atterberg limits, shear parameters and compaction characteristics. Table 1 presents the Geotechnical properties of Red soil and Mn slag individually.

Table 1. Geotechnical properties of Red soil and Mn slag

Property		Red soil	Manganese slag
Specific gravity		2.47	1.89
Gradation characteristics			
i.	Gravel (> 4.75 mm)	9 %	0
ii.	Sand (4.75 mm – 0.075 mm)	77 %	100 %
iii.	Fines (< 0.075 mm)	14 %	0
Consistency limits			
i.	Liquid limit	24 %	NP
ii.	Plastic limit	17 %	NP
iii.	Plasticity index	7 %	-
IS classification		SP	SP
Compaction characteristics			
i.	Optimum moisture content	12 %	6.5 %
ii.	Maximum Dry density	2.68 g/cc	1.55 g/cc
Shear characteristics			
i.	Cohesion	18 kN/m^2	0
ii.	The angle of internal friction (ϕ)	28°	42°

4.2 Geotechnical characterization of soil stabilized with Mn slag

In this study, the Red soil was stabilized with various percentages of Mn slag like 5%, 10% and 15%. The material was added to the soil by its dry weight and the Geotechnical characterization was done by different tests. The optimum % of slag is decided based on the values obtained from various tests.

From the laboratory test results, it is observed that the stabilized Red soil shows valid results when it is stabilized with 10 % of Mn slag. After that at 15 % of Mn slag, the soil specific gravity and dry density were rapidly decreasing. Therefore, the optimum % was selected as 10 %. Figure 1 presents the variation of Geotechnical properties of stabilized material with the addition of Mn slag of various percentages.

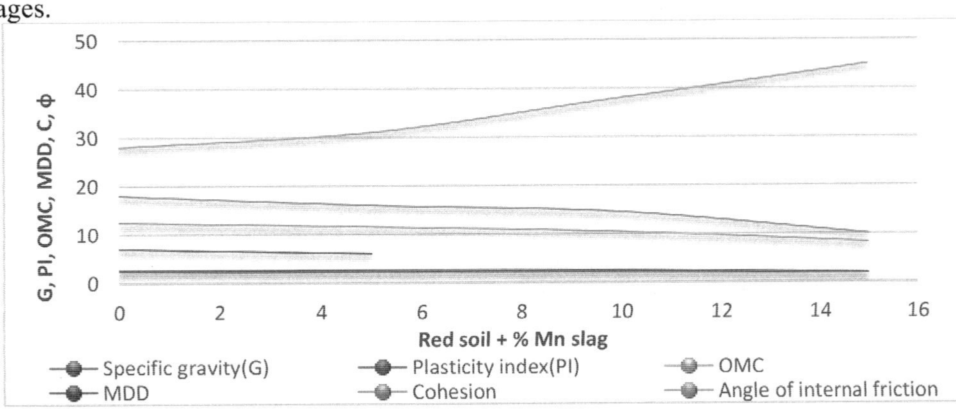

Figure 1. Variation of Geotechnical properties of stabilized material with the addition of Mn slag

Table 2 presents the Geotechnical characterizations of Red soil stabilized with Mn slag of different percentages.

Table 2. Geotechnical characterizations of Red soil stabilized with Mn slag

Property	Soil + 0% Mn slag	Soil + 5% Mn slag	Soil + 10% Mn slag	Soil + 15% Mn slag
Specific gravity	2.47	2.21	2.1	1.8
Gradation characteristics				
Gravel (> 4.75 mm)	9 %	9 %	8 %	8 %
Sand (4.75 mm – 0.075 mm)	77 %	81 %	83 %	85 %
Fines (< 0.075 mm)	14 %	10 %	9 %	7 %
Consistency limits				
Liquid limit	24 %	18 %	NP	NP
Plastic limit	17 %	12 %	NP	NP
Plasticity index	7 %	6 %	NP	NP
IS classification	SP	SP	SP	SP
Compaction characteristics				
Optimum moisture content	12.5 %	11.6 %	10.4 %	8.2 %
Maximum Dry Density (g/cc)	2.25	2.1	1.95	1.5
Shear characteristics				
Cohesion (kN/m^2)	18	16	14.5	10
The angle of internal friction (ϕ)	28°	31°	38°	45°

4.3 Analysis procedure

To meet the requirements of practical issues, a slope with known height and angle were assumed for the analysis. By using the analysis method, called Method of slices, the slope was analysed and value of factor of safety and a most critical circle was determined.

Now, if the slope is constructed with stabilized soil material, the analysis was carried out for the same assumed slope that how the factor of safety was varied.

Analysis problem:

An embankment is to be constructed to a height of 4 m and the side slopes of 1:1 incline 45°. The properties of the soil are unit weight = 22.5 kN/m^3, cohesion = 18 kN/m^3, angle of internal friction is 28°. A minimum factor of safety 2.5 required.

The solution is obtained by assuming different radii of a critical circle using the method of slices and a critical circle is one whose factor of safety is minimum.

Slope of embankment = 45^0

For embankment slope 45^0, $\alpha = 28\ ^0$, $\beta = 37\ ^0$

Trial -1

Analysis by Swedish circle method (Graph 1)

Considering the slip circle radius, r = 6 m

Length of the arc slip surface wedge, L= 9.529 m

Step: 1 Calculation of each slice area by counting the number of squares under the embankment

Step: 2 Weight of each slice is measured by multiplying volume (unit length) by the unit weight of soil

Step: 3 Tangent and Normal are drawn for each slice and angle intended between the tangent and normal was determined

Step: 4 Tangential and Normal Components of Force are determined from the following formulae

Tangential force (T) = W sin α'

Normal force (N) = W cos α'

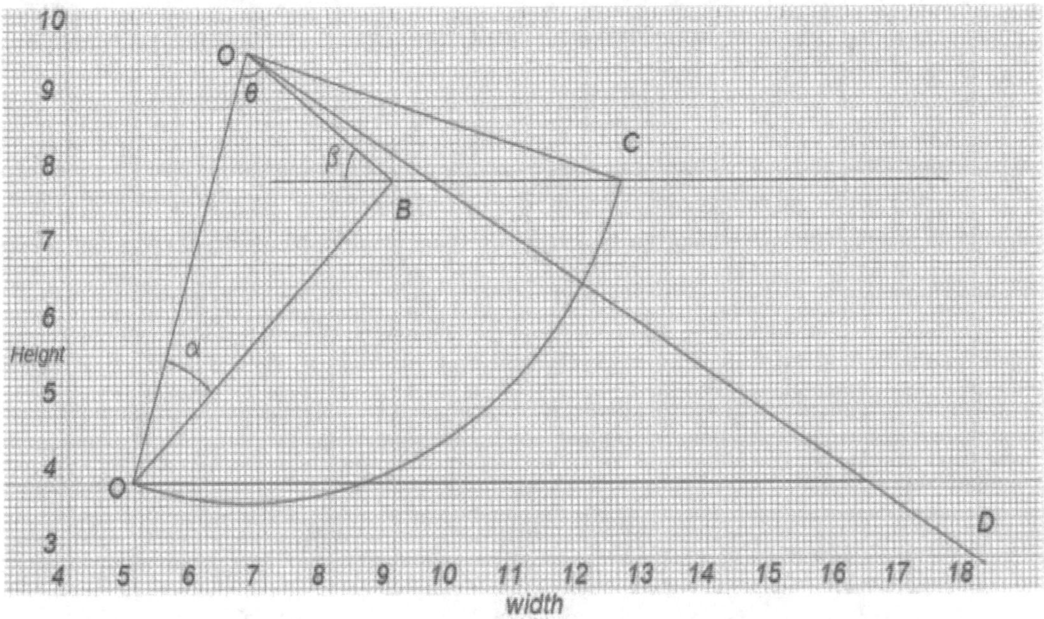

Graph 1: Analysis of failure wedge for slip circle with radius = 6 m (ordinary soil)

The tabulated details are presented in Table 3 for Trail 1.

Table 3. Trail 1 details of Swedish circle method of ordinary soil

Slice no.	Area of each slice (A) m^2	Weight of each slice (W) kN	Angle intended between the tangent and Normal components (α') in °	Tangential force (T) = W sin α'	Normal force (N) = W cos α'
1	0.56	12.6	-10	-2.18	12.4
2	1.7	38.25	-6	-3.99	38.04
3	2.68	60.3	10	10.47	59.38
4	3.45	77.625	20	26.55	72.94
5	3.59	80.775	26	35.40	72.6
6	2.93	65.925	3	40.58	51.94
7	1.92	43.2	50	33.09	27.76
8	0.34	7.65	64	6.83	3.35
				$\sum T$ = 146.75	$\sum N$ = 338.41

Area of each slice is calculated from the graph.
Weight of each slice = Area of each slice x Unit weight of soil.
α′ is an angle between weight acting vertically downwards and normal component of weight resolved.
Now,

$$Factor\ of\ Safety = \frac{Resisting\ Moment}{Overturning\ Moment} = \frac{C\ L + \sum N\ Tan\ \phi}{\sum T}$$

$$= \frac{(18 \times 9.529\ ) + (338.41 \times Tan\ 28)}{146.75} = 2.39$$

In a similar way the analysis is carried out for different radius of a critical circle and the most critical circle with a minimum factor of safety is found out. The key details of all trails are shown in the following table 4.

Table 4. The factor of safety for the slip circle with different radii of ordinary soil

Trail no.	The radius of slip circle (r) m	Length of arc (L) m	The angle at the centre of the arc (°)	Factor of safety
1	6	9.52	91	2.39
2	5.8	10.22	101	2.5
3	5.75	11.34	113	2.69
4	6.5	8.28	73	1.9
5	6.6	7.83	68	2.08
6	5.9	10.19	99	2.2
7	6.2	8.87	82	2.29
8	6.7	9.328	79	2.37

The minimum factor of safety of most critical circle for ordinary soil embankment is 1.9. But the required factor of safety is 2.5. So, the embankment will fail in that most critical surface which results in loss of property and life. So, the soil is stabilized with optimum % of Manganese slag which is 10%. Analysis of the stabilized soil embankment by Swedish circle method is carried out for the same design problem with changed geotechnical engineering properties.

Analysis problem

An embankment is to be constructed to a height of 4 m and the side slopes of 1:1 incline 45°. The properties of the soil are unit weight = 19.5 kN/m^3, cohesion = 14.5 kN/m^3, angle of internal friction is 38°. A minimum factor of safety 2.5 required.

For stabilized soil

The solution is obtained by assuming different radii of a critical circle using the method of slices and a critical circle is one whose factor of safety is minimum. Table 5 presents the details of Swedish circle method of stabilized soil (Trail 1).

Slope of embankment = 45^0

For embankment slope 45^0 $\alpha = 28\,^0$, $\beta = 37\,^0$

Trial – 1

Considering the slip circle radius, r = 6 m and Length of the arc slip surface wedge, L= 9.529 m

Table 5. Trail 1 details of Swedish circle method of stabilized soil

Slice no.	Area of each slice (A) m^2	Weight of each slice (W) kN	Angle intended between a tangent and Normal component (α') in °	Tangential force (T) = W sin α'	Normal force in N W cos α'
1	0.56	10.92	-10	-1.89	10.75
2	1.7	33.15	-6	-3.46	32.96
3	2.68	52.26	10	9.07	51.46
4	3.45	67.275	20	23	63.21
5	3.59	70	26	30.688	62.91
6	2.93	57.135	38	35.17	45.02
7	1.92	37.44	50	28.68	24.06
8	0.34	6.63	64	5.96	2.9
				∑T = 127.618	∑N = 293.27

Now,

$$Factor\ of\ safety = \frac{Resisting\ Moment}{Overturning\ Moment} = \frac{C\,L + \sum N\ Tan\ \phi}{\sum T}$$

$$= \frac{(14.5 \times 9.529\ ) + (293.27 \times Tan\ 38)}{127.618} = 2.88$$

In a similar way the analysis is carried out for different radius of a critical circle and the most critical circle with a minimum factor of safety is found out. The key details of all trails were shown in table 6.

Table 6. Table showing factor of safety for slip circle with different radii of stabilized soil

Trail no.	The radius of slip circle (r) in m	Length of arc (L) in m	The angle at the centre of the arc (°)	Factor of safety
1	6	9.52	91	2.88
2	5.8	10.22	101	2.67
3	5.75	11.34	113	2.6
4	6.5	8.28	73	2.58
5	6.6	7.83	68	2.66
6	5.9	10.19	99	2.7
7	6.2	8.87	82	2.77
8	6.7	9.328	79	2.9

By analysis, the minimum factor of safety of un-stabilized soil is 1.9 at 6.5m radius slip circle which is less than the required factor of safety i.e., 2.5. So, there is a chance of failure at the most critical slip surface. Therefore, the soil is stabilized with the manganese slag material which is freely and abundantly available.

From the analysis of stabilized soil slope analysis, the minimum factor of safety is improved from 1.9 to 2.58. Hence the embankment is safe and the height and slope of the embankment are increased with the same bottom width.

5. Conclusions

Various Geotechnical tests such as specific gravity, compaction, and direct shear tests were conducted for the soil which is taken from the bank of the lake which is just highway extension project of NH 43. From the study, the following observations were made and conclusions were drawn.

1. The soil having less plasticity so that the material can be used as an embankment material. But the shear properties were quite low.
2. To improve shear properties of soil and to decrease the unit weight of soil the soil stabilized with a waste by-product called Mn slag, which is hugely available at the Ferro Alloys Ltd, Garividi, Vizianagaram.
3. The soil was stabilized with 5, 10 and 15 percentage of Mn slag by its dry weight. The geotechnical characterization of stabilized soil at different percentages was found out.
4. From the values, the optimum % of Mn slag was selected as 10 %. With the addition of 10% Mn slag, the soil shear properties were greatly improved and the density was also decreased. And the soil becomes also non-plastic with the addition of 10% slag.
5. A design problem was taken for the analysis concerning field criteria. The slope was analysed and the minimum factor of safety obtained was 1.9. at this minimum factor of safety, the slope has chances to fail.
6. Now, the analysis is carried out for the same designed slope with stabilized soil i.e., higher shear properties and lower density. Now, the minimum factor of safety obtained is 2.7, which safe as we assumed a minimum factor of safety as 2.5.

7. With this improved factor of safety, the slope is safe against the field conditions and we can also increase the top width of the embankment by increasing steepness of the slope.
8. Finally, if the soil was stabilized with Mn slag, it will be safer and more economical for future developments also.

References

[1] Faisal I Shalabi, Ibrahim M Asi, Hisham Y Qasrawi 2017 Effect of by-product steel slag on the engineering properties of clay soils *Journal of King Saud University – Engineering Sciences* **29 (4)** p 394-9
[2] Hamed Niroumand, Khairul Anuar Kassim, Amin Ghafooripour, Ramli Nazir 2012 The role of geosynthetics in slope stability *Electronic Journal of Geotechnical Engineering* **17 (HR)** p 2739-48
[3] Santosh Dhakar, Jain S K 2016 Stabilization of soil: A Review *International Journal of science and research* (IJSR) **5 (6)** p 545–9
[4] Brain O, Oyegblie Benjain, Oyegblie A 2017 Applications of Geo Synthetic membranes in Soil stabilization and Coastal defence structures *International Journal of Sustainable built environment* **6 (2)** p 636-62
[5] Amin Pourkhosravani, Behzad Kalantari 2011 A review of current methods for slope stability evaluation *Electronic Journal of Geotechnical Engineering* **16** p 1245–55
[6] Sakellariou M G, Ferentiou M D 2005 A study of slope stability prediction using Neural network *Springer, Geotechnical and Geological Engineering* **23** p 419–45

Effect of bamboo fibres and lime on engineering properties of expansive soil

D S V Prasad[1], Ch Sivannarayana[2], P Sunitha[3]

[1]Professor & Principal, Department of Civil Engineering, B V C Engineering College, Odalarevu, Andhra Pradesh, India-533210
[2]Department of Civil Engineering, Associate Professor, B V C Engineering College, Odalarevu, Andhra Pradesh, India-533210
[3]Department of Civil Engineering, PG Student, B V C Engineering College, Odalarevu, Andhra Pradesh, India-533210
Email: drdsvp9@gmail.com

Abstract. Shear strength, differential settlements and lofty compressibility, extensive soil usually create a problem for civil engineers and geotechnical engineers. Stabilization is one of the oldest ways to augment the soil quality of problematic soil. In current time, it has been experienced to the totalling of fibres improves the ductility performance of the soil close by tumbling the maturity of fractures throughout the contraction. This paper illustrates the outcome of lime plus bamboo fibres on strength performance black cotton soil blending with different percentages by conducting California Bearing Ratio, Modified Proctor Compaction as well as Unconfined Compressive Strength at various curative periods. The test result indicates that strength characteristics of the most viable combination of black cotton soil-lime sample combined with bamboo fibres are considerably better than un-treated black cotton soil and also the strength of soil increased due to increase in the curing period.
Keywords: Black Cotton Soil, Lime, Bamboo Fibers, Compaction, CBR, Unconfined Compressive Strength.

1. Introduction

In most of the cases, expansive soil is called as problematical soils because of their alteration in nature by fluctuations in moisture content. In the globe, the range of studies (Holtz and Gibbs, 1956; Ganapathy, 1977; Evans and Mc Manus, 1999) reported that many structures constructed on expansive soils are failed. If it is essential to build the structure on such soils, the properties must be improved to bare the loads for any type of structure. In addition to this, the highway position is inhibited to appropriate convenience and connectivity criteria, which perpetually might come across swelling soils on the way and thus it turned into the offing to develop their ability, due to traffic operations, through appropriate practice to the in-situ clay in broad and expansive soils in scrupulous. Several researchers (Dembichi and Jermolowietz, 1988; Rao, 1996) believed that reinforcing the earth could solve the bulk number of civil engineering problems in coming future. Lindh and Eriksson (1991) reported that fibre-reinforced sand in road construction improved the strength and deformation properties of pavements. Due to the expansive soils, there have been continuous efforts around the world to create ways and means to solve the problems. However, this and many other methods are only partially successful and therefore efforts to improve better methods are still ongoing. The use of waste materials and natural fibre helps to improve soil property because they are cheap, locally available and environmentally friendly. The use of natural materials such as jute, coir, sisal and bamboo as a strengthening material in the soil began in the early nineties. At present, it is examined to the accumulation of fibres improves the durability performance of the soil thereby falling the development of fractures during contraction. Navami Chandran B, Veena Vijayan L (2017) are tested the result of rice husk ash (RHA), banana fibre and bamboo fibre on clayey soil by conducting unconfined-compressive-strength, CBR and Compaction tests and from the results, maximum-dry-density decreases and the optimum moisture-content raises with the addition of RHA, Bamboo fibre and Banana fibre and found that the optimum percentage of RHA is 6 %. The optimum percentage obtained by adding soil to different percentages of bamboo and banana fibre is 1% and 0.5% respectively. Dipika Devi and Boken Jempen (2016) Properties of a local soil with and without bamboo fibre reinforcement 1 %, 2 %, 3 %, 4% and 5% are studied and compared. The length of the fibres considered are 20 mm and 30 mm and shear tests are performed with a slow strain rate to simulate drained condition. Fibre-reinforced soils increasing till 4%, shear strength increases. John Paul and Antony Rachel Sneha (2016), Black cotton soil treated with fly ash plus reinforced with dissimilar percentages of bamboo fibres with a mean diameter 0.45 mm and 25 mm pieces be assorted randomly and from results, strength properties of best possible amalgamation of black cotton soil-fly ash reinforced through bamboo fibres be palpably improved than untreated soil and 1% fibre and 20% fly ash is found to be optimum. Fly ash blended with 1% of bamboo fibre raises the strength of the BC soil and reduces the brittle behaviour of the soil sample, while other proportions used fibres show a slight enlarge. The outcome of the studies conducted is that the bamboo fibre fly ash treated BC soil can be used as a strengthening material. Dhanya and Ramya, (2016), Geotechnical characteristics of quarry dust treated soil with bamboo fibres indicated the CBR value goes on rising as a proportion of quarry dust increases up to 25 %, and then decreases on additional adding, enhance in optimum-moisture-content with raise in the proportion of bamboo fibre to the quarry dust treated soil and reduce in maximum dry density with raise in the percentage of the bamboo fibre to quarry dust treated soil and found that 1 % of bamboo fibre was taken as an optimum percentage. Twinkle and Sayida, (2011), experiments were conducted on BC soil blended through the further extent of the lime and of polypropylene fibre, and found that optimum moisture content raises and the maximum-dry-density fallen, adding lime to the soil liquid limit lowered but plastic limit raised. At 6 % lime and 0.75 % polypropylene fibre unconfined compressive strength gained of about 3.8 times and CBR gained of about 3.19 times compared to untreated soil. In the current research work, a trial has been made to calculate the geotechnical properties of BC soil mixing with unlike proportions of lime and bamboo fibres. From the outcome, it is noticed that there is a development in engineering property as a result.

2. Study Design

The present study has been planned in a three-stage process. In the first stage, it is proposed to carry out various tests to be performed in the laboratory for detecting the important characteristics of the materials used during the study. In the second stage, the stabilization technique tried in the laboratory carried out blending with different proportions of lime as shown in figure 1.

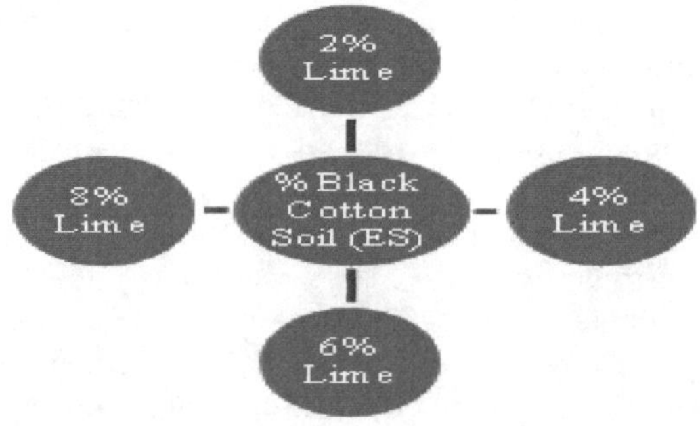

Figure 1. Flow Chart Showing Different Combinations of Lime Mixed with Black Cotton Soil

It is proposed to obtain the strength and other important properties, through laboratory tests. In the third stage, it is proposed to carry out the effect of bamboo reinforcement materials on the soil – lime stabilized mix as shown in figure 2. Based on the observations and assessment results, optimum percentages of lime and bamboo fibres from the laboratory experimentation and relative comparison will be made to know the improvement in geotechnical properties.

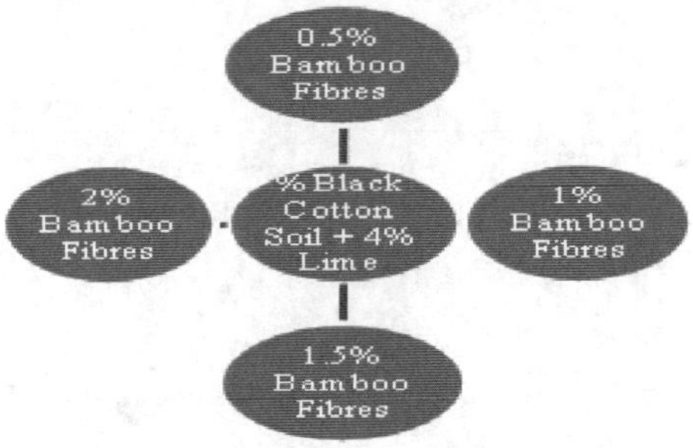

Figure 2. Various Mix Proportions of BC Soil with Lime and % of Bamboo Fibres

3. Materials and Properties

A collection of materials used in the course of the lab tests is described in the subsequent section.

3.1 Black Cotton Soil (ES)

Soil is from Shady grey to black in colour, obtained at Amalapuram, Andhra Pradesh, India. The soil has been taken from a depth of 1.0 to 1.5 m beneath the ground surface. The soil was crushed physically to pass throughout 4.75 mm IS sieve shown in the figure. 3. Soil is graded as inorganic land of high compressibility (CH) as stated by I.S classification. Characteristics of BC soil Code terms are presented in Table 1.

Table 1. Properties of Black Cotton Soil

Laboratory Test	Symbol	Results	Relevant IS Code
Differential Free Swell (%)	PDF	115	IS 2720 Part XI
Atterberg's limits			
Liquid Limit	W_L	82%	IS 2720 Part V
Plastic Limit	W_P	40%	IS 2720 Part V
Plasticity Index	PI	42%	IS 2720 Part V
Specific Gravity	G	2.41	IS 2720 Part III
Grain Size Analysis			
Coefficient of Uniformity	Cu	7.1	IS 2720 Part IV
Coefficient of Curvature	Cc	1.1	IS 2720 Part IV
Compaction Parameters (Modified Proctor)			
Optimum Moisture Content (%)	OMC	22.12	IS 2720 Part VIII
Maximum Dry Density(kN/m^3)	MDD	14.99	IS 2720 Part VIII
California Bearing Ratio (CBR)			
Soaked CBR	CBR	1.79	IS 2720 Part XVI
Un soaked CBR	CBR	2.6	IS 2720 Part XVI
Unconfined Compressive Strength	UCS	350 kN/m^2	IS 2720 Part X

IOP Conf. Series: Materials Science and Engineering 1025 (2021) 012010 doi:10.1088/1757-899X/1025/1/012010

3.2 Lime

Hydrous lime was utilized as a stabilizing agent in this investigation was taken from the local market. 0 % to 8 % amount of lime was mixed by dried weight of soil. The lime's specific gravity is 2.37. Limestone comes in the structure $CaCO_3$, be sieved throughoutby150μ sieve and preserved in strongbox for successive invitations, as revealed in the figure. 4.

Table 2. Properties of Lime

Chemical Formula	$Ca(OH)_2$
Molar Mass	74.093 g/mol
Appearance	White powder
Odour	Odourless
Density	21.58 kN/m^3, solid
Melting Point	580^0c (loses water, decomposes)
Solubility in Water	1.89 g/L (0^0c), 0.66 g/L (100^0c)
Specific Gravity	0.9
Acid Resistance	Very Good
Alkali Resistance	Good
Dispersion	Good
Young's Modulus	3450 MPa

3.3 Bamboo Fibers

Sai Laxman Group from Guntur is the provider of the bamboo shown in figure 5 and physical and chemical characteristics are shown in tables 3 & 4. These fibres are green & biodegradable, strong, UV shielding, flexible, soft and are strong in tension with low modulus of flexibility and also this material available nearby with little cost.

Table 3. Physical Properties of Bamboo Fiber
(Courtesy from Sai Laxman Industrial Group)

Properties	**Units**	**Values**
Length	mm	25
Diameter	microns	20
Specific Gravity	-	0.82
Water Absorption	%	60-85%
Density	kN/m^3	14.45
Tensile Strength	MPa	220

Table 4. Chemical Properties of Bamboo Fibers
(Courtesy from Sai Laxman Industrial Group)

Chemical Name	Percentage (%)
Cellulose	49.1
Lignin	27.7
Hemi-Cellulose	26.1
Starch	2-6
Fat	2- 4
Deoxidized Saccharide	2
Protein	0.8 - 6

IOP Conf. Series: Materials Science and Engineering 1025 (2021) 012010 doi:10.1088/1757-899X/1025/1/012010

Figure. 3. Black Cotton Soil

Figure. 4. Lime

Figure. 5. Bamboo Fibers

4. Laboratory Experimentation

Various experiments are conducted in the lab by a combination of different Proportions of lime and bamboo fibres to black cotton soil to know the optimum value and its result on strength characteristics.

4.1 Index Properties

According to Practice Codes [IS 2720 (Part 5) 1985; IS 2720 (Part 6) 1972], the research has been adopted for the discovery of the plastic and liquid limit for the samples attempted.

4.2 Compaction Properties

Compaction parameter of black cotton soil varied with various proportions of lime and bamboo fibres were determined by using IS Heavy compaction test IS 2720 (Part VIII).

4.3 *California Bearing Ratio (CBR) Tests*

According to the IS Code (IS 2720 (Part 16) 1979), samples have been prepared for CBR testing using a black cotton soil mixture of different percentage of lime and bamboo fibres to influence optimal percentages and geotechnical properties as shown in figure 6.

4.4 Unconfined Compressive Strength (UCS)

These trails are approved out in the laboratory under IS 2720, Part X (1991). These tests are conducted at maximum dry density and optimum moisture content with a strain rate of 1.2 mm/min. Proving Ring used 2 kN capacity for testing models as revealed in the figure. 7.

Figure. 6 California Bearing Ratio Test Apparatus

Figure. 7 Unconfined Compressive Testing Machine

5. Discussion on Test Results

According to the IS Code Rules, various trials were conducted in the lab and the test results are listed below.

5.1 Index Properties

Standard methods recommended by relevant IS codes [IS 2720 (Part 5) 1985; IS 2720 (Part 6) 1972], in the research has followed the findings of the liquid limits and plastic limit values of the samples attempted. The outcome of the liquid limit trail on black cotton soil treated with dissimilar proportions of lime is showing that with raise in the proportion of lime, the liquid limit goes reducing from 82 % to 68 % whereas the plastic limit raises beginning 40 % to 52 % while lime proportion increasing from 0 to 8 % respectively as shown in figure 8.

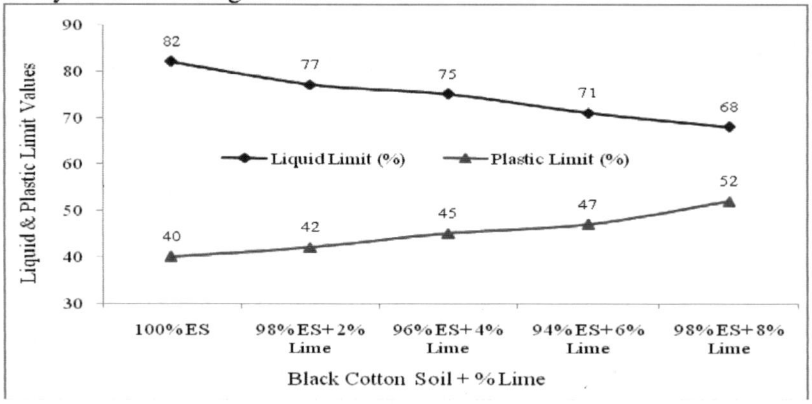

Figure 8. Liquid and Plastic Limit Values of the Expansive soil mixed with various % of Lime

5.2 Compaction Test Results

IS Modified Proctor compaction trails are conducted as per IS: 2720 (Part VIII) in the laboratory mixing with various Proportions of Lime and Bamboo fibres in the pulverized black cotton soil. The graph is drafted between moisture-content & dry-density for each proportion increment of lime and bamboo fibres to Expansive soil and from the results, OMC and MDD values are derived. The results and graph from these tests are presented below. The MDD and OMC values are analysed from the trail results and are shown below. From the outcome the maximum-dry-density values are fallen beginning 14.99 kN/m^3, 14.63 kN/m^3, 14.11 kN/m^3, 13.81 kN/m^{3}and 13.67 kN/m^3; optimum- moisture-content values are raise from 22.12 %, 23.41 %, 24.69 %, 25.93 % and 28.07 % correspondingly when the soil is mixed with 0 %, 2 %, 4 %, 6% and 8 % of lime respectively as shown in figure.9.

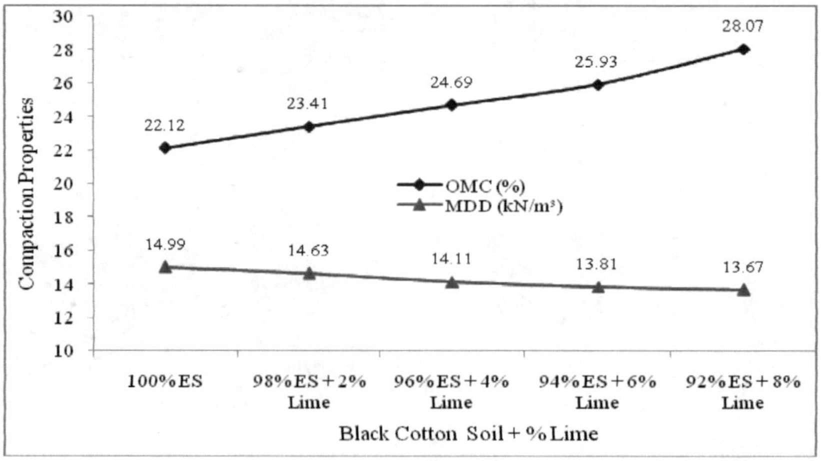

Figure 9. Compaction Parameter of Expansive soil blend with various % of Lime

From compaction test, the MDD values are fallen from14.99 kN/m^3 to 13.67 kN/m^3 and OMC values are raising from 22.12 % to 28.07 % respectively and when the soil –lime mix increases from 0 % to 8 %. The fall in dried unit weight was endorsed to prove that lime responds vastly with black cotton soil ensuing Base-Exchange aggregation and flocculation tends towards amplifying void-ratio of the combination taking to fall in the dry unit weight of the black cotton soil and lime mixture. Optimum 4 % of lime blending with black cotton soil and diverse proportions of bamboo fibres 0 %, 0.5 %, 1.0 %, 1.5 % and2.0 %, the maximum dry density values decreasing from 14.11 kN/m^3, 14.03 kN/m^3, 13.71 kN/m^3, and 13.42 kN/m^{3}and13.23 kN/m^3 whereas optimum moisture content is increasing from 24.69 %, 26.9 %, 28.14 %, 29.33 % and 31.52 % as shown in the figure 10. This increase in OMC can result from flocculation because of the lime reaction of the auxiliary water inside the flux and the OMC because of the greater osmosis of the bamboo fibres.

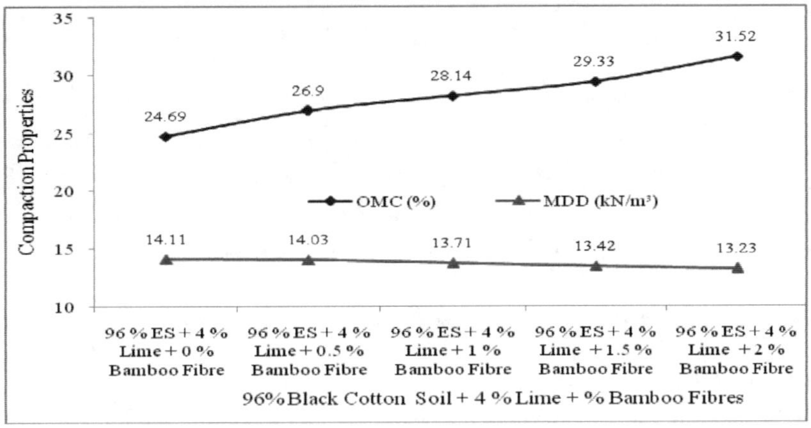

Figure 10. Compaction parameter of BC Soil Blend with 4 % Lime and Various % of Bamboo Fibres

5.3 California Bearing Ratio (CBR) Test

The outcome of soaked and unsoaked CBR trails are performed on black cotton soil blended using assorted proportions of Lime and Bamboo Fibres be presented below. It was found from that black cotton soil added by dissimilar proportions of lime the un-soaked and soaked CBR results are 2.6, 1.79; 4.87, 2.35; 6.87, 3.93; 6.19, 3.37 and 5.37, 3.1 for 0 %, 2 %, 4 %, 6 % and 8 % of lime respectively presented in the figure 11.

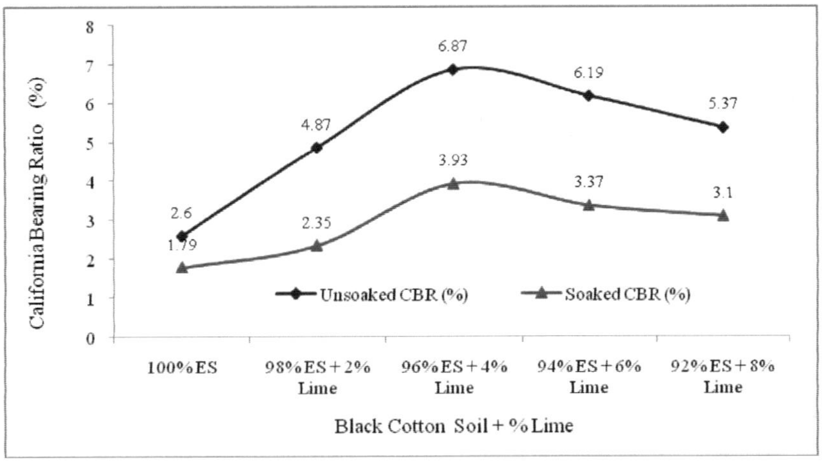

Figure 11. CBR Values of Black Cotton Soil Blend by Various % of Lime

From the above results at 4% lime-soil mix obtained maximum CBR value as contrasted to other samples tested in this work. The unsoaked and soaked CBR values are raised from 6.87, 3.93; 8.29, 4.25; 9.85, 5.78; 8.34, 4.92 and 6.17, 3.9 respectively when the addition of bamboo fibres 0 %, 0.5 %, 1.0 %, 1.5 % and 2 % with the optimum 4% lime – black-cotton soil mix as revealed in figure.12 and the best proportion of lime and bamboo fibres are 4 % and 1 % correspondingly.

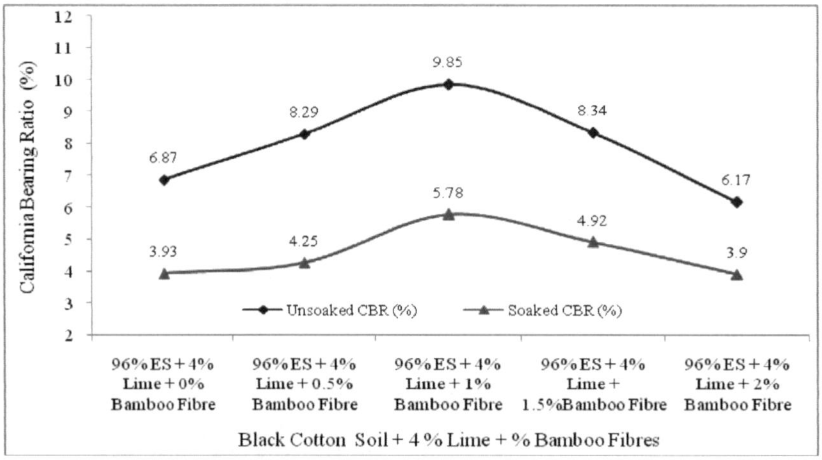

Figure 12. CBR Values of BC Soil Blend by 4% Lime plus Various % of Bamboo Fibres

5.4 Unconfined Compressive Strength (UCS)

UCS testing machine is used for conducting the trails in the laboratory by IS 2720 - Part X at 0, 7, 14- & 28-days curing. Blending lime to the black cotton soil, Unconfined Compressive Strength values raised to 4 % of lime irrespective of curing and beyond it decreases as shown in figure 13.

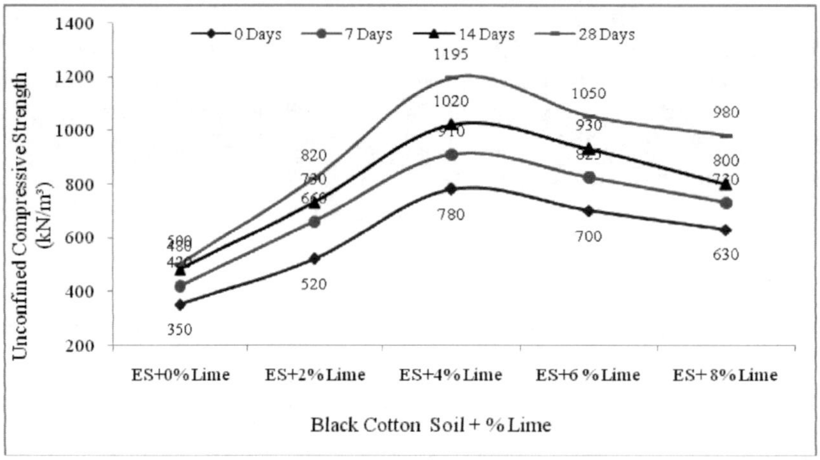

Figure 13. Unconfined Compressive Strength of Black Cotton Soil Blend by Various % of Lime

Similarly, black cotton soil added among 4 % lime plus different percentages of bamboo fibres Unconfined Compressive Strength values are raised to 1 % adding bamboo fibres and further adding fallen the strength as shown in figure 14.

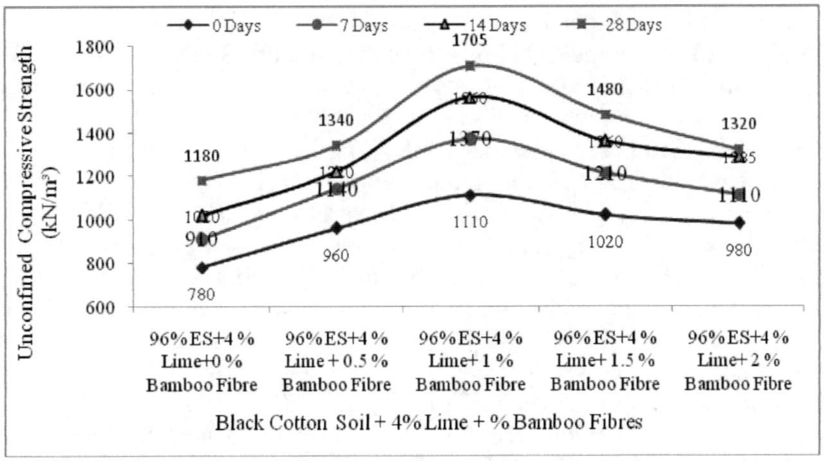

Figure 14. Unconfined Compressive Strength of Black Cotton Soil Blend by 4 % Lime and Various % of Bamboo Fibres

6. Conclusions

The following outcomes were made based on the experimentations carried out.

Adding lime showing falling in liquid limit from 82 % to 68 % and improvement in plastic limit start 40 % to 52 %, as the lime % varies starting 0 % to 8 % with an increment of 2 % added in expansive soil as an output of cation ions from lime reduces in volumetric changes. Addition of lime to the black cotton soil results the decreases the MDD value from 14.99 kN/m^3 to 13.67 kN/m^3 whereas OMC increases from 22.12 % to 28.07 % at 8% of Lime. Compaction characteristics of treated black cotton soil - lime mix at best 4 % of lime, OMC increasing from 24.69 % to 31.52 % and MDD lessening from14.11 kN/m^3 to 13.23 kN/m^3 with the adding up of different proportions of fibres ranges from 0.5 to 2 with an increment of 0.5 % of bamboo fibre. Addition of lime to expansive soil, unsoaked CBR values increases from 2.6 % to 6.87 % and soaked CBR values are increasing from 1.79 % to 3.93 % up to 4 % of lime and beyond the value fallen. Un-soaked and soaked CBR value goes increasing from 6.87 % to 9.85 % and 3.93 % to 5.78 % up to addition of 1% bamboo fibres and beyond it falls with further addition. Unconfined compressive strength increased from 350 kN/m^2 to 780 kN/m^2 up to 4 % adding lime and beyond it decreases and also from 780 kN/m^2 to 1110 kN/m^2 up to 1 % bamboo fibres correspondingly.

Under examination, utilize of lime and bamboo fibres in strengthen soil be able to develop significantly. CBR values raise due to lime significantly bonded with soil particles that resist the entrance of water and adding fibres enhance the strength. Test results proved that the effect of lime and fibres in soil stabilization can develop the strength distinctiveness significantly. Overall, lime and fibre stabilized soil can be used in engineering projects on weak soils with cost-effective concern for land improvement technology. From the trails, the optimum proportion of lime and Bamboo fibres are 4 % and 1 % correspondingly.

7. References

[1] Dembichi E, Jermolowiez P 1988 Model Tests of Bearing Capacity of a Week Subsoil Reinforced by Geotextiles *Proceedings of First Indian Geotextile Conference on Reinforced Soil and Geotextiles* IIT Bombay India **1(1)** p 67-72

[2] Evans R P, Mc Manus K J 1999 Construction of Vertical Moisture Barriers to Reduce Expansive Soil Subgrade Movement *TRR* **1652** *TRB* p 108-112

[3] Ganapathy G V 1977 Experience with Swelling Clay in Western Canada *Proceedings of the First National Symposium on Expansive Soils* HBTI Kanpur India p 3-1 to 3-11

[4] Holtz W G, Gibbs H J 1956 Engineering Properties of Expansive Clays *Transactions of ASCE* **121** p 641-7

[5] Kinney T C, Stone D K, Schuler J 1988 Using Geogrids for Base Reinforcement as Measures by Falling Weight Deflectometer in Full – Scale Laboratory Study *TRR 1611* TRB p 70-7
[6] Lindh E, Eriksson L 1991 Sand Reinforced with Plastic Fibers - A Field Experiment Performance of Reinforced Soil Structures *Proceedings of the International Reinforced Soil Conference held in Glasgow* Scotland September 1991 p 471-3
[7] Rao GV 1996 Geosynthetics in the Indian Environment *Indian Geotechnica Journal* **26** **(1)** p 1-94
[8] Navami Chandran B, Veena Vijayan L 2017 Effect of Natural Fiber on Clayey Soil *International Research Journal of Engineering and Technology (IRJET)* **4** **(4)** p 3044-47
[9] Dipika Devi, Boken Jempen 2016 Shear Strength Behaviour of Bamboo Fiber Reinforced Soil*International Research Journal of Engineering and Technology (IRJET)* **3** **(8)** p 433-7
[10] John Paul V, Antony Rachel Sneha M. 2016 Effect of Random Inclusion of Bamboo Fibers on Strength Behaviour of Flyash Treated Black Cotton Soil *International Journal of Civil Engineering and Technology (IJCIET)* **7** **(5)** p 153–60
[11] Sai Sravya G, Suresh Kommu 2016 Swell and Strength Characteristics of Expansive Soil Reinforced with Synthetic Fibers *Journal on Civil Engineering* **6l** **(4)** p 21-5
[12] Satyam Tiwari, Nisheet Tiwari 2016 Soil Stabilization Using Waste Fiber Materials *International Journal of Innovative Technology and Research* **4** **(3)** p 2927–30
[13] Dhanya P, Ramya K 2016 Strength Characteristics of Kuttanad Soil Treated with Quarry Dust *International Journal of Advanced Research Trends in Engineering and Technology* **3** (**Special Issue 23)** p 36-40
[14] Twinkle S, Sayida M K 2011 Effect of Polypropylene Fibre and Lime Admixture on Engineering Properties of Expansive Soil *Proceedings of Indian Geotechnical Conference* December 15-17 Kochi **Paper No. H-085** p 393-6
[15] Anusha R, Emmanual C Kindo 2011 Behaviour of Bamboo Reinforced Soils – State of Art *Proceedings of Indian Geotechnical Conference* December 15-17 2011 Kochi **Paper No H-247** p 469-72

Subgrade CBR improvement with fly ash as mineral and NaOH as a chemical admixture

Sudheer Ponnada[1*], Rajeswara Rao K[2], SSSV Gopala Raju[3]

[1] Assistant Professor, [2] Professor, Department of civil engineering, MVGR College of Engineering(A), Vizianagaram, Andhra Pradesh, Pin:535005
[3]Professor, Department of Civil Engineering, Rajiv Gandhi University of Knowledge Technologies, Nuzivid, Andhra Pradesh, India 521202
Email: *sudheer.ponnada@mvgrce.edu.in

Abstract. Increase in the rate of traffic volume and heavy vehicle loads gave a revolution in highway construction technology, this increase is due to continuous growth in economic activities. Therefore, it is necessary to construct highly durable and serviceable road structures on any type of soil. In continuation, a good subgrade can influence the thickness of road structure, increases the strength pavement, and reduce the cost of road structure construction. California bearing ratio (CBR) is a generally used in-direct method to estimate the strength of sub-grade and it directly gives the pavement thickness in pavement analysis and design work. So, if increasing in the CBR value results thicknesses of the pavement will decrease and it, in turn, reduces the construction cost of the pavement structure. In the present investigation, the collected soil samples are Black Cotton soils from Dharmapuri, Vizianagaram, AP. Same soil stabilized with various replace percentages of Fly ash (i.e.10, 20, 30, and 40%) and NaOH (i.e. 3, 5, and 7%). The Atterberg limits, Free swell Index, and CBR tests were completed on original and stabilized soils. Finally, authors observed that the considered BC soil stabilized with mineral admixture (Fly ash) and chemical admixture (NaOH) results in more than 400% increase in CBR with 40% decrease in pavement thickness, and 25% decrease in pavement construction cost.
Keywords: Black cotton soil, CBR, Fly-ash and NaOH;

1. Introduction

The highway is a main-road for travel by the public between key destinations, such as cities, large towns. Geotechnical engineering has been critically related to highway construction in case of Bearing-capacity and shear strength of the soil. To construct an economical highway, the engineer must design the best thickness from the lower layer of subgrade to surface course. So, the present study is conducted out to improve the CBR value of the soil by using Fly-ash to decrease the pavement thickness.

1.1. Black Cotton Soil

Soil is generally depicted as silt or another variety of mineral-particles generated by the physical or compound breaking down of rocks in addition to the air, water, natural issue, and different fixings that might be incorporated. Soil is normally a non-homogeneous-, permeable, earthen-material' whose engineering performance is affected by modifications on dampness substance and density. If the rate of montmorillonite is more in soil, then it is said to be Black-cotton soil or Regur soil [8]. This is developed by the weathering of igneous-rocks' and cooling of lava later a volcanic explosion. These do not give any warning and are expansive which is dangerous for construction activity. Normally, lands with black cotton soils are abundant and very great for agriculture, but their bearing capacity is very low (15000 kg/m^2) to lay pavements on them. In general, black-cotton soils discovered in India are a typical example of soft expansive soils. Differential free swell index is one simple method of identifying the expansive soils. Black-cotton soils absorb water, swell heavily loses strength and are easily compressible. This soil shrinks and develops cracks during summer. These soils produce a CBR of 2 to 5% and such soils undergo volumetric modifications leading to pavement distortion, cracking, and general unevenness due to regular wetting and drying [12]. Following are some important reasons for the failure of roads laid on Black-cotton soils, 1) Nature of Black-cotton soils, 2) Poor drainage services, 3) Use of gravel in base and subbase, 4) Supplying of overloaded vehicles and iron wheeled tractors, trolleys etc.,

1.2. Problems of Flexible pavements on Black cotton soils

Flexible pavements having a lesser amount of flexural strength and act as a flexible sheet. These will circulate wheel load stresses to smaller layers by grain-to-grain distribution through the points of contact in the granular section. Embankment design of flexible roads requires to fulfil two important needs on BC soils, they are stability and settlement [13]. The short-term permanence for embankment over soft clay is always extra vital than long term simply because the sub-soil with time under loading and the strength improves.

Some of the failures of flexible pavements built on BC soil are as follows.

- Rutting: Surface recession in the wheel path.
- Corrugation: Occur under dry weather conditions.
- Potholes: These are the end effect of alligator cracking. As alligator cracking turn out to be serious, the interrelated cracks create small lumps on the pavement, the residual hole after the lump is known as potholes.
- Cracking: Cracks may be small and fine (hairline cracks), small interlinked near-circular cracks (fatigue cracks), parallel to pavement centerline longitudinal cracks.

2. Utilization of admixtures in BC soil stabilization

2.1. Using Fly ash as a mineral admixture

Fly-ash' is a solid-waste by-product invention generated by the burndown of coal, it is conducted out of the boiler by flue-gases and removed by electrostatic precipitators or cyclone filters and filter bags. Fly ash used in the current study is gathered from a thermal power station located in Visakhapatnam. Fly-ash brings down the capability of a plastic-soil' to attempt volumetric extension by a physical-cementing' system, which can't be assessed by the plasticity-index (PI). Fly-ash 'self-controls shrink/swell by cementing the soil reviewing all the while much like Portland-cement bonding aggregates simultaneously to make concrete. By holding the soil grains all the while, soil molecule developments are constrained. Fly-ash' is acts as pozzolan [1] [2] [3]. It has been viably utilized with granular and fine-grained materials to enhance soil qualities, proposing sufficient help for pavements and upgrading working situations where unwanted soils are confronted.

2.2. Using NaOH as Chemical admixture

Using chemicals is similarly a conventional soil stabilization technique since it creates a superior characteristic original soil with greater strength and durability than mechanical & physical methods [9] [10] [11]. All chemical methods are reliant on reaction among chemical-additives and soil particles which then generate a solid network that combine the soil grains. Chemically proves mineral polymers are contained of crosslinked units of AlO4 and SiO4-tetrahedral, there charge comparing reactions (Na+, k+, Li+, Ca2+, Ba2+, H3O+) are delivered by alkali metals. In the present investigation the chemical additive used as Sodium-hydroxide' (NaOH). Sodium-hydroxide appears white, feels odorless, non-volatile solution. It may not burn but tremendously responsive. It reacts aggressively with water and several usually encountered materials, producing adequate heat to ignite close to igneous materials. Following key advantages are that it can simply respond with water which results in a strong compaction aid giving a larger density for the same compaction effort [7] [12]. Sodium-hydroxide reacts very effectively with soil abundant in aluminium. The experimental materials are shown in figure 1 (a) and figure 1 (b).

Figure 1(a). NaOH **Figure 1(b).** Fly ash
(Experimental materials)

2.3. Objectives of the present work

The present experimental investigation carried for the following objectives,

- To study the engineering-properties of locally obtainable black cotton soil.
- To improve the CBR value by adding together different % of fly-ash' and NaOH.
- To find the optimal percentage of fly-ash' and NaOH for soil sample that enhances CBR value.
- To determine the suitable structure thickness for the pavement before and after adding fly ash and NaOH.

3. Experimental Investigation and Discussion

For the present study, an experimental investigation was conducted out by two different stages. In the first stage, determine the optimum dosage of fly ash for better CBR among the five design soil mix samples as shown in Table 1. In the second stage, determine the optimum dosage of NaOH with selected soil mix in stage-01 as shown in Table 3.

Table 1. Mix proportions of BC Soil partially replaced with Fly ash in Stage-01

Type of Mix	**M**	**M1**	**M2**	**M3**	**M4**
BC-Soil %	100	90	80	70	60
Fly ash %	-	10	20	30	40

All the five different soil mixes as shown in table 1, with partial replacement of fly ash in stage-01 are tested in the laboratory for Atterberg-limits, Free-swell Index, and CBR value to fix the optimum dosage of fly ash as a replacement in BC soil for stabilization.

Table 2. (a) Atterberg Limits and Free-swell-index (FSI) values for Trial mixes

Type of Mix	Atterberg Limits			FSI (%)
	Liquid Limit (%)	Plastic limit (%)	Plastic Index (%)	
M	89	42	47	74
M1	68	30	38	68
M2	62	25	37	48
M3	55	24	31	19
M4	54	24	30	07
Code of Practice	**IS: 2720 (Part-5)**			**IS: 2720 (Part-40)**

Table 2. (b) Optimum-Moisture-Content, Maximum-Dry-Density and CBR values for Trial mixes

Type of Mix	OMC (%)	MDD (g/cc)	CBR (%)
M	25.80	1.54	2.90
M1	22.20	1.53	5.58
M2	21.80	1.52	5.62
M3	20.45	1.52	5.98
M4	**19.05**	**1.51**	**7.18**
Code of Practice	**IS: 2720 (Part-7)**	**IS: 2720 (Part-7)**	**IS: 2720 (Part-16)**

From Table 2(a) & 2(b), Among all five trial mixes of BC soil partially replaced with fly ash (10 to 40% respectively), M4 mix shows reasonable values. Selected M4 mix is used as an optimum mix for stabilization for further study. In continuation of experimental investigation as a part of stage-02, M4 is considered in determining the optimum dosage of NaOH.

Table 3. Mix proportions of BC Soil partially replaced with Fly ash & NaOH in Stage-02

Type of Mix	M4-01	M4-02	M4-03
NaOH (%)	03	05	07

All the three different soil mixes as shown in table 3, with partial replacement of fly ash and NaOH in stage-02 are tested in the laboratory for CBR value to fix the optimum dosage of NaOH as a replacement in BC soil for stabilization.

Table 4. CBR values for Trial mixes in Stage-02

Type of Mix	OMC (%)	MDD (gm/cc)	NaOH (%) (Replacement)	CBR (%)
M4-01	19.05	1.51	03	9.97
M4-02	19.05	1.51	05	14.16
M4-03	19.05	1.51	07	**15.56**
Code of Practice				**IS: 2720 (Part-16)**

IOP Conf. Series: Materials Science and Engineering 1025 (2021) 012011 doi:10.1088/1757-899X/1025/1/012011

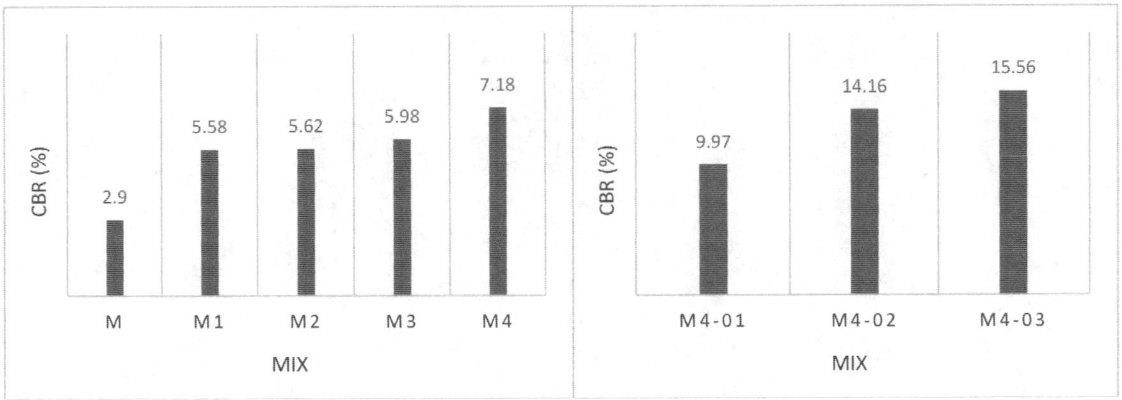

Figure 2. Comparison of CBR for all trail soil mixes

According to the results obtained and presented in figure 2, 7% replacement of NaOH in Mix M4-03 gives better CBR value among all results from the present investigation.

4. Design Criteria for Pavement Structure

According to IRC-37, the thickness of layers for the pavement with CBR 2.9 % (Mix 'M') is 890mm. The individual thickness of all layers for the proposed pavement mentioned in the below table.

Table 5. Thickness of Pavement layers for CBR 2.9 %

CBR 2.9%					
		LAYER COMPOSITION FOR PAVEMENT			
Salient points	**Total pavement thickness (mm)**	**Bituminous surfacing**		**Base(mm)**	**Sub-Base(mm)**
		BC (mm)	**DBM (mm)**		
Cumulative Traffic **150 msa**	890	50	210	250	380
Width of layers for 100 m length		3500	4100	4700	5300

According to IRC-37, the thickness of layers for the pavement with CBR 15.56 % (Mix 'M4-03') is 520mm. The individual thickness of all layers for the proposed pavement mentioned in table 6.

Table 6. The thickness of Pavement layers for CBR 15.56 %

CBR 15.56 %					
		LAYER COMPOSITION FOR PAVEMENT			
Salient points	**Total pavement thickness (mm)**	**Bituminous surfacing**		**Base(mm)**	**Sub-Base(mm)**
		BC (mm)	**DBM (mm)**		
Cumulative Traffic **150 msa**	**530**	30	100	200	200
Width of layers for 100 m length		3500	4100	4700	5300

From the standard rates, the construction cost of the above road layers for CBR 2.9 and 15.56% calculated and compared the final construction cost approximately and presented in the below figure 3.

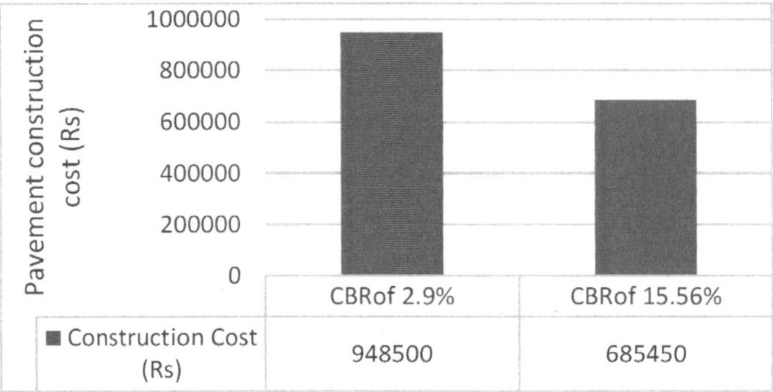

Figure 3. Comparison of the construction cost of road layers

From Table 5 & 6, It is concluded that pavement structure decreases from 890mm to 530mm, means all most more than 25% of cost reduction for the CBR 15.56 % compared to the CBR 2.9 %.

5. Conclusions

From the above study an effort has been made to use fly-ash' and NaOH as mineral and chemical admixtures for improving the properties of the black cotton soil and to reduce the road structure construction cost. The following are the conclusions drawn from the above study,

1. It is noticed that there is a decrease in OMC, PI and FSI with an increase in fly ash.
2. There is an upsurge in CBR value observed with an increase in the percentage of fly ash also with NaOH.
3. Black cotton soil stabilized with 40 % fly ash had resulted in a substantial decline in liquid limit and plasticity index.
4. The effect of the addition of chemical admixture NaOH (3 to 7 %) was studied and it was found that CBR had tremendously increased with mixture M4-03 i.e. natural soil with 40 % fly ash and 7 % NaOH.
5. CBR value of the present soil sample replaced with 40% fly-ash' and 7% NaOH improved considerably up to 400%.
6. The upsurge in CBR had resulted in a decrease in the pavement thickness by 40 % and thereby resulted in more than 25 % decrease in the cost of the road project.

References

[1] Ali M, Sreenivasulu V 2004 An experimental study on the influence of rice husk ash and lime on properties of betonies *Proceedings of Indian Geotechnical Conference* Warangal (India) p 468-71

[2] Brooks R M 2009 Soil stabilization with fly as hand rice husk ash *International Journal of Research and Reviews in Applied Sciences* **1 (3)** p 209-17

[3] Robert M Brooks 2009 Soil stabilization with fly ash and rice husk ash *International Journal of Research and Reviews in Applied Sci*ence **1 (3)** p 209-17

[4] IS 2720 1987 (Part 16) Laboratory determination of CBR *Bureau of Indian standards* New Delhi

[5] IS 2720 1977 (Part 40) Determination of free swell index of soil *Bureau of Indian standards* New Delhi

[6] IRC 37 2001 Guidelines for the Design of flexible pavements *The Indian Rods Congress* New Delhi

[7] Islam M Rafizul, Md. Assaduzzaman, Muhammed Alamgir 2012 The effect of chemical admixtures on the geotechnical parameters of organic soil *International Journal of Applied Sciences and Engineering Research* **1 (4)** p 623-34

[8] Laxmikant Yadu, Rajesh Kumar Tripathi, Dharamveer Singh 2011 Comparison of Fly Ash and Rice Husk Ash Stabilized Black Cotton Soil *International Journal of Earth Sciences and Engineering* **4 (6 Spl.)** p 42-5

[9] Sudheer P, Munireddy M G, Adiseshu S 2017 Study on Strength of Hybrid Mortar Synthesis with Epoxy Resin, Fly Ash and Quarry Dust under Extreme Conditions *IOP Conference Series: Materials Science and Engineering* **225** **(1)** 012169

[10] Sudheer P, Munireddy M G, Adiseshu S 2016 An experimental study on strength of hybrid mortar synthesis with epoxy resin, fly ash and quarry dust under mild condition *Journal of Advances in Materials Research, An International Journal Techno-press* **5** **(3)** p 171-9

[11] Sudheer P, Munireddy M G, Adiseshu S 2016 Study on Strength of Innovative Mortar Synthesis with Epoxy Resin, Fly Ash and Quarry Dust *American Journal of Engineering Research* **5** **(5)** p 221-6

[12] Pankaj R Modak, Prakash B Nangare, Sanjay D Nagrale, Ravindra D Nalawade, Vivek S Chavhan 2012 Stabilization of Black Cotton Soil Using Admixtures *International Journal of Engineering and Innovative Technology* **1** **(5)** p 1-3

[13] Surendra P Jadhav, R M Damgir 2011 Use of Jute Geo Textile for Strengthening of Sub Grade of Road Work *Innovative Systems Design and Engineering* **2** **(4)** p 40-8

Effect of various mineral and chemical admixtures on the improvement of subgrade CBR

Sudheer Ponnada[1], K Rajeswararao[2], P Markandeya Raju[2]

[1]Assistant Professor, Department of Civil Engineering, MVGR College of Engineering (Autonomous), Vizianagaram, Andhra Pradesh, 535005
[2]Professor, Department of Civil Engineering, MVGR College of Engineering (Autonomous), Vizianagaram, Andhra Pradesh, 535005
Email: sudheer.ponnada@mvgrce.edu.in, markandeyaraju@mvgrce.edu.in

Abstract. Increase in the rate of traffic volume and heavy vehicle loads gave a revolution in highway construction technology, this increase is due to continuous growth in economic activities. Therefore, it is necessary to construct highly durable and serviceable road structures on any type of soil. In continuation, a good subgrade can influence the thickness of road structure, increases the strength pavement, and reduce the cost of road structure construction. California bearing ratio (CBR) is a generally utilized in-direct technique to evaluate the strength' of subgrade and it directly gives the pavement thickness in pavement analysis and design work. So, if increasing in the CBR value results thicknesses of the pavement will decrease and it, in turn, reduces the construction cost of the pavement structure. In the present investigation, the collected soil samples are Black Cotton soils from Dharmapuri, Vizianagaram, AP. Same soil stabilized with various replace percentages of Fly ash and Ground Granulated Blast Furnace Slag (GGBS) (i.e. up to 10%) and NaOH (i.e. 5%). The Atterberg limits, Free swell Index, and CBR tests were completed on original and stabilized soils. Finally, authors observed that the considered BC soil stabilized with mineral admixture (Fly ash and GGBS) and chemical admixture (NaOH) results in more than 400% increase in CBR with 35% decrease in pavement thickness, and 23% decrease in pavement construction cost.
Keywords: Black cotton soil, CBR, Fly-ash, GGBS and NaOH

1. Introduction

The highway is a main-road for travel by the public between key destinations, such as cities, large towns. Geotechnical engineering has been critically related to highway construction in case of Bearing capacity and shear strength of the soil. To construct an economical highway, the engineer must design the best thickness from the lower layer of subgrade to surface course. So, the present study is conducted out to enhance the CBR value of the soil by utilizing Fly-ash, GGBS and NaOH to reduce the pavement thickness.

1.1. Black Cotton Soil

Soil is depicted as silt or another assortment of mineral-particles generated by the physical/compound breaking down of rocks in addition to the air, water, natural issue and different fixings that might be incorporated. Soil is usually a non-homogeneous, permeable, earthen material whose engineering performance is affected by variations on dampness substance and density. If the rate of montmorillonite is more in soil, then it is said to be Black-cotton-soil or Regur soil [1]. This is developed by the weathering' of igneous rocks and cooling of lava after a volcanic explosion. Mentioned do not give any warning and are expansive which is dangerous for construction activity. Usually, homelands with black cotton soils are productive and very great for agriculture, but their bearing capacity is very low (15000 kg/m^2) to lay pavements on them. In general, black-cotton soils discovered in India are a typical example of soft expansive soils. Differential free swell index is one simple method of identifying the expansive soils. Black-cotton soils absorb water, swell heavily loses strength and are easily compressible. This soil shrinks and develops cracks during summer. These soils produce a CBR of 2 to 5% and such soils undergo volumetric modifications leading to pavement distortion, cracking, and general unevenness due to regular wetting and drying [2]. Following are some important reasons for the failure of roads laid on Black-cotton soils, 1) Nature of Black-cotton soils, 2) Poor drainage services, 3) Use of gravel in base and subbase, 4) Supplying of overloaded vehicles and iron wheeled tractors, trolleys etc.,

1.2. Problems of Flexible pavements on Black cotton soils

Flexible pavements having a lesser amount of flexural strength and act as a flexible sheet. These will scatter wheel load-stresses' to smaller layers by grain-to-grain transmission through the points of contact in the granular section. Embankment design of flexible roads requires to fulfil two important needs on BC soils, they are stability and settlement [3]. The short-term permanence for embankment over soft clay is always extra vital than long term simply because the sub-soil with time under loading and the strength improves.

Some of the failures of flexible pavements built on BC soil are as follows.

- Rutting: Surface recession in the wheel path.
- Corrugation: Occur under dry weather conditions.
- Potholes: These are the end effect of alligator cracking. As alligator cracking turn out to be dangerous, the interrelated cracks generate small lumps on the pavement, the residual hole after the lump is known as potholes.
- Cracking: Cracks may be small and fine (hairline cracks), small interlinked near-circular cracks (fatigue cracks), parallel to pavement centerline longitudinal cracks.

2. Utilization of admixtures in BC soil stabilization

2.1. Using Fly-ash' and Ground Granulated Blast Furnace Slag (GGBS) as a mineral admixture

Fly-ash' is a strong waste development created by the consuming of coal, it is led out of the boiler by flue gases and evacuated by electrostatic precipitators or cyclone filters and filter bags [4] [5] [6]. Ground Granulated Blast Furnace Slag (GGBS) is acquired by sating molten iron slag from a blast furnace [7]. Flash and GGBS combinedly used in the current study are gathered from a thermal power station and steel plant located in Visakhapatnam. Fly-ash and GGBS bring down the capability of a plastic soil to attempt volumetric extension by a physical cementing system, which cannot be assessed by the plastic index (PI). Fly-ash' and GGBS controls shrink/swell by cementing the soil reviewing all the while much like portland cement bonding aggregates simultaneously to make concrete. Through holding the soil grains all the while, soil molecule developments are constrained. Fly-ash' and GGBS are pozzolans'. It has been viably utilized with granular and fine-grained materials to enhance soil qualities, offering sufficient help for pavements and upgrading working situations where unwanted soils are confronted.

2.2. Using NaOH as Chemical admixture

Using chemicals is similarly a conventional soil stabilization technique since it creates a superior characteristic original soil with greater strength and durability than mechanical & physical methods [8] [9] [10]. All chemical methods are reliant on reaction among chemical-additives and soil particles which then generate a solid network that combine the soil grains. Chemically proves mineral polymers are contain crosslinked units of AlO_4 and SiO_4-tetrahedral, there charge comparing reactions (Na+, k+, Li+, Ca2+, Ba2+, H3O+) are delivered by alkali metals. In the present investigation the chemical additive used as Sodium-hydroxide' (NaOH). Sodium-hydroxide appears white, feels odourless, non-volatile solution. It may not burn but tremendously responsive. It reacts aggressively with water and several usually encountered materials, producing adequate heat to ignite close to igneous materials. Following key advantages are that it can simply respond with water which results in a strong compaction aid giving a larger density for the same compaction effort [11]. Sodium-hydroxide reacts very effectively with soil abundant in aluminium.

Figure 1. **(a)** NaOH **Figure 1.** **(b)** Fly ash (Experimental materials) **Figure 1.** **(c)** GGBS

2.3. Objectives of the present work
Present experimental research carried with the following objectives,

- To determine the engineering properties of locally available black cotton soil.
- To improve the CBR value by adding up various % of fly-ash', GGBS and NaOH.
- To determine the optimum percentage of fly-ash', GGBS and NaOH for soil sample that enhances CBR value.
- To determine the suitable structure thickness for the pavement before and after adding fly ash, GGBS and NaOH.

3. Experimental Investigation and Discussion
For the present study, an experimental investigation was conducted on five different soil trail mixes excluding original black cotton soil sample. As cited, determine the optimum dosage of fly ash, GGBS and NaOH for better CBR among the all five design soil mix samples as shown in Table 1.

Table 1. Mix proportions of BC Soil partially replaced with Fly ash, GGBS and NaOH

Sl. No.	Mix Symbol	Soil %	Fly ash %	GGBS %	NaOH (%)
1.	M	100	-	-	-
2.	M1	90	8	2	5
3.	M2	90	6	4	5
4.	M3	90	4	6	5
5.	M4	90	2	8	5
6.	M5	90	-	10	-

All the five different soil mixes as shown in table 1, with partial replacement of fly ash, GGBS and NaOH are tested in the laboratory for Atterberg-limits, Free swell index, and CBR value to fix the optimum dosage of fly-ash', GGBS and NaOH as a replacement in BC soil for stabilization.

From table 2 and 3, among all five trial mixes of BC soil partially replaced with fly ash, GGBS and NaOH (i.e. 10% and 5% respectively), M4 and M5 mixes are shows reasonable values. Here M4 mix is used as an optimum mix for stabilization because of combine use of these waste by-products effectively. According to the above-obtained results 5% replacement of NaOH and 10% replacement of fly ash and GGBS combinedly i.e. M4 mix gives more than 400% of CBR value among all results from the present investigation compared with mix M.

Table 2. Atterberg Limits and Free sell index values for Trial mixes

Type of Mix	Atterberg Limit			FSI (%)
	Liquid Limit (%)	Plastic Limit (%)	Plasticity Index (%)	
M	43	30	13	25
M1	34	26.6	6.4	22
M2	33	27.2	5.8	19
M3	32	26.3	7.7	18
M4	20	15	5	16
M5	18	13	4	14.5
Code of Practice		**IS: 2720 (Part-5)**		**IS: 2720 (Part-40) [12]**

Table 3. Optimum moisture content, maximum dry density and CBR values for Trial mixes

Type of Mix	OMC (%)	MDD (g/cc)	CBR (%)
M	20	2.0	2.70
M1	16	1.81	5.08
M2	15	1.83	5.99
M3	14	1.84	8.99
M4	13	1.84	14.21
M5	13	1.85	27.18
Code of Practice	**IS: 2720 (Part-7)**	**IS: 2720 (Part-7)**	**IS: 2720 (Part-16) [13]**

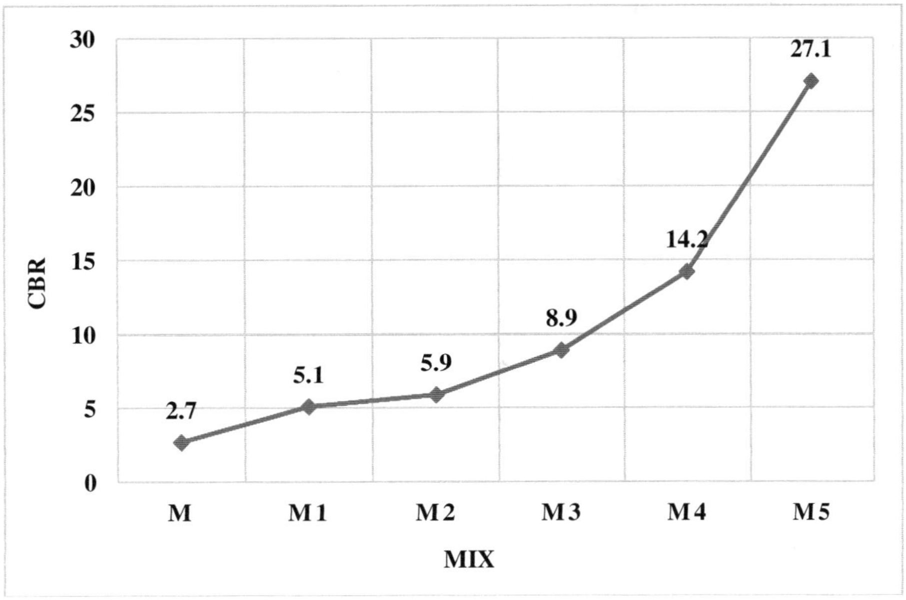

Figure 2. Comparison of CBR for all trail soil mixes

4. Design Criteria for Pavement Structure

According to IRC 37 [14], the thickness of layers for the pavement with CBR 2.7 % (Mix 'M') is 970mm. The individual thickness of all layers for the proposed pavement mentioned in the below table 04.

Table 4. The thickness of Pavement layers for CBR 2.7 %

CBR 2.7%					
Salient points	Total pavement Thickness(mm)	LAYER COMPOSITION FOR PAVEMENT			
		Bituminous surfacing		Base (mm)	Sub-base (mm)
		BC (mm)	DBM (mm)		
Cumulative Traffic **150 msa**	**970**	50	210	250	460
Width of layers for 100 m length		3500	4100	4700	5300

According to IRC 37, the thickness of layers for the pavement with CBR 14.21 % (Mix 'M4') is 630mm. The individual thickness of all layers for the proposed pavement mentioned in the below table 05.

Table 5. The thickness of Pavement layers for CBR 14.21 %

CBR 14.21 %					
Salient points	Total pavement Thickness(mm)	LAYER COMPOSITION FOR PAVEMENT			
		Bituminous surfacing		Base (mm)	Sub-base (mm)
		BC (mm)	DBM (mm)		
Cumulative Traffic **150 msa**	**630**	30	150	250	200
Width of layers for 100 m length		3500	4100	4700	5300

From standard rates, the construction cost of the above road layers for CBR 2.7 and 14.21% calculated and compared the final construction cost approximately and represented in the below figure 3.

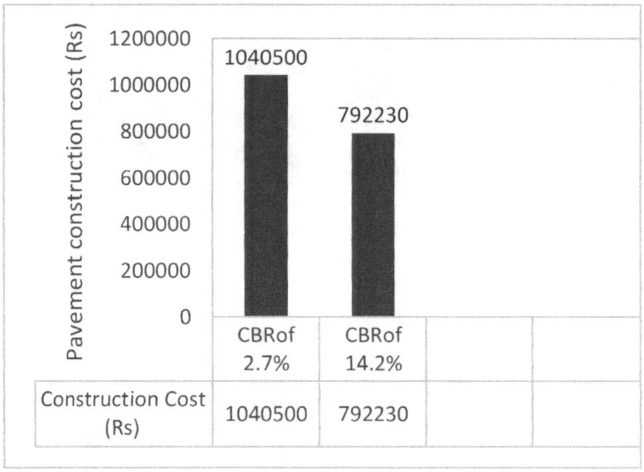

Figure 3. Comparison of the construction cost of road layers

From Table 4 & 5, it is concluded that pavement structure decreases more than 30% from 970mm to 630mm, means almost more than 20% of pavement construction cost reduction for the CBR 14.21 % compared to the CBR 2.7 % as per figure 3.

5. Conclusions

On the globe, we have a limited quantity of good quality of earth is available. It is beyond the realm of imagination to expect to provide the quality soil to each enhancement; in such circumstances, we must go for the modification practices. In this experimental investigation, an attempt has been made to use fly ash, GGBS and NaOH as mineral and chemical admixtures for improving the properties of the soil and to reduce the pavement construction cost.

According to the present investigation, the conclusions are made and are as follows,

1. Natural soil mix (M) stabilized with 2 % fly ash, 8 % GGBS with 5 % NaOH had shown promising results. Results outcome is a substantial reduction in the liquid limit (LL), plasticity index (PI) and free swell index (FSI).
2. Total pavement thickness required for natural soil mix (M) is 970 mm and the same was reduced to 630 mm when mix M4 was used. The reduction in the pavement is due to more than 400% increase in CBR of stabilized soil (M4) when compared to un-stabilized soil M.
3. The reduction in pavement thickness had resulted in a decrease in the cost of the road project construction cost by about 23 %.

References

[1] Laxmikant Yadu, Rajesh Kumar Tripathi, Dharamveer Singh 2011 Comparison of Fly ash and Rice Husk Ash Stabilized Black Cotton Soil *International Journal of Earth Sciences and Engineering* **4 (6 SPL)** p 42-5

[2] Pankaj R Modak, Prakash B Nangare, Sanjay D Nagrale, Ravindra D Nalawade, Vivek S Chavan 2012 Stabilization of Black Cotton Soil Using Admixtures *International Journal of Engineering and Innovative Technology IJEIT* **1 (5)** p 1-3

[3] Surendra P Jadhav, Damgir R M 2011 Use of Jute Geo Textile for Strengthening of Subgrade of Road Work *Innovative Systems Design and Engineering* **2 (4)** p 40-47

[4] Ali M., Sreenivasulu V 2004 An experimental study on the influence of rice husk ash and lime on properties of bentonites *Proceedings of Indian Geotechnical Conference* Warangal India p 468-71

[5] Brooks R M 2009 Soil stabilization with fly as hand rice husk ash *International Journal of Research and Reviews in Applied Sciences* **1 (3)** p 209-17

[6] Robert M Brooks 2009 Soil stabilization with fly ash and rice husk ash *International Journal of Research and Reviews in Applied Sci*ence **1 (3)** p 209-17

[7] Oormila T R, Preethi T V 2014 Effect of Stabilization Using Flyash and GGBS in Soil Characteristics *International Journal of Engineering Trends and Technology (IJETT)* **V11 (6)** p 284-9

[8] Sudheer P, Munireddy M G, Adiseshu S 2017 Study on Strength of Hybrid Mortar Synthesis with Epoxy Resin, Fly Ash and Quarry Dust under Extreme Conditions *IOP Conference Series: Materials Science and Engineering* **225 (1)** p 012169

[9] Sudheer P, Munireddy M G, Adiseshu S 2016 An experimental study on strength of hybrid mortar synthesis with epoxy resin, fly ash and quarry dust under mild condition *Techno-Press Journal of Advances in Materials Research, An International Journal (AMR)* **5 (3)** p 171-9

[10] Sudheer P, Munireddy M G, Adiseshu S 2016 Study on Strength of Innovative Mortar synthesis with Epoxy Resin, Fly Ash and Quarry Dust *American Journal of Engineering Research (AJER)* **5 (5)** p 221-6

[11] Islam M Rafizul, Md. Assaduzzaman, Muhammed Alamgir 2012 The effect of chemical admixtures on the geotechnical parameters of organic soil: A new statistical model *International Journal of Applied Sciences and Engineering Research* **1 (4)** p 623-34

[12] IS 2720 Part 40 1977 Determination of free swell index of soil *BIS* New Delhi India

[13] IS 2720 Part 16 1987 Laboratory determination of CBR *BIS* New Delhi India

[14] IRC 37 2001 Guidelines for the Design of flexible pavements *The Indian Rods Congress* New Delhi

Sustainable materials in the production of pavement base and subbase courses

B A V Ram Kumar[1,2], G Ramakrishna[3]

[1]Assistant Professor, Civil Engineering Department, GMRIT, Rajam, India.
[2] PhD. Scholar at Pondicherry Engineering College, Pondicherry, India.
[3]Professor, Department of Civil Engineering, Pondicherry Engineering College, Pondicherry, India.
Email: ramkumar.bav@gmrit.edu.in, grkv10@pec.edu

Abstract. The use of sustainable materials in construction is the need of the hour amidst the depletion of natural resources at a rapid rate. The significance of green roads is increasing day by day and green approach for road construction can be only achieved by the use of sustainable construction materials. In the present study, an effort has been made to make base and subbase courses sustainable using Reclaimed Asphalt Pavement (RAP) and Ground Granulated Blast Furnace Slag (GGBS) along with a very less quantity of cement. RAP used in the present study was obtained from the in situ milling of NH-16 at Srikakulam. To meet the gradation requirements as per IRC Ground Granulated Blast Furnace Slag (GGBS) was used along with RAP. Upon initial trial tests, it was concluded that unbound specimens made with RAP and GGBS have failed to meet the minimum unconfined compressive strength(UCS) requirements as per IRC: SP-89:2010 due to which cement by weight of 2% of total mix was added to meet the UCS requirement of 4.5 MPa at seven days. The optimum proportion of RAP that can be used to replace virgin aggregate (VA) was found to be 30% based on UCS test results. The durability studies of GGBS stabilized RAP/VA mixes were found to be satisfactory. The retained strength of the samples after durability studies is 91.43% which is greater than 80% as specified by IRC: SP-89:2010.

1. Introduction

Reducing the damage caused during the extraction and manufacturing of conventional construction materials and conserving them for future generations without compromising the present needs is the key to sustainable development. The damage reduction due to the extraction and manufacturing of conventional construction materials can only be achieved after an extent by reducing their usage in construction. For reducing the use of conventional construction materials and thereby reducing their implications on environment recycling is one of the ways forward. The use of recycled materials in construction is not a new development but solutions for overcoming the ill effects due to the use of recycled materials as a partial substitution for conventional materials is certainly an area where significant research is required. This kind of research will promote the use of recycled materials in construction thereby contributing to less use of conventional materials thereby reducing pollution created by extraction and manufacture of conventional construction materials, saving cost, giving an answer to their disposal problems, etc. RAP is the material obtained during the process of milling an old pavement surface. It consists of aggregates coated with bitumen fragmented by the process of milling. The major challenge for using RAP as a construction material is the presence of an aged and oxidized bitumen layer over the RAP aggregates. This layer prevents establishing a bond between RAP and any other material. GGBS, on the other hand, is an industrial by-product generated during the manufacture of steel in the blast furnace at steel plants. RAP can be an excellent pavement construction material when properly blended with suitable material for its application in various courses of flexible pavements [3]. RAP is an inferior material compared to VA and can be only used and partial replacement for VA or has to be used besides binders like cement [8]. If cement is not an economically viable solution mineral admixtures like Fly ash and other slags can also be used along with RAP for improving its properties [11]. Utilization of RAP and its extraction not only saves the economy but also keeps in limit the rise in height of the pavement over years due to resurfacing on the old pavement [1]. Durability is a property that defines the resistance offered by material for damage caused by the surrounding environment [5].

2. Materials used

2.1. Reclaimed Asphalt Pavement material

RAP used in the present study was obtained from in situ-controlled milling of 50 mm thick surface course NH-16 at Srikakulam where the existing four lanes are being upgraded to six lanes national highway. RAP consists of aggregates both coarse and fine coated with bitumen which in most cases is aged and oxidized. The major challenge in using RAP is this oxidized and aged layer of bitumen which prevents bonding. Both fine and coarse fractions of RAP are used in the present study for casting the specimens. Sieve analysis of RAP was performed to check its suitability as per the gradation requirements of base and subcase courses of pavements specified in MORTH Section-400. It was observed from the grain size distribution of RAP that certain fractions specified in MORTH gradation for base and subbase courses were missing in the RAP collected. The grain size analysis is presented in the following table 1.

Table 1. The grain size distribution of RAP.

Sieve size	Retained weight	% of retained	Cumulative % of retained	% of passing
26.4	0.9	2.3	2.3	97.7
20.0	1.2	3.0	5.3	94.7
4.8	31.0	76.8	82.2	17.8
600.0	6.0	14.7	96.9	3.1
75.0	1.3	3.1	100.0	3.6

2.2 Cement

Ordinary Portland cement of grade 43 in compliance to IS 8112 was used in the study. Basic tests like initial setting time, final setting time, soundness, consistency, fineness are determined to ensure that it qualifies for the specified limits as per IS 4031. The results were presented in the following table 2.

Table 2. Test results of cement.

Sl. No.	Name of test	Value Obtained	Specified Limit
1.	Initial Setting Time	42 min	>30 min
2.	Final Setting Time	290 min	<10 hrs
3.	Consistency of Cement	30%	27% – 32%
4.	Specific Gravity of Cement	3.1	3.0 -3.2
5.	Soundness of Cement	3 mm	<10 mm
6.	Fineness of Cement	2%	<10%

2.3 Virgin Aggregates

The natural aggregates used in the present study were collected from a quarry near Palakonda, Andhra Pradesh of maximum nominal size 26.5mm as specified in MORTH Section-400 for base and subbase courses. Aggregates were tested for abrasion, impact, crushing and specific gravity as per IS 2386 PART IV-1963 and were presented below in the following table 3.

Table 3. Test results for virgin aggregate and RAP.

Sl. No.	Name of Test	Virgin Aggregate	RAP
1	Aggregate impact test	28.19%	16.94%
2	Aggregate crushing test	20.3%	16.65%
3	Los Angeles abrasion test	23.68%	26.12%
4	Aggregate specific gravity test	1.96	1.97
5	Aggregate Water Absorption	1.01%	0.39%

Figure 1 shows the reclaimed asphalt pavement material.

Figure 1. Reclaimed Asphalt Pavement Material.

It can be observed from the test results that except abrasion value RAP perform better than virgin aggregates in terms of mechanical properties which can be attributed to the presence of hard oxidized and aged bitumen coating on the aggregate surface. Water absorption of RAP also was observed to be less than of the conventional aggregate because of the same reason.

2.4 GGBS

GGBS collected from Auto Nagar, Visakhapatnam with the source of steel plant located at Vishakhapatnam is used in the study. GGBS is a by-product of iron ore mixed with limestone and coke and burned in the blast furnace. The slag is then quenched from blast furnace and results in the formation of granules which are then ground to granules to a desired fineness and then used in concrete. Some properties of GGBS and Cement are similar. So, it was used to replace some percentage of cement in construction. It acts as a sustainable material and decreases pollution. GGBS reduces the heat of hydration and also the formation of early aged cracks. It also reduces chlorine and sulfate attacks. In this study, we tested the mix of GGBS, RAP by adding 2% to 4% of cement for using in base and subbase courses of pavements. Grain size analysis was performed for GGBS and is presented in figure 2 and table 4.

Figure 2. Ground Granulated Blast Furnace Slag at the site.

Table 4. The grain size distribution of GGBS.

Sieve size	Retained weight	% retained	Cumulative % retained	% passing
4.75mm	0.00	0.00	0.00	100.00
2.36mm	0.00	0.00	0.00	100.00
1.18mm	0.00	0.00	0.00	100.00
600microns	0.06	6.02	6.02	93.98
425microns	0.20	20.08	26.10	73.90
300microns	0.27	27.11	53.21	46.79
150microns	0.12	11.65	64.86	35.14
90microns	0.28	27.91	92.77	7.23
75microns	0.07	7.23	100.00	0.00

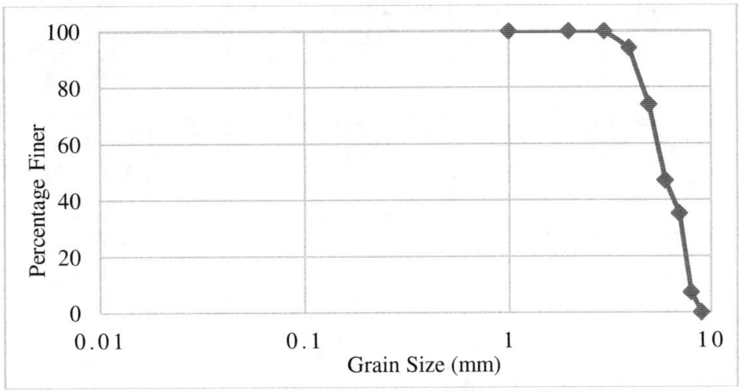

Figure 3. Ground Granulated Blast Furnace Slag gradation curve

2.5 Research Objectives
To determine the optimum proportion of RAP to replace with virgin aggregate for base and subbase course of pavements.
To assess the durability of cement-treated RAP and GGBS for base and subbase courses.

3. Methodology
The specimens of size 100 mm × 200 mm cylinders were cast and cured through moist curing for 7, 14 and 28 days and were tested for UCS are shown in the following figure 4. The flow chart shown in figure 5 depicts how this experimental study is conducted. As per the guidelines formulated for base and subbase courses in IRC 37-2012 the specimens should attain a minimum unconfined compressive strength of 4.5 MPa after curing for 28 days.

Figure 4. Moist curing of UCS samples by wet cloth wrapping

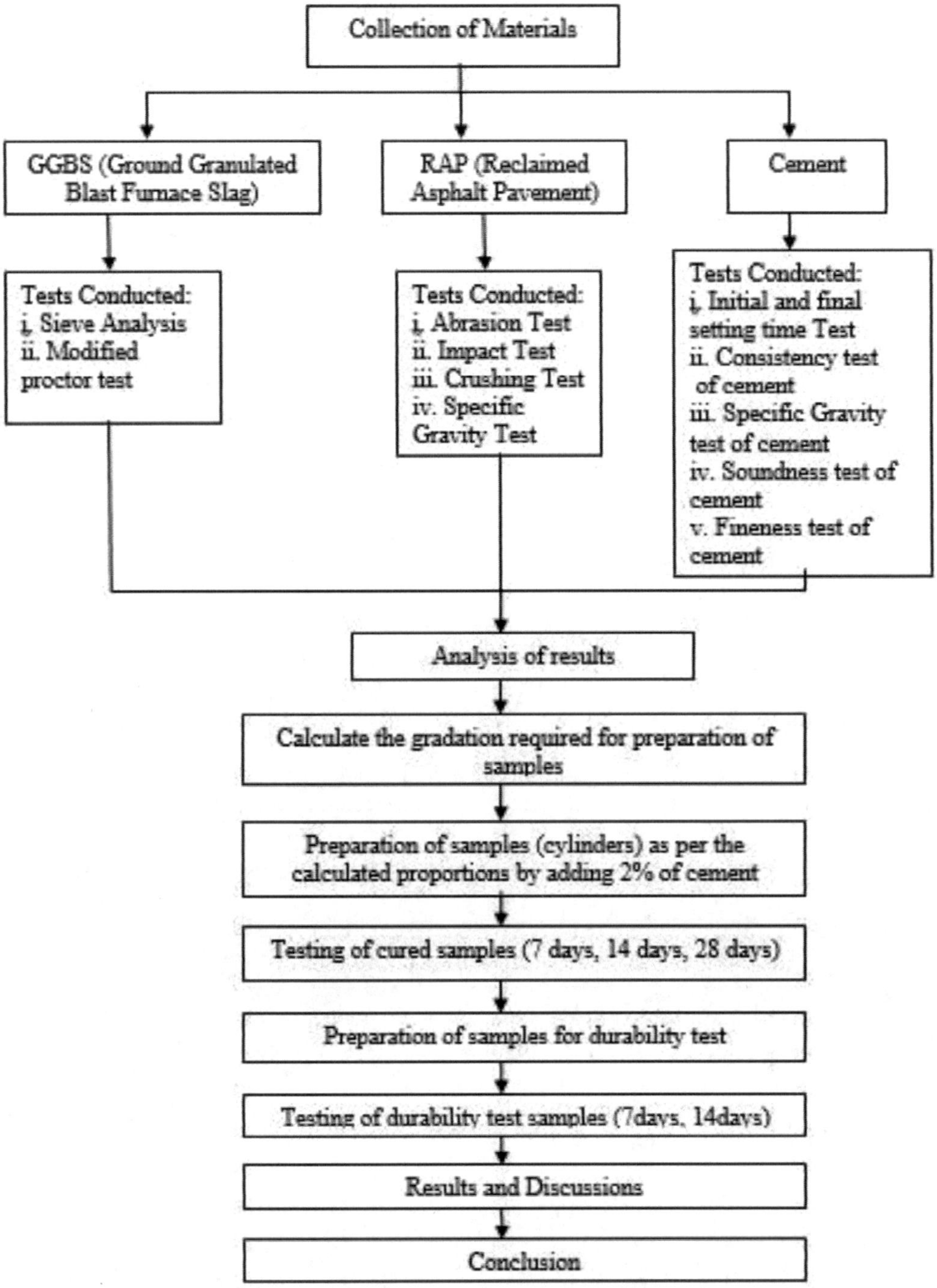

Figure 5. Methodology flow chart.

For blending RAP and GGBS to achieve the gradation specified in the MoRTH section -400 trial and error method was employed and a ratio of 70:30 for RAP to GGBS is fixed in this study. This trail and error method also satisfied with the 80:20 ratio of RAP to GGBS and results of the trial and error method of gradation are given in the following table 5. Initially, the trial test has been made for unbound base courses and subbase courses without cement using 70:30 ratio of RAP and GGBS but

IOP Conf. Series: Materials Science and Engineering 1025 (2021) 012013 doi:10.1088/1757-899X/1025/1/012013

they failed to reach the minimum UCS of 4.5MPa so cement was added starting with 2%in increments of 1% up to 4% by weight of the total mix. The UCS values for 2% addition of cement itself have satisfied the minimum UCS value hence it was adopted for further study. Initially, a compaction test was conducted for virgin aggregates and GGBS mix to determine optimum moisture content. Later RAP was used to replace virgin aggregates in the mix starting with 100% in decrements of 10% up to 60%.

Table 5. Gradation of RAP and GGBS.

SI No.	Sieve Size	Grad. % as per MORTH	RAP Gradation %	GGBS Gradation %	70% RAP	30% virgin Aggregate	Combined Gradation %	80% RAP	20% Virgin Aggregate	Combined Gradation %
1.	26.5 mm	100	97.7	100	68.3	30	98.3	78.1	20	98.2
2.	22.4 mm	50-100	94.6	100	66.2	30	96.2	75.7	20	95.7
3.	4.75 mm	35-55	17.8	100	12.4	30	42.4	14.2	20	34.3
4.	600µ	10-30	3.1	93.97	2.1	28.1	30.3	2.4	18.7	21.3
5.	75µ	2-5	3.6	3.11	2.5	0.9	3.4	2.8	0.622	3.5

4. Results and discussion

4.1 Durability Test

The durability of specimens was calculated based on the procedure outlined in IRC SP 89-2010. There are two methods outlined for assessing the durability of base and subbase courses in which method 1 was adopted in this study. Two sets of specimens each set coning three UCS samples were considered for durability testing. Initially, both sets of specimens were cured at constant moisture content for 7 days after which one set was continued to cure at constant moisture content and another set was submerged in water till 14 days. At the end of 14 days both sets were tested for UCS and the percentage of unconfined compressive strength of set submerged in water to that of constant moisture curing is calculated and this parameter is treated as a measure of the durability of that respective mix. The test results of durability for different proportions of RAP replacing VA were presented in table 6. The resistance to water on strength was determined by calculating the percentage of strength achieved by constant moisture curing to that of immersion curing. As per the recommendations of IRC SP 89-2010, this percentage should be greater than 80% and the mix with only 80% RAP and 20% VA exhibited slightly less than 80% and other mixes exhibited good durability.

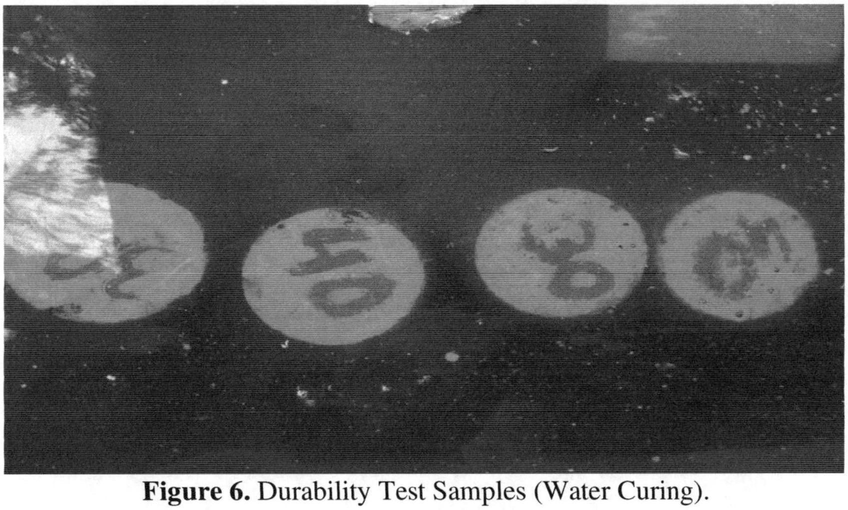

Figure 6. Durability Test Samples (Water Curing).

Table 6. Durability test results.

Mix ID	% of RAP	% of conventional aggregate	Compressive Strength for Moisture Curing (N/mm^2)	Compressive Strength for Water Curing (N/mm^2)	Percentage of UCS immersed in water to that cured at a constant moisture content
1.	80	20	7.32	5.72	78.14
2.	70	30	9.12	7.64	83.77
3.	60	40	10.33	9.22	89.25

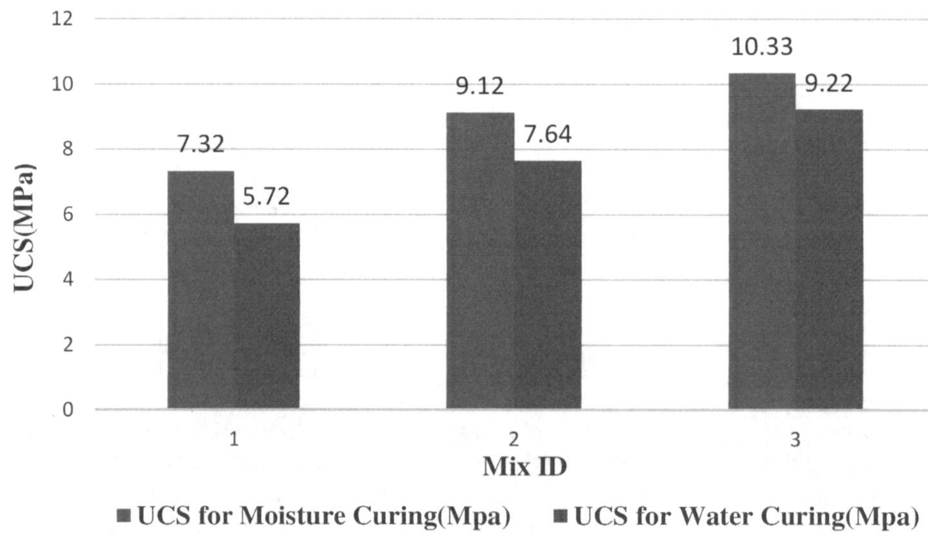

Figure 7. Bar chart showing durability test results.

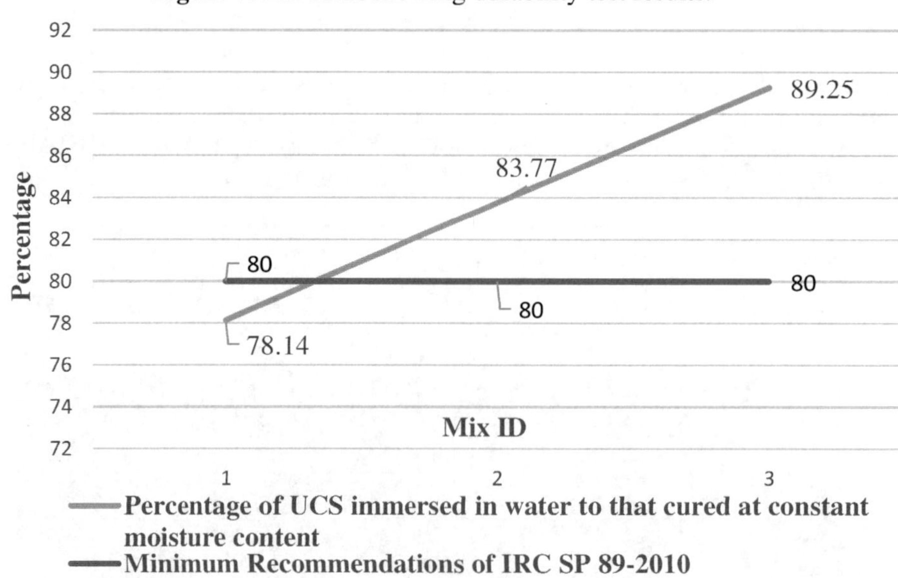

Figure 8. Line showing percentage of UCS immersed in water to that cured at constant moisture content.

IOP Conf. Series: Materials Science and Engineering 1025 (2021) 012013 doi:10.1088/1757-899X/1025/1/012013

4.2 Unconfined compressive strength test

UCS test was performed on cylindrical specimens of size 100mm×200 mm following the procedure outlined in IRC SP 89-2010. The results of the UCS test were presented in the following Table 7. ALL the mixes after the addition of cement have exhibited the minimum UCS value of 4.5MPa at 28 days but unbound mixes have failed to reach the minimum UCS value. Although cement addition was done up to 4% in the bound mixes, the mixes with 2% cement addition itself have satisfied the minimum UCS requirement and hence that was adopted as an optimum percentage of cement in the mix. With a decrease in the percentage of RAP used for replacing VA unconfined compressive strength has increased which can be observed in the results presented in table 7. The maximum percentage of RAP that can replace VA is found to be 80 % since the UCS value exhibited by this proportion of RAP has crossed the minimum UCS value of 4.5 MPa needed for base and subbase courses.

Table 7. UCS test results.

SI. No.	% of RAP	% of conventional aggregate	% of cement added	UCS (MPa)		
				7 days	14 days	28 days
1.	-	100	0	3.12	3.73	4.12
2.	-	100	4	10.72	14.1	20.56
3.	-	100	3	10.59	13.33	16.82
4.	100	-	0	2.41	2.82	3.21
5.	100	-	2	3.40	3.92	4.32
6.	80	20	2	4.72	5.34	7.32
7.	70	30	2	5.79	8.32	9.12
8.	60	40	2	6.32	8.56	10.33

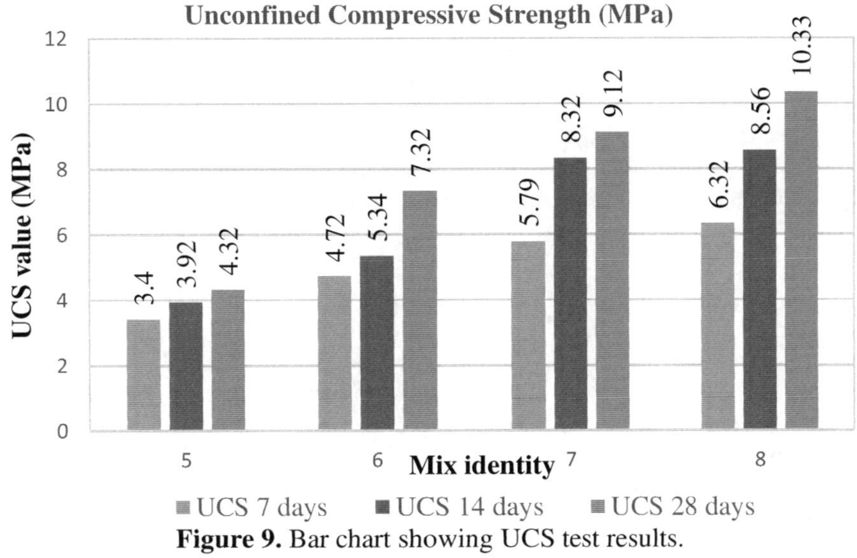

Figure 9. Bar chart showing UCS test results.

5. Conclusions

1. The mechanical properties of RAP like crushing value, impact value are better than Virgin Aggregate (VA) due to the presence of a layer of oxidized and aged bitumen on the aggregate surface.
2. Based on initial trial test results it was found that unbound RAP and GGBS samples did not meet

the minimum UCS of 4.5 MPa at 28 days and this made cement addition mandatory for RAP-GGBS mixes to attain the required UCS as per IRC SP 89-2010.

3. The durability studies of GGBS stabilized RAP/VA mixes performed satisfactorily except the mix with 80% RAP and 20% VA. The retained strength of other mixes after durability studies is greater than 80% as specified by IRC: SP-89:2010.
4. The maximum proportion of replacement of RAP with Virgin Aggregate is 80% since the mix at this percentage of replacement attained the minimum UCS required for base and subbase courses.
5. Overall, it was observed that GGBS treated RAP and Virgin Aggregate with 2% addition of cement can be utilized effectively for base and sub-base courses of pavements for making them economical as well as sustainable.

References

[1] Kandhal P S, Mallick R B 1997*Pavement recycling guidelines for state and local Governments* FHWA-SA-98-042 National Center for Asphalt Technology Washington D.C

[2] MoRTH Section -400 2000 Specifications for base and subbase courses *Indian Roads Congress* on behalf of the Government of India Ministry of Road Transport and Highways.

[3] Taha R, Ali G, Basma A, Al-Turk O 2007 Evaluation of reclaimed asphalt pavement aggregate in road bases and sub-bases *Transp. Res. Rec.* **1652** p 264–69

[4] Li L, Benson H C, Edil T B, Hatipoglu B 2008 Sustainable construction case history: fly ash stabilization of reclaimed asphalt pavement materials *Geotech. Geol. Eng.* **26** p 177–87

[5] Ganne V K 2009 Long-term durability studies on chemically treated reclaimed asphalt pavement materials *Master thesis report* University of Texas Arlington

[6] IRC SP-89 2010 Guidelines for Soil and Granular material stabilization using Cement, Lime and Fly ash *Indian Roads Congress* New Delhi

[7] Rupnow T D, Icenogle P, Reech S 2011 Evaluation of cement and fly ash treated recycled asphalt pavement and aggregates for base construction *Louisiana Transportation Research Centre* **68** FHWA/LA.11/481 p 1–68

[8] Puppala A J, Hoyos L R, Potturi A K 2011 Resilient moduli response of moderate cement-treated reclaimed asphalt pavement aggregates *J. Mater. Civ. Eng.* **23** p 990–8

[9] IRC 37 2012 Guidelines for the design of flexible pavements *Indian Roads Congress* New Delhi India

[10] ASTM D15573-12 2012 Standard test methods for laboratory compaction characteristics of soil using modified effort *American Society for Testing and Materials* West Conshohocken PA

[11] Arulrajah A, Piratheepan J, Disfani M 2013 Reclaimed asphalt pavement and recycled concrete aggregate blends in pavement subbases: Laboratory and field evaluation *J. Mater. Civil Eng.* **26** p 349–57

[12] Mohammadinia A, Arulrajah A, Sanjayan J, Disfani M, Bo M, Darmawan S 2014 Laboratory evaluation of the use of cement-treated construction and demolition materials in pavement base and subbase applications *J. Mater. Civ. Eng.* **27** 04014186

[13] Saride S, Avirneni D, Javvadi S C P 2015 Utilization of reclaimed asphalt pavements in Indian low volume roads *J. Mater. Civ. Eng.* **28 p** 1-10

[14] Saride S, Avirneni D, Javvadi S C P, Puppala A J, Hoyos L R 2015 Evaluation of fly ash treated reclaimed asphalt pavements for base/subbase applications *Ind. Geotech. J.* **45 (4)** p 401–11

[15] Mohammadinia A, Arulrajah A, Sanjayan J, Disfani M, Win Bo M, Darmawan S 2016 Stabilization of demolition materials for pavement base/subbase applications using fly ash and slag geopolymers: laboratory investigation *J. Mater. Civ.* **28** 04016033

[16] Mohammadinia A, Arulrajah A, Sanjayan J, Disfani M, Bo M, Darmawan S 2016 Strength development and micro fabric structure of construction and demolition aggregates stabilized with fly ash-based geopolymers *J. Mater. Civ. Eng* **28** 04016141

[17] Avirneni D, Peddinti R T P, Saride S 2016 Durability and long term performance of geopolymer stabilized reclaimed asphalt pavement base courses *Constr. Build. Mater.* **121** p 198–209

Economic and environmental benefits of thin bituminous surface pavements over thick bituminous surface pavements

Venkata Jogarao Bulusu[1], Sudhakar Reddy Kusam[2], Amarnatha Reddy Muppireddy[2]

[1]Research Scholar, Indian Institute of Technology, Kharagpur, 721302, India
[2]Professor, Indian Institute of Technology, Kharagpur, 721302, India
Email: j4jogs@gmail.com

Abstract. The design of bituminous pavements, using mechanistic-empirical design method, involves the selection of thicknesses of the bituminous layer and the granular layers to keep the strains at critical locations of the pavement structure within the allowable limits arrived from performance equations. The tensile strain under the bituminous layer defines the fatigue life, and compressive strain on top of the subgrade defines the rutting life of the pavement based on the performance equations adopted. It is generally considered that increasing the thickness of the bituminous layer decreases the strain at the bottom of bituminous (binder) layer. So, most of the pavement designers increase the thickness of the bituminous layer to restrict the strains. However, this is not always true. In fact, the tensile strain at the bottom of the bituminous layer increases with the increase in the thickness of bituminous layer up to a critical thickness and then slowly starts decreasing as the thickness further increases. The critical thickness of bituminous layer ranges between 50 mm and 100 mm. In the majority of the cases, the critical tensile strain under the bituminous layer corresponding to the thicknesses of the bituminous layers of 25 mm and 200 mm are very close. Apart from the economy, the problem of mix rutting is greatly minimized in the case of thin bituminous surfacing (TBS). The issues in rehabilitation and recycling of pavement materials are very less as the granular layer can be reused without much processing requirements.
Keywords: critical locations, tensile strain, fatigue, rutting;

1. Introduction

Most of the Indian highways and the majority of the urban roads are bituminous pavements, due to their ease of construction and repair. The urban roads are designed for 10- 20 MSA traffic depending on the requirements. The flexible pavement design is done, in India, as per IRC: 37 guidelines [1]. All the revisions of IRC: 37 in the past two decades were based on mechanistic-empirical pavement design principles [1, 2, 3]. In the IRC method, the design of pavement structure involves the selection of thicknesses of different layers to keep the strains at critical locations within the limits as calculated using the performance equations adopted for predicting the fatigue life and rutting life and the failure criteria. However, the selection of thicknesses of different layers can have many possible combinations satisfying the design criteria and hence selecting one combination of layers' thicknesses by trial and error method is difficult. So, the pavement engineers generally try to fix the thickness of granular layers and vary the thickness of the bituminous layer to match the limiting strains. A similar approach was presented in the design charts given in IRC: 37-2001 onwards, only for the convenience of the pavement engineer. This approach leads to the pavement designs with varying thickness of bituminous layers depending on the traffic levels, keeping the granular layers' thickness constant.

2. Problem definition

Majority of the pavement designers are giving higher importance to the bituminous layer and lower importance to the granular layer in designing a flexible pavement. This is leading to pavements being designed with the thicker bituminous layer. However, the same pavement performance can be obtained with a thin bituminous layer and thick granular layer at an economical cost of construction. This type of pavement has better qualities in terms of scope for future strengthening and material reusability. The present paper attempts to show that thin bituminous pavements are economical and ideal in places where sufficient quantity of aggregate is available for use. The environmental benefits of the thin bituminous surfacing are also discussed.

3. Methodology

To highlight the advantages of thin bituminous surface pavements, strains at critical locations were calculated for a set of bituminous layer and granular layer thickness combinations, and fatigue and rutting lives were estimated based on these strain values. A subgrade layer's elastic modulus of 70 MPa and bituminous layer's elastic modulus of 1700 MPa (corresponding to dense bituminous macadam mix with VG30 binder as per IRC: 37 – 2012 [2]) are considered for the study. The granular layer modulus varies depending on the thickness of the granular layer for a given subgrade modulus. The critical strain locations and the typical pavement structure with dual wheel load are presented in Figure 1 along with the commonly used notations representing the layer thickness and modulus values.

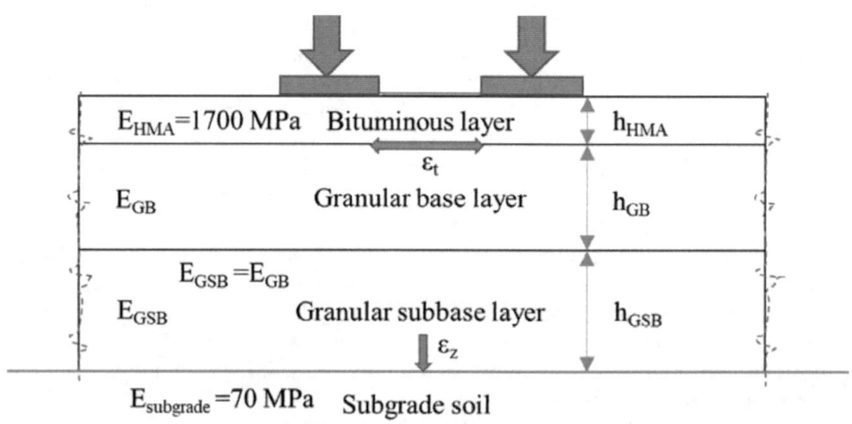

Figure 1. Critical strain locations considered in flexible pavement design

The sub grade's modulus of resilience is considered to be 70MPa (CBR is close to 9%) and elastic modulus of the granular layer is estimated based on the sub grade's modulus of resilience using Equation 1 given in IRC: 37 [1,2].

$$E_{granular\ layer} = 0.2\, h^{0.45}\, E_{Subgrade} \quad (1)$$

The thickness values of granular layer considered are 350 mm, 500 mm and 650 mm, and thickness of bituminous layer varied from 25 mm (1 inch) to 200 mm (8 inches) to study the influence of the thickness of these layers on the fatigue and rutting performance of bituminous pavements.

For example, the design charts are given in IRC: 37 -2012 [2] for 5 MSA load repetitions for a CBR of 8% show a granular layer thickness of 400 mm and bituminous layer thickness of 75 mm. Similarly, for 10 MSA load repetitions, the suggested granular layer thickness is 450 mm and a bituminous layer thickness of 100 mm for the CBR of 8%.

However, there can be many combinations of the granular layer and bituminous layer thicknesses to satisfy the critical strain criteria for 5 MSA and 10 MSA traffic. This present work is an attempt to show different combinations of layer thicknesses which can satisfy the strain criteria and to show that thin bituminous surface pavements have more advantages over thick bituminous layer pavements.

The strains were calculated using the IITPave [2] software with the inputs like elastic moduli of layers, thicknesses and Poisson's ratio values along with the wheel load configuration and tire pressure. Poisson's ratio value of 0.35 was considered for all the layers. The strains and deflection values calculated for different combinations of layer thicknesses were represented in a set of graphs to better understand the trends in the variation of values.

4. Observations

4.1 Estimation of strains at critical locations

The tensile strain under the bituminous layer is observed to vary non-linearly with the change in thickness of the bituminous layer. This trend has a peculiar nature. The tensile strain increases with increase in the thickness of bituminous layer up to a critical thickness and then starts decreasing with further increase in thickness. This trend can be seen in Figure 2 for different thicknesses of the granular layer (sub-base and base combined). In all the cases, the tensile strain under the bituminous layer for a bituminous layer thickness of 25 mm is less than that for a 200 mm thick bituminous layer. The reason attributed to this behaviour is how the bituminous layer distributes the loads on to the granular layer in the case of thick bituminous layers. If the flexural rigidity of the bituminous layer is high (in the form of thickness and/or bituminous layer's modulus) compared to that of the granular layer, the load distribution will be done predominantly by the bituminous layer and hence may lead to higher tensile strains.

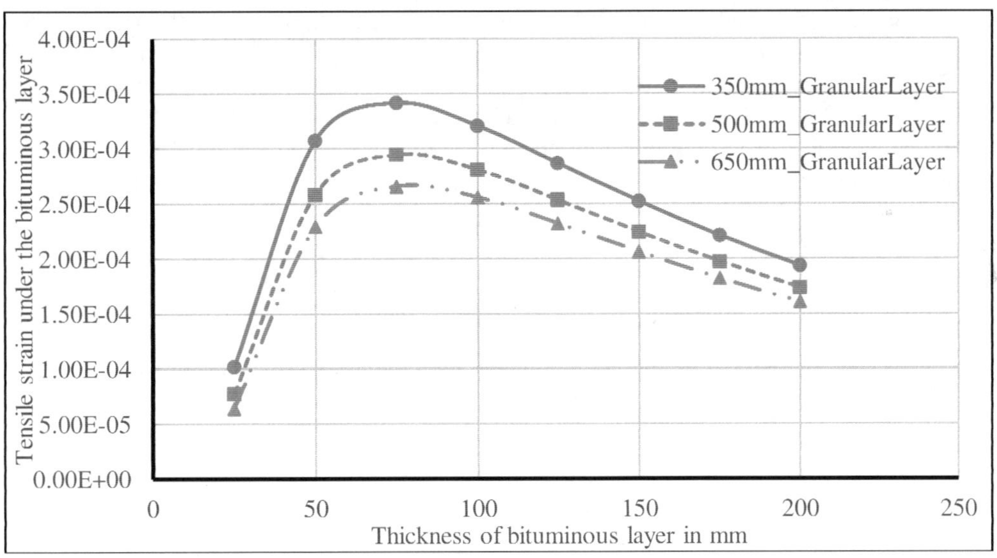

Figure 2. Variation of tensile strain under the bituminous layer with its thickness

On the other hand, the variation of compressive strain on top of the subgrade with a thickness of bituminous layer follows a uniform trend. The compressive strain decreases with increase in the thickness of the bituminous layer as shown in Figure 3. The compressive strain on top of subgrade decreases with increase in bituminous layer thickness and also with the increase in the granular layer thickness. However, the effect of an increase in the thickness of the bituminous layer is less for a higher thickness of the granular layer as can be seen in Figure 3. This can be attributed to the ability of a thick granular layer in distributing the vertical load before transmitting them onto the subgrade soil. Though the granular layers cannot be considered for carrying tensile loads, their ability to distribute the load through compression is very effective with aggregate interlocking aiding in the increase of the area of load distribution on the subgrade. The effect of bituminous layer thickness on compressive strain on subgrade decreases with increase in the thickness of the granular layer.

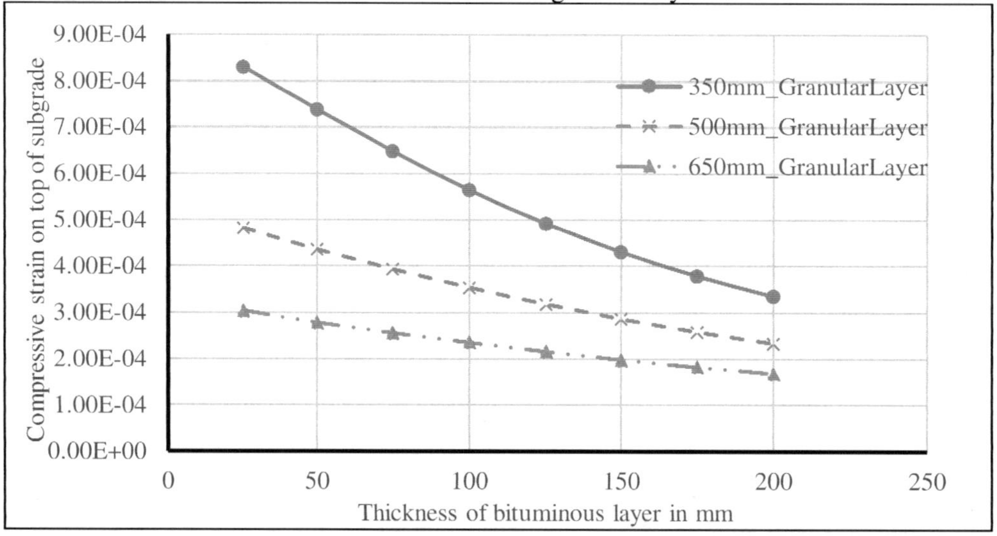

Figure 3. Variation of compressive strain on subgrade with the thickness of the bituminous layer

The variation of surface deflection of the pavement with the thickness of the bituminous layer is similar to the trend followed by the compressive strain on top of subgrade as shown in Figure 4. In the surface deflection, the compressive strain within the top layers also gets compounded along with the compressive strain on top of the subgrade. So, the trends in the variation of surface deflection and vertical compressive strain on top of the subgrade with a thickness of bituminous layer are not the same, though there is a similarity in the trend.

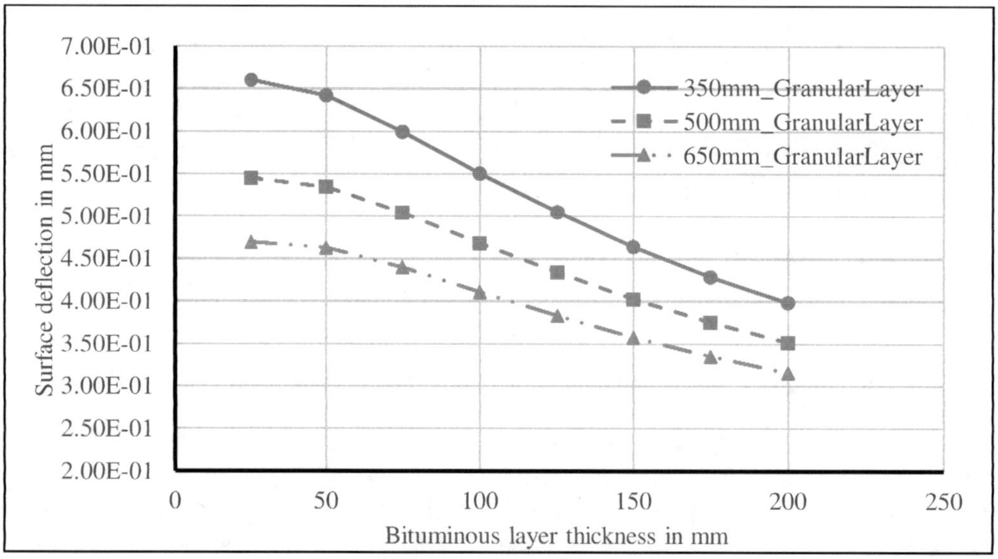

Figure 4. Variation of surface deflection with a thickness of the bituminous layer

Though the pavement surface deflection is not considered in the design, it gives a good indication of the pavement structure's ability to distribute the load. AASHO road tests and the AASHTO's pavement design guidelines are based on the pavement surface deflection measurements made with Benkelman beam [4].

4.2 Estimation of pavement service life

The mechanistic-empirical design methods use the mechanistic parameter, like strain at a critical location, to predict the life of a pavement using a performance equation similar to the correlations presented in Equation 2 and Equation 3. Equation 2 is used to predict the fatigue life of the pavement based on the initial tensile strain at the bottom of a bituminous layer as per IRC: 37 – 2018 [3] guidelines published by Indian roads congress (IRC). Similarly, Equation 3 is used to predict the rutting life in terms of the number of standard axle load repetitions the pavement can withstand before undergoing a surface rutting of a predefined magnitude as per IRC: 37 -2018 guidelines [3]. The strains are calculated using the IITPave [1, 2] software which is based on the linear elastic layer analysis, similar to the KENLAYER software [5]. The performance equations proposed in IRC: 37 -2018 are similar to the equations proposed in the asphalt institute method of flexible pavement design. These equations were incorporated into IRC: 37 guidelines from the revision made in the year 2001. Later on, the concept of reliability was associated with the performance equations and the Equations 2&3 were considered for 80% reliability. A new set of coefficients were proposed for designing with 90% reliability. However, for the sake of simplicity only the 80% reliability equations are shown here.

$$N_{fatigue} = 2.21 * 10^{-4} * \left[{}^{1}/_{\varepsilon_t}\right]^{3.89} * \left[{}^{1}/_{M_R}\right]^{0.854} \quad (2)$$

$$N_{Rutting} = 4.1656 * 10^{-8} * \left[{}^{1}/_{\varepsilon_v}\right]^{4.5337} \quad (3)$$

The inputs for the fatigue equation are the tensile strain at the bottom of the bituminous layer (ε_t) and resilient modulus (M_R) of the bituminous layer. For a particular bituminous mix selected, M_R will be a constant and hence only the tensile strain controls the fatigue life. The only input for the rutting performance equation is the vertical compressive strain on top of subgrade (ε_v).

The fatigue life of the bituminous layer can be estimated using the performance equation (Equation 2) when the tensile strain under the bituminous layer is known. The estimated fatigue life values (in number) for different bituminous layer and granular layer thicknesses are presented in the graph shown in Figure 5. This graph indicates the fall in fatigue life by 100 times when the bituminous layer thickness is increased from 25 mm to 50 mm as there is a shift in the load distribution mechanism between the thickness values of 25 mm and 50 mm. The lowest fatigue life value is corresponding to a bituminous layer thickness of approximately 60 mm and from there the fatigue life increased with increase in the thickness of the bituminous layer. This 60 mm thickness associated with the lowest fatigue performance is specific to the combination of subgrade strength and bituminous layer modulus selected in this study.

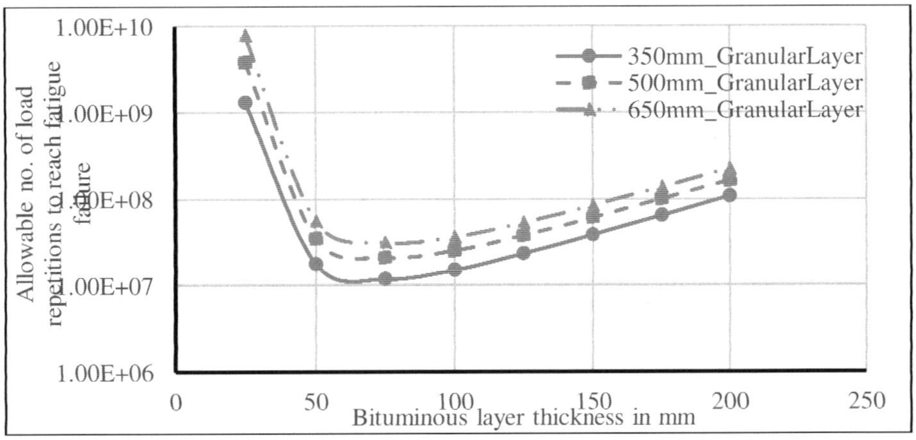

Figure 5. Variation of estimated fatigue life with a thickness of the bituminous layer

From the figure 5, it can be observed that the design for 50 MSA traffic can have either a bituminous layer thickness of less than 50 mm or greater than 125 mm depending on the thickness of the granular layer to satisfy the fatigue criteria.

The number of repetitions to reach rutting failure or simply rutting life of the pavement can be estimated from the value of vertical compressive strain on top of subgrade using Equation 2 as discussed earlier. Figure 6 shows the variation of rutting life with a thickness of the bituminous layer and it shows a fairly uniform trend of increase in rutting life with an increase in the thickness of bituminous layer for all the granular layer thicknesses considered in the study. Figure 6 also shows the degree of variation in the rutting life with an increase in the thickness of the granular layer and bituminous layer.

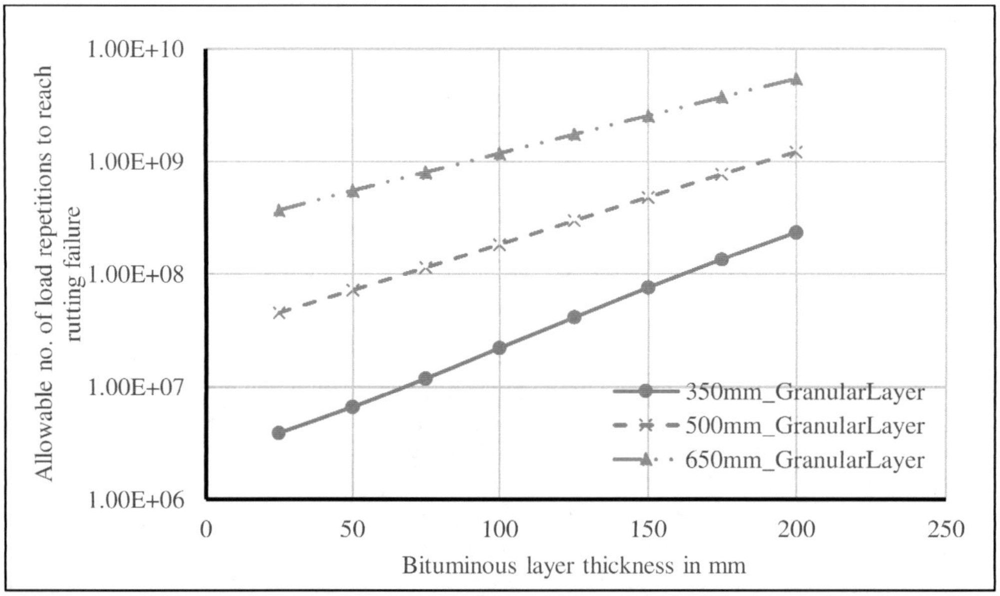

Figure 6. Variation of estimated rutting life with a thickness of the bituminous layer

Similar to the earlier considered case of design for 50 MSA traffic, the granular layer thickness of 500 mm for 25 mm thick bituminous surface layer and 350 mm for 125 mm thick bituminous layer. Though these two options sound similar, the cost difference will be very high as the cost of bituminous mix is nearly four times the cost of granular material per unit volume (DSR 2018) [6].

The combined plot of variation in fatigue life and rutting life with a thickness of bituminous layer for different thicknesses of granular layers is presented in Figure 7. This plot gives a better understanding of both fatigue and rutting lives of the pavement structure as a whole. The letter 'R' indicates 'Rutting' and the letter 'F' indicates 'Fatigue' in the plot. Majority of the pavement designs implemented in the field do not have a balance between the fatigue and rutting lives of the pavement structure and it may lead to an uneconomical pavement section where some material property or geometry of the pavement is under-utilized.

In the places where the aggregate is available in plenty, having a thicker granular layer is economical and also eco-friendly when future overlays and reconstructions are taken into account. The granular layer consists of granular sub-base (GSB) of a minimum thickness of 150 mm and granular base, which is usually, wet mix macadam (WMM), for the remaining thickness. Though the material requirements are different for GSB and WMM materials, the IITPave software considers the thickness of granular layer as a single entity as the modulus of resilience or elastic modulus of the unbound granular base is dependent on the elastic modulus of the underlying soil.

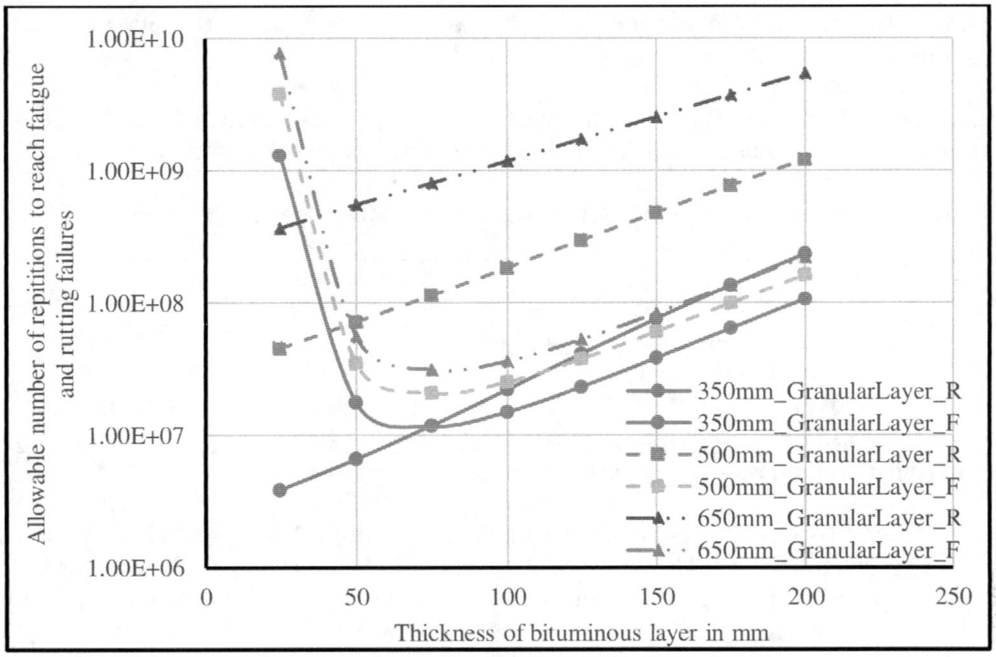

Figure 7. Variation of fatigue and rutting lives with a thickness of the bituminous layer

The data presented in Figure 7 is also shown in tabular format in Table 1, as the exact numbers can be seen in the table data which are not easy to decipher in the graphical representation. The layer combinations suitable for a minimum of 100 MSA traffic are highlighted in the table for different granular layer thicknesses chosen in the study. From the data, it can be seen that a 25 mm bituminous layer and 650 mm granular layer has the best fatigue and rutting life combination.

Table 1. Number of load repetitions for fatigue and rutting failure

Asphalt layer thickness in mm	Rutting life for 350mm Granular Layer	Fatigue life for 350mm Granular Layer	Rutting life for 500mm Granular Layer	Fatigue life for 500mm Granular Layer	Rutting life for 650mm Granular Layer	Fatigue life for 650mm Granular Layer
25	3.87E+06	1.30E+09	4.53E+07	3.80E+09	**3.69E+08**	**7.78E+09**
50	6.59E+06	1.77E+07	7.19E+07	3.48E+07	5.52E+08	5.57E+07
75	1.19E+07	1.17E+07	1.14E+08	2.09E+07	8.02E+08	3.12E+07
100	2.22E+07	1.50E+07	1.84E+08	2.52E+07	1.18E+09	3.63E+07
125	4.13E+07	2.33E+07	2.99E+08	3.77E+07	1.73E+09	5.29E+07
150	7.58E+07	3.84E+07	4.82E+08	6.06E+07	2.55E+09	8.38E+07
175	1.36E+08	6.42E+07	**7.69E+08**	**9.94E+07**	3.73E+09	1.36E+08
200	**2.36E+08**	**1.07E+08**	1.21E+09	1.63E+08	5.44E+09	2.22E+08

5. Discussion on results

Based on the fatigue life and rutting life of different combinations of the granular layer and bituminous layer thicknesses, some cost comparisons were presented in this section to highlight the advantages of thin bituminous surface pavements. The cost comparisons and environmental benefits are presented in this section. However, the environmental benefits are not quantified for consideration in the decision making process.

5.1 Cost comparisons

The costs of material and construction of pavement layers are presented here in Table 2 as per the rates specified in the Delhi schedule of rates (DSR, 2018) [6] published by the CPWD.

Table 2. Rates for construction of different pavement layers as per DSR, 2018

Sl. No.	Material description	Cost per cubic meter in Rs.	As per clause
1	Granular sub-base (GSB) with min. CBR of 30	2517.70	16.78, page 284
2	Wet mix macadam (WMM)	2641.60	16.79, page 285
3	Dense Bituminous macadam (DBM)	10064.25	16.54, page 279

*Rs. was the older symbol used for representing Indian currency of Rupees

For ease of comparison, the cost of both the granular base and sub-base is considered to be Rs. 2,550 and that of dense bituminous macadam material is considered to be Rs. 10,050. Taking into account the data from Table 1 and Table 2, the cost estimate for the selected layer combinations to withstand 100 MSA of traffic is presented in Table 3 per 1 m length and 9 m width of the pavement section. So, a typical volume estimate for pavement layers will be length X width X thickness, where length is considered as 1 m and width as 9 m. Hence, the volume of mix required is 9.h where h is the thickness of the layer in m.

Table 3. Cost estimates for the three alternative pavement compositions

Combination number	Bituminous layer thickness (m)	Granular layer thickness (m)	Cost of bituminous layer (Rs.)	Cost of granular layer (Rs.)	Total cost (Rs.)
1	0.2	0.35	18090	8032.5	26122.5
2	0.175	0.5	15828.75	11475	27303.75
3	0.025	0.65	2261.25	14917.5	17178.75

Alternative 3 has nearly 30 % saving in the cost compared to alternative 1, as can be seen from the data presented in Table 3. The alternative 3, known as thin bituminous surfacing, has environmental benefits along with the cost-saving offered as recycling of the pavement at a later date is easy as the granular layer is an un-bound layer.

AUSTROADS pavement design guide [7] provides the flexible pavement design charts for thin bituminous surfacing and thick bituminous surfacing separately. Hence, this is not a new approach and has been in use for a long time [8].

5.2 Environmental benefits of thin bituminous surfacing

Majority of the flexible pavements, at the end of their service life, are either overlaid or scrapped to construct a new pavement structure. The scrapped bituminous pavement is known as recycled asphalt pavement (RAP) material and it is very difficult to dispose of it or utilize it in the new construction. In case of reconstruction, due to the difficulties in re-heating of RAP material is used only to a limited extent and the remaining material is dumped at a disposal site. The dumping of RAP material is environmentally not acceptable as this does not allow the vegetation to grow there.

On the other hand, disposing or reusing of thin bituminous surfacing is much easier as the quantity of RAP generated is less. The process of overlaying to structurally strengthen the pavement is also advantageous in case of thin surfacing, as the existing granular layer provides strong base support.
The cost of milling will be less for thinner bituminous surfaces.

6. Concluding remarks

The thin bituminous surface pavements are cost-effective and have a great scope for future strengthening and reusability of aggregate. The cost of construction of thin bituminous surface pavement is only 70% of the cost of a thick bituminous pavement, in addition to the various advantages it has in terms of environmental friendly construction practices. The air pollution due to preparation of hot bituminous mix will also be less in the case of a thin bituminous surface.

The data presented here is for only one subgrade condition and one bituminous mix considered. However, the trends in the variation of fatigue life and rutting life of pavements remain the same with little offsets. The example shown for cost comparison is corresponding to 100 MSA traffic as the numbers are better matched for the combination of layer thicknesses selected for the study.

However, a very thin bituminous layer of the order of 25-30 mm thickness may not offer high shear resistance under heavy axle loads with frequent application of brakes. So, these thin bituminous surface pavements may not be appropriate for highway pavements. However, these pavements are best suited where the traffic is limited in terms of heavier axle loads and frequent brake applications. A binder rich surfacing will have better durability as the modulus of the bituminous layer has little significance on fatigue and rutting lives at thicknesses of the order of 25 – 30 mm.

Acknowledgments

The authors would like to acknowledge Late Prof. B B Pandey, Advisor, SRIC, Department of Civil Engineering, Indian Institute of Technology Kharagpur for his initiation and efforts in developing the IRC: 37 guidelines into mechanistic-empirical design guidelines for the past two decades and for introducing many layer combinations in the recent revisions of the IRC: 37 guidelines.

References

[1] IRC 37 2001 Tentative guidelines for the design of flexible pavements *Indian Roads Congress* New Delhi India
[2] IRC 37 2012 Tentative guidelines for the design of flexible pavements *Indian Roads Congress* New Delhi India
[3] IRC 37 2018 Guidelines for the design of flexible pavements *Indian Roads Congress* New Delhi India
[4] AASHTO 1993 AASHTO guide for design of pavement structures *American Association of State Highway and Transportation Officials* USA
[5] Yang H Huang 2004 Pavement analysis and design Second edition *Pearson -Prentice Hall*
[6] DSR 2018 Delhi schedule of rates *CPWD* New Delhi https://cpwd.gov.in/Publication/DSR_Vol_2_2018.pdf, Accessed on 14th October 2019
[7] Jameson G 2008 Technical basis of Austroads guide to pavement technology Part 2 Pavement Structural Design No. **AP-T98/08**
[8] Freeme C R, Marais C P 1973 Thin Bituminous Surfaces: Their Fatigue Behavior and Prediction *Highway Research Board* Special Report **140** p 158–79

Shear critical Reinforced Concrete beams with Recycled Coarse Aggregate

G Papa Rao[1], N Sunil[2], D Sree Ramachandra Murty[3]

[1]Professor, Department of Civil Engineering, Gayatri Vidya Parishad College of Engineering (Autonomous), Visakhapatnam, India.
[2]Post Graduate Student, Department of Civil Engineering, Gayatri Vidya Parishad College of Engineering (Autonomous), Visakhapatnam, India.
[3]Professor, Department of Civil Engineering, College of Engineering (Autonomous), Andhra University, Visakhapatnam, India.
E-mail: gprao_74@yahoo.co.in

Abstract: Various kinds of structures are made of premier construction material, concrete. Concrete is extensively used, only second for water; world consumption of concrete is staggering at 4.5 billion tons a year. All the constituent raw materials that go in the production of concrete are Ordinary Portland cement, coarse aggregate and sand which are valuable natural resources; these are facing fast depletion, as the demand for concrete is rising by leaps and bounds. In this context, conservation of concrete materials is inevitable at all costs. The world concrete industry consumes over ten billion tons of coarse aggregate annually. To conserve the natural coarse aggregate, industrial by-products must be increasingly substituted for natural coarse aggregate. In big and overpopulated cities, old and dilapidated structures are demolished for new constructions; the waste can serve as an alternative source, named recycled coarse aggregate (R.C.A). The current research is directed to investigate the suitability of R.C.A as natural coarse aggregate (N.C.A), in shear critical reinforced concrete beams.
Keywords: Demolition waste, Natural coarse aggregate (NCA), Recycled coarse aggregate (RCA), Shear critical beams, Service strength, Ultimate strength, Deflection and Crack width.

1. Introduction

Endless utilization of coarse aggregate in the extensively used concrete leads to the shortage of natural resource, Natural Coarse Aggregate (NCA). Therefore, industrial by-products must be increasingly substituted for NCA. Non–availability or non-accessibility in some parts of the country, like Gangetic belt, ever-increasing demand rate and consequent depletion of NCA, fast urbanization and the resulting elimination of NCA quarries and enforcement of environmental regulations have increased demand and decrease in the supply of NCA for concrete production. Since the needs of NCA is expected to keep increasing, present shortages become more acute and incessant effort for an alternative material, has found concrete waste as a new alternative, beneficial source. In big and overpopulated cities, old and dilapidated structures are demolished for new constructions. The waste disposal becomes a serious threat to the environment, besides becoming an unmanageable burden involving time, effort, non-availability of space and money. This disposal crisis can be converted into an advantage by using hard and inert materials like demolished concrete as CA, for the preparation of new concrete, solving disposal crisis as well as the meagre availability of CA simultaneously.

The method of preparation of RCA (Recycled Coarse Aggregate) from demolished concrete waste originated in Europe, at the end of world war -II and it is the need of the hour anywhere. Xiaoa et al. 2012 [1] presented an Overview of Study on Recycled Aggregate Concrete in China (1996–2011). Rao et al. 2011 [2] studied on the influence of Field Recycled Coarse Aggregate on Properties of Concrete. Fonteboa and Abella 2007 [3], Ignjatović et al. 2017 [4], Katkhuda and Shatarat 2016 [5], Rahal and Alrefaei 2017 [6] studied on the shear behaviour of Recycled Concrete Beams. Kim et al. 2013 [7] studied the size effect in shear failure of reinforced concrete beams with recycled aggregate. Marinkovica et al. 2010 [8] compared the Environmental impact of Natural and Recycled Aggregate Concrete. Murthy et al. 2019 [9] experimented on the high volume of fly ash cement recycled coarse aggregate and stone crusher dust combined for sustainable structural concrete. Tayfun et al 2014 [10] studied the use of Waste Marble and Recycled Aggregates in Self-Compacting Concrete for Environmental Sustainability. Ponnada 2014 [11] studied the combined effect of flaky and elongated aggregates on strength and workability of concrete.

The tests established that RCA from demolition concrete waste can serve as a useful substitute for NCA. It is predicted by them that the momentum for recycling of concrete and demolition waste for new concrete will culminate in its being a routine procedure. This occurrence is a non-distant future in India. Most of the work on RCA by researchers, related to the determination of mechanical properties of concrete made of RCA. RCA needed more water than NCA for the same slump, as water absorption was more for RCA. Compressive strength of concrete with RCA was lower than that of concrete with NCA. Impurities and type of manufacture of RCA influenced the behaviour of resulting concrete with RCA. Inferior physical properties of RCA, such as lower density, higher water absorption and low workability, compared to NCA are principally attributable to the adhesion of cement paste to the demolished CA. Resistance to mechanical actions such as impact, crushing and abrasion are lower for RCA concrete than conventional concrete. Impurities such as brick pieces in demolished CA, lower the performance of RCA. Crushed concrete waste has been successfully used as sub-base for roads, sidewalks and parking lots. In some countries, up to 90% of CA in concrete is from recycled waste. RCA can be shaped to suit individual requirements. A formal framework for the use of recycled construction waste is in different stages of development in different parts of the world, but it is virtually non-existent in India. The construction boom in India, combined with depleted sources of NCA, increasing infrastructure development, are bound to bring the recycling of concrete waste system to a stage where it will culminate in its being an unavoidable routine procedure.

2. Experimental investigation

2.1 Experimental program

An experimental program was carried out in the present study to evaluate the strength and deformation behaviour of six shear critical reinforced concrete beams, with concrete, having NCA and RCA in different proportions. All the six beams tested had the same longitudinal reinforcement in the bottom and top. Except for the beam NAB1, all the other five beams had the same shear reinforcement, 6mm diameter two-legged stirrups at 140 mm spacing. The details of the test specimens are shown in Figure 1 and Table 1. All the beams were 150mm wide 250mm deep and 1500mm long. The properties of steel reinforcement used are listed in Table 2.

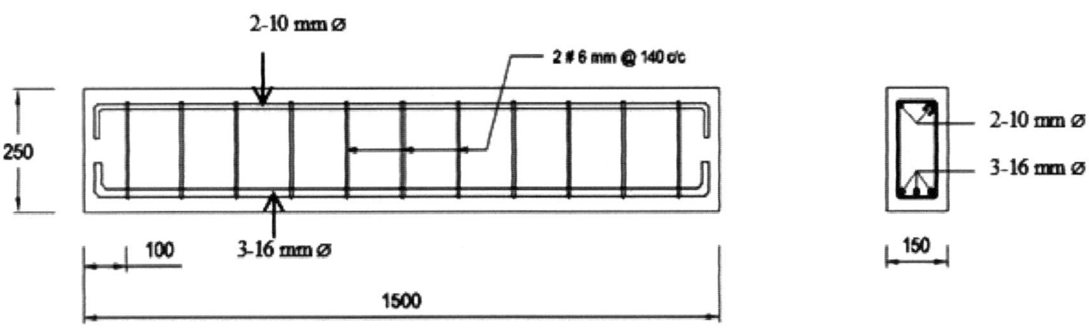

Figure 1. Reinforcement Details of Beams NAB-1, NAB-2, RCAB-3, RCAB-4, RCAB-5 & RCAB-6 (All dimensions are in mm)

Table 1. Properties of Test Specimens

Specimen label	Effective span (mm)	Shear span to effective depth ratio	Flexural reinforcement		Water-cement ratio by weight	Cube compressive strength (MPa)	Split tensile strength (MPa)	Shear Reinforcement
			Bottom	Top				
NAB-1	1300	1.99	3-#16	2-#10	0.5	34.40	2.38	Without shear reinforcement
NAB-2	1300	1.99	3-#16	2-#10	0.5	34.40	2.38	6mm– 2 legged stirrups @140mm c/c
RCAB-3	1300	1.99	3-#16	2-#10	0.5	34.36	2.19	
RCAB-4	1300	1.99	3-#16	2-#10	0.5	33.34	2.13	
RCAB-5	1300	1.99	3-#16	2-#10	0.5	30.68	2.00	
RCAB-6	1300	1.99	3-#16	2-#10	0.5	26.60	2.00	

Table 2. Properties of steel reinforcement

Sl. No.	**Property**	**Results**		
		16 mm	**10 mm**	**6 mm**
1	Yield stress (N/mm2)	571.96	572.95	327.14
2	Ultimate stress (N/mm2)	678.89	910.41	419.10
3	Weight per meter length	1.58	0.61	0.22
4	% Elongation	20.00	20.00	22.00

2.2 Materials and methods

The cement used for the investigation was normal Portland cement. In the present study river sand conforming to zone II as per, IS 383 1970 [12] was used. Normal coarse aggregate (NCA) used was from an established quarry satisfying the requirements IS: 383 1970 [12], 20mm and 10mm size aggregates in combination were used in a proportion. RCA was prepared from the concrete elements that were used in the earlier laboratory tests. The crushed concrete was screened by using sieve analysis. Specific gravity, water absorption, and impact value were performed on NCA and RCA; the details are furnished in Table 3. The concrete mix details are given in Table 4.

Table 3. Properties of Coarse Aggregate used in specimens

Sl. No.	Property	RCA	CA
1	Specific Gravity	2.63	2.7
2	Water Absorption %	3.15	0.7
3	Impact Value %	21	12

Table 4. Concrete Mix Design Details (kg/m^3 of concrete)

Mix proportions	W/C ratio	Cement	Water	Fine Aggregate	Coarse Aggregate	
					NCA	RCA
M1 (0% RCA- 100% NCA) (Without Shear Reinforcement)	0.5	360	180	876.86	1067.60	-
M2 (0% RCA- 100%NCA)	0.5	360	180	876.86	1067.60	-
M3 (25% RCA- 75%NCA)	0.5	360	180	876.86	800.70	266.90
M4 (50% RCA- 50%NCA)	0.5	360	180	876.86	533.80	533.80
M5 (75% RCA- 25%NCA)	0.5	360	180	876.86	266.90	800.70
M6 (100% RCA- 0%NCA)	0.5	360	180	876.86	-	1067.60

2.3 Nomenclature
The six beams are named NAB-1, NAB-2, RCAB-3, RCAB-4, RCAB-5 and RCAB-6. The first beam NAB-1 had 0% of RCA and 100% NCA, NAB-2 had 0% RCA and 100% NCA, beam NAB-3 contained 25% RCA and 75% NCA, beam NAB-4 was provided with 50% RCA and 50% NCA, beam NAB-5 contained 75% RCA and 25% NCA and finally NAB-6 had 100% RCA and 0% NCA.

2.4 Casting of specimens
Laboratory floor was used as the bed for casting the beams, brick masonry moulds were used for casting the beams. The moulds were made of bricks and plastered with mortar. The inner dimensions of the moulds were equal to the dimensions of the beams (0.15 m × 0.25 m × 1.5 m).
In the moulds cover blocks of 20mm depth were placed over which beam steel grill was positioned. The concrete materials were weigh-batched and the concrete mixer was used for mixing the concrete. Needle vibrator was used for the vibration of the beams. The top surface was smoothened with a trowel. Simultaneously from each beam mix, corresponding cubes and cylinders were cast in a standard way. Curing of the specimen started after 24hours of casting and beams were shifted from the moulds to curing tank after 7days of casting. The testing of the beams started 28days after casting day. The test setup is shown in Figure 2. Test specimens were instrumented to obtain the measurement of transverse load on the beam, mid-span vertical deflections and crack width. A dial gauge was used for recording deflections and a hand-held microscope capable of measuring a minimum value of 0.05mm was used to note the crack width.

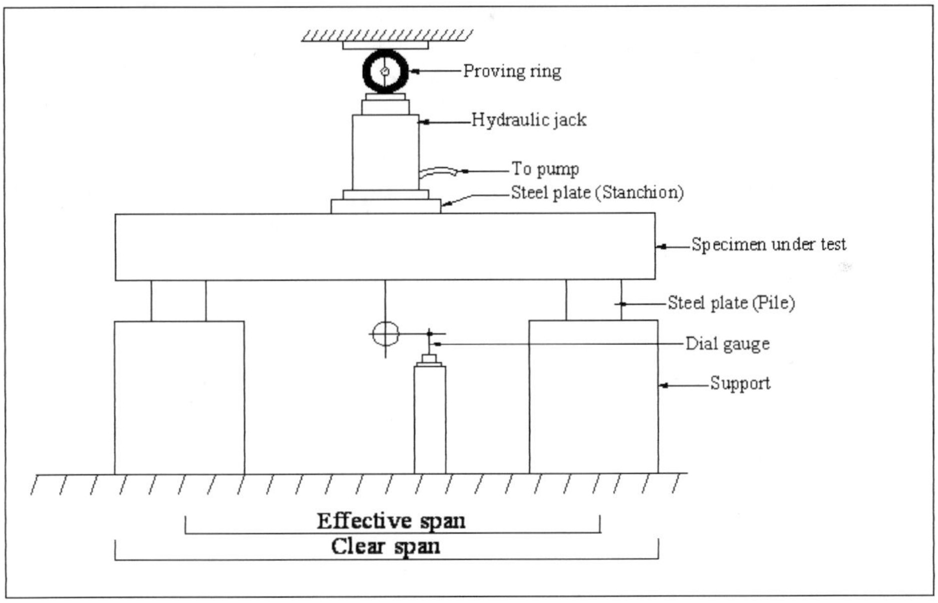

Figure 2. Schematic diagram of the load test set-up for beams

2.5 Testing procedure
The beams are tested under monotonic increasing static load, using a hydraulic jack of 1000kN size for load application. To measure the load applied a load cell of 1000kN capacity was used. The beams were placed under the jack that was on a steel reaction frame. The hydraulic jack, load cell for load measurement and deflection dial gauge were located at the mid-span of the beam.

A distributing steel beam was used to apply the load at two points on the top of the beam. The beams were simply supported on a span of 1300mm. In testing, the initial cracks were flexural at the bottom of the side face. As the loading increased diagonal cracks started appearing. With further load increases, the diagonal cracks extended to the top face and touched the load application point. At ultimate, the diagonal crack widened and extended from the bottom face and touched the load application point. All the beams failed in shear suddenly with wider shear cracks. The tested specimens were photographed after the test. Tested beam specimens are shown in Figure 5. All the beams failed in shear. The load detection and moment–crack width curves are shown in Figure 3 and Figure 4 respectively. The principal test results are shown in Table 5.

3. Presentation of test results

Beam NAB–1 contained NCA only, but was provided with no shear reinforcement; in this beam shear resistance of stirrups was nil; the ultimate test shear strength was 70kN. Beam NAB-2 was identical to NAB-1 but contained shear reinforcement which enhanced the shear resistance of the beam, permitting shear failure at a higher load of 80kN than the companion beam NAB-1.

Beams RCAB3, RCAB 4 and RCAB 5 contained RCA 25%, 50% and 75% respectively; the shear resistance decreased as the percentage of RCA increased. With 25% RCA, shear capacity was 75kN; with 50% RCA shear capacity was same at 75kN. But when the RCA increased to 75%, the shear strength further decreased to 70kN. RCAB6 contained 100% RCA; the loss in shear strength was maximum and the shear capacity was 65kN. The strength reduction in an RC beam with RCA is more in shear than in flexure, because of lesser resistance offered by RCA in shear mechanism than in flexure mechanism. Load-deflection curves are shown in Figure 3. The sudden failure of an RC beam in shear is typified in the load-deflection curves in Figure 3. At service load bending moment, 23.11 kN-m in all the beams, deflection increased as the RCA percentage in the concrete increased. Similarly, as RCA percentage in the concrete increased, the crack width at service load increased.

Table 5.a. Principal test results at service load

Specimen label	The moment at first crack (kN-m)	The moment at 0.3mm crack width (kN-m)	At service load					
			Moment (kN-m)	Shear force (kN)	Shear stress (MPa)	Deflection (mm)	Crack width (mm)
NAB-1	13.00	22.87	20.22	46.67	1.43	2.85	0.12
NAB-2	13.00	26.62	23.11	53.33	1.63	2.54	0.10
RCAB-3	10.83	22.96	23.11	53.33	1.63	3.22	0.25
RCAB-4	13.00	25.63	23.11	53.33	1.63	3.30	0.30
RCAB-5	13.00	18.88	23.11	53.33	1.63	3.78	0.40
RCAB-6	10.83	18.77	23.11	53.33	1.63	4.42	0.57

Table 5.b. Principal test results at ultimate load

Specimen label	The moment at first crack (kN-m)	The moment at 0.3mm crack width (kN-m)	At ultimate load				
			Moment (kN-m)	Shear force (kN)	Shear stress (MPa)	Deflection (mm)	Crack width (mm)
NAB-1	13.00	22.87	30.33	70	2.15	5.98	15
NAB-2	13.00	26.62	34.67	80	2.45	6.03	15
RCAB-3	10.83	22.96	32.50	75	2.30	6.45	15
RCAB-4	13.00	25.63	32.50	75	2.30	6.64	10
RCAB-5	13.00	18.88	30.33	70	2.15	7.15	15
RCAB-6	10.83	18.77	28.17	65	1.99	7.50	12

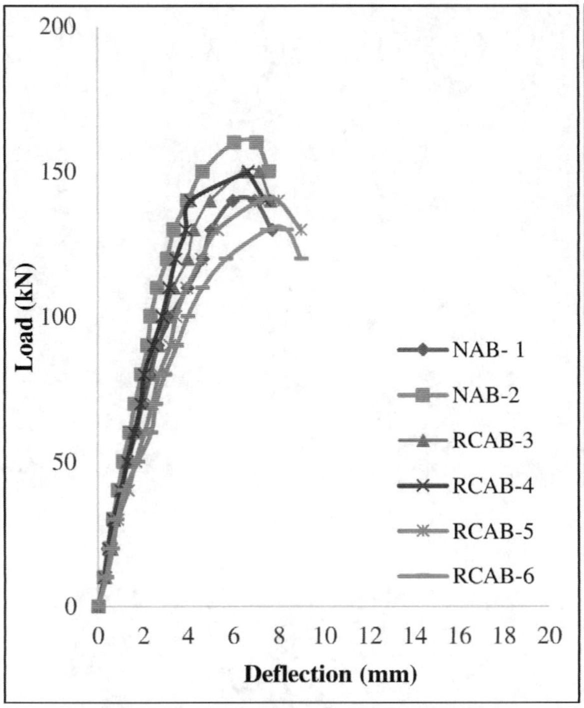

Figure 3. Load vs Deflection Curves

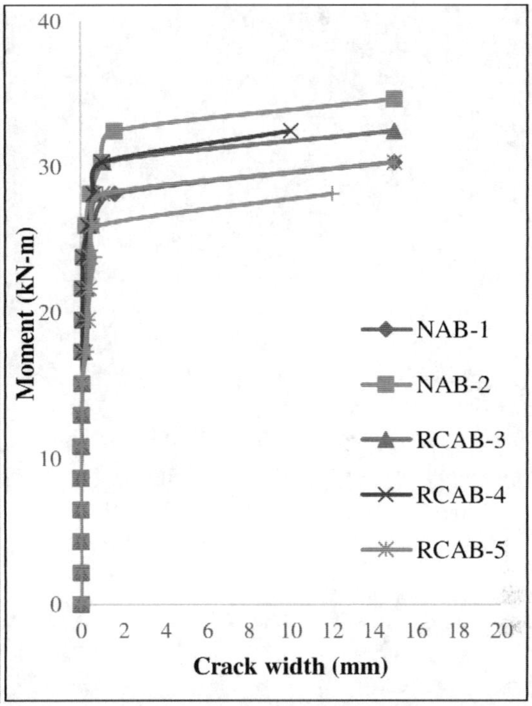

Figure 4. Moment vs Crack width Curves

Figure 5.a. BEAM NAB- 1 (100% NCA (without shear reinforcement)

Figure 5.b. BEAM NAB- 2 (100% NCA)

IOP Conf. Series: Materials Science and Engineering 1025 (2021) 012015 doi:10.1088/1757-899X/1025/1/012015

Figure 5.c. BEAM RCAB -3 (25% RCA)

Figure 5.d. BEAM RCAB -4 (50% RCA)

Figure 5.e. BEAM RCAB -5 (75% RCA)

Figure 5.f. BEAM RCAB – 6 (100% RCA)

Figure 5. Beams NAB-1, NAB-2, RCAB-3, RCAB-4, RCAB-5 & RCAB-6 after testing

4. Conclusions

Uncontrolled demand and use for several years of natural coarse aggregates in concrete making and consequent significantly reduced availability of natural coarse aggregates, forced the concrete profession to use substitutes in place the primary natural resource of coarse aggregate. The current investigation replaces in different percentages natural coarse aggregate with recycled coarse aggregate (NCA with RCA) in shear critical structural concrete beams and investigates the desirability of the use of RCA in structural concrete. Resourceful ideas gleaned from the study are shown in the following conclusions.

1. In the present context of fast depletion of concrete making materials, the need for the alternative sources for cement, sand and coarse aggregate is urgent.
2. As an alternative to natural coarse aggregate (NCA), recycled coarse aggregate (RCA), having its origin in Europe during World War II, is most suitable and rewarding, notwithstanding initial setbacks, in the preparation of RCA.

3. In the investigation undertaken, use of RCA up to 50% in reinforced concrete shear critical beams is appropriate.
4. Use of RCA in shear critical RC beams results in small percentage reduction, in shear strength.
5. At service load, the increase in deformations, deflection and crack width increased with increasing the percentage of R.C.A.
6. Adoption of RCA up to 50% of total coarse aggregate appears reasonable based on the current study. The use of a higher percentage of RCA becomes possible in the country, as the refining mechanism of RCA reaches its peak.

References

[1] Xiaoa J, Lia W, Fana Y, Huanga X 2012 An Overview of Study on Recycled Aggregate Concrete in China (1996–2011) *Journal of Construction and Building Materials* **31** p 364-383

[2] Rao M C, Bhattacharyya S K, Barai S V 2011 Influence of Field Recycled Coarse Aggregate on Properties of Concrete *Journal of Materials and Structures* **44 (1)** p 205-220

[3] Fonteboa B G, Abella F M 2007 Shear Strength of Recycled Concrete Beams *Journal of Construction and Building Materials* **21** p 887-893

[4] Ignjatović I S, Marinković S B, Tošić N 2017 Shear behaviour of recycled aggregate concrete beams with and without shear reinforcement *Engineering Structures* **141** p 386-401

[5] Katkhuda H, Shatarat N 2016 Shear behaviour of reinforced concrete beams using treated recycled concrete aggregate *Construction and Building Materials* **125** p 63-71

[6] Rahal K N, Alrefaei Y T 2017 Shear strength of longitudinally reinforced recycled aggregate concrete beams *Engineering Structures* **145** p 273-282

[7] Kim S W, Jeong C Y, Lee J S, Kim K H 2013 Size effect in shear failure of reinforced concrete beams with recycled aggregate *Journal of Asian Architecture and Building Engineering* **12 (2)** p 323-330

[8] Marinkovica S, Radonjaninb V, Malesevb M, Ignjatovica I 2010 Comparative Environmental Assessment of Natural and Recycled Aggregate Concrete *Journal of Waste Management* **30 (11)** p 2255-64

[9] Murthy R D, Suresh D, Dinakar P, Lava Kumar S K V S T 2019 High volume of fly ash cement recycled coarse aggregate and stone crusher dust combined for sustainable structural concrete *Indian Concrete Institute Journal* **19 (4)** p 29-36

[10] Tayfun U, Ilker B T, Atila G C 2014 Use of Waste Marble and Recycled Aggregates in Self-Compacting Concrete for Environmental Sustainability *Journal of Cleaner Production* **84** p 691-700

[11] Ponnada M R 2014 Combined effect of flaky and elongated aggregates on strength and workability of concrete *International Journal of Structural Engineering* **5 (4)** p 314-325

[12] Indian Standard Specification for coarse and fine aggregates from natural sources for concrete IS 383: 1970 Bureau of Indian Standards, New Delhi

Compressive strength of concrete with partial replacement of Fine Aggregate with crushed Electronic waste plastic

Kandalam Veerabhadram[1], B L Satya Bhavya[2], S Mani Bhaskar[3]

[1]Professor at Civil Engineering Department, GITAM (Deemed to be University), Visakhapatnam, INDIA
[2]UG Student, Department of Civil Engineering, GITAM (Deemed to be University), Visakhapatnam, INDIA
[3]Research Scholar, Department of Civil Engineering, GITAM (Deemed to be University), Visakhapatnam, INDIA
E-mail: veerabhadram.kandalam@gitam.edu

Abstract: Electronic waste (E-waste) of about 53.6 MT is generated globally with 7.3 kg per capita, wherein India generates about 1.72 MT. Out of the total generated e-waste, only 17.4% is recycled properly. The e-waste constitutes both organic and inorganic compounds while handling. Plastic waste is one of the inorganic compounds of e-waste, which has majorly become a massive problem owing to its non-degradable property. Moreover, a very less percentage of this is being reused/recycled every year. This also has many negative impacts on the environment so there is a necessity to study its uses alternately. Therefore, it is preliminarily desired to attempt in replacing the fine aggregate to certain percentage viz 0%, 10%, 15%, 20% by using available plastic components of e-waste available from circuit boards, body cabinets of a computer after crushing. In this paper, an attempt has been made to observe the effects of partially replacing fine aggregate with e-waste (plastic). The impact of alternative material to fine aggregate in material compositions is determined by testing the physical (Density) & strength (Compressive and Flexural) properties of concrete. Testing for these impacts helps in understanding the extent up to which these wastes can be used in concrete. Thereby, the usage of natural materials in construction practices can be reduced. Finally, it is inferred that the usage of e-waste plastic particles as a part of the reuse mechanism helps in improving sustainability due to less usage of sand.
Keywords: e-waste; Plastic waste; Construction; Engineering properties of concrete.

1. Introduction

It is inevitable that due to globalization, there is widespread economic development, which is prone to mobility and Urbanization. This has made the modern societies to enhance their living standards, which ultimately linked to production and consumption of Electrical and Electronic Equipment (EEE). As per the Euro Monitor report 2020 [1], e-waste of 53.6 MT was generated globally at an average rate of 7.3kg per capita. Out of the generated quantity, the informal sector handled the waste and resulted in only 17.4% was recorded as re-cycled quantity. In the Indian context, 1.72 MT of e-waste was expected to generate with CAGR ranging from 7 % to 8 %. On perspective view, out of all the states in INDIA, Maharashtra contributes the highest share of about 20% in generating e-waste. This e-waste predominantly contributed by large household appliances (42%), Electrical and Electronic equipment (34%) and consumer electronics & others (24%).

Many researchers across the globe worked on the properties of concrete with aggregate replaced by E-waste. Lakshmi and Nagan (2010) [2] studied the properties of concrete containing E plastic waste.

Kale and Pathan (2015) [3] researched on recycling of Demolished Concrete and E-waste. Amalu et al (2016) [4] reported the uses of waste plastic as a fine aggregate substitute in concrete. Raghatate Atul (2012) [5] researched the uses of plastic in concrete to improve its properties. Khilesh Sarwe (2014) [6] studied on the strength of Concrete using waste plastics and steel fibres. Zainab and Enas (2008) [7] uses of waste plastic as aggregate in concrete. Panneer Selvam and Gopala Krishna (2016) [8] researched on the recycling of E-waste in concrete. Ashwin Manjunath (2016) [9] studied on the partial replacement of E-plastic Waste as Coarse-aggregate in concrete. Markandeya Raju (2015) [10] studied the properties of concrete made with flaky and elongated aggregates. Naga Rajesh et al (2019) [11] reviewed on sustainable micro-concrete. Naveen Kumar et al (2019) [12] reviewed the utilization of crumb rubber in various ingredients of concrete.

The e-waste constitutes of organic material viz Lead, Cadmium, Beryllium, Gold, Copper, Silver, Mercury etc. and inorganic material viz PBC and Flamed retarders majorly of Plastic. Each material has adverse effects on environmental sustainability and health. Hence there are different methods viz Incineration, Landfill, recycle for disposing of the e-waste. Based on this the quantification can be measured. But in this paper, it is preliminarily desired to attempt in replacing the fine aggregate to certain percentage viz 0%, 10%, 15% and 20% by using available e-waste of plastic nature which is available from circuit boards, body cabinets of the computer. This study shall pave a new path in handling e-waste and also for deriving new composition in material science. Figure 1 shows e-waste in GITAM campus, Visakhapatnam.

Figure 1. E-waste in GITAM campus

2. Objectives of the study
The objective of this paper is to study the impact of the usage of alternative material viz e-waste plastic in the place of fine aggregates in concrete.

3. Materials used
It involves in the preparation of new material compositing with – Cement (53 grade OPC confirming to IS 12269: 2013 [13]), fine aggregates, coarse aggregates, e-waste plastic granules & potable water. The material composition mix design suiting to M 20 grade is carried and derived as follows Cement (485 kg), Fine Aggregate (610.27 kg), Coarse Aggregate (1147.50 kg) having a mix proportion of 1:1.258:2.36. Various tests were conducted to find out engineering properties of hardened concrete.

4. Experimental setup and results

As per the significance of the study the strength of concrete containing e-waste plastic as partial replacement of fine aggregate is determined by Compression test. Hence the concrete cubes (150 × 150 × 150 mm^3 size) were tested at the age of 7 days, 28 days like compressive strength on the cube, flexural strength on the beam of size 100 × 100 × 500 mm^3. As per IS 516: 1959 [14], a total of 72 specimens were tested and results were studied.

4.1 Compressive strength

The methodology involved adhering to IS 10262-2019 [15] is adopted. Compressive strength tests were performed on cube samples using a compression testing machine. The test is carried out by placing the cube in the compression testing machine and load is applied at the rate of 140kg/cm^2/minute. The load at which the cube fails is its failure load of compression three samples per batch were tested with the average strength values reported in table 1.

Table 1. Average Compressive strength of M 20 grade concrete

Sl. No.	e-waste (%)	Compressive Strength at 7 days (N/mm^2)	Compressive Strength at 28 days (N/mm^2)
1	0	16.75	27.00
2	10	17.5	28.5
3	15	19.33	31.00
4	20	17.667	29.16

4.2 Flexural strength

The Beam specimen of size 10 cm × 10 cm × 50 cm are cast and cured for 28 days and marked for 40 cm i.e. 20 cm from the centre on both sides. UTM is used for applying loading at a rate of 180 kg/min, using two rollers placed along the length, within the 40cm span for determining the flexure test, on the beam specimen. The load was increased until the specimen failed and the failure load was recorded in table 2. The prism specimen and specimen under testing in UTM (Universal Testing Machine) are shown in figure 2.

Table 2. Average Flexural Strength (N/mm^2) at 28 days of M20 Grade Concrete

Sl. No.	% of e-waste	Flexural Strength
1	0	3.13
2	10	4.806
3	15	4.968
4	20	2.212

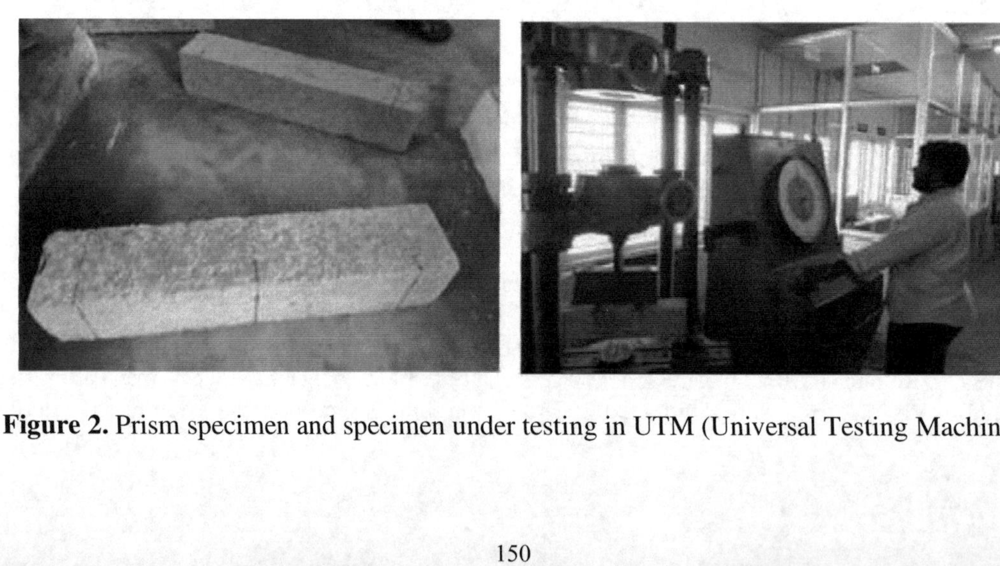

Figure 2. Prism specimen and specimen under testing in UTM (Universal Testing Machine).

4.3 Density test

It is carried out to study the change in density on replacement of fine aggregate with the waste. The weight of cube is measured and divided by volume and this gives the density. This is again compared with the average density of concrete of that mix type. The below values represent an average density of 3 specimens for each replacement %., which results in reducing the density of the material by 2.78%, 4.78% & 7.72% respectively. The densities of the concrete with all four combinations are presented in table 3.

Table 3. Densities of Material composition

Sl. No.	e-waste (%)	Density of Material (kN/cu.m)
1	0	25.1
2	10	24.4
3	15	23.9
4	20	23.16

5. Results and discussion

From the above table 4, it is observed that the compressive strength of the material composition increases from 0% to 15% addition of e-waste plastics, thereafter there is a trend of downfall. Similarly, flexural strength increases from 0 % to 15 %. Thereafter there is a decreased trend. Moreover, the density of material compositions is decreasing from 25. kN/cu.m to 23.16 kN/cu.m, which resembles a lightweight material.

6. Applications

This may be used as Lightweight concrete since there is a reduction in Self-weight of the new material composition.

7. Conclusions

From this study, the following conclusions can be drawn.

1. Partially replacing E-waste (plastic) with fine aggregates has given good compressive strength. The compressive strength was retained even after replacing fine aggregate with E-waste (plastic). Out of 10%, 15% and 20% replacements; 15% replacement improved the compressive strength in a better way, whereas 20% replacement was not able to give a good impact.
2. The density of concrete which is responsible for dead load has decreased by an appreciable value on the partial replacement. 20% replacement gave the best result of decreasing the density.
3. The flexural strength of the concrete has improved to a great extent on the addition of E-waste (plastic) as a partial replacement. 15% replacement escalated the flexural strength by a good value.

References

[1] Forti V, Baldé C P, Kuehr R, Bel G 2020 The Global E-waste Monitor 2020: Quantities, flows and the circular economy potential. *United Nations University (UNU)/United Nations Institute for Training and Research (UNITAR) co-hosted SCYCLE Programme, International Telecommunication Union (ITU) & International Solid Waste Association (ISWA)* Bonn/Geneva/Rotterdam p 1-120

[2] Lakshmi R, Nagan S 2010 Studies on Concrete containing E plastic waste *International Journal of Environmental Sciences* **1 (3)** p 270-81

[3] Kale S P, Pathan H I 2015 Recycling of Demolished Concrete and E-waste *International Journal of Science and Research (IJSR)* **4 (1)** p 789-92

[4] Amalu R G, Azeef Ashraf, Muhammad Hashim, Rejith K U, Vijitha V, Nithya Kurup 2016 Use of waste plastic as Fine aggregate substitute in concrete *International Journal of Scientific & Engineering Research* **7 (4)** p 172-7

[5] Raghatate Atul M 2012 Use of plastic in concrete to improve its properties *International Journal of Advanced Engineering Research and Studies* **1 (3)** p 109-11
[6] Khilesh Sarwe 2014 Study of Strength Property of Concrete Using Waste Plastics and Steel Fibers *The International Journal of Engineering and Science IJE)* **3 (5)** p 09-11.
[7] Zainab Z Ismail, Enas A AL Hashmi 2008 Use of waste plastic in concrete mixture as aggregate replacement *Waste Management* **28 (11)** p 2041-47
[8] Panneer Selvam. N, Gopala Krishna G V T 2016 Recycle of E-waste in concrete *International Journal of Science and Research* **5 (4)** p 1590-3
[9] Ashwin Manjunath B T 2016 Partial replacement of E-plastic Waste as Coarse-aggregate in Concrete *Procedia Environmental Sciences* International Conference on Solid Waste Management, 5IconSWM 2015 **35** p 731-9
[10] Ponnada Markandeya Raju 2014 Combined effect of flaky and elongated aggregates on strength and workability of concrete *International Journal of Structural Engineering* **5 (4)** p 314-25
[11] Naga Rajesh K, Rath M K, Markandeya Raju P 2019 A research on sustainable micro-concrete *International Journal of Recent Technology and Engineering* **8 (2 Special Issue 3)** p 1137-9
[12] Naveen Kumar S, Rath M K, Markandeya Raju P 2019 A review on utilization of crumb rubber in various ingredients of concrete *International Journal of Recent Technology and Engineering* **7 (5)** p 414-8
[13] IS 12269 2013 Indian Standard Ordinary Portland Cement, 53 Grade - Specification (First Revision) *Bureau of Indian Standards* New Delhi, India
[14] IS 516 1959 Indian Standard Methods of Tests for Strength of concrete *Bureau of Indian Standards* New Delhi, India
[15] IS 10262 2019 Recommended guidelines for concrete mix design, *Bureau of Indian Standard,* New Delhi.

Strength properties of ceramic waste concrete

Venkata Krishna Atkuri[1], G V Rama Rao[2]

[1]Research Scholar, Department of Civil Engineering, Andhra University College of Engineering (A), Visakhapatnam – 530 003, Andhra Pradesh, India.
[2]Professor, Department of Civil Engineering, Andhra University College of Engineering (A), Visakhapatnam – 530 003, Andhra Pradesh, India.
E-mail: atkurivenkatakrishna@gmail.com

Abstract. Today significant research study is ongoing into the use of waste materials like, fly ash, GGBS, waste glass powder etc. as Portland cement replacements. A ceramic waste powder is used as a binder with partial replacement of cement which takes some part of reaction at the time of hydration, and also it acts as a filler material. This study is focused on the waste ceramic powder used as a replacement to the concrete ingredient i.e. cement and the strength properties like compressive, split tensile and flexural strengths were measured and the size effect of ceramic waste powder on strength of concrete. To check the strength effect of replacement of cement by ceramic waste powder, the cement ingredient in concrete is replaced at 10%, 20%, 30%, 40% and 50%. This paper focuses on the size effect of ceramic waste powder, having a size less than 90 micron and another size ranges from 90 microns to 150 microns. For this study, concrete mixtures were prepared, for low-grade M20 and medium grade M40. Strength studies like compressive, split tensile and flexural strengths were conducted and comparisons have to be made with the conventional concrete.

1. Introduction

Ceramic Industries are generating 15% -30% waste materials from the total production. Ceramic wastes are produced as a result of ceramic processing. These wastes cause soil, air, and groundwater pollution. This waste is not recycled in any form at present [1]. Ceramic waste powder coming from industries is dumping in nearby vacant spaces or pits. As the ceramic waste is piling up every day, there is a pressure on ceramic industries to find a solution for its disposal. [2-4]. The possibility of pozzolanic reactivity in ceramic products is responsible for long term strength and good durability due to Ceramic production conducts on temperature 200°C to 2000°C.[5-9].To get various properties of concrete, Ceramic waste powder can be used as a partial replacement of cement.

1.1 Objectives of the Study

- Carry out the mix design for low-grade M20 and medium grade M40 for both Conventional concrete and ceramic waste concrete.
- Study the strength characteristics of concrete-like compressive, split tensile and flexural strengths.
- Arrive the optimum percentage replacement of cement with ceramic waste.

1.2 Need for Present Study

The use of ceramic waste generated from the ceramic industry showing that, prospective application in construction as an alternative to the cement. The following problems of increasing magnitude are pressurizing the construction industries to use ceramic waste in the new concrete.

1.2.1 Sustainability in the concrete industry. Design of concrete mixtures with optimum content of cement and enhancement of concrete durability are the main issues towards sustainability in the concrete industry.

1.2.2 Shortage of Dumping Ground and Disposal Problems. In the present days, there has been an enormous increase in the quantity of ceramic waste. Therefore, disposal of such waste is posing a threat to the environment besides the shortage of dumping area.

2. Literature Review

The optimum percentage of waste ceramic tile to be used within the concrete mix with a water/cement ratio of 0.5was about 30%. Their findings reveal that using waste ceramic tile leads to enhancing the properties of concrete [10]. The optimum percentage of replacement of OPC cement with ceramic waste achieved was 30% by weight for M20 grade concrete and decreases the compressive strength of the concrete by the further replacement [11]. The optimum percentage of replacement of OPC cement with ceramic waste achieved was 20% by weight for M30 grade concrete and decreases the compressive strength of the concrete by the further replacement [12].

3. Experimental Methods

Series of test specimens comprising, cubes, cylinders and prisms of standard dimensions were cast for concrete mixes produced with different percentage replacement (0%, 10%, 20%, 30%, 40% and 50%) of cement with ceramic waste to study the strength characteristics. The strength characteristics of concrete like compressive strength, split tensile strength and flexural strength are established experimentally.

3.1 Experimentation

In this study, cement was replaced by the ceramic waste powder of particle size less than 90 microns and 90 microns to 150 microns with replacement from 10% to 50%. For this study, low-grade M20 and medium grade M40 of concrete was adopted. Cubes, cylinders and prisms were cast to arrive at the strength parameters. The strength parameters were studied at the ages 7, 28, 56 and 90 days. Concrete mix with 0% ceramic waste powder forms the basic reference mix to compare the results of mixes with replacement from 10% to 50%.

3.2 Materials

The materials used in the present study namely cement, fine aggregate and coarse aggregate have been tested in the laboratory. The specifications and properties of these materials were presented in the following sections.

3.2.1. Cement. Ordinary Portland Cement of 53 Grade conforming to Bureau of Indian standards was used and tested for various properties as per IS 12269 1987 [13].

Table 1. Physical Properties of Cement

Property	Value	IS Standards
Normal consistency	34%	33-35%
Fineness of cement	6.50%	<10%
I.S time	105 Min.	$\geq$ 30 Min.
F.S time	300 Min.	$\leq$ 600 Min.
Specific Gravity	3.1	3.1-3.2

3.3 Coarse Aggregates

20 mm size coarse aggregate were used for all mixes and tested as per IS 383 1970 [14]. The aggregate used here is brought from nearby quarries which were breached from the stone available near Godavari regions.

Table 2. Physical Properties of Coarse Aggregate

Property	Value	IS Standards
Bulk density gm/cc	1.451 (loose state) 1.576 (dry rodded)	1.45-1.80
Specific gravity	2.85	2.6-2.9
Water absorption (%)	0.6	0.5-1.0
Fineness Modulus	7.21	6.5-8

IOP Conf. Series: Materials Science and Engineering 1025 (2021) 012017 doi:10.1088/1757-899X/1025/1/012017

3.4 Fine Aggregate

The river sand conforming to Zone II as per IS 383 1970 [14] was used as fine aggregate in this study.

Table 3. Physical Properties of Fine Aggregate

Property	Results	IS Standards
Specific Gravity	2.69	2.5-2.8
Bulk Density gm/cc	1.58 (loose state) 1.64 (dry state)	1.45-1.65
Fineness Modulus	2.5	2.2-2.6
Zone	II	I – IV

3.5 Ceramic waste:

Ceramic waste is gathered from nearby SILICA CERAMICS PRIVATE LIMITED situated near Narayanapuram, West Godavari District, Andhra Pradesh, India.

Table 4. Chemical Properties of Ceramic waste

Materials	Ceramic powder %
Sio_2	68.11
Al_2O_3	16.48
Fe_2O_3	0.59
K_2O	3.14
Na_2O	3.78
CaO	0.85
MgO	1.61
TiO_2	0.02
L.O.I	4.75

3.6 Casting and Curing

For preparing the specimens materials were weighed and mixed as per IS 516-1959 (1989) [15]. The demolded specimens after 24 hours are immersed in the potable water for curing period of 28 days at room temperature of 27°C ± 2°C.

Table 5. Concrete Design Mix (M20 Mix) Proportion

Type of Concrete	Concrete Mix Design Proportion					
	Water/ Cement ratio	Cement	Fine Aggregate	Coarse Aggregate	Ceramic Waste Powder	Water
L0	0.55	1	1.89	3.21	0	0.55

Table 6. Concrete Mix Design (M40 Mix) Proportion

Type of Concrete	Concrete Mix Design Proportion					
	Water/ Cement ratio	Cement	Fine Aggregate	Coarse Aggregate	Ceramic Waste Powder	Water
M0	0.4	1	1.79	3.04	0	0.4

Table 7. Design Mix Proportion for Different Concrete Mixes

Grade of the Concrete	Type of the Concrete	Cement Replacement with Ceramic waste powder
M20	L0	Standard Concrete
M20 (<90-micron size of ceramic waste powder)	LA10	10% -replacement
	LA20	20% -replacement
	LA30	30% -replacement
	LA40	40% -replacement
	LA50	50% -replacement
M20 (90-150 Microns size of ceramic waste powder)	LB10	10% -replacement
	LB20	20% -replacement
	LB30	30% -replacement
	LB40	40% -replacement
	LB50	50% -replacement
M40	M0	Standard Concrete
M40 (<90-micron size of ceramic waste powder)	MA10	10% -replacement
	MA20	20% -replacement
	MA30	30% -replacement
	MA40	40% -replacement
	MA50	50% -replacement
M40 (90-150 Microns size of ceramic waste powder)	MB10	10% -replacement
	MB20	20% -replacement
	MB30	30% -replacement
	MB40	40% -replacement
	MB50	50% -replacement

4. Results and Analysis

4.1 General

To examine the effect of replacement of cement with ceramic waste, five different concrete mixtures using different percentages of ceramic waste (10%, 20%, 30%, 40% and 50%) with the same compressive strength were produced. All the concrete mixtures with different percentage of replacement were compared with the reference concrete mix produced with conventional concrete i.e. 0% replacement. Compressive, split tensile and Flexural strengths tests were performed on the concrete specimens to obtain the strength characteristics of concrete made with ceramic waste and conventional concrete.

4.2 Strength Characteristics

4.2.1 Compressive Strength. The compressive strength test was conducted on test specimens at the ages of 7, 28, 56, 90 days after proper curing till the day of the test. The compressive strength values obtained from the test for Conventional Concrete, Concrete with different percentage (10, 20, 30, 40 and 50) of ceramic waste were presented in table 8. The results were plotted below in figure 1.

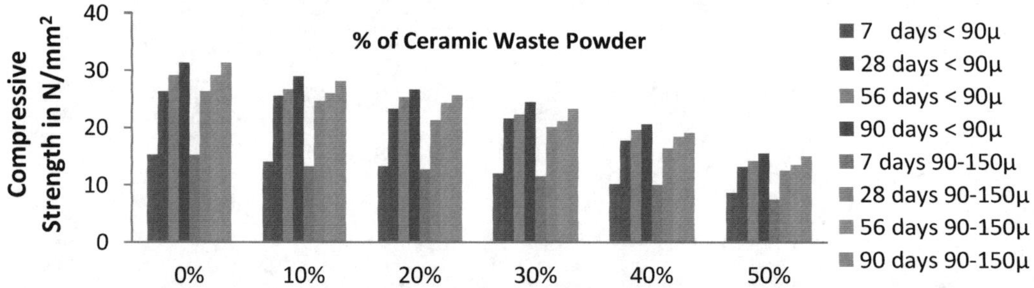

Figure 1. Compressive Strength (MPa): M20 grade for < 90µ and 90-150µ for 7 to 90 days

Table 8. Compressive Strength (M Pa) Results of Cubes of M20 grade for < 90μ and 90-150μ for 7, 28, 56 and 90 days

% of Ceramic Powder	7 days < 90μ	28 days < 90μ	56 days < 90μ	90 days < 90μ	7 days 90-150μ	28 days 90-150μ	56 days 90-150μ	90 days 90-150μ
0%	15.24	26.33	29.1	31.33	15.24	26.33	29.1	31.33
10%	14.02	25.52	26.67	28.89	13.2	24.55	26.02	28.06
20%	13.23	23.24	25.21	26.55	12.65	21.33	24.22	25.55
30%	12.05	21.55	22.24	24.45	11.52	20.05	21.15	23.33
40%	10.16	17.72	19.54	20.56	10.01	16.41	18.44	19.04
50%	8.65	13.15	14.23	15.55	7.55	12.55	13.55	15.11

Table 9. Compressive Strength (M Pa) Results of Cubes of M40 grade for < 90μ and 90-150μ for 7, 28, 56 and 90days

% of Ceramic Powder	7 days < 90μ	28 days < 90μ	56 days < 90μ	90 days < 90μ	7 days 90-150μ	28 days 90-150μ	56 days 90-150μ	90 days 90-150μ
0%	30.36	50.55	53.25	55.24	30.36	50.55	53.25	55.24
10%	31.24	48.52	50.52	52.11	29.54	47.55	49.54	51.24
20%	26.67	43.24	46.87	49.15	24.02	43.04	45.87	48.24
30%	23.33	39.55	43.22	45.14	23.12	38.64	41.14	42.88
40%	21.11	35.06	38.65	39.54	19.55	33.54	35.24	36.22
50%	16.45	29.44	31.98	30.88	13.23	27.65	29.14	30.14

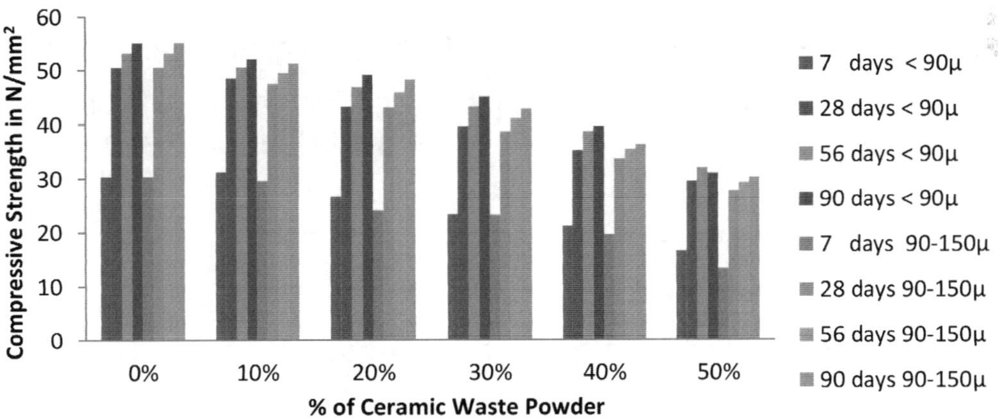

Figure 2. Compressive Strength (MPa): M40 grade for < 90μ and 90-150μ for 7, 28, 56 and 90days

4.2.2 Split Tensile Strength. This test was conducted on test specimens at the ages of 7, 28, 56, 90 days after proper curing till the day of the test. The tensile strength values obtained from the test for Conventional Concrete, Concrete with different percentage (10, 20, 30, 40 and 50) of ceramic waste were presented in table 10. The results were plotted below.

Table 10. Splitting Tensile Strength (MPa): M20 grade for < 90μ and 90-150μ for 7, 28, 56 and 90 days

% of Ceramic Powder	7 days < 90μ	28 days < 90μ	56 days < 90μ	90 days < 90μ	7 days 90-150μ	28 days 90-150μ	56 days 90-150μ	90 days 90-150μ
0%	3.66	4.45	4.51	4.66	3.66	4.45	4.51	4.66
10%	2.54	3.46	3.57	3.82	2.75	3.6	3.67	3.84
20%	2.6	3.67	3.74	3.86	2.19	3.72	3.84	3.91
30%	2.73	3.81	3.88	3.96	2.19	3.25	3.44	3.64
40%	2.05	3.53	3.6	3.67	2.05	3.1	3.26	3.34
50%	1.41	2.82	3.04	3.32	1.82	3.04	3.1	3.25

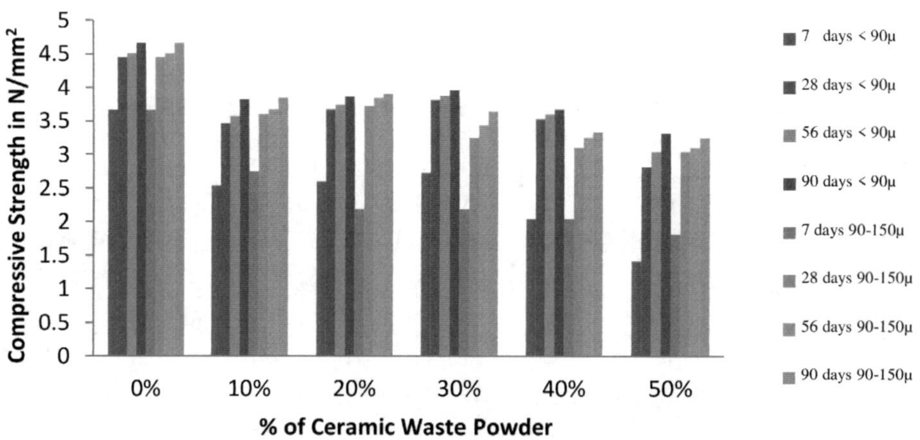

Figure 3. Split Tensile Strength (MPa): M20 grade for < 90μ and 90-150μ for 7, 28, 56 and 90days

Table 11. Split tensile Strength (MPa): M40 grade for < 90μ and 90-150μ for 7, 28, 56 and 90days

% of Ceramic Powder	7 days < 90μ	28 days < 90μ	56 days < 90μ	90 days < 90μ	7 days 90-150μ	28 days 90-150μ	56 days 90-150μ	90 days 90-150μ
0%	3.53	4.24	4.3	4.81	3.53	4.24	4.3	4.81
10%	3.18	4.03	4.38	4.81	3.46	4.03	4.03	4.52
20%	3.32	4.17	4.52	4.69	3.32	4.38	4.43	4.57
30%	2.61	3.81	4.38	4.38	3.11	3.88	3.99	4.39
40%	2.61	3.53	4.38	4.24	3.11	3.03	3.31	4.17
50%	1.76	2.75	3.53	3.67	3.47	2.88	3.17	4.02

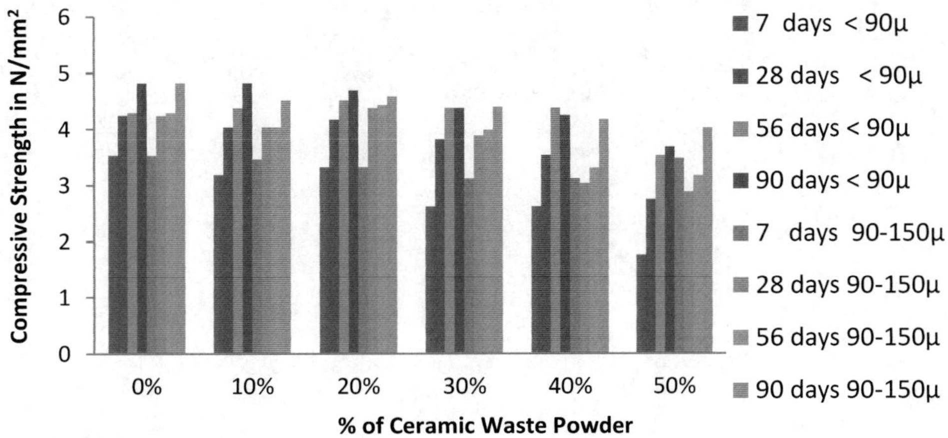

Figure 4. Split tensile Strength (MPa): M40 grade for < 90μ and 90-150μ for 7, 28, 56 and 90days

4.2.3 Flexural Strength. This test was conducted on test specimens at the ages of 7, 28, 56, 90 days after proper curing till the day of the test. The Flexural strength values obtained from the test for Conventional Concrete, Concrete with different percentage (10, 20, 30, 40 and 50) of ceramic waste were presented in table 12. The results were plotted below.

Table 12. Flexural strength (MPa): M20 grade for < 90μ and 90-150μ for 7, 28, 56 and 90days

% of Ceramic Powder	7 days < 90μ	28 days < 90μ	56 days < 90μ	90 days < 90μ	7 days 90-150μ	28 days 90-150μ	56 days 90-150μ	90 days 90-150μ
0%	3.67	4.46	4.89	5.09	3.67	4.46	4.89	5.09
10%	3.18	4.2	4.54	5.08	3.04	4.93	4.97	4.29
20%	3.23	4.23	4.59	4.72	3.06	4.85	5.25	5.19
30%	3.49	4.36	4.66	4.77	2.52	3.11	3.48	3.65
40%	2.7	4.04	4.22	4.55	2.42	3.02	3.06	3.12
50%	1.94	3.69	3.81	3.85	2.41	2.93	2.96	2.98

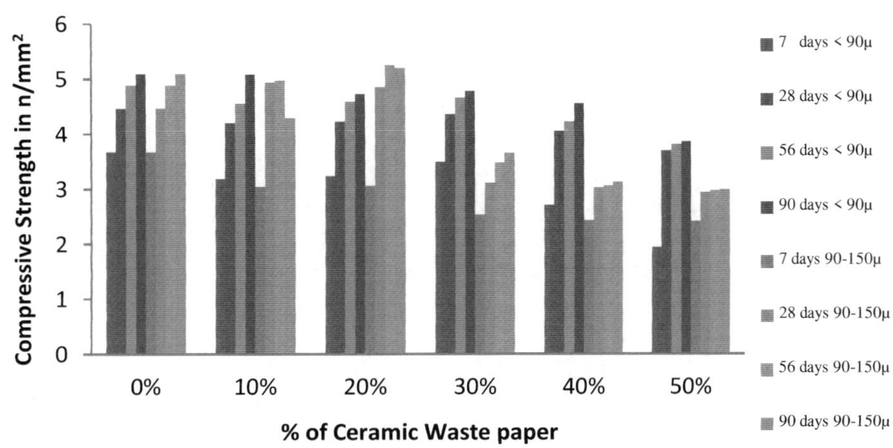

Figure 5. Flexural strength (MPa): M20 grade for < 90μ and 90-150μ for 7, 28, 56 and 90days

Table 13. Flexural Strength (MPa): M40 grade for < 90µ and 90-150µ for 7, 28, 56 and 90days

% of Ceramic Powder	7 days < 90µ	28 days < 90µ	56 days < 90µ	90 days < 90µ	7 days 90-150µ	28 days 90-150µ	56 days 90-150µ	90 days 90-150µ
0%	4.5	5.36	5.86	5.92	4.5	5.36	5.86	5.92
10%	3.13	4.63	5.07	5.97	4.07	4.13	4.24	4.92
20%	3.16	4.15	4.58	5.52	4.01	4.09	4.13	4.39
30%	3.65	4.18	4.38	5.02	3.25	3.76	3.77	3.82
40%	2.75	4.18	4.33	5.01	2.69	3.42	3.58	4.09
50%	2.2	3.39	3.65	4.64	3.14	3.38	3.46	3.52

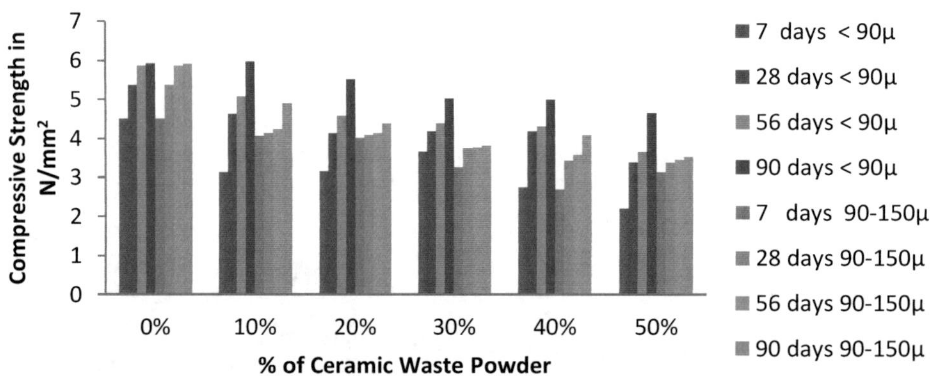

Figure 6. Flexural Strength (MPa): M40 grade for < 90µ and 90-150µ for 7, 28, 56 and 90days

5. Conclusions

By the experimental study, the following observations are made:

1. In M20 grade concrete, Compressive strength decreases when the replacements of cement with ceramic waste powder (of size < 90 microns and 90-150 microns) up to 30% by weight of cement and decrease the strength by the further replacement. Compressive strength loss obtained is 18.15% from L0 to LA 30 and 23.85% from L0 to LB 30.
2. In M20 grade concrete, split tensile strength decreases when the replacement of cement with ceramic powder of size < 90 microns up to 30% and ceramic powder of size for 90-150 microns up to 20% by weight of cement and decreases the strength by the further replacement.
3. In M20 grade concrete, Flexural strength decreases when the replacement of cement with ceramic powder of size < 90 microns up to 30% and ceramic powder of size for 90-150 microns up to 20% by weight of cement and decreases the strength by the further replacement.
4. In M40 grade concrete, Compressive strength decreases when the replacements of cement with ceramic powder (of size < 90 microns and 90-150 microns) up to 20% by weight of cement and decrease the strength by the further replacement. Compressive strength loss obtained is 14.46% from M0 to MA 20 and 14.85% from M0 to MB 20.
5. In M40 grade concrete, split tensile strength decreases when the replacement of cement with ceramic powder (of size < 90 microns and 90-150 microns) up to 20% by weight of cement and decreases the strength by the further replacement.
6. In M40 grade concrete, flexural strength decreases when the replacement of cement with ceramic powder (of size < 90 microns and 90-150 microns) up to 10% by weight of cement and decreases the strength by the further replacement.

With the increase of grades, the percentage of replacement of ceramic waste is decreased. The observations made during the present investigation are in agreement with the results reported by earlier investigators. The experimental study has helped to investigate the various properties of Ceramic waste concrete as a partial replacement of cement with ceramic waste in the production of structural concrete. Because of the other advantages such as conservation of natural resources, free the ceramic waste material from landfills and elimination of disposal problems, the ceramic waste Concrete can be considered as a potential and suitable alternative material with a bright future.

References

[1] Amit Kumar Raval, Indrajit N Patel, Jayesh Kumar Pitroda 2013 Ceramic Waste: Effective Replacement of Cement For Establishing Sustainable Concrete *International Journal of Engineering Trends and Technology* **(4) 6** p 2324-29
[2] Senthamarai R M, Devadas Manoharan P 2005 Concrete with ceramic waste aggregate *Cement and Concrete Composites* **27 (9-10)** p 910–13
[3] Jay Patel P, Shah J, Patel B K 2014 Ceramic powder in concrete by partial replacement of cement- A literature analysis *Journal of International Academic Research for Multidisciplinary* **2 (3)** p 712-7
[4] Moho Mustafa Al Bakri Abdullah, Kamarudin Hussin, Che Moho Ruzaidi, Shamsul Baharin Rozaimah Ramly, Nur Khairiatun Nisa 2006 Concrete Ceramic Waste Slab (CCWS) *Journal of Engineering Research and Education* **3** p 139-45
[5] Binici Hanifi 2007 Effect of crushed ceramic and basaltic pumice as fine aggregates on concrete mortar properties *Construction and building materials* **21(6)** p 1191-97
[6] J.de Brito, Pereira A S, Correia J R 2005 Mechanical Behavior of non- structural concrete made with recycled ceramic aggregates *Cement and Concrete Composites* **27 (4)** p 429-33
[7] Correia J R, J.de Brito, Pereira A S 2005 Effects on concrete durability of using recycled ceramic aggregates *RILEM publications Materials and structures* **38** p 1-7
[8] Daniyal Md., Ahmad Shakeel 2015 Application of Waste Ceramic Tile Aggregates in Concrete *International Journal of Innovative Research in Science, Engineering and Technology* **4 (12)** p 12808-15
[9] Hardik Patel, Arora N K, Shradda R Vania **2015** Use of Ceramic waste powder in Cement Concrete *International Journal for Innovative Research in Science & Technology* **2(1)** p 91-7
[10] Chandana Sukesh, Bala Krishna Katakam, P Saha, K Shyam Chamberlin 2012 A Study of Sustainable Industrial Waste Materials as Partial Replacement of Cement *IACSIT Coimbatore Conferences IPCSIT, IACSIT Press, Singapore* **28** p 161-6
[11] Vimal Kumar N Patel, Chetan Kumar D Modhera, Panseriya M 2017 Utilization of Sanitary Ceramic Wastes in Concrete *International Journal of Civil Engineering and Technology* **8 (4)** p 711-8
[12] Zimbili O, Salim W, Ndambuki J M 2014 A Review on the Usage of Ceramic Wastes in Concrete Production *World Academy of Science, Engineering and Technology International Journal of Civil, Environmental, Structural, Construction and Architectural Engineering* **8 (1)** p 91-5
[13] IS 12269 1987 Indian Standard Specification for 53 Grade Ordinary Portland Cement *Bureau of Indian Standards* New Delhi
[14] IS 383 1970 Indian Standard Specification for Coarse and Fine aggregates from natural sources for concrete *Bureau of Indian Standards* New Delhi
[15] IS 516 1959 (1989) Indian Standard methods of tests for strength of concrete *Bureau of Indian Standards* New Delhi

IOP Conf. Series: Materials Science and Engineering 1025 (2021) 012018 doi:10.1088/1757-899X/1025/1/012018

Study on early age and ultimate compressive strength of M30 grade self-curing concrete

George Vincent Jasmine[1], Madisetti Pavan Kumar[2], Ponnada Markandeya Raju[3,*]

[1]Assistant Professor, Department of Civil Engineering, Raghu Engineering College, Visakhapatnam, INDIA
[2]Assistant Professor, Department of Civil Engineering, SVP Engineering College, Visakhapatnam, INDIA
[3]Professor and Head, Department of Civil Engineering, MVGR College of Engineering (Autonomous), Vizianagaram, INDIA
Email: vincent.jasmine2227@gmail.com, pavanidea@gmail.com, markandeyaraju@mvgrce.edu.in

Abstract: Curing concrete is one of the important stages in making quality concrete. Research has not given due attention to this area. Due to the scarcity of potable water and subsequent need for conservation of water, the construction industry is facing a challenge in ensuring proper and adequate curing of concrete elements. Concrete is one of thelargest used construction materials across the world, which can easily be accepted as it has changed the face of rural areas. But in the localities where there is a water insufficiency and in locations of higherelevations, water curing is difficult. There is also a need for reduced water usage for curing for the conservation of potable water for environmental sustainability. These problems can be rectified by using Self-Curing or Internal Curing technique. Self-curing admixturesconsiderablyassist in the preservation of water in concrete, by bringing down the evaporationand transpiration during the hydration of concrete. The present paper is focused to study the performance of M30 grade concrete by using different self-curingcompounds like Poly-vinyl alcohol, Liquid paraffin wax, and Poly-ethylene glycol 400 in comparison with conventional immersion curing.

1. Introduction

Concrete is one of the largest used construction materials across the world, which can easily be accepted as it has changed the face of rural areas. But in the localities where there is a water insufficiency and in locations of higher elevations, water curing is difficult. A lot of water is used for curing concrete across the globe leaving a carbon foot-print on the earth. There is also a need for reducingthe curing water usage for the conservation of potable water for environmental sustainability.

Proper curing is necessary to meet the strength and durability requirements of any structure or structural component. This can be achieved by conventional external curing applied after mixing, placing and finishing of structural components made by concrete. Conventional curing is broadly categorised into five types.

a) Immersion curing: Submersion of concrete normally done during testing of concrete specimens in the laboratory.
b) Ponding: Used where water can be effortlessly held on the surface of the concrete elements like slabs and pavements. A layer in the form of a small bund of sand or earth is kept on the border of the surface to retain water on it.

c) Fogging: Misting or Fogging is adopted in conditions where temperatures are less than the freezing point. Fogging raises and maintains the moistness over the concrete surface by showering a fine fog of water.
d) Wet Covering: Curing of hardened concrete while it is wet with a wet plastic sheet to cover it adequately without harming the solid's surface is wet covering.Covering can be done by a plastic sheet, sand, canvas, burlap or straw. The cover has to be kept wet during this entire process.
e) Plastic Sheeting: Curing concrete with plastic sheeting requires covering every uncovered region of the concrete as quickly as time permits after casting without harming the concrete shape. When the plastic sheeting is used over level surfaces, like concrete pavements or surfaces, it should be ensured that no edge is left uncovered.
f) Layer Forming Curing Compounds: Curing compounds are either mixed or sprayed on the surface of the concrete to retain moisture in concrete thereby ensuring effective hydration of cement without the need for an additional supply of water. One type of these compoundsforms an impermeable layer that impedes the loss of dampness from the concrete. Another type retains the moisture in concrete by restricting the loss of moisture due to evaporation and transpiration.

Hence, Self-curing is a method that can be used to giveextra moisture in concrete for more effective hydration of cement and reduce self-desiccation. The concrete in which the water is used for mixing is restricted inside the concrete by using some chemical admixture called “Self-Curing Concrete”.Self-curing compounds are Polymers that are added to the mix to form a hydrogen bond with water molecules that reduce the chemical potential of the water molecules. The differential free energy between vapour and liquid phases results in the uninterrupted vanishing of water content from the concrete surface exposed to the atmosphere. This results in a reduction of vapour pressure and hence the rate of evaporation and transpiration from the surface leading to adequate availability of water for cement hydration.

2. Literature review

Many researchers have worked on studying the effect of using curing compounds on various properties of concrete. Amal Viswam (2016) [1] done their study on Self Curing Concrete and observed that concrete mixes with self-curing admixtures have higher water retention when compared to those with conventionally cured mixes. The weight loss was considered as a measure of loss of water due to evaporation and transpiration. Concretemixes with self-curing admixtures show the effective hydration with time under drying state compared to concrete with conventional curing. Nithya et al (2017) [2] studied compressive strength property of self- curing concrete using a self-curing agent PVA. Maximum compressive strength of the considered mix obtained at optimum dosage of 0.5%. The slump of M30 grade concrete mix was increased with an increased percentage of PVA. From their study, it was concluded that concrete mixes with self-curing agent show good results and it can be used in the place of conventional concrete. Rajeswari Unnithan and Sarah Anil (2017) [3] conducted an experimental study on self-curing concrete with lightweight aggregate using polyethylene glycol as an admixture. From the results of the mechanical properties of concrete, it was observed that self-curing concrete resulted in higher strength than that of conventionally cured concrete mixes. Shikha Tyagi (2017) [4] compared the compressive strength property of self-cured concrete and conventionally cured concrete. For the study, they considered different concrete grades i.e., M20, M25, M30 and M40. By using curing compound LPW, maximum compressive strength was achieved which was nearly 99% of that achieved through normal curing method for all considered concrete grades. Ahmad Mustafa Sabaoon and Navinderdeep Singh (2018) [5] studied the performance of self-curing concrete made using self-curing agent, polyethylene glycol-400. From the test results, it was observed that maximum compressive strength for M20 and M25 grade concrete occurs at 1% dosage. A decrease in strength was observed at 2% dosage for both grades. Ankita Arsude and Dhawale (2018)[6] studied the characteristics of self-curing concrete by using self-curing agent Polyethylene Glycol-400 (PEG-400). They found maximum compressive strength, modulus of rupture and split tensile strength at 1% dosage of the self-curing agent. Karthick and Amrin Sulthana

(2018) [7] studied the properties of self-curing concrete using various self-curing admixtures. From the result, they found that self-curing agent Sodium Polyacrylate shows better results than Paraffin wax coating. Gaddam Sai Krishna (2018) [8] performed an experimental study on self-curing concrete by partial replacing cement by red mud and fine aggregate by quarry dust in various percentages. PEG 400 was used as a self-curing admixture. Self-curing agent dosageof 0.5% to the weight of cement recommended for M40 grade of concrete. Vaseem Akram and Balachandiran (2018) [9] studied the mechanical properties of self-curing concrete by using PVA at different percentages i.e., 0.5%, 1%, 1.5% and 2%. Mechanical properties like compressive strength and split tensile strength were studied. From the study, they found that the optimum percentage of PVA was 1.5% which gives higher strength. Soorya Tharshini and Belciya Mary (2018) [10] examined the compressive strength property of self-curing concrete. Compressive strength of self-curing concrete (SC) was increased by 15.9% when compared to conventional curing concrete (CC). Lidiya and Preethi (2018) [11] performed an experimental investigation on self-curing concrete using self-curing agent PEG-600. From the obtained results, it was found that self-curingconcrete performs better when compared to concrete mix with conventional curing. At 1.5 % dosage of PEG-600 in M50 grade concrete, obtained maximum compressive strength. Compressive strength of self-curing concrete was increased by 13% when compared to concrete mix with conventional curing. Vikram et al (2018) [12] done their study on self-curing concrete with the utilization of chemical agent. The optimum dosage of self-curing agents PEG-400 and PEG-600 was 1% weight of cement. Zameer and Karthik (2018) [13] examined the strength characteristics of self-curing concrete included with fly ash. At an optimum percentage of self-curing agents, it provides good results. An unnecessary increase in dosage of self-curing agent leads to a decrease in the strength of concrete.

In the present study, an attempt is made to compare the performance of M30 grade self-curing concrete in terms of its compressive strength by using different self-curing admixtures i.e. polyvinyl alcohol, Liquid paraffin wax and Poly-ethylene glycol 400. Early age and ultimate compressive strength of self-curing concrete with above mentioned self-curing agents are compared with conventional concrete to know the suitability.

3. Experimental Program

All the ingredients of concrete are tested for their suitability of making concrete. Table 1, 2 and 3 present the properties of the concrete ingredients used in this study.

Cement- Ordinary Portland Cement (53 grade)

Table 1. Results of tests on Cement- Ordinary Portland Cement (53 grade)

Sl. No.	Property	Test Result
1	Specific gravity	3.15
2	Fineness modulus	96%
3	Initial setting time	47 minutes
4	Final setting time	225 Minutes

Fine aggregate -Locally available river sand

Table 2. Results of tests on fine aggregate

Sl. No.	Property	Test Result
1	Specific gravity	2.79
2	Fineness modulus	2.75%
3	Bulking of sand	25%

Coarse aggregate - Crushed angular coarse aggregate

Table 3. Results of tests on coarse aggregate

Sl. No.	Property	Test Result
1	Specific gravity	2.78
2	Nominal max. Size	20mm

Water - Locally available potable water

Chemical Admixtures - Self-curing admixtures used for the present study are as follows.

1. Poly Ethylene Glycol (PEG) [0.5%]
2. Liquid Paraffin Wax (LPW) [1%]
3. Poly Vinyl Alcohol (PVA) [1.5%]

Mix proportions of Self Curing Concrete

Table 4 presents the mix proportions for 'M30 grade concrete with a workability of 30 mm slump' in terms of quantities of materials in kg per 1 cum

Table 4. Mx proportions

Cement	Fine aggregate	Coarse aggregate	Water	Water/cement ratio
438.13 kg	603.84 kg	1149.48 kg	189 litres	0.4

4. Results and Discussions

This section presents the comparison of compressive strength and workability results for all the four mixes

4.1 Workability

Minimum workability of about 30 mm is required assuming the concrete is used for making VRCC (Vibrated Reinforced Concrete). Table 5 shows the workability test results from the slump cone test. From the results, it is observed that there is an increase in the slump value by 33.33% for M30 grade concrete with PVA [0.5%] and PEG400 [1.5%] when compared to the conventional concrete mix. In the case of M30 grade concrete with LPW [1%], there is no variation of the slump. The improvement in workability after adding the curing compounds can be attributed to the restricted loss of moisture in the fresh concrete mix.

Table 5. Results of the slump cone test

Mix Designation	Mix Description	Slump (mm)
CM	M30 Conventional mix	30
SCC-PVA	M30 mix with Poly Vinyl Alcohol [0.5%]	40
SCC-LPW	M30 mix with Liquid Paraffin Wax [1.0%]	30
SCC-PEG	M30 mix with Poly Ethylene Glycol - 400 [1.5%]	40

4.2 Compressive Strength

The standard cube samples of 15 cm size are tested in 3000 kN Compression Testing Machine. The cube samples each for the various percentage of LPW, PVA, PVG-400 were prepared to determine the 7 days,14 days, and 28 days compressive strength. Table 6 shows the compressive strength results for

7, 14 and 28 days by using PVA, LPW and PEG-400. Based on the results graphs are plotted and shown in Figures 1, 2 and 3.

Table 6. Compressive strength results of various mixes at 7, 14- and 28-days curing period

Mix Designation	Mix Description	Compressive strength N/mm²		
		7 days	14 days	28 days
CM	M30 Conventional mix	26.16	35.55	44.29
SCC-PVA	M30 mix with Poly Vinyl Alcohol [0.5%]	33.77	35.73	40.54
SCC-LPW	M30 mix with Liquid Paraffin Wax [1.0%]	28.32	30.44	40.85
SCC-PEG	M30 mix with Poly Ethylene Glycol - 400 [1.5%]	31.77	35.2	40.48

Figure 1. presents the compressive Strength of M30 grade concrete with conventional curing and self-curing agents at 7 days curing period.

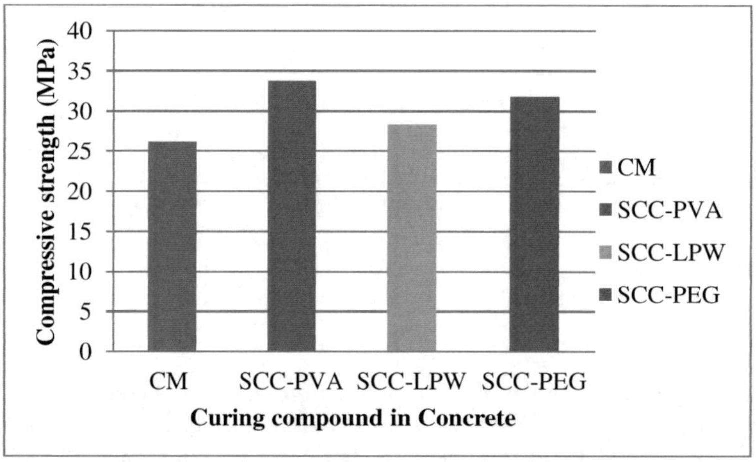

Figure 1. Compressive Strength of M30 grade concrete with conventional curing and self-curing agents at 7 days curing period

From figure 1, it is observed that when compared to the conventional method of curing, all three curing compounds improve the compressive strength of concrete after 7 days. There is an increase in the compressive strength by 22% for Poly-vinyl alcohol, 7% for Liquid paraffin wax and 17% for poly-ethylene glycol 400. Hence, adding Poly-vinyl alcohol as curing compound to concrete will result in highest early age strength followed by Poly Ethylene Glycol - 400 and Liquid Paraffin Wax respectively.

Figure 2. presents the compressive Strength of M30 grade concrete with conventional curing and self-curing agents at 14 days curing period.

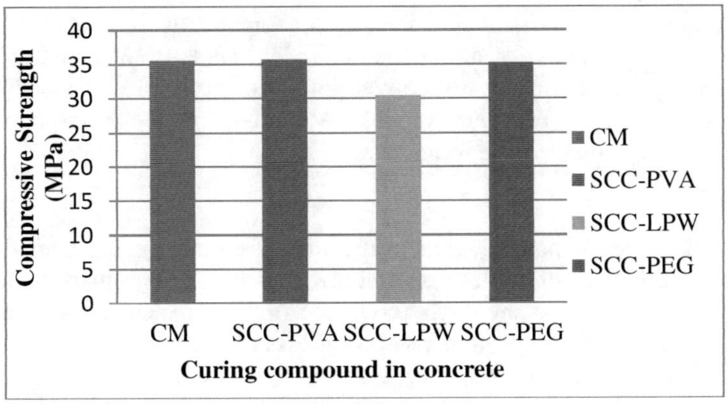

Figure 2. Compressive Strength of M30 grade concrete with conventional curing and self-curing agents at 14 days curing period

From the figure 2, it is observed that when compared to the conventional method of curing there is an increase in the compressive strength by 0.5% for Poly-vinyl alcohol, and decrease in the compressive strength by14% for Liquid paraffin wax and 0.98% for poly-ethylene glycol 400. The marginal increase in 14 days strength for Poly-vinyl alcohol is observed because it has already attained maximum strength in 7 days. Further, although there is a decrease in the compressive strength by14% for Liquid paraffin wax and 0.98% for poly-ethylene glycol 400 when compared with compressive strength of conventional mix at 14 days, their strengths are more than their respective 7 days strength. This means that the curing compound is working and the mixes have been continuously gaining strength.

Figure 3. presents the compressive Strength of M30 grade concrete with conventional curing and self-curing agents at 28 days curing period.

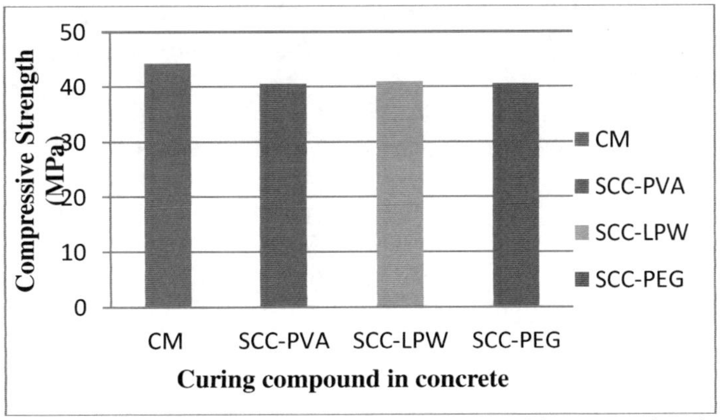

Figure 3. Compressive Strength of M30 grade concrete with conventional curing and self-curing agents at 28 days curing period

From the figure 3, it is observed that when compared to the conventional method of curing there is a slight decrease in the compressive strength by 8% for Poly-vinyl alcohol, by 7.76% for Liquid paraffin wax and 8.6% for poly-ethylene glycol 400.The conventional method of curing i.e. immersion of specimens in potable water at room temperature results in maximum ultimate compressive strength when compared to the concretes with various curing compounds used in this study. However, the strengths of concrete with all the three curing compounds resulted in the attainment of the almost same strength and beyond target mean strengths. Further, in the construction industry, immersion curing of concrete elements likebeams, columns and slabs is not practically possible and water for curing is sprayed twice or thrice a day on gunny bags wound around these elements. Hence, the curing compound is a very good alternative to conventional curing and will surely result in strengths acceptable. The use of curing compounds should also be encouraged in the construction industry in light of the urgent need for potable water conservation.

5. Conclusions

From the study, it has been concluded that usage of self-curing agents doesn't show considerable reduction in the compressive strength property. Hence, it is recommended to adopt self-curing concrete to conventional curing concrete where we can find an inadequacy of water and the pumping of water is difficult (i.e., in the case of high-rise buildings).

Even though the observed compressive strength values are reduced for the considered self-curing agents when compared to the conventional concrete, but these values achieved the target compressive strength. Among all the considered self-curing agents, Liquid paraffin wax (LPW) performs better at 28 days curing period. The use of curing compounds should also be encouraged in the construction industry in light of the urgent need for potable water conservation.

References

[1] Amal Viswam, Arjun Murali 2016 Review on the Study of Self Curing Concrete *International Journal of Advanced Research and Innovative Ideas in Education* **4 (1)** p 271-4

[2] K Nithya, K Ranjitha, P Kumar 2017 Experimental Study on Self-Curing Concrete *International Research Journal of Engineering and Technology* **4 (3)** p 1758-62

[3] Rajeswari S Unnithan, Sarah Anil 2017 Review on Self Curing Concrete *International Research Journal of Engineering and Technology* **4 (5)** p 1022-4

[4] Shikha Tyagi 2017 Comparative Study on Compressive Strength of Self Cured Concrete and Normally Cured Concrete *International Research Journal of Engineering and Technology* **4 (8)** p 2190-3

[5] Ahmad Mustafa Sabaoon, Navinderdeep Singh 2018 A Review Paper on Self Curing Concrete *International Research Journal of Engineering and Technology* **5 (1)** p 745-7

[6] Ankita Arsude, Ass W Dhawale 2018 An Overview on Characteristics of Self-Curing Concrete Using Polyethylene Glycol - 400 (PEG) *International Journal of Engineering Science Invention* **7 (3)** p 8-11

[7] Karthick R, Amrin Sulthana I 2018 An Experimental Investigation on Self-Curing Concrete Using Different Curing Agents *International Research Journal of Engineering and Technology* **5 (3)** p 429-32

[8] Gaddam Sai Krishna 2018 Experimental Study on SelfCuring Concrete with Red Mud and Quarry Dust *International Journal of Engineering Research and Management* **5 (3)** p 7-13

[9] Vaseem Akram N, Balachandiran P 2018 Experimental Study of Self Curing Concrete by Using Poly Vinyl Alcohol *International Journal of Scientific & Engineering Research* **9 (3)** p 102-5

[10] Soorya Tharshini C, Belciya Mary B 2018 Strength Characteristics of Selfcuring Concrete *International Journal of Civil Engineering and Technology* **9 (4)** p 1612-17

[11] Lidiya, Preethi M 2018 Mechanical Properties of Self Curing Concrete incorporating Polyethylene Glycol-600 *International Research Journal of Engineering and Technology* **5 (6)** p 534-89

[12] Vikram M B, Vijay B V, Deepak Kumar, Rashmi R Ghali 2018 Utilization of Chemical agent in Concrete for Self Curing*International Research Journal of Engineering and Technology* **5 (5)** p 1897-901

[13] Zameer K, Karthik N M 2018 Study on Strength Characteristics of Self Curing Concrete incorporated with Fly Ash*International Research Journal of Engineering and Technology* **5 (5)** p 1429-34

A review on sustainable concrete mix proportions

Kanta Naga Rajesh[1, 2], Ponnada Markandeya Raju[3], Kapileswar Mishra[4], Pavan Kumar Madisetti[5]

[1]Research Scholar, Centurion University of Technology &Management, Paralakhemundi, Odisha-761211, India
[2]Assistant Professor, Department of Civil Engineering, GMRIT, Rajam, Andhra Pradesh-532127, India
[3]Professor & Head, Department of Civil Engineering, MVGR College of Engineering (Autonomous), Vizianagaram, Andhra Pradesh-535005, India
[4] Professor, Department of Civil Engineering, Centurion University of Technology & Management, Jatni, Odisha-752050, India
[5]Assistant Professor of Civil Engineering, SVP Engineering College, Visakhapatnam
Email: markandeyaraju@mvgrce.edu.in, rajeshkanta@gmail.com

Abstract. The need for reducing the carbon footprint and effect of hazardous gases, etc., released from the construction sectors and other product manufacturing industries on the environment is the need of the hour. In this regard, the construction sector has been utilizing some of the wastes generated from various sources in making concrete. Notable among them is the Ground granulated blast furnace slag (i.e. GGBS) as a partial substitution of cement. Further, waste foundry sand, a by-product of the foundry industry is observed to be an appropriate alternative to partial replacement of fine aggregate and demolition waste as a partial replacement of coarse aggregate. These waste utilizations have addressed some of the global environmental concerns like solid waste management, scarcity of landfilling, sustainability etc. but there is still is inefficient, given the rate at which we are producing waste. This paper presents an overview of some of the recent research published on the utilization of wastes in the production of sustainable RMC/SMC. It also presents the effect of waste utilization in concrete on mechanical, microstructural and durability factors.
Keywords: Ready-mixed concrete, Green concrete, recycled aggregates, Demolition waste, Sustainable concrete.

1. Introduction

Ready Mix Concrete (RMC) is concrete made by simultaneously mixing the ingredients of concrete and transporting the same to work location while maintaining the required workability by adding admixtures and superplasticizers. The time of haul is restricted to 2 hours of the time of loading [1]. RMC is the next widely used construction material when compared with conventional SMC. Worldwide there is a huge concern on reducing the environmental pollution viz. greenhouse gases and the contribution of the construction sector in this aspect is important as most of the industrial wastes can be used as a partial replacement of various ingredients of concrete. Cement Fine Aggregate (FA) and Coarse Aggregates (CA) along with water and admixtures (if necessary) are basic ingredients on concrete. Cement hydration that results in strength attainment of concrete which is accountable in emitting CO_2 into the environment and atmosphere. It is established that 1 Ton production and use of cement in the construction activities emit CO_2 of about 0.9 Ton [2, 3]. The second place in releasing of CO_2 is the construction industry next to the automotive sector. Natural aggregates like fine aggregates (FA)such as Sand and Coarse Aggregate (CA) are reducing at a faster rate on account of abundance

IOP Conf. Series: Materials Science and Engineering 1025 (2021) 012019 doi:10.1088/1757-899X/1025/1/012019

consumption in concrete manufacturing. Due to this, there is an immense outlook for research in looking for substitutes of concrete making constituents. As per the literature study, industrial wastes such as slag (GGBS), waste glass powder, fly ash, silica fume, etc. other than above rice husk ash, sawdust ash, pulverized fuel ash, etc which are known as Supplementary Cementitious Materials (SCM's). Similarly, FA can be partially replaced with industrial secondary-products like Used Foundry Sand (UFS), M sand (MS), Marble Powder (MP) etc. CA can be partially replaced with Plastic aggregates (HDPE), Rubber Waste (RW), construction debris, Demolition Waste (DW), Recycled Concrete Aggregates (RCA) etc [4,5]. The suitability of replacing an ingredient of concrete with a waste material depends on the extent of compatibility of their physical and chemical properties. The results of research on the performance of concrete with these replacements initiated the utilization of green materials in the manufacture of cement concrete so that subsidizing to the decrease of environmental contamination. This review paper aims to assess the feasibility of manufacturing RMC with maximum use of green materials.

2. Background

The first RMC industry was established in Germany in the year 1903, whereas the first product of RMC was produced in Turkey in the year 1906 [6]. The first commercial product was supplied in Baltimore, the USA in the year 1913. In the year 1926, a first of its kind small transit mixer of revolving type was used for supplying RMC. Later in the 20th century, Europe and the USA have seen remarkable growth in this industry. Currently, USA is consuming 75% and Europe is consuming 47% of cement through RMC plants[7]. In India, RMC utilization was started in the year 1950 for only mega or large-scale projects. Due to rapid growth in industrialization, infrastructure and housing projects, the need for quality, quantity and faster supply of concrete was increasing. Meeting this demand was possible only through RMC. The first commercial RMC plant in India was set up in Pune in 1992 [7]. Over the years, the RMC industry has expanded beyond metropolitan cities to smaller cities and towns. Presently it is estimated that in India, around 1000 RMC plants are producing annually 35-40 mm^3 of concrete. The projected growth of this industry for the next five years is expected to be above 7%. The existing RMC plants in India are modernized with the automated control system as shown in figure 1 [8].

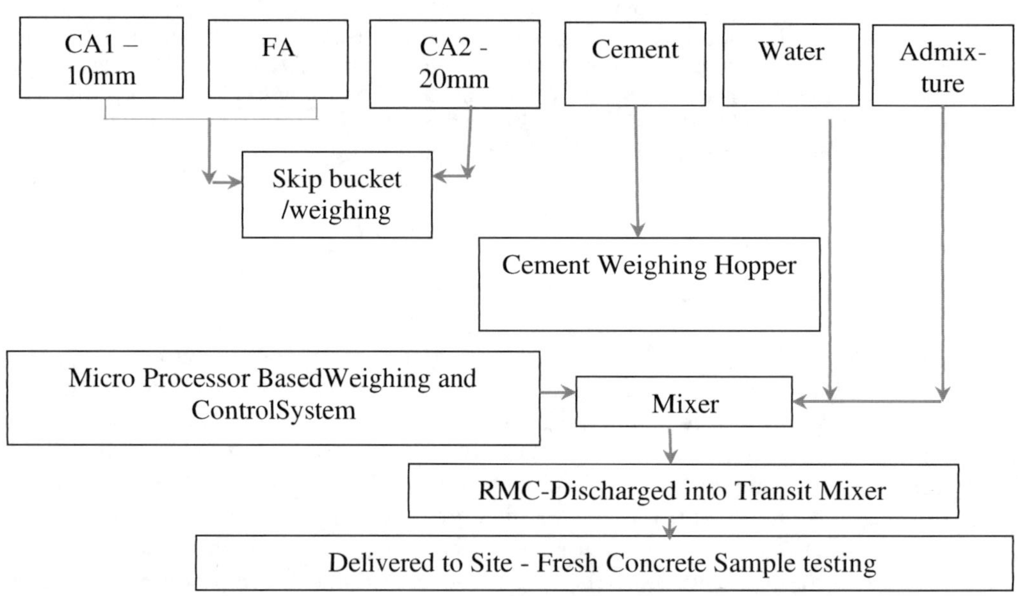

Figure 1. Flow Chart-Manufacturing Process of RMC.

IOP Conf. Series: Materials Science and Engineering 1025 (2021) 012019 doi:10.1088/1757-899X/1025/1/012019

3. Review of Literature

A review of some recent research on utilization of industrial by-products and manufacture of RMC is presented in this section.

3.1. Fly ash, GGBS and Bottom ash as Partial Replacement of Cement

Figure 2. Fly ash.

Figure 3. GGBS.

Some of the by-products in the steel manufacturing industry are fly ash and ground granulated blast furnace slag (GGBS)and shown in figures 2 and 3 respectively. In the Netherlands, due to Greendale initiation, Geopolymer concrete (GC) was manufactured in RMC plant with a combination of GGBS and fly ash, FA and CA which were activated separately with powdered alkalis of sodium silicate and sodium hydroxide called as dry mix and with liquid alkalis called as wet mix. Both were compared with the control mix. It was found that dry binder has good results when compared with the wet binder. After 91 days, the dry binder has shown 1.12 times higher strength than the wet binder which has shown 0.04 times lesser compressive strength than reference concrete. The modulus of elasticity of dry binder is 0.81 times less and wet binder is 0.75 times less than reference concrete. Production of GC in conventional RMC plant by mixing, transporting and casting was successful with minor changes. It was also found that the adopted design mix was workable up to 90 minutes. However, even a single minute delay resulted in difficulty in placing the same. Adjustments may be required in the conventional durability factors while adopting them for special concretes like GC. European codes have not yet standardized the regulations for the designs and construction of structures with concrete made with geopolymer as the binder. It is suggested to use this new type of concrete for the product-based elements like precast slabs, precast fence posts, Pavement blocks, tiles etc [9].

GGBFS was added as a part of binder in geopolymer concrete based fly ash. The optimum mixtures of concrete specimens containing 10% GGBFS + 90% Fly ash activated by 40% Na_2SiO_3 (of total binder content) and 14M NaOH alkali activator solution (with a ratio of Na_2SiO_3/NaOH 1.5-2.5 without adding any extra water) showed moderate compressive strength under ambient curing. The setting time of binder was comparably same to that of conventional concrete. When Fly ash was replaced with Slag up to 30%, the compressive strength of this geopolymer concrete was nearly 55 N/mm^2, and mortar strength was up to 63 N/mm^2 after 28 days curing. Beyond this replacement, reduction in strength compared to conventional concrete was observed [10]. Cement with 'GGBS & bottom ash' and sand with 'foundry sand' (figures 4a & 4b)were partially replaced and tested for mechanical and durability properties. Low concentrations of NaOH i.e., 4M showed satisfactory results under ambient curing. With an increase in density, the grade of concrete, replacements percentages, splitting tensile and flexural strength increased whereas water absorption, weight loss for sulphate and chloride attack reduced [11]. The ready-mix mortar was prepared with three types of cement to study the effect of cement and water content on the plastic or fresh state properties. For maintaining the longer duration of workability for up to 72 hours, the mortar was mixed with hydration stabilizing admixture and air-entrained admixture. The ratio of water to dry material had a substantial influence on the air entrainment but the cement types in ready mix mortar and its storage duration have not shown any effect on the same. The consistency of the mix increases with increase in the ratio of water to dry materials. There was no effect for the type of cement used on the consistency index for 12% water to dry materials ratio. The consistency of the mix decreases with increase in storage time. With the increase in cement fineness and storage of the mix, the maximum displacement observed by squeeze flow test increased. The rheological properties of the mix were changed with storage time and cement types. It was also noticed that there was no change in air entrainment and consistency properties [13].

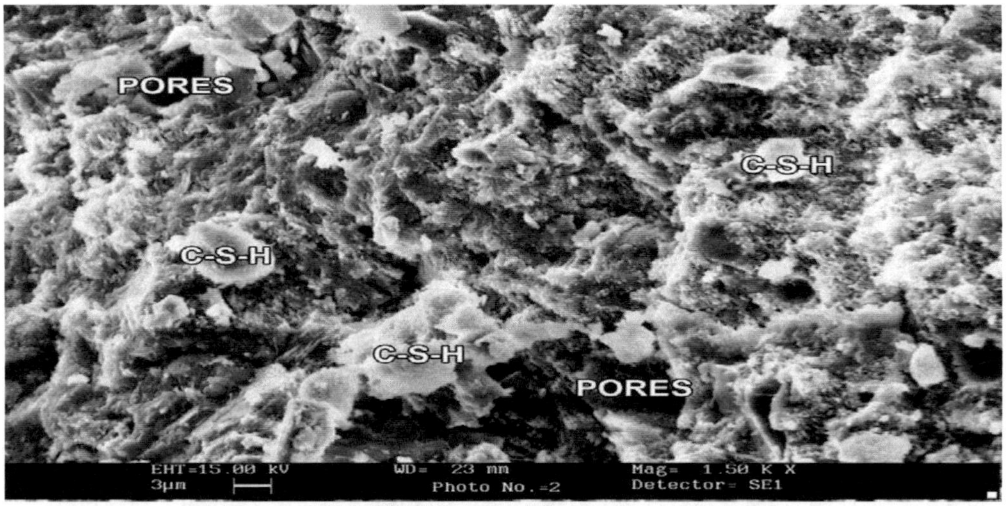

Figure 4. (a) SEM Image ofConventional Mix [12].

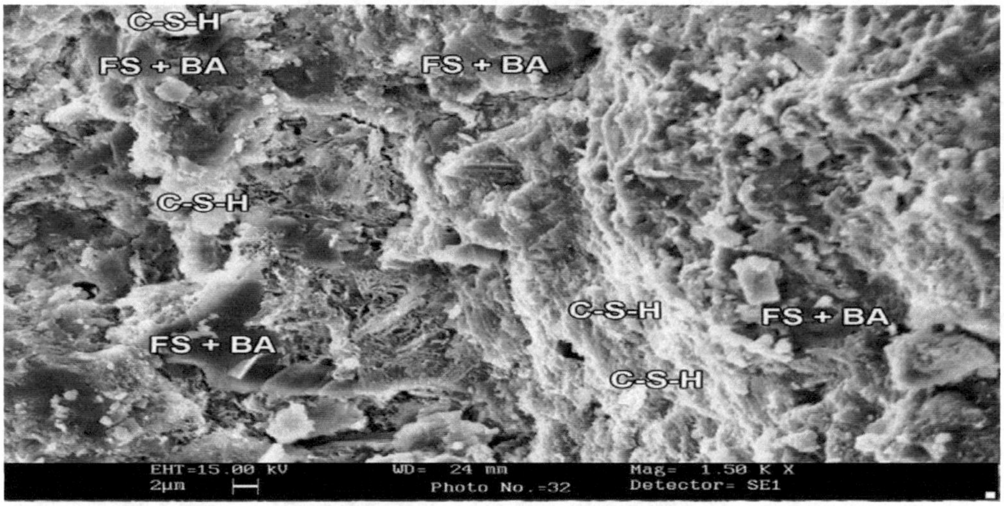

Figure 4.(b) SEM Images of Non-Conventional Mix [12].

Ready-mix alkali-activated cement was produced by treating construction waste thermally up to 1200°C and activating it with NaOH solution. This thermally treated construction waste is very stable and can be converted into amorphous form, which when reacts with water forms hard material similar to conventional cement with appreciable compressive strength. With the increase in temperature and NaOH%, the performance of ready-mix alkali-activated cement also increases. It is an eco-friendly material and white in colour. Due to this, it can be used for decorative works and in insignificant projects [14].

3.2.Waste Foundry Sand, Dredged Material and Fine Recycled aggregate as Partial Replacement of Fine Aggregate

Waste foundry sand (WFS) is the by-product of the foundry industry. After repeated use of high-quality silica in the foundry process, it is discarded and sometimes used as landfills. It was showed that it can also be used as a replacement of fine aggregate [15]. WFS was used in the manufacturing of RMC by replacing sand at various percentages by its weight. Solidification and stabilization were done for all concrete mixtures by which the properties were improved. The mechanical, leaching and microstructural properties of WFS-RMC's final product was studied as per the standards of Turkey and environmental concerns. When compared with control mix, it was observed that as per the Turkey standards "TS 706 EN 12620+A1:2009-Coarse aggregates" [16], WFS has shown similar physical, geometrical and chemical properties on par with conventional sand. When WFS was partially replaced with regular sand, there was a decrease in strength properties when compared with the control mix. Optimum results in compression and splitting tensile strengths were observed up to 20% replacement. With the increase in replacement percentages, water absorption increased, but the density of the mix decreased. Non-conventional mixes cured for 56 days showed results similar to 90 days. The eluate concentrations (Ni, Zn, Cr, F, TDS, DOC, TOC)of WFS treated with solidification and stabilization under different pH conditions were incompatible with EULFD limits. 20% WFS mix and control mix are almost same in the microstructural and morphological properties. To utilize the WFS in RMC, its replacement should be 20% by weight of regular sand. Further, it should not show any adverse effect on mechanical, microstructural and environmental factors. It was also observed that WFS properties may vary from foundry to foundry. It was also found that corrosion effect has to be studied when WFS is used [17]. Dredging is the process of removing silt and other sediment materials from the water bodies. The removed material is called as Dredged Material (DM). DM samples were dredged from Istanbul Ambarlı Port (DM1), Mersin Erdemli Fishery Harbour (DM2), _Izmir PETK_IM Container Port (DM3), and Samsun Port (DM4). This DM was used as fine aggregate by treating and without treating the same at various percentages (0%, 25%, 75%, 100%) in RMC as shown in figures 5a & 5b. As per the Turkey standards (17 05 06), DM was identified as "non-hazardous waste". The leaching characteristics of DM according to "ADDDY – Appendix 2: The acceptance criteria of landfilling waste" [18], were above the limits of Class III – inert waste as landfill material. The fresh and hardened state properties of treated and untreated DMs of concrete were compared and found that treated DM1 and DM2 performed better than treated DM3 and DM4, the untreated DMs also showed the results like treated DMs. Due to this, the treated and untreated DMs were transformed to form composite treated and composite untreated DMs. The composite treated (untreated) DMs consists of treated (untreated) DMs of 40% each of DM1 and DM2 and 10% each of DM3 and DM4 were mixed as per Turkey's gradation distribution. As per "TS 706 EN 12620 + A1:2009" [16], the composite treated DMs showed that the fine aggregate properties are similar to silica sand. From the experimental results, it was observed that RMC2 (50% Composite untreated) and RMC8 (100% composite treated) had shown minimum strength as per the requirement.

The composite treated DMs had shown better physical and mechanical properties than composite untreated. On testing the RMC0 (control RMC), RMC2 (SEM showed in figures 6a & 6b) and RMC8 when tested for durability, it was found that higher w-c ratio results in higher permeable voids and permeability.

The leachable properties (eluate concentrations) of RMC2 and RMC8 are observed to be under limits according to "ADDDY-Appendix 2" [18]. The RMC2 had shown similar chemical and mineralogical properties when compared with RMC0 than RMC8. This may be due to the low percentage of replacements. From this study, the author concluded that treated DM can be used as a partial replacement of fine aggregate in RMC or concrete manufacturing locations [19].

IOP Conf. Series: Materials Science and Engineering 1025 (2021) 012019 doi:10.1088/1757-899X/1025/1/012019

Figure 5(a). SEM image of silica sand [19].

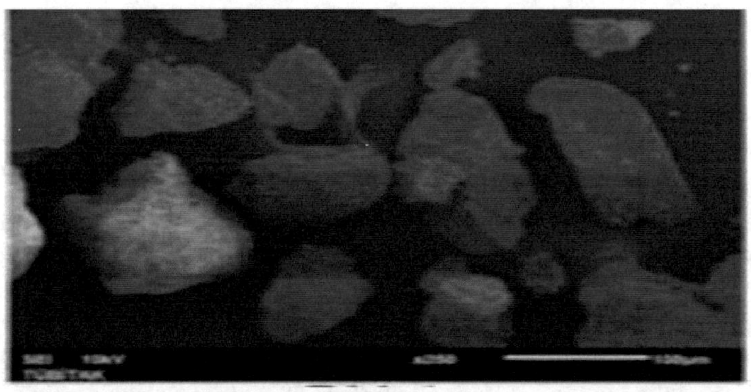

Figure 5(b). SEM image of Dredged material Samples [19].

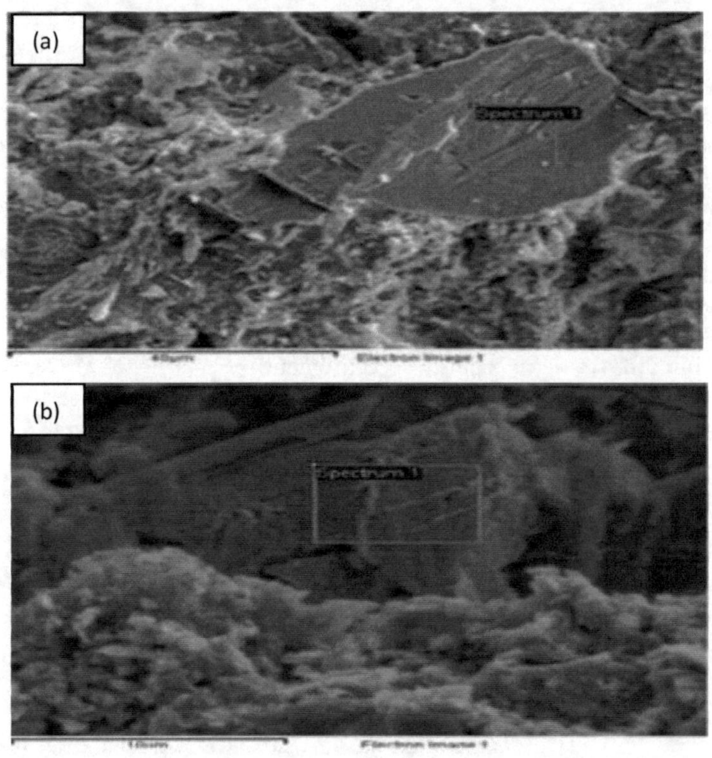

Figure 6. SEM images/EDS spectra of (a) RMC-0(b) RMC 2 concrete [19].

Fine Recycled aggregate (FRA) was collected from the demolition and construction wastes of the structures. In this study, two types of FRA were used as shown in figures 7a & 7b. One is FRAL (Fine Recycled Aggregate with Lower water absorption ratio of 5.83%) and FRAH (Fine Recycled Aggregate with Higher water absorption ratio of 7.95%). Conventional sand was replaced with FRAL at various percentages (0, 30, 60, 100) and FRAH with 35 and 70% and their effect was studied when used in the RMC. From the experimental study, it was observed that with the increase in replacements, the slump of the RMC was reduced and regardless of FRA replacements and its types, the air content of the RMC was almost similar to that of conventional concrete. With an increase in the replacement percentages of FRA, there was a decrease in compressive strength. As per the Korean standards "KS F 2573" [20], and replaced with 100% FRA, there was a 13% decrease in compressive strength when compared with conventional concrete. With 100% FRAL and 70% FRAH, no significant effect was observed on the splitting tensile strength and modulus of rupture when compared with the conventional mix. With an increase in replacements when compared with conventional concrete, the shear and flexural capacity of RCC beams have not shown any significant decrease. However, deflection values noted during flexure test increased. There was a decrease in 28 days of bond strength also. It was concluded that a 30% replacement of FRA in RMC gives moderate compressive strength [21].

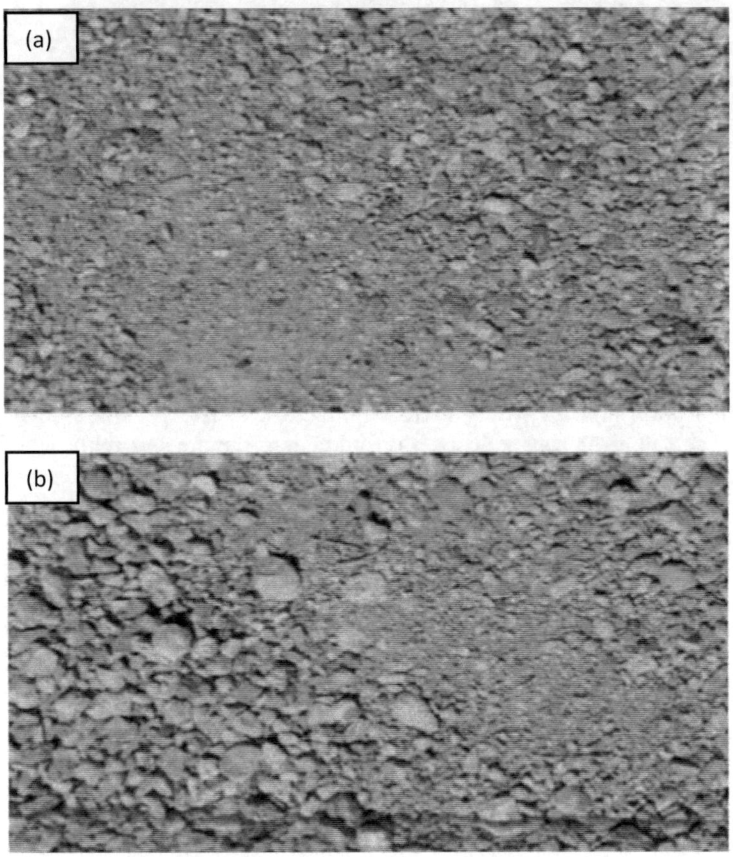

Figure 7. Images of Recycled Fine Aggregates (a) FRAL (b) FRAH [21].

In this study, Fly ash bricks were manufactured with cement, sand, sludge lime and gypsum. NCF50 bricks showed better results when compared with NCF100 in the compressive strength, and water absorption. There is no sign of efflorescence in NCF50 and NCF100. With the help of these results, conventional sand can be partially replaced with waste foundry sand up to 50% [22].

3.3. Construction Demolition Waste, RMC Waste as Partial Replacement of Coarse Aggregate

In 1999, an earthquake occurred in Taiwan and generated approximately 20 million tons of construction demolition wastes (CDW). The demolished clay brick and tiles (CBT) out of this waste were used as a replacement of coarse aggregate. The RAC (Recycled Aggregate Concrete) made with washed CBT has shown higher values of the slump, compressive strength and flexural strength when compared with unwashed CBT. The low slump value of unwashed CBT is due to impurities present in it[23]. Conventional coarse aggregate was replaced with two kinds of waste generated, one is the initial waste of RMC plant collected from Hormigones Ebro (HE) Company and other was demolition waste from Hormigones Rioja (HR) as shown in figure 8. It was found that HE wastes concrete has yielded lesser strength, whereas HR demolition waste concrete was similar to conventional coarse aggregate concrete. With the prior treatment of HR aggregates, it was shown that up to 50% can be utilized in concrete production by replacing natural aggregates [24].

Figure 8. Image of Recycled Coarse Aggregate derived from concrete waste [24].

To estimate the environmental impact of concrete structures, a wide-range study was carried out on the life cycle assessment of the same grade of recycled concrete and ordinary concrete. Both are commercial products. It was found that recycled concrete has less impact on the environment when compared with conventional concrete. Carbon emissions are slightly less for 1cum of recycled concrete when compared with control concrete [25]. The durability parameters viz. volume of permeable voids (VPV), water absorption and sorptivity of ready mixed concrete of grades G40, G45 and G50 are studied. It was reported that the water absorption percentage increased with decrease in grade of concrete with respect to its curing condition and age. G40 concrete that underwent 28 days S2 (site cured) curing on top zone has 7.12% water absorption value. In the case of G50 concrete that underwent 90 days S1 (sealed cured) curing, the bottom zone recorded the lowest value of 3.59%. G40 has 13.50% VPV when compared with G50s VPV that has 6.17% on the bottom zone. G40 left zone has the highest sorptivity of 0.123 $g/mm^2 /\sqrt{t}$ for 28 days curing and 0.038 $g/mm^2 /\sqrt{t}$ on the bottom zone of G50 concrete cured for 90 days. So, the Performance indicators play a vital role in RMC transported to the site [26]. The effect of sustainable materials in the production of micro-concrete is reviewed [27]. Nowadays conventional concrete is manufacturing with wastes generated from demolition and construction which in turn reduces the environmental pollution [28].

4. Conclusion

1. The combination of GGBS and fly ash can partially replace cement in the production of sustainable concrete mixes.
2. For adopting in mix design, the workability of geopolymer Ready Mix Concrete (RMC) is to be retained to up to 90 minutes.
3. When Fine aggregate was partially replaced with Waste Foundry Sand (WFS), it adversely affected the workability and water absorption in the Site Mix Concrete (SMC).
4. WFS up to 20% replacement resulted in similar strength properties as that of control conventional concrete.

5. For minor works where strength required is nominal, WFS can be used as a partial replacement to produce low-strength materials without any major modification or alteration to the SMC.
6. Treated dredged material (DM) can be used as a partial replacement of sand without significant the effect in physical, mechanical, environmental, microstructural properties of RMC.
7. For moderate compressive strength, construction demolition waste (CDW) can be used up to 30% as coarse aggregate replacement in RMC.
8. CDW's washed clay brick tile when used as a coarse aggregate replacement showed a higher slump and improved mechanical properties when compared with unwashed clay brick tiles.
9. Natural coarse aggregates can be replaced with Hormigones Rioja (HR) aggregates up to 50% in concrete production.
10. Using these wastes in concrete production, showed a significant reduction in carbon emissions.

5. Scope of the Study

Many of the research studies on sustainable green materials exhibited acceptable performance in terms of strength and durable properties of concrete. On the other hand, there is a requirement for further research work in finding the impact of utilizing sustainable materials on dynamic properties of new-age concretes like sustainable green concrete. Also, the life cycle assessment has to be done based on real-time data on specific and existing market products. Furthermore, the scope of sustainable materials in the production of RMC and SMC needs to be studied extensively.

References

[1] IS 4926 2003 Ready-Mixed Concrete - Code of Practice Second Revision *Bureau of Indian Standard New* Delhi India:
[2] EPA (Environmental Protection Agency) 2010 *Available and Emerging Technologies for Reducing Greenhouse Gas Emissions from the Portland Cement Industry* Washington D.C.
[3] USGS (US Geological Survey) 2005 *Background Facts and Issues Concerning Cement and Cement Data* Reston VA
[4] Vinita Vishwakarma, Ramachandran D 2018 Green Concrete mix using solid waste and nanoparticles as alternatives – A review *Construction and Building Materials* **162** p 96–103
[5] Liew K M, Sojobi A O, Zhang L W 2017 Green concrete: Prospects and challenges, *Construction and Building Materials* **156** p 1063–95
[6] Karakule F, Akakın T, Ucar S 2004 Ready-mixed concrete industry in Turkey and in the world, *In: Proceedings of concrete congress*-Istanbul
[7] *Ready Mixed Concrete*-PWD Handbook Chapter No. 41 2011 Maharashtra India
[8] *Criteria for Production Control of Ready-Mix Concrete for RMC Capability Certification* 2013 Building Materials & Technology Promotion Council, Ministry of Housing & Urban Poverty Alleviation *Government of India* New Delhi
[9] Peter De Vries Geopolymer Concrete Ready Mixed-A Challenge 2015 Springer International Publishing AG D.A. Hordijk and M. Luković (eds.), *High Tech Concrete: Where Technology and Engineering Meet*
[10] Pradip Nath, Prabir Kumar Sarker 2014 Effect of GGBFS on setting, workability and early strength properties of fly ash geopolymer concrete cured in ambient condition *Construction and Building Materials* **66** p 163-71
[11] Janardhanan Thaarrini, Venkatasubramani Ramasamy 2016 Properties of foundry sand, ground granulated blast furnace slag and bottom ash based geopolymers under ambient conditions, *Periodica Polytechnica Civil Engineering* **60 (2)** p 159-68
[12] Yogesh Aggarwal and Rafat Siddique 2014 Microstructure and properties of concrete using bottom ash and waste foundry sand as partial replacement of fine aggregates *Construction and Building Materials* **54** 210–23

[13] Casali J M, Melo F D, Serpa V C, Oliveira A L de, Betioli A M, Calçada L M L 2018 Influence of cement type and water content on Fresh state properties of ready-mix mortar *Ambiente Construído* Porto Alegre **18 (2)** p 33-52

[14] Abdel-Gawwad H A, Heikal E, El-Didamony H, Hashim F S, & Mohammed A H 2018 Recycling of concrete waste to produce ready-mix alkali-activated cement *Ceramics International* **44 (6)** p 7300–04

[15] Bhardwaj B, Kumar P 2017 Waste foundry sand in concrete: A review *Construction and building Materials* **156** p 661–74

[16] TS 706 EN 126202009 *Aggregates for concrete* Ankara: Turkish Standards Institution (In Turkish)

[17] Merve Basar H and Nuran Deveci Aksoy2012 The effect of waste foundry sand (WFS) as partial replacement of sand on the mechanical, leaching and microstructural characteristics of ready-mixed concrete *Construction and Building Materials* **35** p 508–15

[18] Anonymous*The Regulation on the Landfilling of Waste*, Official Gazette no.: 27533 2010 *Ministry of Environment and Urbanization Ankara* Turkey (In Turkish)

[19] Pembe Ozer-Erdogan, Merve Basar H, Ibrahim Erden and Leyla Tolun 2016 Beneficial use of marine dredged materials as a fine aggregate in ready-mixed concrete: Turkey example *Construction and Building Materials* **124 p** 690–704

[20] KATS (*Korea Agency for Technology and Standards*) 2011 http://www.kssn.net/english Seoul Korea

[21] Jang S J, Yun H D 2015 Mechanical properties of ready-mixed concrete incorporating fine recycled aggregate *Magazine of Concrete Research* **67 (12)** 621–32

[22] Naga Rajesh K, Srinivasa Rao R 2017 Effect on Replacement of Conventional Sand with Used Foundry Sand in Fly ash Bricks *International Journal for Research in Applied Science & Engineering Technology* **5 (XI)** p 1670-79

[23] Sheen Y N, Wang H Y, Juang Y P, Le D H 2013 Assessment on the engineering properties of ready-mixed concrete using recycled aggregates *Construction and Building Materials* **45 p** 298–305

[24] Fraile-Garcia E, Ferreiro-Cabello J, López-Ochoa L, López-González L 2017 Study of the Technical Feasibility of Increasing the Amount of Recycled Concrete Waste Used in Ready-Mix Concrete Production *Materials* **10** p 817

[25] Kleijer A L, Lasvaux S, Citherlet S, Viviani M 2017 Product-specific Life Cycle Assessment of ready-mix concrete: Comparison between a recycled and an ordinary concrete Resources *Conservation and Recycling* **122** p 210-18

[26] Matarul J, Mannan M A, Safawi M Zb M I, Ibrahim A, Jainudin N A, Yusuh N A 2016 Performance-based Durability Indicators of Different Concrete Grades Made by the Local Ready Mixed Company: Preliminary Results *Procedia - Social and Behavioral Sciences* **224** p 620 – 25

[27] Naga Rajesh K, Manoj Kumar Rath, Markandeya Raju P 2019 A Research on Sustainable Micro-Concrete *International Journal of Recent Technology and Engineering* **8** p 1137-39

[28] Markandeya Raju Ponnada, Kameswari P 2015 Construction and Demolition Waste Management – A Review *International Journal of Advanced Science and Technology* **84 p** 19-46

Influence of glass fibres in stone mastic asphalt

V Udaya Bhanu[1], NVLN Pavan Kumar[2]

[1]Assistant Professor, Civil Engineering, SVCET, Srikakulam, Andhra Pradesh, India.
[2]Assistant Professor, Civil Engineering, SVCET, Srikakulam, Andhra Pradesh, India.
E-mail: bhanu.12256@gmail.com

Abstract. The study in this paper examines the effect of fibres on the stability of Stone Mastic Asphalt (SMA) mixture in flexible pavement. Two types of fibres namely Class-C Glass fibres(Glass – C), and Alkali Resistant Glass Fiber(ARGF) were used in this investigation. From the results, it was found that the addition of glass fibres showed the improvement in the properties of the stone mastic asphalt mix. The addition of fibres at 0.4% with 6% binder content results in increased stability values and low draindown values.
Keywords: Class-C Glass Fiber; Alkali Resistant Glass Fiber; Draindown; Stability.

1. Introduction

Stone Matrix Asphalt (SMA) is a mixed proportion of the coarse aggregate with fine aggregate and the bitumen to obtain the strong and durable mix to withstand the traffic loads. As the SMA offers rut resistance, durable and textured surface course, it has been successfully utilized in many of the developing countries. SMA consists major portion of coarse aggregates. SMA has a typical composition consisting of coarse aggregate of 70–80%, filler of 8–12%, a binder of 6.0–7.0%, and fibre of 0.3 - 0.5%. Aggregate grading, filler, binder, proportion and the type of stabilizer additive will determine the stability of the mix. The interlocking and binding properties of the SMA Mix are influenced by the type of fibre used. The binding, as well as the property of interlocking, are influenced by the type of fibre used in SMA. Fibre affects the asphalt in its properties like Penetration, Ductility of the Mix, also affect the rut Resistance.

The Fibres utilization in the Stone Mastic Asphalt helps in the reduction of the Draindown, which also affects the stability of the mix. In this, the Paving Mixes were examined in terms of Stability, Void filling Resistance to Displacement and its Consistency (Bhanu and Venkateswara Rao (2016) [1]). The performance of SMA with various types of fibres was studied in the earlier investigations. Bindu et al (2010) [2] has conducted the experimental investigation with waste plastic as the stabilizer in the SMA from 0-12%. Laboratory tests like Marshall Stability and draindown were conducted and concluded that 10% of plastic waste increased stability and also reduced the draindown. Ahmadinia et al (2011) [3] has studied on Stone Mastic Asphalt properties using plastic waste bottles as additive. From the experimental investigation, it was come to know that up to 0.4% addition of additive increased the stability and further addition decreases significantly. Moghaddam et al (2012) [4] has studied regarding the properties of SMA containing waste polyethylene terephthalate. From the test results indicated that 0.4% addition of waste polyethylene terephthalate has increased the stability. Suresh et al (2013) [5] studied on the performance of modifying strength parameters using glass fibres of lengths 0.3 millimeters and 0.6 millimeters. From the test results of stability and penetration, it was found that fibres with 0.3 mm lengths performed better results compared to the mix with 0.6 mm lengths. Thanh et al (2013) [6] has studied the influence of three different types of fibres in stone mastic asphalt. The fibres used are lignin. Mineral and polyester fibres respectively. From the Marshall stability and rutting tests, it was identified that mineral fibre has increased the properties compared to the lignin and polyester fibres.

2. Properties of Different Materials Using Different Test Methods

2.1 Identification of Material Properties

In this experimental investigation, the coarse and fine aggregates used were used from the near quarry. The aggregate particle size is ranging from between 75μ to 13.2 millimeters. Different test methods are conducted to the collected aggregates to determine their properties are shown below.

Table 1. Different Aggregate Properties (as per IS 2386 1963 [7])

Characteristic Property	Parameter	Result Obtained (%)
Strength	Impact Value	17.7
	Abrasion Value	18
	Crushing Value	25.3
Particle shape	Combined Index of Elongation and Flakiness	26.7
Water Absorption	Water Absorption	0.5

The grade of bitumen is VG – 30, which is brought from HPCL, Vizag. 1.02 is the obtained specific gravity. The properties of bitumen are mentioned below.

Table 2. Bitumen properties

Characteristic Property	Method Obtained	Result
Test for Penetration	IS 1202 [8]	64
Test for Softening Point	IS 1205 [9]	50
Test for Ductility	IS 1208 [10]	Greater than 100

The glass fibres used in this investigation are obtained from Ashwin ceramics and reliance industries and their properties are mentioned below.

Table 3: Properties of Fibres

Property	Glass – C Fibre	Alkali resistant glass Fibre
Density	2.70	2.60
Tensile strength	3600 MPa	3500 MPa
Elastic modulus	760 GPa	740 GPa
Abrasion	High resistance	High resistance

2.2 Aggregate Gradation

Stone mastic asphalt mixes were prepared using IRC SP 79 2008 [11] of SMA gradation. It is represented in figure 1.

2.3 Procedure for the Preparation of SMA Specimens

As per standard specifications of AASHTO MP8 [12], the stone mastic asphalt mixes were designed. Two different fibres were used to prepare two different mixes. The percentage of fibre addition is obtained from the draindown test. For the first mix Glass - C type fibre is used as a stabilizer additive. In case next mix alkali-resistant glass fibre is added as the stabilizer additive. For the determination of optimum bitumen content, aggregate and filler of 1200 gm are heated to 175 – 190^{0}C. Then the fibres are added and uniformly mixed. Bitumen was heated and maintained temperature from 121^{0}C to 125^{0}C for trail 1, the content of bitumen was taken as 5% by weight of aggregates.

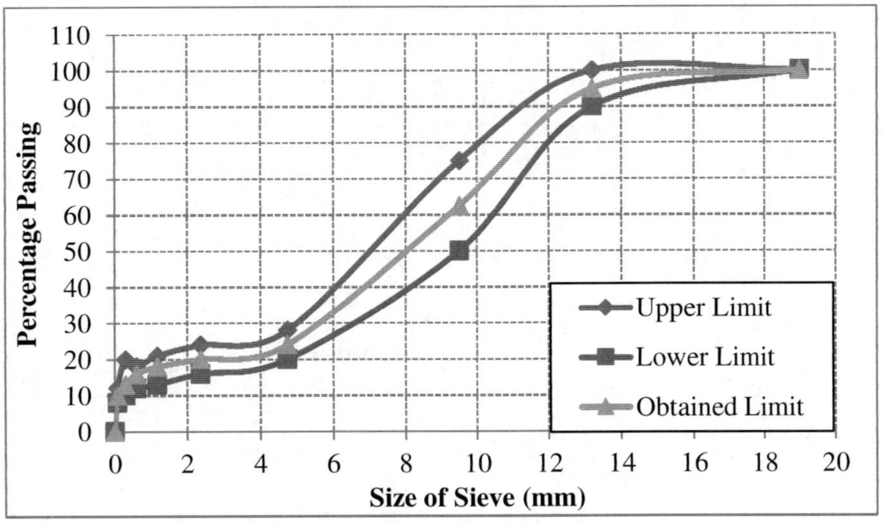

Figure 1. Grading of Aggregates

The aggregate and bitumen were mixed thoroughly at 154 - 160^0 temperature. The mix was placed in the mould which is preheated at 110^{0}C. the mix is prepared by compacting the mix for 50 blows of either side by maintaining the temperature at 138^{0}C – 149^{0}C. The thickness of the compacted specimen must be 63.5$\pm$3 mm. Bitumen content for the next mixes is varied by 0.5% and the above procedure is repeated.

2.4 Test Procedures

2.4.1Binder Runoff Test (Draindown)

To identify the runoff of the binder from the mix, the draindown test was used. fibres of different percentages were added to the mix. The percentage fibre addition giving the draindown value less than 0.3 is considered as the optimum additive and used in further analysis.

2.4.2 Test for Marshall Stability

This is to analyze Stability and Flow values, with different mixes of bitumen. Here in this mix Stability, Flow, Percentage of Air Voids (%Vv), Percentage of Voids filled with Bitumen (%VFB), Percentage of voids in Mineral Aggregates (%VMA). At 4% Air voids (%Vv), the Optimum Binder Content (OBC) was considered.

3. Analysis of Test Results

Initially, the test of draindown was performed to obtain the optimum content of fibre added with the bitumen. Later the Marshall Stability tests were performed on the specimen with optimum fibre content and the results were discussed below.

3.1 Draindown Test for Optimum Fibre Content

The Draindown test was performed using the basket drainage test as per ASTM D 6390 (2005) [13] for the obtained gradation mix with 7% bitumen content. It is conducted at a temperature of 160^{0}c with fibre addition in percentage of 0.05%, 0.1%, 0.2%, 0.3%, 0.4%, and 0.5%.

It is found that the optimum stabilizer additive content for both the fibres was obtained at 0.4%. The draindown values obtained for the Glass - C and ARGF are 0.199 and 0.159 respectively. Hence it satisfies the minimum condition of draindown as per IRC: SP 79 – 2008, the draindown test value should be less than 0.3.

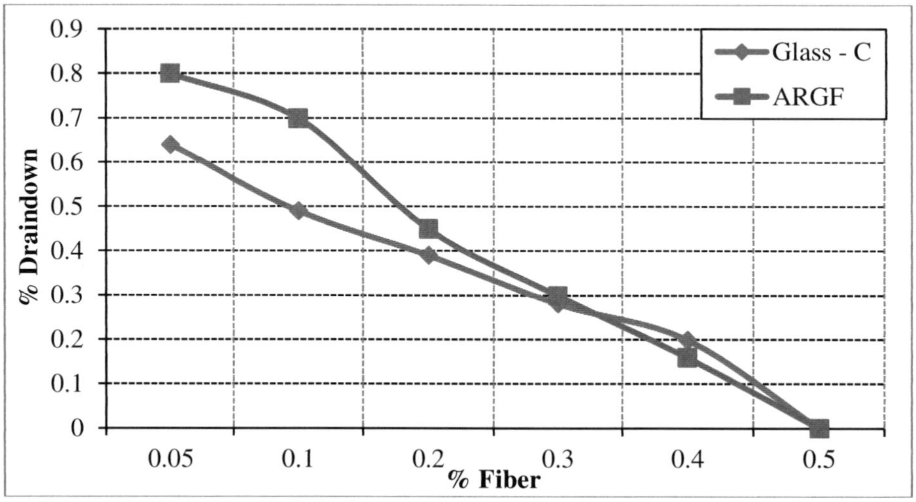

Figure 2. Variation of draindown value for glass fibres as stabilizer additive.

Figure 2 represents the draindown of the SMA mix with the fibre fraction. As the fibre fraction increases the draindown decreases gradually. The reason may be attributed to the presence of fibres due to which void space will be reduced in the bitumen mix. The better interlocking of fibre and aggregate helps in attaining the maximum stability. Comparatively Glass - C showed the lesser draindown than the ARGF. The draindown values obtained for both types of fibres were less than 0.3 at the optimum fibre content of 0.4 % fibre.

3.2 Preparation of Stone Matrix Asphalt Samples with Optimum Stabilizer Additive

The SMA specimens were prepared by varying the binder content from 5.5% to 7.0% i.e., 5.5%, 6.0% 6.5% and 7.0% respectively at optimum stabilizer additives (0.4% obtained by draindown test). Then the Marshall properties are identified and represented below.

3.2.1 Stability

Figure 3 represents the comparison of the stability value to percentage binder varying from 5.5% to 7.0% at 0.5% increment at optimum stabilizer additives (glass - C, ARGF). Both the mixes gradually increased the stability and obtained maximum stability at 6% and thereafter stability decreases gradually with the increased binder content.

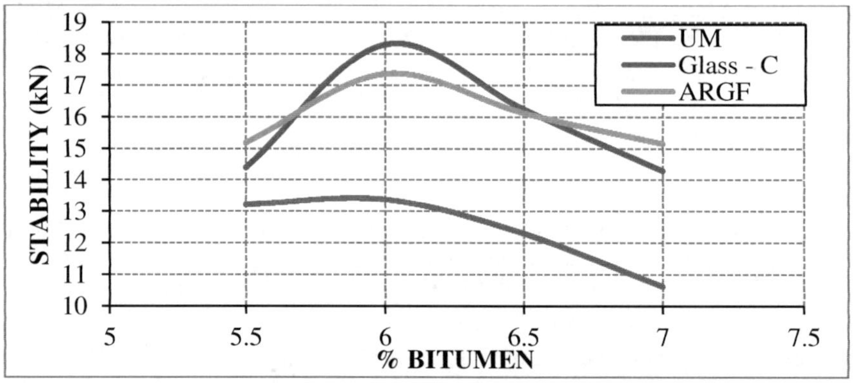

Figure 3. Comparison of Value of Stability with binder content for stabilizer additives.

The maximum stability obtained for Un-modified mix is 13.38 kN. For the Glass - C and ARGF are 18.30 kN and 17.36 kN. For glass - C, an increment of 36.8% in the stability to unmodified mix was observed and 5.4% increment in its stability was found compared to the mix with ARGF.

3.2.2 Flow

Figure 4 represents the comparison of the value of the flow of the mixes at optimum stabilizer additives (glass - C and ARGF) with the increased percentage binder. The flow value for glass – C fibre is ranging from 2.32mm to 3.12mm and the flow value of glass fibre is ranging from 2.31mm to 4.28mm. The results showed that the increase in the flow value is due to the increase in the binder content. Due to the elastic nature of the binder and the fibres used in the mixes increased the flow values. Mix with ARGF fibre showed a higher flow value compared to the mixes with unmodified and glass –C fibre.

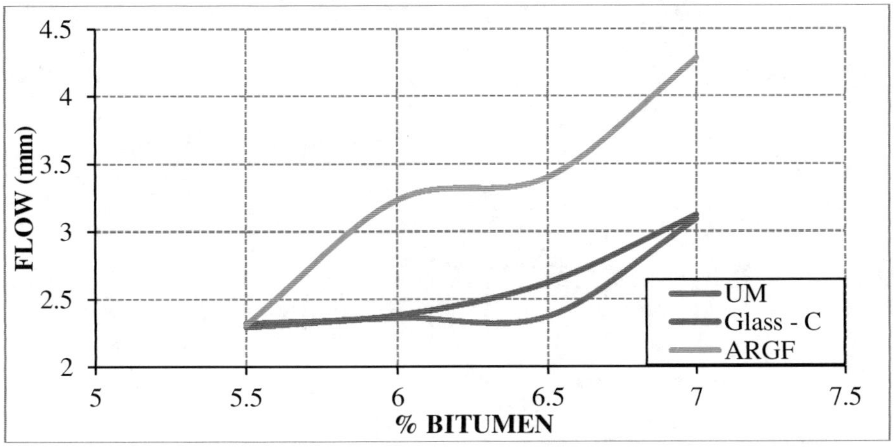

Figure 4. Comparison of the Flow Value with the content of fibre for stabilizer additives.

3.2.3 Air voids

Figure 4 represents the comparison of the value of % air voids with the % increment in the binder content in the mixes with stabilizer additives. The results show the increase in the content of fibre reduced air voids percentage for the mixes. The reason may be considered as the fibres coated with the binder helps in filling the vacant space in the specimens, thereby resulted in a decrease in the air voids. By the addition of glass – C fibre at different binder percentages resulted in the reduction of air voids from 5.74% to 1.68% and by the addition of ARGF, air voids reduced from 5.33% to 2.52%. In the case of glass – C fibres at a lower percentage of binder high amount of air voids were observed. Beyond 6% binder content the less percentage of air voids were observed in the case of glass – C fibre compared to the unmodified mix and mix with ARGF.

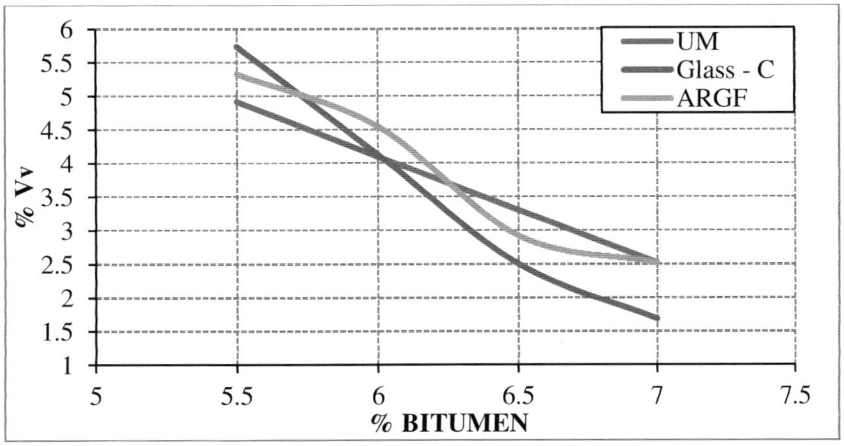

Figure 5. Comparison of the value of % Air Voids with the content of binder for stabilizer additives.

3.2.4 Percentage of Voids in Mineral Aggregates

Figure 6 shows the comparison of the value of % Voids in Mineral Aggregates for the mixes to the increase in the percentage binder. In case of glass – C fibre as stabilizer additive the voids in mineral aggregates gradually decreases from 18.14% to 17.74%. In case of ARGF as stabilizer additive, it was observed that from 5.5% to 6.0% there is decrement from 17.79 to 17.31% and thereafter there is an increment up to 7% binder by obtaining 18.44% in the % VMA. Mix with ARGF attained the lower VMA at 6% binder and thereafter increased up to 7%. For glass – C fibre there is a continuous decrement in VMA.

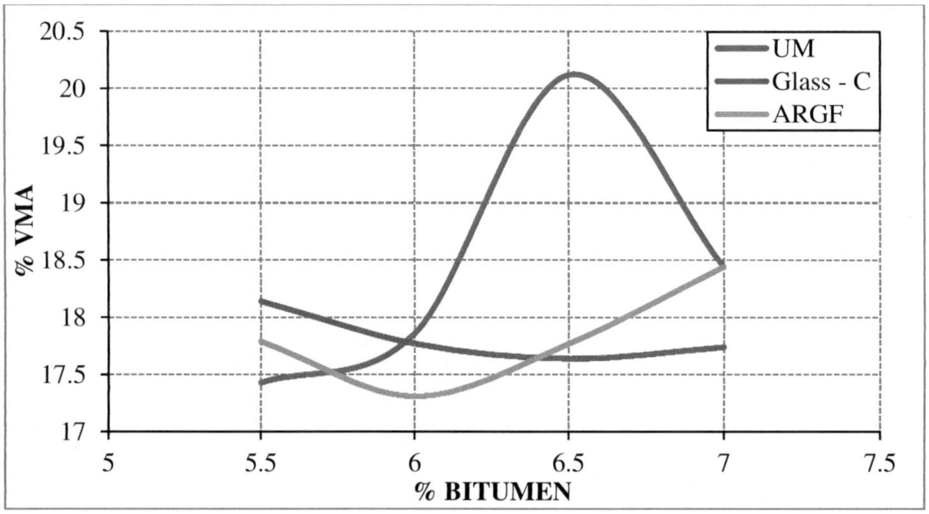

Figure 6. Comparison of Percentage VMA with the Content of Binder for stabilizer additives.

3.2.5 Percentage of Voids filled with Bitumen (VFB)

Figure 7 shows the comparison of percentage VFB with the % for the mixes to the increase in the percentage binder.

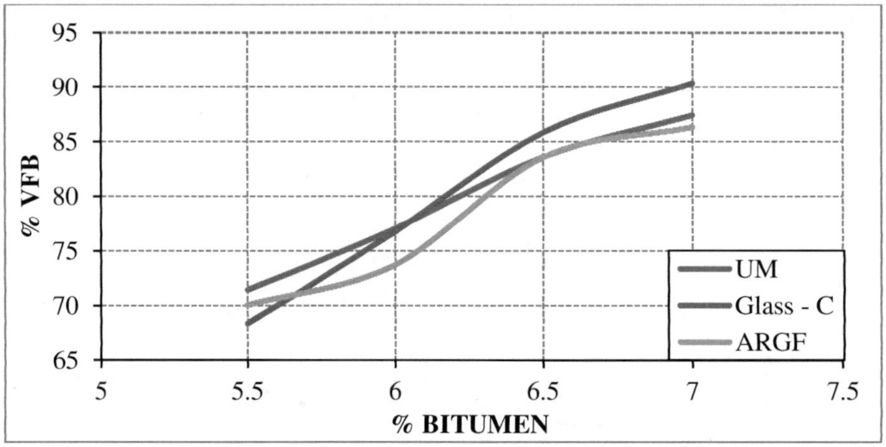

Figure 7. Comparison of Percentage VFB with the content of Binder for stabilizer additives.

The result shows that the increase in the Voids filled with Bitumen is due to the increase in the percentage Binder. As a result, there is a decrement in percentage air voids, there must be an increment in the percentage Voids Filled with Bitumen. The range of increment in Glass – C fibre is from 68.36 to 90.36% and for ARGF the increment is in the range of 70.14% to 86.33%. For both the fibres binder there is an increment in VFB with the increase in the binder content. Beyond 6% of binder content, the glass – C fibre showed a higher value of VFB. In the case of mix with ARGF, there is no appreciable increase in VFB with the increase in binder content.

4. Conclusions

From the laboratory tests, the influence of the glass fibres as the stabilizer additive in the SMA mix, which indicates that Marshall Stability Properties has been examined and compared for different properties. The following conclusions were suggested from the above tests conducted.

1) From the results of the draindown test, it was concluded that 0.4% fibre addition is optimum fibre addition for both the glass fibres.
2) The maximum stability values obtained for mixes with glass - C and ARGF are 18.30kN and 17.36kN.
3) Stability of the mix with glass – C showed an increment of 36.8% to unmodified mix and 5.4% increment when compared to the mix with ARGF.
4) Flow value is higher in case of mix with ARGF and lowers in case of glass – C fibre comparing with the unmodified mix.
5) Utilization of fibres in SMA increased the VFB compared to the unmodified mix.
6) Glass – C fibre showed better Marshall properties compared to the alkali-resistant glass fibre.

References

[1] Udaya Bhanu V, Venkateswara Rao J 2016 Influence of Glass Powder and Quartzite Rock Dust as Fillers in Stone Mastic Asphalt *International Journal of Engineering Research & Technology* **5 (11)** p 222–6

[2] Bindu CS, Beena KS 2010 Waste plastic as a stabilizing additive in Stone Mastic Asphalt *International Journal of Engineering and Technology* **2 (6)** p 379-87

[3] Esmaeil Ahmadinia, Majid Zargar, Mohamed Rehan Karim, Mahrez Abdelaziz, Payam Shafigh 2011 Using waste plastic bottles as additive for stone mastic asphalt *Materials and Design* **32 (10)** P 4844–9

[4] Taher Baghaee Moghaddamand, Mohamed Rehan Karim 2012 Properties of SMA Mixtures Containing Waste Polyethylene Terephthalate *International Journal of Chemical and Biological Engineering* **6 (2)** p 170-5

[5] Suresh T, Gobinath R, Kannan T, Saranya S, Sruthy Radhakrishnan, Neelima Singhvi, Chitravel V 2013 Studies on Modifying Strength Parameters of Bitumen Using Fibres *International Journal of Engineering Research & Technology (IJERT)* **2 (10)** p 1085-93

[6] Dang Van Thanh, Cheng Pei Feng 2013Influence of Different Fibres on High Temperature and Water Stability of Stone Matrix Asphalt *International Journal of Applied Engineering and Technology* **3 (1)** p 1-10

[7] IS 2386 Part I 1963 Methods of test for aggregates for concrete *Bureau of Indian Standards,* New Delhi, India p 11-4

[8] IS 1203 1978 Methods for testing tar and bituminous materials: Determination of penetration *Bureau of Indian Standards* New Delhi, India p. 19-22

[9] IS 1205 1978 Methods of testing tar and bituminous material: Determination of softening point *Bureau of Indian Standards* New Delhi, India p 33-8

[10] IS 1208 1978 Methods of testing tar and bituminous material: Determination of ductility *Bureau of Indian Standards* New Delhi, India

[11] IRC SP 79 2008 Tentative specifications for stone mastic asphalt *Indian Roads Congress*, India

[12] AASHTO M 325-08 2012 Standard specification for Stone Mastic Asphalt (SMA) *AASHTO*
ASTM D 6390 05 Standard test method for determination of draindown characteristics in uncompacted asphalt mixtures *ASTM*

Hardened properties of pervious concrete using mixed aggregate blended with admixtures

V Vinay[1], S Adiseshu[2], S Chandramouli[3]

[1]Research Scholar, Department of Civil Engineering, Andhra University College of Engineering (Autonomous), Visakhapatnam Andhra Pradesh
[2]Professor, Department of Civil Engineering, Andhra University College of Engineering (Autonomous), Visakhapatnam
[3]Professor, Department of Civil Engineering, MVGR College of Engineering (Autonomous), Vizianagaram 535005
Email: vinayv@mvgrce.edu.in

Abstract: In the present study, a new combination of materials for making pervious concrete have been obtained based on compressive strength and permeability. The investigation is done for finding hardened properties and its main aim is to know the improvement in compressive strength and permeability by different combinations of materials. Also, different tests for void ratio, porosity and water absorption are conducted. The different sizes of aggregate used are 20mm, 12.5mm and 10mm with cement replacement with fly-ash 10%, 20%, 25% and 30% and with maximum cement content 450kg/m^3, 500kg/m^3 of 53 grade. It is observed that the strength and other properties are highly fluctuating. The maximum compressive strength obtained is 21.3MPa with 450kg/m^3 cement content of aggregate combination 60-20-20 with 30% of fly ash. The maximum permeability obtained is 2.87 cm/sec with 450kg/m^3 cement content of aggregate combination 20-40-40 with 20% fly-ash. The maximum water absorption obtained is 4.087% with500kg/m^{3}of aggregate combination 50-20-30 with 30% fly-ash. The maximum void ratio obtained is 37.545% with 500kg/m^3 cement content of aggregate combination 40-30-30 with 25%fly-ash. The maximum porosity obtained is 68.912% with 500kg/m^3 cement content of aggregate combination 40-30-30 with 25% fly-ash.
Keywords: Pervious concrete, Compressive strength, Permeability, Water absorption, Porosity, Void ratio

1. Introduction

"Pervious concrete" it is a special type of concrete in which fine aggregate is not used, it consists of a high volume of voids, it is divided into interconnected pores, capillary pores, dead-end pores, which will increase porosity from, permeability. It is mainly used to improve groundwater recharge by the action of percolation. Pervious concrete is used for the development of transportation infrastructure, to avoid water pollution, waterlogging, to control flood run-off and also to improve the skid resistance of roads in rainy seasons. Advantages of pervious concrete are low expansive, low density, it will have good isolating properties. Pervious concrete is made by using ordinary Portland cement which is designed for various applications as per requirements. However, it is used in several parts around the world but still, it is not practised in some countries may be due to some technical issues, material preparation etc. Maintenance of pervious concrete is somewhat difficult due to clogging problems. Research is going on to improve compressive strength, the effect of water to cement ratio, aggregate to cement ratio, aggregate sizes, binding material type, its quantity on strength of pervious concrete and use of recycled aggregate in pervious concrete, use of cement replacing materials like fly-ash, silica fume, GGBFS.

Omkar Deo et al. (2010) [1] studied the improvement of permeability which is reduced due to clogging effect. To avoid this clogging several experiments and models are done with different proportions of aggregate to improve pore structure. Permeability generally is reduced when fine sand is used as clogging material. Here the volumetric porosity, porosity is resolved by image analysis and hydraulic conductivity by permeability cell. Modelling has been done for clogging behaviour. Here the modelling approach is based on the geometrical model and probabilistic particle which was captured in pores. And then by all these modellings and testing the clogging is reduced up to some extent. Lian et al. (2011) [2] developed a mathematical model to characterize the relationship between compressive strength and porosity for porous concrete by analysing empirical results and theoretical derivations. The suitability of existing equations for porous concrete is assessed and a new model is then proposed. Bradley and Andrew (2011) [3] found different techniques for preparation test specimen of pervious concrete. As that specimen should be similar to the pervious pavement which is compared to each other by taking cores from pavements. This comparison is based on porosity, infiltration capacity etc. For this. cylinders of 150mm dia and 300mm height and slabs of 150mm thickness with three different sizes are cast. Correlation between porosity and infiltration is done. The flow chart of the experimental approach is clearly given for 3 cylinders, pavements, and slabs for determining the best specimen. Rasiahsriravindra rajah et. al. (2012) [4] prepared a mix design for pervious concrete which is made of recycled aggregate and to replace the cement 70% of its weight with Ground Granulated Blast Furnace Slag (GGBS). Here the relationship developed between permeability, porosity and compressive strength. Based on this the mix design was prepared. The compressive strength relies on type, size, shape of materials and porosity. As the porosity decrease, the strength will be increased. By using this recycled aggregate strength was reduced and porosity improved and there is no change in permeability. The empirical equations for compressive strength and porosity were given for recycled aggregate. Daniel and Jorge (2014) [6] observed that stormwater management can be done by proper drainage systems by involving Sustainable urban drainage systems which impact on the economy, environment, and society. Pervious concrete pavements are the solution for managing urban stormwater runoff. The authors further opined that there are many possibilities based on integrated value models to facilitate the proper selection. For this accurate information is given to know the alternatives. Here a series of variables are modelled by stochastic simulations based on Monte Carlo methods and opinions of experts related to water management about selected criteria comparing with Analytical Hierarchy Process (AHP). Three types of pervious pavement are figured to determine the favourable model.

Ahmed et al. (2014) [7] developed a linear regression relationship to establish relationships between density and porosity, compressive strength and permeability, tensile strength and permeability, and compressive strength and porosity. The results showed that properties such as permeability, porosity, are significantly affected by using either one or two coarse aggregate sizes in all concrete mixtures. Furthermore, density can be an effective factor for predicting compressive strength, and porosity. In this study, the maximum compressive strength was 6.95 MPa, which obtained by using one aggregate size of 9.5 mm with 250 kg/m^3 cement content. Ruizhong and Kay (2015) [8] developed an ultra-high-performance cement-based matrix with compressive strengths over 150 MPa and high durability properties are designed and applied to the mixture design concept of pervious concrete. The research results show that compressive strength and elastic modulus increase by up to 150% and 100%, respectively, without sacrificing the hydraulic conductivity of the concrete. Furthermore, freeze-thaw tests have been carried out to compare the durability performance of conventional pervious concrete with high-performance pervious concrete.

Hariyadi and Hiroki (2015) [9] observed the relationship between mechanical properties of concrete with volcanic pumice explored and compared them with those of other researchers' works and ACI standard. They concluded that utilizing volcanic pumice on porous concrete mixtures resulting in a high porosity (void content) and a low modulus of elasticity of porous concrete. Moreover, by using volcanic pumice a higher tensile strength (flexural strength) than that of ACI standard can be obtained. This volcanic pumice porous concrete is potential for future structure with adequate strength and good impact energy-absorbing. Anush and Krishna (2016) [10] observed that the water to cement ratio is a very important variable and is lower compared to those used in the conventional concrete mix and has been historically varied over the range of 0.28 to 0.4 with the main intention to provide sufficient cement coating for the aggregates. Tejas Joshi and Urmil Dave (2016) [11] studied the effect of different aggregate size and water-cement ratio on pervious concrete. The experiments are done for determining void ratio, permeability, strength and density. As the strength increased the density also increased was noticed. Rui Zhong and Kay Wille (2015) [12] observed that pervious concrete has remarkable potential to counteract the adverse impacts on the environment while providing necessary structural integrity, thus supporting continued urbanization. Broader application of pervious concrete could be achieved through increased ravelling resistance and enhanced durability performance. This research emphasizes the development and characterization of high-performance pervious concrete aiming at improved mechanical resistance and advanced durability properties. Jiusu et al. (2017) [13] aimed at a solution towards the current deficient strength, durability and inconvenient maintenance in pervious concrete. Substantial performance evaluation indicated that HSPC possesses ideal compressive and flexural strength while maintaining excellent water permeability. Furthermore, an efficient drainage system and clogging-preventing paths from bottom to top were combined in the precast concrete pavement systems to overcome other technical bottlenecks in current pervious concrete technology.

In the present study, an attempt has been made to find out the hardened properties of pervious concrete such as compressive strength, permeability, water absorption, voids ratio and porosity made with different combinations of materials. Also, to know the improvement in compressive strength and permeability by using different combinations of materials.

2. Materials and Methods

2.1 Mix design:

The practice of mix design of pervious concrete is similar to traditional concrete. In our country, there is no suitable or perfect mix design for pervious concrete. As we know IS 12727 1989 [14] it is the code of practice for no fines concrete. This code gives us various proportions of materials which are used in pervious concrete. But this code is not in practice because the maximum compressive strength by using this code is 10MPa. So, to improve the compressive strength based on both codes IS 456 2000 [15] and IS 10262 2009 [16] the mix design of pervious concrete is followed by neglecting fine aggregate. The mix proportions of materials used are shown in tables 1 to 3.

Table 1. Mix proportions for 450 kg/m^3 and 500 kg/m^3

Mix	Cement / fly ash (kg/m^3)	Coarse aggregate (kg/m^3)	Water (kg)	Proportion Cement: CA	W/binder ratio
1	450	1861.75	180	1: 4.15	0.4
2	500	1828.96	197.16	1: 3.65	0.4

Table 2. Details of aggregate proportions

Sl. No	Mixed Proportions	Aggregate sizes 10mm	12.5mm	20mm	W/C ratio
1	A1	50%	20%	30%	0.4
2	A2	60%	20%	20%	0.4
3	A3	40%	30%	30%	0.4
4	A4	20%	40%	40%	0.4

IOP Conf. Series: Materials Science and Engineering 1025 (2021) 012021 doi:10.1088/1757-899X/1025/1/012021

Table 3. Percentages of fly-ash and cement

Designation	Cement (%)	Fly-ash (%)
MFC1	90	10
MFC2	80	20
MFC3	75	25
MFC4	70	30

2.2 Experimental Program

With the above mix proportions, the cubes and cylinders are cast and shown in figure 1.

Figure 1. Casted cubes and cylinders

2.2.1 *Compressive strength test*

Tested as per IS 516 1958 [17], IS 456 2000 [15] after 28 days curing of the cube (15 cm × 15 cm × 15 cm)

2.2.2*Water absorption test*

The water absorption of pervious concrete samples was evaluated by calculating the difference between the dry weight of the cylinder and the saturated weight of the cylinder. By using the equation.

$$W\% = \left(\frac{w1 - w2}{w1}\right) \times 100$$

Where,

w1 is the wet weight of cylinder in kg

w2 is the dry weight of cylinder in kg

2.2.3 *Falling head permeability test to find out permeability*

Figure 2. Falling head permeability apparatus

IOP Conf. Series: Materials Science and Engineering 1025 (2021) 012021 doi:10.1088/1757-899X/1025/1/012021

This is one of the methods to find out permeability. The following formula is adopted in the present study for falling head permeability test to calculate permeability.

$$K = \left(\frac{a \times L}{A \times T}\right) \log\left(\frac{h0}{h1}\right)$$

Where,
K : coefficient of permeability (cm/sec)
a : Cross-sectional area of pipe (cm^2)
L : Length of the cylinder (cm)
A : is the cross-sectional area of the cylinder (cm^2)
T : Time taken for head to fall from h_0 to h_1
h_0 : Initial water head (cm)
h_1 : Final water head (cm)

2.2.4 *Void ratio*

The void ratio of the cylinder is calculated according to ASTM C1688 [18]. Void ratio is explained as the total percentage of voids present divided by the volume of the specimen. The equation is given below.

$$\text{Void ratio \%} = \frac{T - D}{T} \times 100$$

Where,
$D = (M_c - M_m)/V_m$ (Density)
M_c = Mass of measure filled with concrete
M_m = Net mass of concrete by subtracting the mass of measure
V_m= Volume of measure
$T = M_s/V_s$ (Theoretical Density)
M_s = Total mass of materials batched
V_s = Total absolute volume of materials

2.2.5 Porosity

The total porosity was calculated by measuring the difference between the weight of the dry cylinder and weight of the submerged cylinder and dividing with the volume of the cylinder. Using formula

$$P = 1 - \left(\frac{\left(\frac{W1 - W2}{\rho w}\right)}{V}\right) \times 100$$

Where,
W_1 = Weight of dry cylinder in (kgs)
W_2 = Weight of submerged cylinder (kgs)
P_w = Density of water (kg/m^3)
V = Volume of cylinder (m^3)

3. Results and Discussions

With the proportions of the materials as obtained in materials and methods section, the cubes of 15 cm × 15 cm × 15 cm are cast in the laboratory and tested for their 28-days compressive strength with 450 kg/m^3 and 500 kg/m^3 as maximum cement content. The various results such as compressive strength, permeability, porosity, voids ratio and water absorption obtained are presented graphically and then discussed. The strength and permeability characteristics of various combinations of aggregate with the weight of cement 450 kg/m^3 and 500 kg/m^3 are shown graphically in figures 3 to 6.

From figure 3, it is observed that the maximum compressive strength obtained for 450 kg/m^3 of cement content is 21.3 MPa for MFC4A2450. It is varying in the range of 7.8 to 21.3 MPa. Also, it is observed that the increasing trend of compressive strength from 10% to 30% fly-ash for all combinations of aggregates.

From figure 4, it is observed that the maximum permeability obtained for 450 kg/m^3 of cement content is 3.01 cm/sec for MFC4A4450. It is varying in the range of 3.01 to 1.51 cm/sec. Also, it is observed that the decreasing trend of permeability from 10% to 30% fly-ash for all combinations of aggregates.

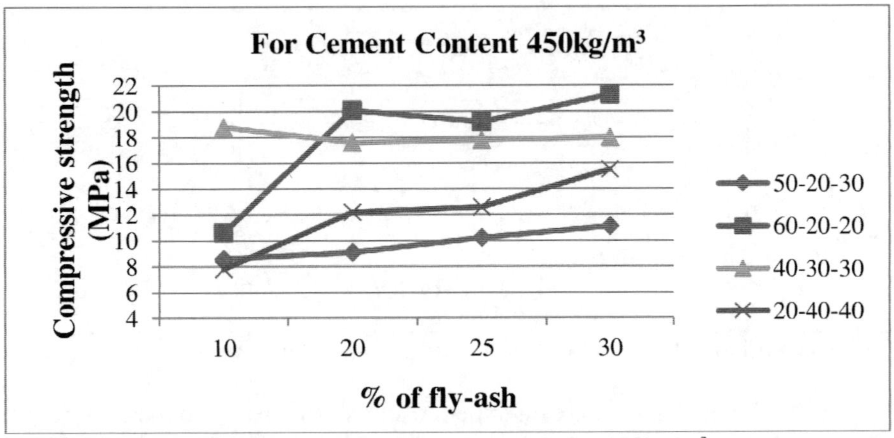

Figure 3. %of Fly-ash Vs Compressive Strength for 450kg/m^3 cement content

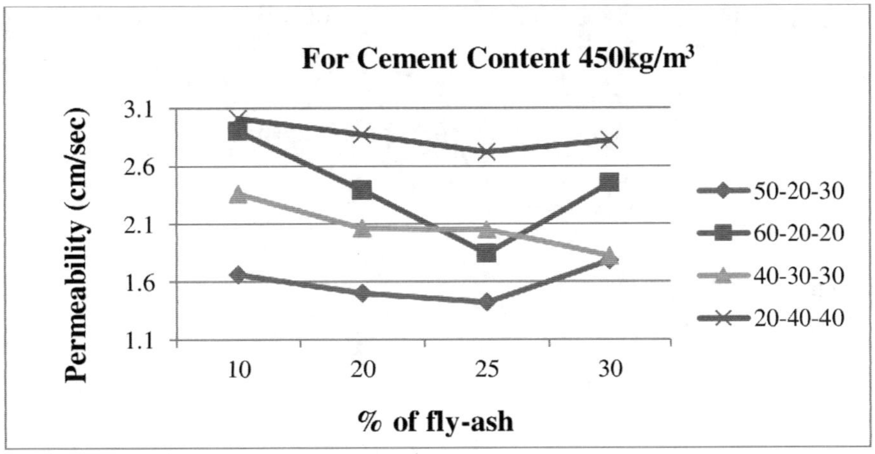

Figure 4. % of Fly-ash Vs Permeability for 450 kg/m^3 cement content

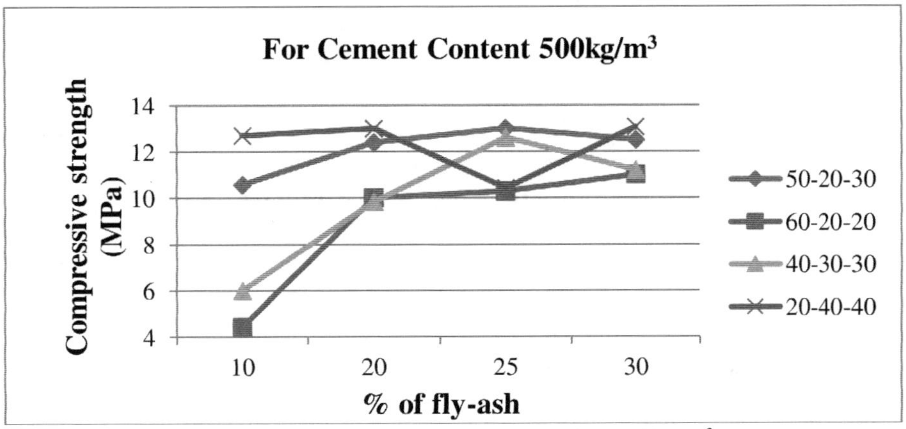

Figure 5. %of Fly-ash Vs Compressive Strength for 500kg/m^3 cement content

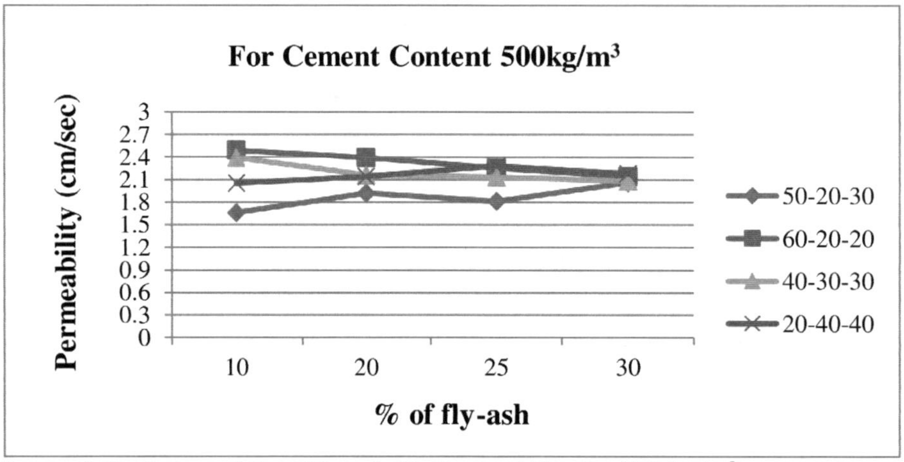

Figure 6. Variation of % of Fly-ash Vs Permeability for 500 kg/m^3 cement content

From figure 6, it is observed that the maximum permeability obtained for 450 kg/m^3 of cement content is 2.49 cm/sec for MFC4A4450. It is varying in the range of 2.49 to 1.66 cm/sec. Also, it is observed that the decreasing trend of permeability from 10% to 30% fly-ash for all combinations of aggregates. The water absorption, porosity and void ratio characteristics of various combinations of aggregate with the weight of cement 450 kg/m^3 and 500 kg/m^3 are shown graphically in figures 7 to 12.

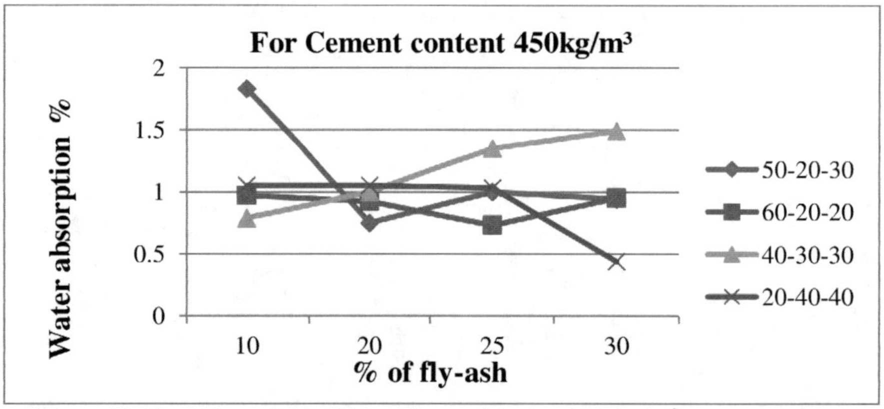

Figure 7. % of Fly-ash Vs Water absorption for 450 kg/m^3 cement content

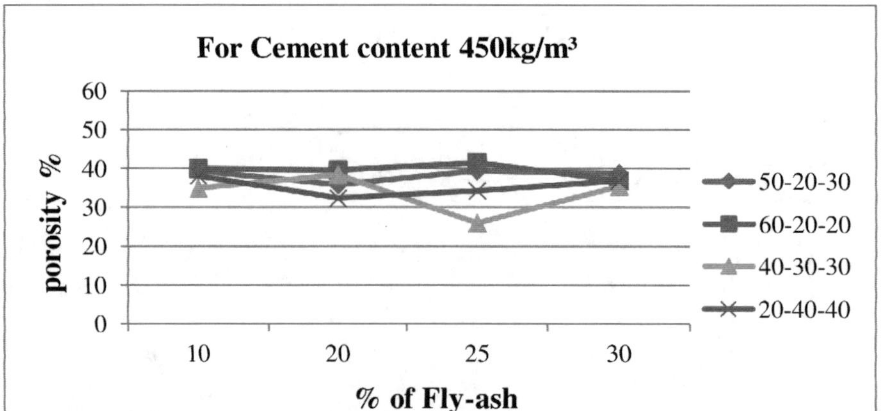

Figure 8. % of Fly-ash Vs Porosity for 450 kg/m^3 cement content

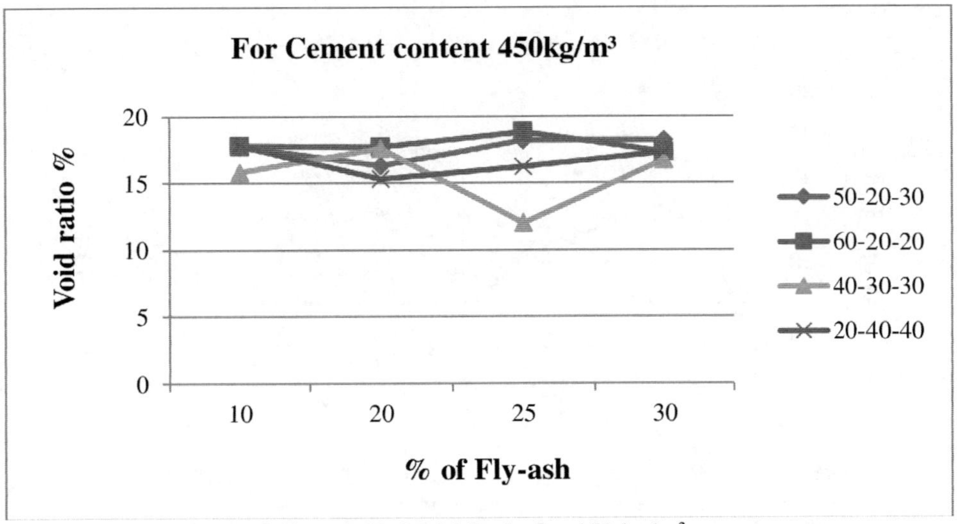

Figure 9. % of Fly-ash Vs Void Ratio for 450 kg/m^3 cement content

From figure 7, 8 and 9, it is observed that for 450kg/m^3 cement content, the maximum values of water absorption, void ratio and porosity obtained are of aggregate combination 50-20-30 with fly-ash 10% is 1.83%, an aggregate combination of 60-20-20 with fly-ash 25% is 18.83% and aggregate combination of 60-20-20 with fly-ash 25% is 41.50% respectively.

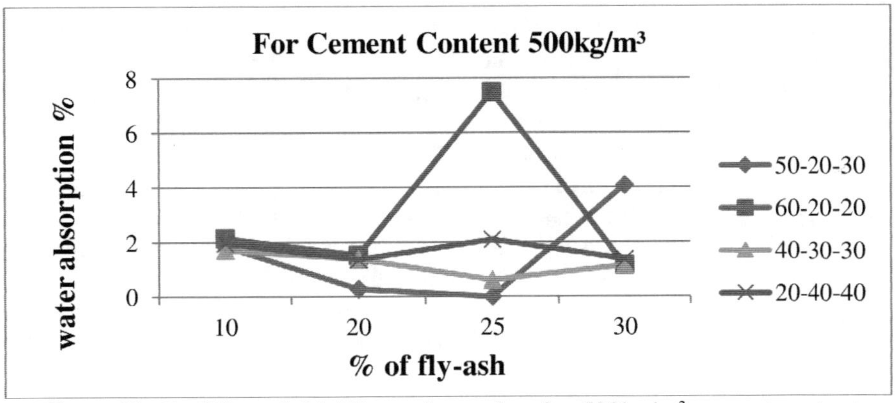

Figure10. %of Fly-ash Vs Water absorption for 500kg/m^3 cement content

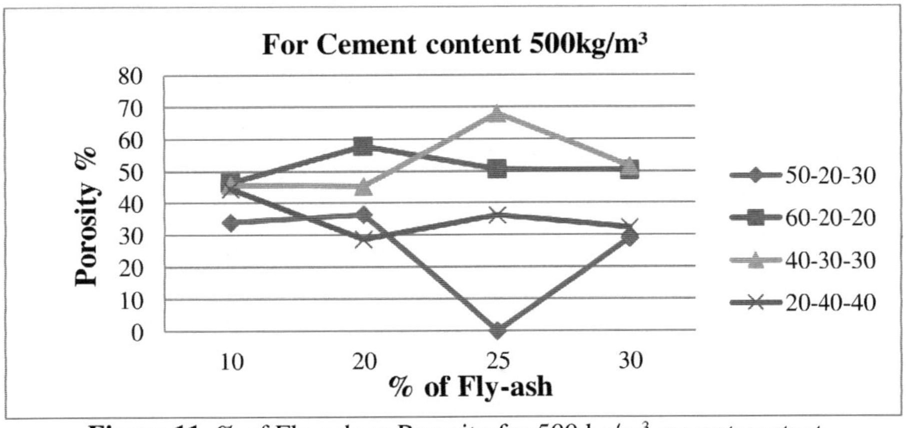

Figure 11. % of Fly-ash vs Porosity for 500 kg/m^3 cement content

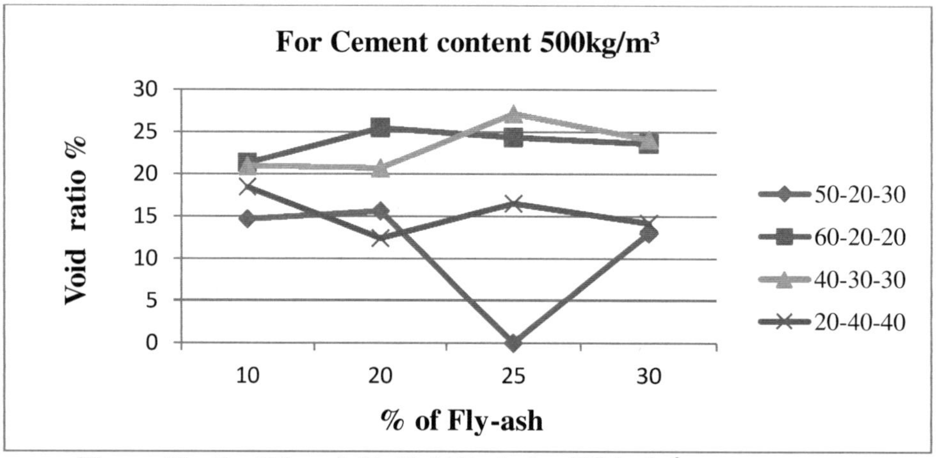

Figure 12. % of Fly-ash Vs Void Ratio for 500 kg/m^3 cement content

From figure 10, 11 and 12, it is observed that for 500kg/m^3 cement content, the maximum values of water absorption, void ratio and porosity obtained are of aggregate combination 50-20-30 with fly-ash 30% is 4.087%, an aggregate combination of 60-20-20 with fly-ash 25% is 37.54% and aggregate combination 40-30-30 with fly-ash 25% is 68.91% respectively.

4. Conclusions

1. The maximum values of compressive strength and permeability for a combination of water-cement ratio 0.4 and cement content 450kg/m^3 obtained are 10.43 MPa, at 30% fly-ash and 2.90 cm/sec at 10% fly-ash for 60-20-20 aggregate combination. For this combination of materials, the maximum values of water absorption, void ratio and porosity obtained are of aggregate combination 50-20-30 with fly-ash 10% is 1.83%, an aggregate combination of 60-20-20 with fly-ash 25% is 18.83% and aggregate combination of 60-20-20 with fly-ash 25% is 41.50% respectively.
2. The maximum values of compressive strength and permeability for the combination of water-cement ratio 0.4 and cement content 500kg/m^3 obtained are 13.06 MPa at 30% fly-ash for 20-40-40 aggregate combination and 2.49cm/sec at 10% fly-ash for 60-20-20 aggregate combination. For this combination, the maximum values of water absorption, void ratio and porosity obtained are of aggregate combination 50-20-30 with fly-ash 30% is 4.087%, an aggregate combination of 60-20-20 with fly-ash 25% is 37.54% and aggregate combination 40-30-30 with fly-ash 25% is 68.91% respectively.
3. It is observed from the graph of %of fly ash vs compressive strength drawn for different combinations of aggregates, that, there is a rising trend of compressive strength with increase in fly ash up to 30 %.
4. It is observed from the graph of %of fly ash vs permeability drawn for different combinations of aggregates, that, there is a decreasing trend of permeability with an increase in fly ash up to 30%.
5. It is observed from the graphs drawn for %of fly ash vs, water absorption, voids ratio and porosity that there is no specific trend either increasing or decreasing.

Acknowledgements
The authors are highly thankful to the Postgraduate students of structural Engineering who helped in conducting various tests in the laboratory.

References

[1] Omkar Deo, Milani Sumanasooriya, Narayanan Neithalath 2010 Experimental study on permeability reduction due to clogging and modelling *ASCE Journal of Materials in Civil Engineering* **22 (7)** p 741-751

[2] C. Lian, Y. Zhuge, S. Beecham 2011 Relationship between porosity and strength for porous concrete *Journal of construction and building materials* **25** p 4294-4298
[3] Bradley J. Putman, Andrew I, Neptune 2011 Comparison of test specimen preparation techniques for pervious concrete pavements *Journal of construction and building materials* **25** p 3480-3485
[4] Sriravindrarajah R, Wang N D H, Ervin L J W 2012 Mix Design for Pervious Recycled Aggregate Concrete *International Journal of Concrete Structures and Materials* **6 (4)** p 239–246
[5] Daniel Jato-Espino, Jorge Rodriguez-Hernandez, Valerio Carlos Andres-Valeri 2014 Fuzzy stochastic multi-criteria model for urban pavement selection *Journal of expert systems with applications* **41** p 6807-6817
[6] Willam D. Martin, Nigel B, Bradley J. Putman 2014 Impact of vertical porosity distribution on the permeability of pervious concrete *Journal of construction and building materials* **59** p 78-84
[7] Ahmed Ibrahim, Enad Mahmoud Yamin, Varun Chowdary Patibandla 2014 Experimental study on Portland cement pervious concrete mechanical and hydrological properties *Journal of construction and building materials* **50** p 524-529
[8] Ruizhong, Kay Wille 2015 Material design and characterization of high-performance concrete *Journal of construction and building materials* **98** p 51-60
[9] Hariyadi, Hiroki Tamai 2015 Enhancing the performance of porous concrete by utilizing the pumice aggregate *Journal of Procedea Engineering* **125** p 732-738
[10] Anush K. Chandrappa, Krishna Prapoorna Biligiri 2016 Comprehensive investigation of permeability characteristics of pervious concrete, It's a hydraulic approach *Journal of Construction and Building Materials* **123** p 627-637
[11] Tejas Joshi, Urmil Dave 2016 Evaluation of permeability, void ratio and strength of pervious concrete with changing the water-cement ratio and aggregate size *Journal of International Civil Engineering and Technology* **7 (4)** p 276-284
[12] Rui Zhong, Kay Wille 2016 Linking pore system characteristics to the behaviour of pervious concrete *Journal of Cement and Concrete Composites* **70** p 130-138
[13] Jiusu Li, Yi Zhang, Guanlan Liu, Xinghai Peng 2017 Preparation and performance evaluation of an innovative pervious concrete pavement *Journal of Construction and Buildings Materials* **138** p 479-485
[14] IS 12727 1989 Indian standard Code of practice for no-fines cast-in-situ cement concrete *Bureau of Indian Standards* New Delhi
[15] IS 456 2000 Indian standard Code of practice for Plain and reinforced cement concrete *Bureau of Indian Standards* New Delhi
[16] IS 10262 2009 Indian standard Concrete mix design proportioning – Guidelines *Bureau of Indian Standards* New Delhi
[17] IS 516 1958 Indian standard methods of tests for strength of concrete *Bureau of Indian Standards* New Delhi
[18] ASTM C1688 Standard Test Method for Density and Void Content of Freshly Mixed Pervious Concrete *American Society for Testing Materials* USA

IOP Conf. Series: Materials Science and Engineering 1025 (2021) 012022 doi:10.1088/1757-899X/1025/1/012022

Mechanical properties of hybrid polypropylene-steel fibre-reinforced concrete composite

Sagi Murali Sagar Varma[1], Ponnada Markandeya Raju[2]

[1]Assistant Professor, Department of Civil Engineering, MVGR College of Engineering (A), Vizianagaram – 535 005, INDIA,
[2]Professor and Head, Department of Civil Engineering, MVGR College of Engineering (A), Vizianagaram – 535 005, INDIA*(Corresponding author)
E-mail: [2]markandeyaraju@mvgrce.edu.in, [1]muralisagariitr@gmail.com

Abstract. Fibres are materials with axial resistance to load that can withstand tensile stresses also when they are mixed in concrete. Recent research shows that the use of fibres improves the mechanical properties in terms of tensile strength and use of hybrid fibres, the mechanical properties further get improved. This paper presents an experimental investigation to quantify the improved Mechanical Properties for different amounts of mono fibres as hook end steel fibres and hybrid fibres that is a combination of hook end steel and polypropylene fibres. Compressive and splitting tensile strengths, modulus of rupture, and modulus of elasticity are studied. Steel fibres are used in the volume fractions of 0%, 0.1%, 0.3%, 0.5%, 0.7%, 0.9%, 1%, 1.25% and polypropylene fibres in a fixed quantity of 400 gm/m^3 in M35 grade concrete. From the results, all the Mechanical properties improve with an increase in the fibre content up to an optimum volume fraction of 0.9%. Addition of steel fibres and polypropylene fibres improves flexural strength and split tensile strengths significantly. However, compressive strength and elastic modulus are slightly improved. It is also observed that the addition of fibres increases the post-peak behaviour of the concrete.

1. Introduction

Concrete is a Strong, reliable, easily mouldable and most sought-after construction. Over seven billion tons of concrete are manufactured on this planet yearly. Fiber Reinforced Concrete (FRC) is a cement composite made with cement, fine aggregate, coarse aggregate, water and fibres. The idea of adding fibres to concrete and other building materials is there since olden days when horsehair has been utilized in mortar and straw in mud blocks. In the 1900s, asbestos strands have been used in concrete. In the 1950s, the idea of utilizing composite materials came into construction practice and FRC had begun to be one of the research areas of interest for concrete technologists. As it was discovered that using asbestos as fibre in concrete is resulting in health hazards for construction workers, the focal point of exploration moved towards finding a substitute for the substance in concrete and other structural materials in place of asbestos. In the 1960s, researchers started studying the use of steel, glass and other manufactured fibres like synthetic, polypropylene and glass in concrete.

Concrete is weak in tension and its tensile strength is around one-tenth of its compressive strength. To overcome this disadvantage, concrete can be reinforced with little randomly dispersed fibres. The collaboration of fibres and concrete ingredients improves numerous properties of concrete like flexibility and reduced crack initiation and propagation because of plastic shrinkage and drying shrinkage. They also reduce permeability and void ratio of concrete. The value of FRC in different Structural Building applications is consequently unquestionable.

The use of SFRC (Steel Fiber Reinforced Concrete) has opened numerous new areas of applications especially in precast, cast in situ, and shotcrete applications. The advantages of SFRC have now been perceived and used in a precast application where designers are searching for more slender segments and progressively complex shapes. Applications include sea- defence, walls and blocks, blast-resistant storage cabins, piles, coffins Pipes, prefabricated storage tanks, highway kerbs, building panels, composite panels and ducts etc. Precast fibre reinforced solid covers and frames are effectively being utilized in India, Europe and USA.

In the ongoing research, Hybrid Fiber Reinforced Concrete (HFRC) in which more than one type of fibre is added to concrete is prominently being utilized in various applications with promising results. The utilization of at least two kinds of fibres resulting in an appropriate blend is found to improve the general properties of concrete thereby bringing about a superior concrete. The process of combination of fibres is sometimes called hybridization.

To optimise overall system behaviour, the mixing of at least two distinct kinds of fibres (diverse in material and or geometry) is getting increasingly normal. The expectation is that the adoption of two or more fibres in concrete would improve the behaviour much superior to that resulted by individual fibres due to synergy between the fibre types. Banthia and Gupta, 2004 [1] characterized this synergy in three categories depending on the mechanisms involved.

1. Fibre constitutive response: One fibre is stiffer and stronger providing strength, while the other is more ductile and provides toughness at high strains.
2. Fibre dimensions: One fibre is very small and controls the formation of microcracks at early stages of loading; the other fibre is larger and provides a bridging mechanism across macrocracks to control the propagation of cracks formed if any.
3. Fibre function: One type of fibre improved fresh mix properties suitable for the required application while the second improved the strength or toughness in the hardened composite.

2. Literature review

The interaction of many fibres and their combinations has already been studied by many researchers. Zhihong and Dorel [2] studied the characteristics of various fibres and their effect on the behaviour of the concrete composites when they are used as reinforcements in concrete. Song and Hwang [3] evaluated some of the mechanical properties of concrete for the various percentage of steel fibres in the volume fraction of 0.5%, 1%, 1.5%, 2%. The Compressive strength is observed to be maximum at 1.5% of volume fraction and is about 15.3% increment. But the Split tensile strength and Modulus of Rupture have got maximum values at 2% of volume fraction, and the improvements are observed to be around 98% and 126% respectively. John et al. [4] observed that the microfibres have been delayed the development of macro cracks so the composite exhibited greater strength and resistance to cracks than the similar matrix which is reinforced with macro fibres only. Yurtseven et al. [5] prepared nine mixes to study the mechanical properties of HFRC. Four types of different fibres have been used in combinations. Among them, two are macro steel fibres and two are microfibres. The volume percentage of fibres has been kept constant at 1.5%. for HFRC the percentage of macro fibres is 1% and the remaining percentage is contributed by microfibres. the compressive strength, split tensile strength, modulus of rupture has been evaluated after 28 days. and these values are compared with the numerical model values to find the effect of the fibre inclusion. The analysis is given that the steel fibres contribute to the strength of the concrete and the microfibres contribute to the ductility of the concrete. Thomas and Ramaswamy [6] conducted an experimental investigation and analytical models have been derived based on the regression analysis to evaluate the influence of adding the fibres on mechanical properties of concrete. Hsie et al. [7] used two types of polypropylene fibres to prepare the HFRC. One is stapled fibres and other is monofilament fibres. The monofilament fibres are used in the quantity of 3 kg/m^3, 6 kg/m^3, 9 kg/m^3. The stapled fibres are used in the quantity of 0.6 kg/m^3. From the study, it is observed that the HFRC is showing improved properties when compared with the single fibre reinforced concrete. Staple fibres have good dispersion capacity and fineness so, they can control the premature cracks. The monofilament fibres have high Young's Modulus and stiffness and these are similar to steel fibres. so, they can withstand more stress during loading. As, the combination can disperse throughout the mixture the drying shrinkage strain of concrete can be reduced. Bencardino et al. [8] prepared the standard cylindrical specimens of plain and steel fibre reinforced concrete for the volume fractions of 1%, 1.6%, 2%. Tests have been done under compression for stress-strain curves and compared with the numerical model results to evaluate the actual stress-strain relations of fibre reinforced concrete. Padmanabha Rao et al. [9] used two types of hooked steel fibres and one type of twisted steel fibres as reinforcement in the concrete. The various mechanical properties like compressive strength, split tensile strength, first crack load, ultimate load, Elastic modulus, toughness is evaluated to know the type and content of the fibre. From the study, it is found that the compressive strength and Elastic modulus are slightly increased with the steel fibre reinforcement in the concrete but the flexural strength is significantly increased with the steel fibre reinforcement. Singh et al. [10] used HFRC in combinations of steel fibre and polypropylene fibres in the volume fraction of 1-0, 0.75-0.25, 0.5-0.5, 0.25-0.75, 0-1 respectively. The mechanical properties like compressive strength, split tensile strength and flexural toughness have been evaluated. From the results, it is been observed that the combination of 0.75% of steel fibres with 0.25% of polypropylene fibres have given the good improvement in compressive strength, flexural toughness as well as split tensile strength. Gencel et al. [11] added monofilament polypropylene fibres to the self-compacting concrete. The study was done with two cement contents of 350 and 450 kg/m^3 as well as with four fibre contents of 3, 6, 9 and 12 kg/ m^3. The results have shown that the compressive strength, split tensile strength, Elastic modulus, flexural strength have increased significantly up to 9 kg/m^3 of fibres. Amir et al. [12] studied the hardened properties of concrete like compressive strength, split tensile strength, flexural strength with the addition of steel fibres and synthetic fibres separately. From the results, it is observed that the steel fibres were found to be effective for flexural strength at an early stage. But at 28 days, the compressive strength, split tensile strength and flexural strength are almost same for two types of fibre reinforced concretes.

2.1. Conclusions from literature

Most of the research projects investigating mechanical properties by mixing of either steel fibres or polypropylene fibres.

A limited number of mostly experimental studies were conducted to investigate the mechanical properties of hybrid fibre reinforced concrete with a combination of micro polypropylene as well as macro steel fibres. Hence the present study has been done by using the combination of steel fibres and polypropylene fibres in concrete. When compared to other micro fibres, polyester is more available fibres as they are the by-product in the textile industry.

3. The objective of the work

The main objective of the work is to evaluate the improvement in various Mechanical Properties (Compressive Strength, Split Tensile Strength, Modulus of Rupture, Elastic Modulus) of concrete by adding macro steel and micro polypropylene fibres in different proportions.

4. Scope of the work

In this study, two types of fibres are used, one is steel fibres and other is polyester fibres. For HFRC, the polyester fibres are taken in fixed quantity as 400gm/m^{3}of concrete. For SFRC and HFRC the steel fibres are varied in percentages (0%, 0.1%, 0.3%, 0.5%, 0.7%, 0.9%, 1%, 1.25%).

For SFRC for each percentage of steel fibres, 5 specimens of cubes (15 cm × 15 cm × 15 cm), 6 specimens of cylinders (3 for split and 3 for the stress-strain curve of size 15 cm × 30 cm), 3 specimens of prisms (50 cm × 10 cm × 10 cm) are cast.

For HFRC also the same specimens are cast but for 0.7%, 0.9%, 1%, 1.25% of steel fibres only. By studying the results, an attempt is made to know the optimum percentage of fibres to be added to achieve good strength as well as the improvement in the mechanical properties by altering the percentage of steel fibres.

IOP Conf. Series: Materials Science and Engineering 1025 (2021) 012022 doi:10.1088/1757-899X/1025/1/012022

5. Experimental program

5.1. *Material details*: Ordinary Portland cement conforming to IS 12269 1983 [13] with specific gravity 3.15, Zone II fine aggregate with specific gravity 2.74 and crushed coarse aggregate of a combination of 4.75 mm size passing and 10 mm retained proportion and 10 mm passing -20 mm retained proportion with specific gravity 2.74 was used in the mix. A combination of Duraflex hook end Steel fibres and Recron 3S polyester fibres are used in this study. Potable water was used for mixing.

5.1.1. Fibbers

a) *Duraflex Hook ends steel fibres* (figure 1).

Duraflex (Hook End Steel) Fibbers are unglued and loose fibres made of high strength steel (> 100 MPa) developed for convenient and cost-effective mixing in concrete as reinforcement. The Fibers are as per ASTMA820 [14] M04 Type I Standards. These fibres in concrete enhance one or more of the following properties of concrete.

- ductility
- overall efficiency
- load-bearing capacity
- flexural Strength
- microcracking Mechanism
- mechanical anchorage substantially
- impact Resistance
- fatigue strength
- flexibility with various types of Portland cement, admixtures and shotcrete mix designs

These fibres are loose and unglued and hence requires less labour and time to place in concrete than with glued steel fibre. It also results in even distribution in the concrete mix in a fast and convenient manner thus effective in avoiding balling, it also reduces overall cost significantly. Table 1 presents the properties of Duraflex hooked end Steel fibres

Table 1. Properties of Duraflex hooked end Steel fibres

Length	Diameter	Aspect ratio	Young's modulus (N/mm^2)
50 mm	0.9 mm	55	210000

Applications. These fibres are presently being used in Shotcrete in Mines/Tunnels, Industrial Floorings, Pavements, Highways, Roadways and Bridges, Precast Structures and elements, Seismic Resistant structures (Dams), Explosive and Impact resistance Structures.

Figure 1. Duraflex hook end steel fibres

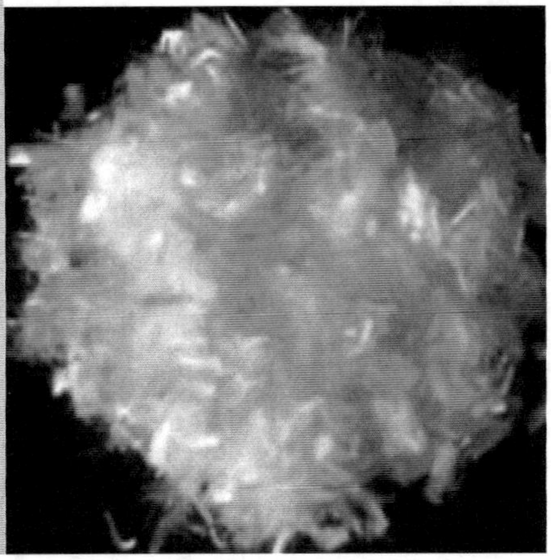

Figure 2. Recron 3S polyester fibres

b) Recron 3S Polyester Fibres (figure 2)
Recron 3S polyester fibres used in this study belong to the thermoplastic polymer group and are available in monofilament form. They are a by-product from the textile industry and are temperature sensitive and hydrophobic in nature. These non-biodegradable fibres unlike glass, steel or carbon fibres and belong to the low-modulus fibres group. They are added to the concrete as 0.1 % by volume to

- control cracking due to plastic and drying shrinkage in concrete.
- improve the abrasion resistance of concrete.
- reduce the water permeability,
- reduce rebound loss
- increase ductility.

The extent of improvement in any property depends on the length, diameter and dosage of the fibre. Table 2 presents the properties of Recron 3S Polyester Fibres

Table 2. Properties of Recron 3S Polyester Fibres

Property	Value
Cut length	6-12mm
Shape	Special for the improved holding of cement aggregates
Tensile strength	4000-6000 kg/cm^2

5.1.2. Mix Design
The mix proportions adopted in this study are shown in Table 3.

Table 3. Mix design

Description	Mix proportion (by weight)
Cement	1
Sand (Fine aggregate)	2
Coarse aggregate	2
Water	0.5

5.1.3 Specimen making procedure:
Locally available potable water was used for mixing concrete. Iron cube, cylinder and beam moulds of standard size are used for casting specimens. Both table and needle vibrator were used to thoroughly compacting the concrete in the moulds. The moulds were sealed by plaster of Paris avoid leakage of cement slurry through the gaps of the joints. The inside edges of all the moulds are oiled before pouring concrete the miller mixed concrete ingredients as per mix design into them. After compacting the concrete in the moulds, the top surface of the specimens was levelled with a trowel. After demoulding the concrete specimens after 24 hours of casting, the specimens are immersion cured in potable water for a period of 28 days for curing.

6. Test setup and testing procedure

6.1. Testing compressive strength
Cubes of size 150 mm are placed under the 3000 kN CTM as shown in figure 3 and the rate of loading is applied at a rate of 5kN/sec. The maximum load that the cube can take under compression will be considered as the compressive strength of the cube.

6.2. Testing for split tensile strength
For testing the split tensile strength of the concrete standard cylinders of diameter 15cm and height 30cm are used. The cylinder specimens are placed under CTM as shown in figure 3. The load is applied at a rate of 5kN/sec. and the maximum load at which the cylinder fails will be noted. The splitting tensile strength of concrete is calculated by using the following formula

Split tensile strength T =2P/ΠDL
Where,
P=the maximum load taken by the cylinder at failure
D=diameter of the specimen
L=length of the specimen.

6.3. Testing of cylinders for Stress-strain curve

For obtaining the Modulus of Elasticity of concrete, the stress-strain curve for the standard cylinder specimen of diameter 15cm and height 30cmare used. Longitudinal compresso-metre with an attached dial gauge is attached to the cylinder as shown in figure 3 and then is placed under the CTM. The load is applied and the corresponding deflection is noted in the dial gauge. From the recorded values Stress-Strain curves are plotted. The slope of the Stress-Strain curve will give the Young's Modulus.

6.4. Testing of prisms

Standard prisms of square cross-section 10cm and length 50cm are tested by4-point loading as shown in figure 3. The load is applied in the UTM and the corresponding deflection is noted. The modulus of rupture will be calculated as $3FL/4BD^2$

Where F is the maximum load at the fracture point, L is the length between the supports, B is the width of specimen and D is the depth of specimen

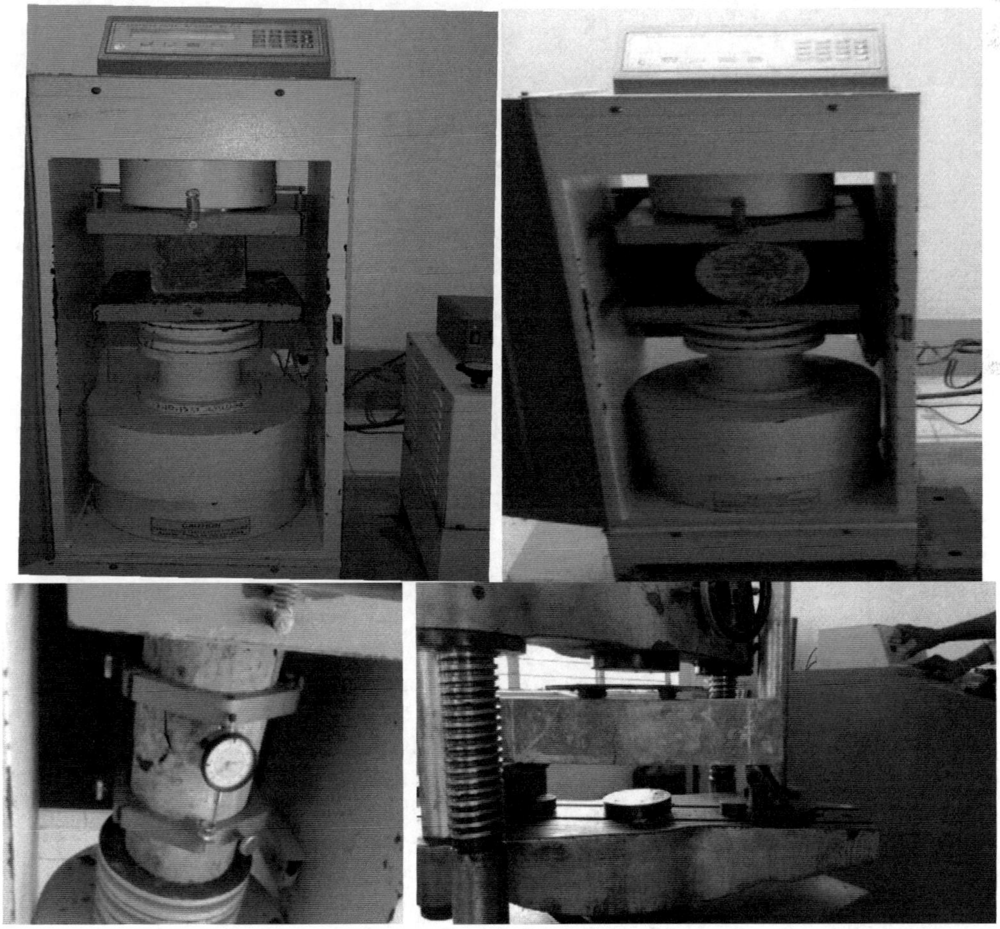

Figure 3. Specimens under testing for Compressive strength, Splitting tensile strength, Modulus of Elasticity and Flexural strength (respectively)

IOP Conf. Series: Materials Science and Engineering 1025 (2021) 012022 doi:10.1088/1757-899X/1025/1/012022

7. Results and Discussions

The experimental results of all specimens with different combinations of Steel fibre contents are presented in this section. Their behaviour throughout the test is described using recorded data of all the specimens that are tested for their ultimate strengths. 0% of steel fibres specimens are taken as the control specimens for comparison. It is observed that the control specimens had less load carrying capacity values compared to that of the other specimens. The study is to evaluate various mechanical properties. They are compressive Strength, Split Tensile Strength, Modulus of Rupture, Stress-Strain Curves, Young's Modulus. The recorded test results in graphical form are presented along with discussion as follows.

7.1. Compressive strength

It can be observed from figure 4 that for concrete with addition of steel fibres alone, the Compressive strength is following a slight increment profile with the increase in the percentage of steel fibres up to 0.9% of steel fibres. After that a slight decrease in the compressive strength is observed i.e., the addition of steel fibres to the plain concrete is increasing its compressive strength by 9.17% at 0.9% of steel fibres after that no significant improvement even fibres are added.

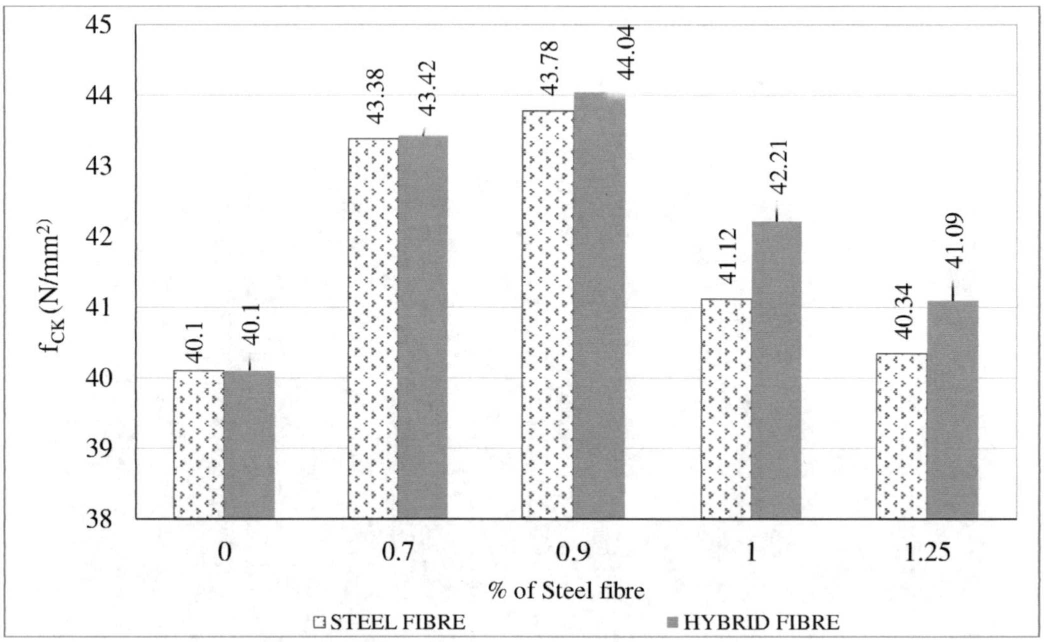

Figure 4. Comparison of compressive strength

7.2. Split tensile strength

It can be observed from figure 5 that for concrete with addition of steel fibres alone, the Split Tensile Strength is following a slightly increasing profile up to 0.9% of fibres after that a slight decreasing profile i.e., the addition of steel fibres is slightly increasing the split tensile strength when compared with the plain concrete. The split tensile strength is increased by 52% at 0.9% of steel fibres when compared with the 0% of steel fibres. After 0.9% of steel fibres, the increase in the percentage of steel fibres is not significant.

The split tensile strength is slightly increased for HFRC when compared with the SFRC. The split tensile strength is increased by 58.24% for HFRC at 0.9% of steel fibres which is slightly greater than 52% for SFRC i.e., the addition of micro-fibres improving the split tensile strength of concrete.

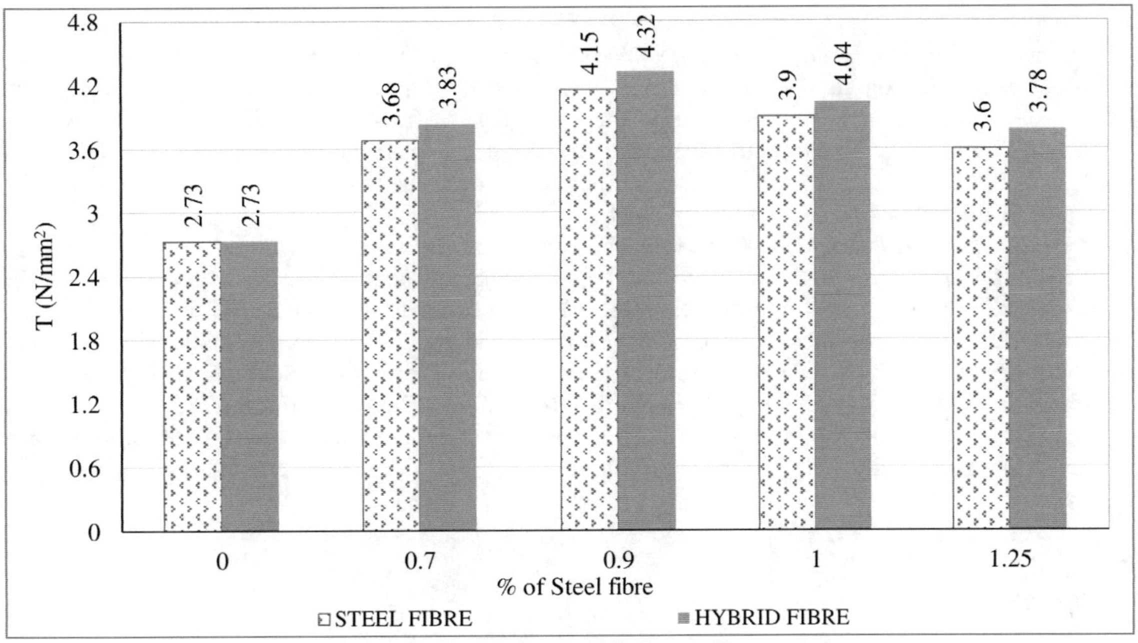

Figure 5. Comparison of Split tensile strength

7.3. Comparison of peak load

From figure 6, the peak load in the flexure test is increasing with the increase in the percentage of steel fibres. It is maximum at 0.9% of steel fibres. When comparing SFRC with the HFRC the increase in the peak load is more for HFRC. For HFRC the peak load is increased by 218.41% at 0.9% of fibres, which is greater than 174.88% for SFRC at 0.9% of steel fibres when compared with the 0% of steel fibres. i.e. the addition of microfibres will increase the load-carrying capacity in flexure.

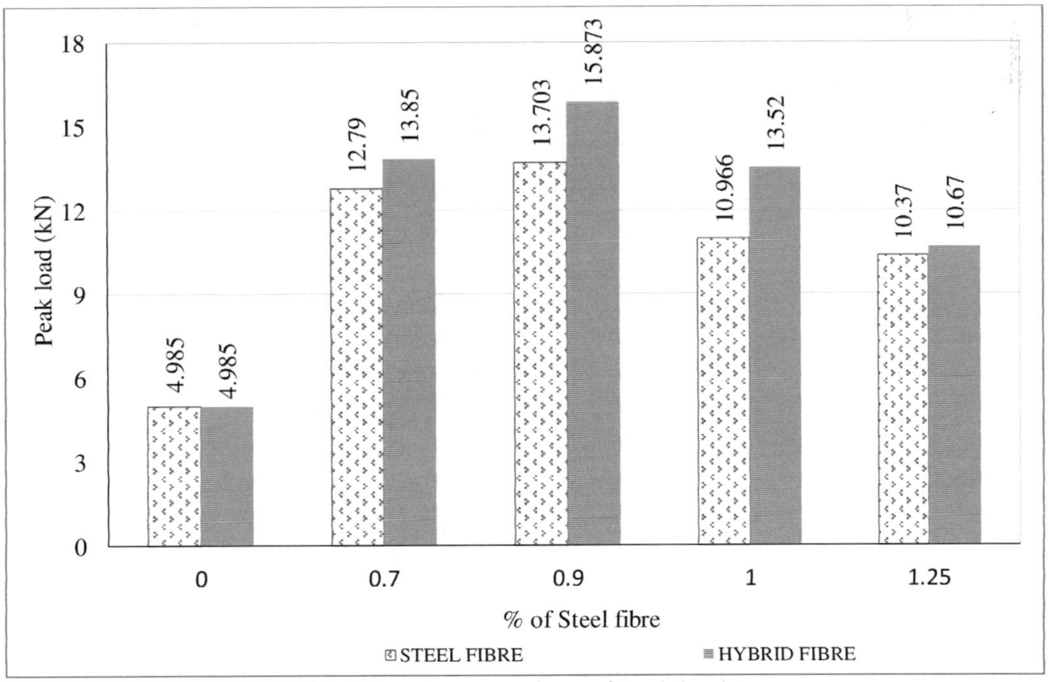

Figure 6. Comparison of Peak load

7.4. Individual Load-Deflection behaviour of Hybrid and Steel Fiber Reinforced concrete

From the load-deflection curves shown in figure 7, it is clear that the peak load, as well as the peak deflection, are increasing with an increase in the percentage of steel fibres for both SFRC as well as HFRC when compared with the control specimen. The peak load is increased by 171.28% and deflection by 51.72% at 0.9% of steel fibres for HFRC. Whereas the peak load is increased by 218.41% and the deflection by 60.68% for HFRC at 0.9% of steel fibres.

7.5. Comparative Load-Deflection behaviour of Hybrid and Steel Fiber Reinforced concrete

From all four graphs in figure 8, it is clear that the load-carrying capacity, as well as the deflection, are increasing with the addition of microfibres in HFRC when compare with the SFRC. For 0.7% of steel fibres, HFRC is exhibiting increment of 8.28% in peak load and 7.6% in deflection when comparing with the SFRC. Similarly, at 0.9%of steel fibres HFRC is exhibiting an increment of 15.8% in peak load and 15.9% in deflection, at 1%of steel fibres HFRC is exhibiting increment of 15.3% in peak load and 13.6% in deflection, at 1.25% of steel fibres HFRC is exhibiting an increment of 14.65% in peak load and 12.26% in deflection when compare with the SFRC at the same percentage of steel fibres respectively.

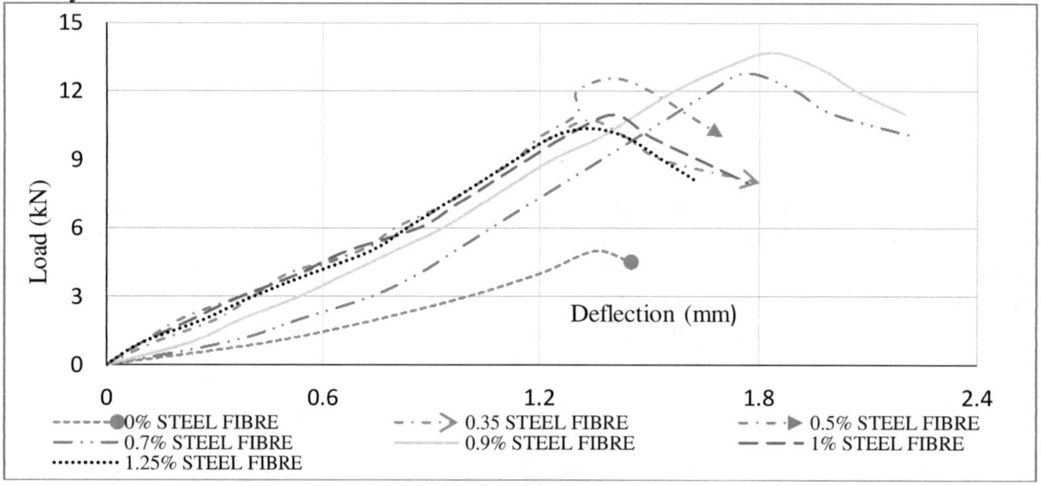

a) Steel fibre

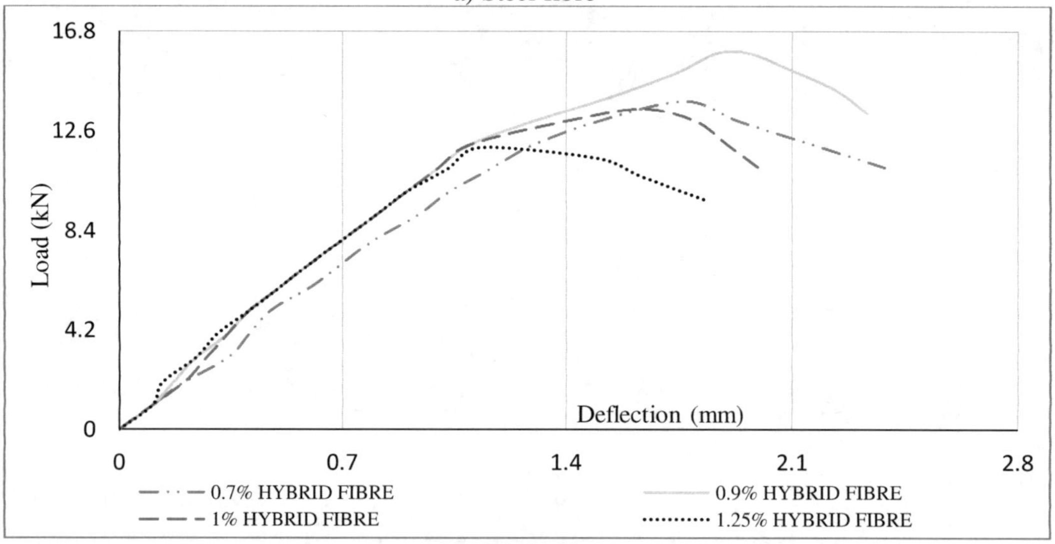

b) Hybrid fibre

Figure 7. Individual Load – deflection behaviour of Hybrid and Steel Fiber Reinforced concrete

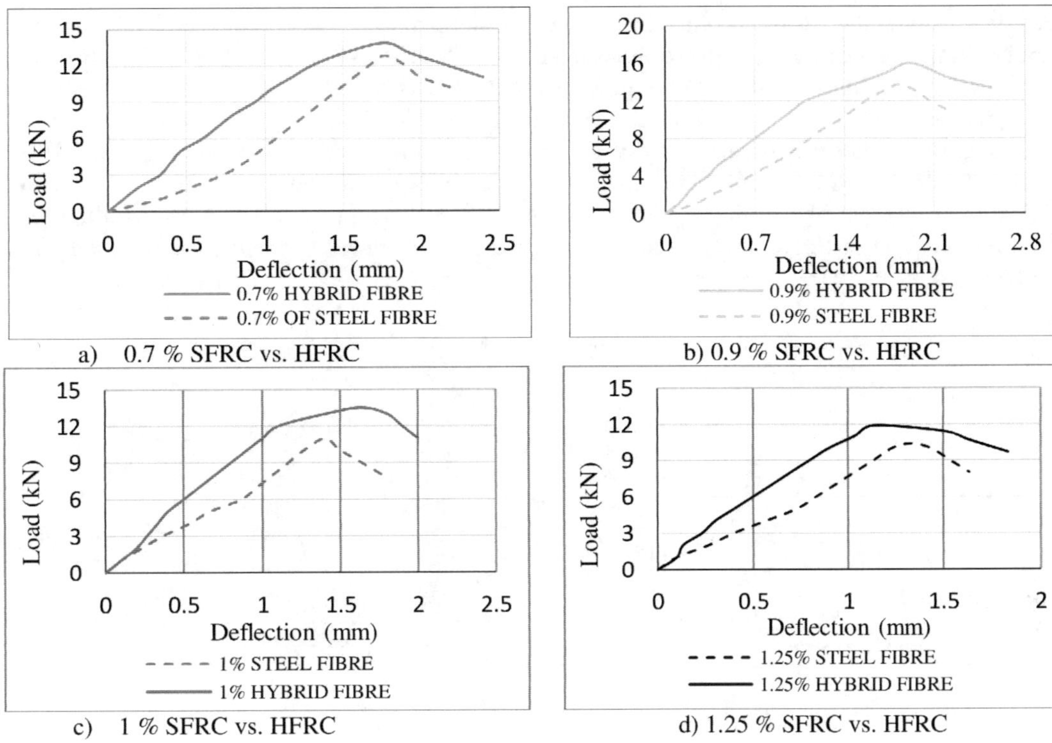

Figure 8. Comparative Load-Deflection behaviour of Hybrid and Steel Fiber Reinforced concrete

7.6. Comparison of Modulus of rupture

From figure 9, it is observed that for concrete with addition of steel fibres alone, the Modulus of Rupture is following an increment profile up to 0.9% of steel fibres after that it is slightly decreasing.

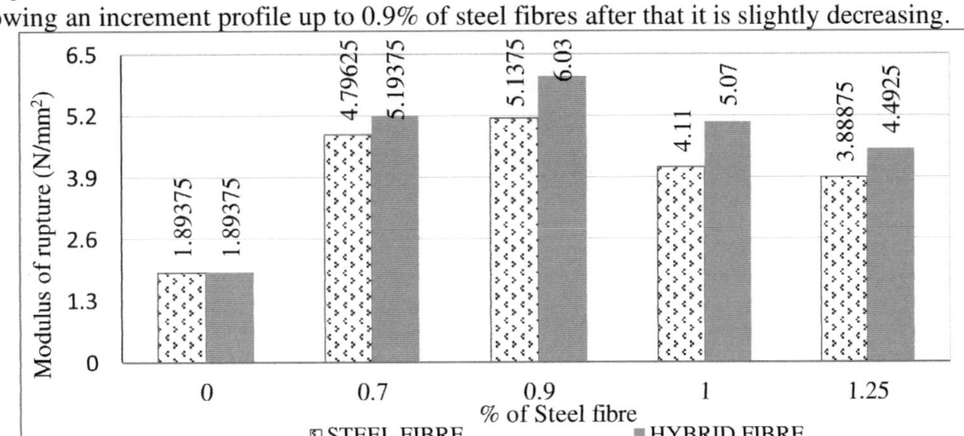

Figure 9. Comparison of Modulus of Rupture for SFRC and HFRC

The Modulus of Rupture is increased by 171.28% at 0.9% of steel fibres when compared with the 0% of steel fibres. The modulus of rupture is increasing with the increase in the % of steel fibres for both HFRC and SFRC. For SFRC the Modulus of Rupture is increased by 171.28% and for HFRC it is increased by 218.41%. Adding of microfibres is improving the Modulus of rupture by 47.13%.

7.7. Comparison of Young's modulus

Young's modulus is calculated as the slope of the initial tangent drawn on the stress-strain curves as shown in figures 11 and 12. The value of Young's modulus is calculated and compared for different amounts of fibres proportions and shown in figure 10. It is observed that for concrete with the addition

of steel fibres alone, the Elastic Modulus is following the increasing profile up to 0.9% of steel fibres after that a slight decrease in the profile observed. After 0.9% of steel fibres, the Elastic Modulus became constant. i.e., the addition of steel fibres greater than 0.9% is not effective. The Elastic Modulus is increased by 10.15% at 0.9% of steel fibres when compared with the 0% of steel fibres.
The modulus of Elasticity is increasing with the increase in the % of steel fibres for both HFRC and SFRC. For SFRC the Modulus of Elasticity is increased by 10.15% and for HFRC it is increased by 12.65% at 0.9% of steel fibres when compared with the 0% of steel fibres. The stress-strain curves for all the combinations of SFRC are presented in figure 11 and for all the combinations of HFRC are presented in figure 12.

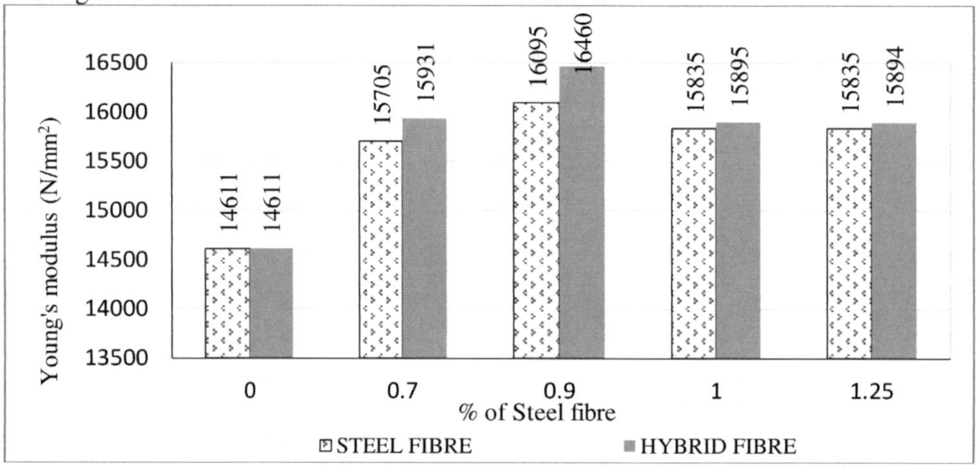

Figure 10. Comparison of Young's Modulus for SFRC and HFRC

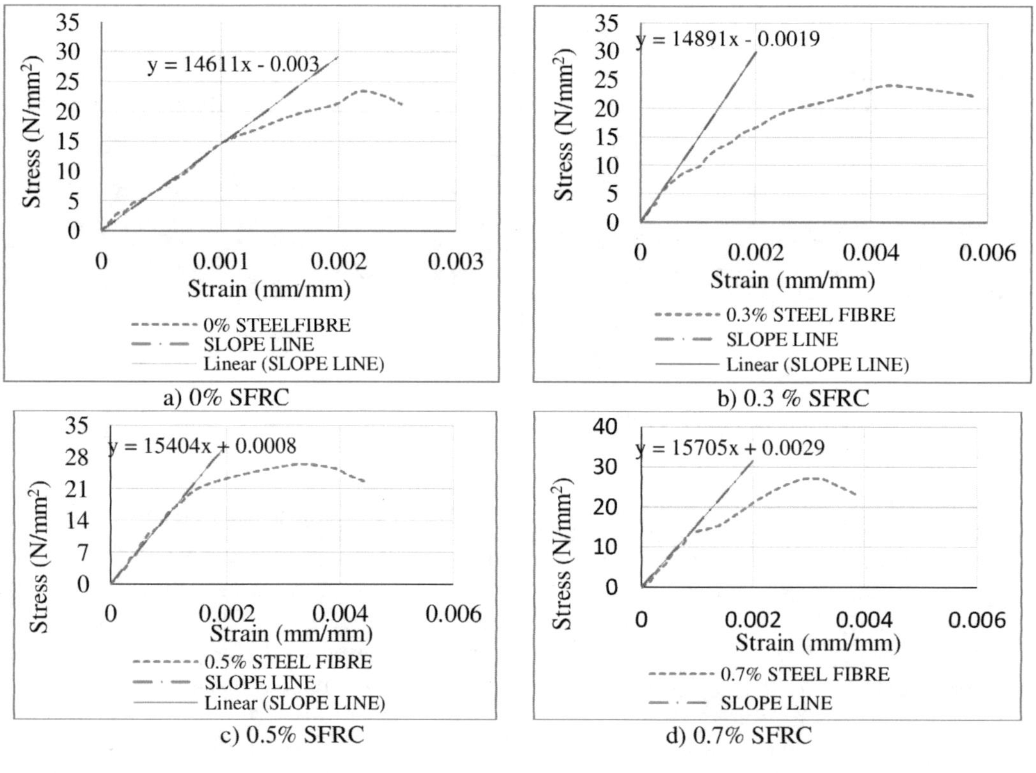

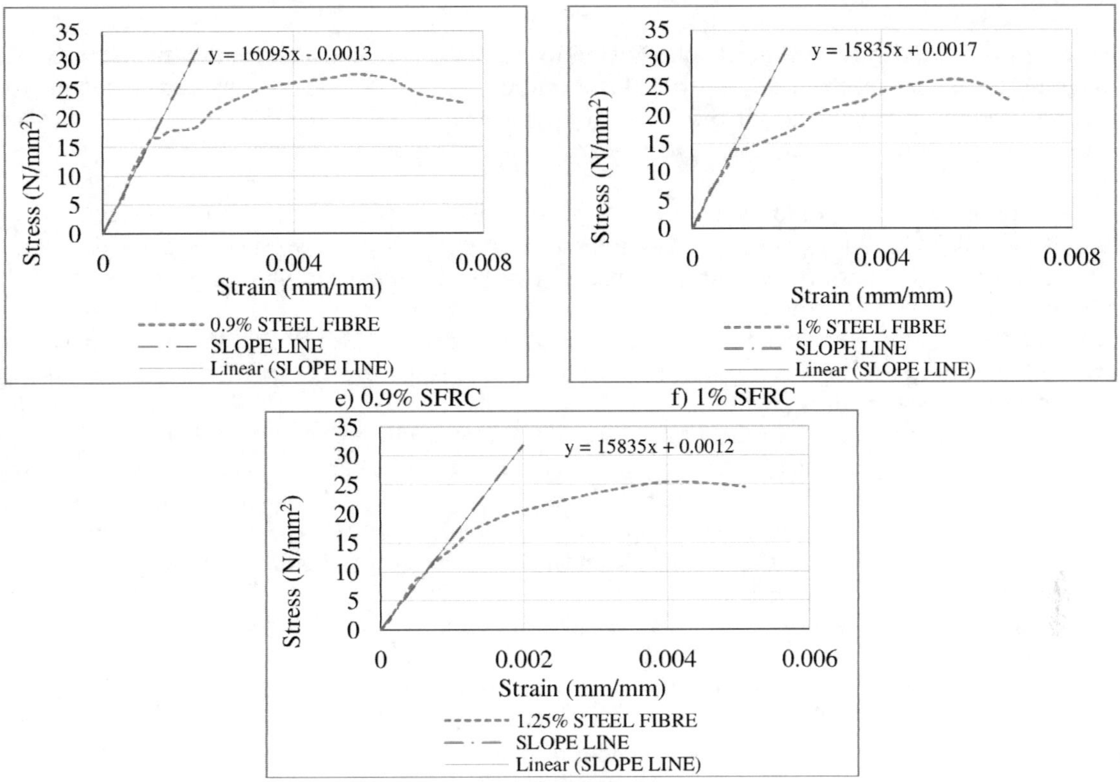

Figure 11. Stress-strain curves for finding Young's modulus of SFRC

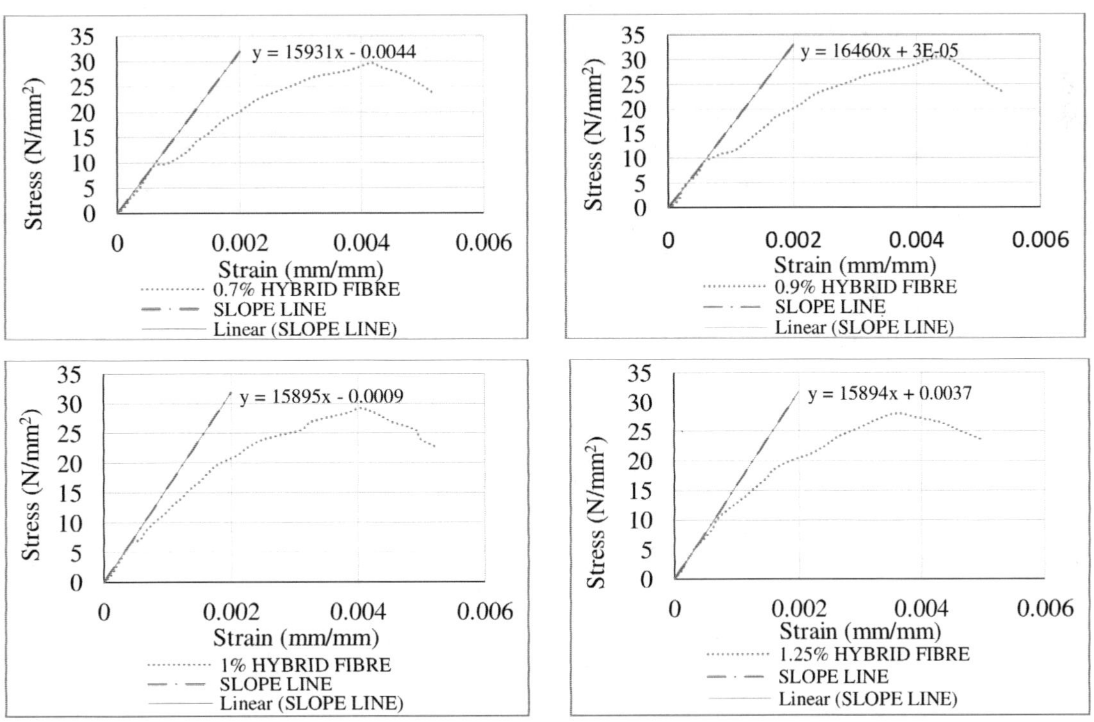

Figure 12. Stress-strain curves for finding Young's modulus of HFRC

8. Conclusions

The Mechanical properties of concrete play a vital role in improving the applicability of concrete. The change in the various mechanical properties has been reported for variation in the percentage of steel fibres and various combinations of hybrid fibres from the analysis of the recorded data. After analysing the results, the following conclusions have been made.

1. The addition of steel fibres to the plain concrete is increasing its compressive strength by 9.17% at 0.9% of steel fibres in SFRC. After that no significant improvement even when fibres are added.
2. For HFRC the compressive strength is increased by 9.82% which is slightly higher than 9.17% at 0.9% of fibres.
3. The split tensile strength is increased by 52% at 0.9% of steel fibres for SFRC. When compared with the 0% of steel fibres. After that, no significant improvement is observed even after fibres are added.
4. The split tensile strength is slightly increased for HFRC when compared with the SFRC. The split tensile strength is increased by 58.24% for HFRC at 0.9% of steel fibres which is slightly greater than 52% for SFRC.5. The peak load in flexure test is increasing with the increase in the percentage of steel fibres. It is maximum at 0.9% of steel fibres. When comparing SFRC with the HFRC the increase in the peak load is more for HFRC. For HFRC the peak load is increased by 218.41% at 0.9% of fibres, which is greater than 174.88% for SFRC at 0.9% of steel fibres when compared with the 0% of steel fibres.
6. The peak load is increased by 171.28% and deflection by 51.72% at 0.9% of steel fibres for HFRC. Whereas the peak load is increased by 218.41% and the deflection by 60.68% for HFRC at 0.9% of steel fibres.
7. The load-carrying capacity, as well as the deflection, are increasing with the addition of microfibres in HFRC when compare with the SFRC. For 0.7% of steel fibres, HFRC is exhibiting increment of .28% in peak load and 7.6% in deflection when comparing with the SFRC. Similarly, at 0.9%of steel fibres HFRC is exhibiting an increment of 15.8% in peak load and 15.9% in deflection, at 1%of steel ibres HFRC is exhibiting increment of 15.3% in peak load and 13.6% in deflection, at 1.25% of steel fibres HFRC is exhibiting an increment of 14.65% in peak load and 12.26% in deflection when compare with the SFRC at the same percentage of steel fibres respectively.
8. The Modulus of Rupture is increased by 171.28% at 0.9% of steel fibres for SFRC when compare with 0% of steel fibres.
9. For SFRC the Modulus of Rupture is increased by 171.28% and for HFRC it is increased by 218.41%. Adding of microfibres is improving the Modulus of rupture by 47.13%.
10. The Elastic Modulus is increased by 10.15% at 0.9% of steel fibres for SFRC when compare with 0% of steel fibres.
11. For SFRC the Modulus of Elasticity is increased by 10.15% and for HFRC it is increased by 12.65% at 0.9% of steel fibres when compared with the 0% of steel fibres.
12. From all the results at 0.9% of steel fibres the maximum increment in all properties will be achieved. So, 0.9% of steel fibres can be taken as an optimum percentage of steel fibres.

9. Future scope

The present study is done on varying the percentage of steel fibres and constant quantity of micro(polyester) fibres as 400gm/m^{3}of concrete. So, the study can be extended to variation of the mechanical properties by varying the number of microfibres along with the steel fibres. The optimum combination of steel fibres, as well as micro-fibres, can be derived. A lot of scope is there for the present study by using various types of micro as well as macro fibres

References

[1]. Nemkumar Banthia, Rishi Gupta 2004 Hybrid fiber reinforced concrete (HyFRC): Fiber synergy in high strength matrices *Materials and Structures* **37 (10)** p 707-716
[2]. Zheng Z Feldman D 1995 Synthetic Fibre-Reinforced Concrete *Progress in Polymer* **20** p 185-210

[3]. Song P S, Hwang S 2004 Mechanical properties of high-strength steel fiber-reinforced concrete Elsevier *Construction and Building Materials* **18 (9)** p 669-673

[4]. John S, Lawler, Davide Z, Surendra P and Shah2005 Microfiber and Macrofiber Hybrid Fiber-Reinforced Concrete *ASCE Journal of Materials in Civil Engineering* **17 (5) p** 595-604

[5]. Yurtseven A E, Yaman I, Tokyay M 2006 Mechanical Properties of Hybrid Fiber Reinforced Concrete Springer *Journal of Measuring, Monitoring and Modelling Concrete Properties* Springer, Dordrecht.

[6]. Thomas J, Ramaswamy A 2007 Mechanical Properties of Steel Fiber-Reinforced Concrete *ASCE Journal of Materials in Civil Engineering* **19 (5)** p 385-392

[7]. Hsie M, Chijen Tu, Song P S 2008 Mechanical Properties of Polypropylene Hybrid Fiber-Reinforced Concrete *Elsevier Materials Science and Engineering A* **494** p 153–157

[8]. Bencardino F, Rizzuti L, Spadea G, Ramnath N Swamy 2008 Stress-Strain Behaviour of Steel Fiber-Reinforced Concrete in Compression *ASCE Journal of Materials in Civil Engineering* **20 (3)** p 255-263

[9]. Padmanabha Rao T, Mo Y L Thomas T C, Hsu, John V 2009 Mechanical Properties of Steel Fiber Reinforced Concrete Beams *ASCE Structures Congress* **2009** p 1-10

[10]. Singh S P, Singh A P, Bajaj V 2010 Strength and Flexural Toughness of Concrete Reinforced with Steel – Polypropylene Hybrid Fibres *Asian Journal of Civil Engineering (Building and Housing)* **11 (4)** p 495-507

[11]. Gencel O, Ozel C, Brostow W, Martı´nez-Barrera G 2011 Mechanical properties of self-compacting concrete reinforced with polypropylene fibres *Materials Research Innovations* **15 (3)** p 216-225

[12]. Amir M, Alani, Aboutalebi M 2013 Mechanical Properties of Fibre Reinforced Concrete- A Comparative Experimental Study *International Journal of Civil, Environmental Structural Construction and Architectural Engineering* **7 (9)** p 646-651

[13]. IS 12269 1983 Indian Standard Specification for 53 grade ordinary Portland cement *Bureau of Indian Standards* New Delhi

[14]. ASTM A820/A820M - 16 Standard Specification for Steel Fibers for Fiber-Reinforced Concrete *American Society for Testing and Materials* USA

Effect of Fibres on the behaviour of Bottle shaped strut

Murali Sagar Varma Sagi[1], S Suriya Prakash[2]

[1]Research Scholar, Department of Civil Engineering, IIT Hyderabad, Hyderabad (and)
[1]Assistant Professor, Department of Civil Engineering, MVGR College of Engineering (A), Vizianagaram – 535 005, INDIA,
[2]Professor, Department of Civil Engineering, IIT Hyderabad, Hyderabad, India
Email: muralisagariitr@gmail.com

Abstract: Bottle-shaped struts are the critical elements in the design of D-regions using Strut-and-Tie method. The transverse tension develops due to dispersion of compression load which further leads to splitting crack in bottle shaped strut. Due to this, the strut fails before reaching its ultimate compression capacity. Higher resistance to the transverse tension can improve the strut capacity and its efficiency in load transfer. Transverse tensile stress can be resisted by providing the steel reinforcement or through the addition of discrete fibres in the concrete. Many international codes like ACI, AASTHO, and CSA have suggested guidelines about the addition of steel fibre reinforcement in the bottle-shaped struts for resisting the transverse tension. Still, the influence of the discrete fibres on the performance of the bottle-shaped strut is not well established. The performance of the bottle-shaped strut in terms of efficiency factors, crack pattern and failure mode for different amounts of macro steel fibres and micro polypropylene fibres are studied using experimental investigation. Specimens of 600mm x 600mmx 100mm size were tested under compression. Steel fibres are added in the proportions of 0.7%, 0.9% and 1.1% volume fractions to the concrete. Effect of fibre hybridization also studied by adding micro polypropylene fibres in the proportions of 1% and 2% in addition to the steel fibres. Experimental results showed that adding discrete fibres in concrete significantly improved the resistance to the transverse tension in bottle-shaped struts and led to the increased load-carrying capacity of the specimens. A 75% improvement in the efficiency factor is observed at 0.9% volume of steel fibre addition. Addition of micro polypropylene fibres to the macro steel fibres further enhanced the load-carrying capacity of the bottle-shaped struts. Microfibres in the concrete effectively arrested the micro-cracks and delayed the occurrence of a first splitting crack in the strut region. Due to this the mode of failure changed to ductile through the formation of a greater number of small cracks with less crack width at the ultimate load. Results of this study clearly show that the addition of discrete fibres to the concrete is an effective solution to improve the performance of the bottle-shaped struts in terms of ultimate strength and serviceability.
Keywords: Bottle-Shaped Struts, Steel fibres, Hybrid fibres, Cement and Concrete

1. Introduction

Structural members can be categorized into two portions namely Bernoulli regions (B-regions) and Disturbed regions (D-regions). B-regions are the portion of members where beam theory applies i.e. Bernoulli's hypothesis is valid. In D-regions, the strain profile is significantly non-linear and hence the basic assumption of 'plane section remains plane in an initially straight beam remains plane after bending' is not valid. D-regions forms at the geometric discontinuities, which are adjacent to openings, abrupt changes in the cross-section, or at the statical discontinuities, which are regions near the concentrated loads and support reactions. Structural members with significant extents of D-regions are called Non-flexural members, examples Deep beams, Pile caps, Corbels, Dapped-ends etc. D-regions cannot be designed using elastic theory due to invalidity of Bernoulli's hypothesis. The Strut-and-Tie method (STM) is a rational method which is useful in the design of both B & D-regions in concrete members. In STM, the complex D-region is modelled as a truss carrying the applied loads to the supports or the adjacent B-regions. This truss is known as the StrutandTie model. The Strut and Tie model consists of compression members and tension members which are connected at nodes. Struts carry compressive forces and denoted using broken lines whereas ties carry tensile forces and denoted with solid lines. Struts represent compressive stress fields with principal stresses are predominantly in the direction of the strut axis. Struts are categorized based on the geometric shapes and load transfer mechanism as prismatic, fan-shaped and bottle-shaped as shown in figure 1. Prismatic struts (figure 1a) have uniform cross-section throughout the length. Fan-shaped struts are formed when compressive stresses flow from a large area to smaller area as in the case of a deep beam subjected to uniformly distributed load as shown in figure 1c. Bottle-shaped strut is diagonal strut which is formed between the concentrated load and adjacent support as shown in figure 1b.

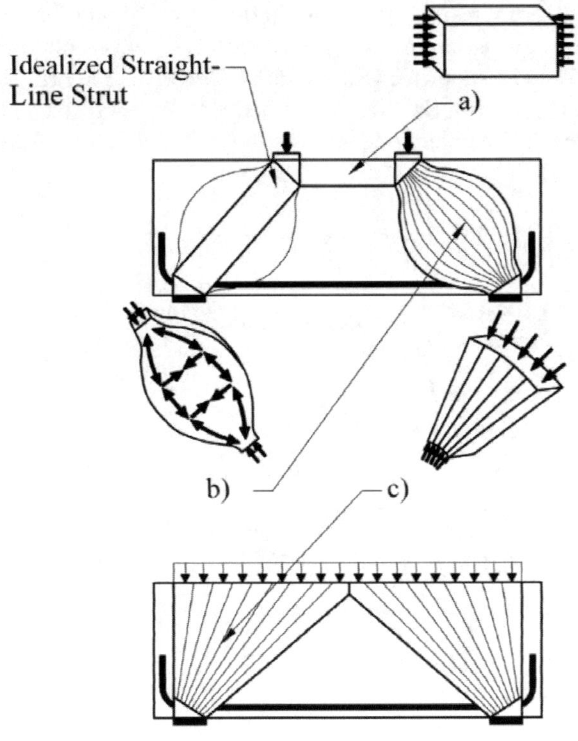

Figure 1. Types of Struts a) Prismatic strut b) Bottle- shaped strut c) Fan-shaped strut

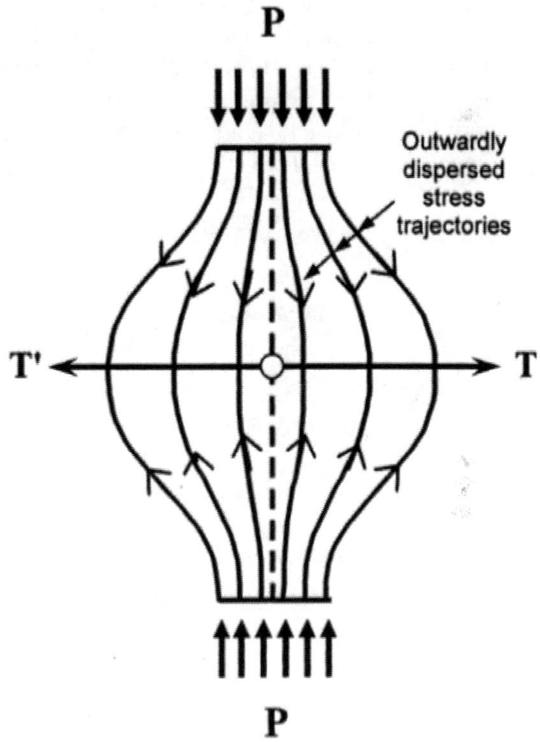

Figure 2. Dispersion of compressive forces in the bottle-shaped strut

1.1 Bottle-shaped strut

Bottle-shaped strut forms when the external load is applied in a relatively small area and the compression disperses as they flow through the member as shown in figure 2. During this dispersion, tension perpendicular to the strut axis develops to counteract the lateral component of the angled compression. Concrete being weak tension, the developed transverse tension in the strut cause longitudinal splitting and failure of the bottle-shaped struts much before the compressive stress reaches the compressive strength of concrete. The behaviour of the cracked concrete strongly influences the compressive strength of bottle-shaped struts in the strut. Compared to prismatic and fan-shaped struts, the bottle-shaped strut is critical due to the existence of transverse tension in the strut region. Resistance to transverse tension is essential in bottle-shaped struts to not only increase its strength but also to control cracking at service loads. The transverse reinforcement provisions in bottle-shaped struts vary from one code of Practice to another. The ACI 318-11 [1] provisions for evaluating the strength and minimum reinforcement requirements of a bottle-shaped strut are discussed here. As per ACI 318-11 [1], the theoretical load-carrying capacity of the bottle-shaped strut is

$$ß_s 0.85 f_c' A_S \quad (1)$$

where
$ß_s$= strut efficiency factor suggested by ACI and is listed in Table 1
f_c' = Compressive strength of cylinder
A_s = Minimum cross-section area of the strut (loaded area)

The ACI (American Concrete Institute) identifies two types of bottle-shaped struts namely reinforced and unreinforced struts and the efficiency factors specified for these two types of struts for normal concrete are 0.75 and 0.60 respectively for cylinder compressive strengths up to 40 MPa. The larger efficiency factor of 0.75 can be used for bottle-shaped struts having minimum reinforcement as per (A-4) of ACI 318-11 [1]. It means ACI allows un-reinforced struts with lesser factors. The efficiency factor of the strut is a very important parameter to design the strut in Strut-and-Tie method. Actual resistance to the transverse tension in the bottle-shaped strut can be understood indirectly from the value of the efficiency factor obtained experimentally.

Table 1. Strut and Nodal efficiency factors in ACI 318-11[1] (American Concrete Institute 2011)

Type of strut	**Efficiency factor ($β_s$)**	**Type of Node**	**Node Efficiency factor ($β_n$)**
Prismatic strut	1.00	CCC node	1.00
Bottle-shaped strut with reinforcement as per equation (A-4) of ACI 318	0.75	CCT node	0.80
Bottle-shaped strut without reinforcement as per equation (A-4) of ACI 318	0.60 λ	CTT node or TTT node	0.60
Struts in tension members	0.40		
Other types of struts	0.60 λ		

Note: λ=1.0 for normal weight concrete, 0.85 for sand light weight and 0.75 for all-light weight concrete.

Experimental strut efficiency factor is calculated by equating the theoretical strut capacity to the actual load resisted before it fails. In equation 1, 0.85 f_c' is the design compressive strength of the concrete. So while calculating the experimental efficiency factor, only f_c' will be considered as the compressive strength of concrete. Therefore, the experimental efficiency factor can be calculated as per Equation 2.

$$ß_{se} = P_u / (A_s \times f_c') \quad (2)$$

1.2 Research Motivation

Sahoo et al. [2] found that bottle-shaped struts should be provided with some reinforcement to resist transverse tension. Provision of minimum reinforcement is essential to meet the serviceability requirements. Also, the unreinforced struts will fail in brittle mode with a single splitting crack along the length of the strut, which is not at all advisable. In some structural members like pile caps, beam-column joints etc., providing the steel reinforcement in the strut region may not be possible due to congestion and inaccessibility. Therefore, there is a need to find an alternative solution to steel reinforcement in the strut region for better performance of bottle-shaped strut. Thomas et al. [3] concluded that discrete steel fibres in the concrete improves the tensile strength and develop bridging action after the first crack for better post crack behaviour also. Hsie et al. [4] suggested that microfibres in the concrete help to arrest the micro-cracks and crack widths. Addition of macro and microfibres in the concrete may get the combined advantages of improved tensile strength, reduced crack width and better post-peak behaviour with a ductile mode of failure. Therefore, in this study, it is attempted to use the discrete macro and microfibres in the concrete to study the performance of bottle-shaped struts. Manu other researchers worked on the behaviour of bottle-shaped struts in the recent years ([5], [6], [7], [8], [9] and [11]).

2. Experimental Program

Concrete panels of size 600 mm × 600 mm × 100 mm with different proportions of fibres are tested with in-plane loading as shown in figure 3. Isolated bottle-shaped struts will form in the panel for the adopted loading arrangement. It is observed from the literature that optimum dosage of hooked end steel fibres is around 1% for better tensile strength of concrete and therefore steel fibres are varied around 1% only. Polypropylene fibres are added in 0.1% and 0.2% volume fraction, to mix with optimum steel fibres for higher efficiency factor of a bottle-shaped strut. Table 2 presents the specimen details.

Table 2. Specimen details

Sl. No.	Grade of Concrete	% V_f of macro Steel fibres	% V_f of micro PP fibres
1		0	0
2		0.7	0
3		0.9	0
4	M35	0.9	0.1
5		0.9	0.2
6		1	0
7		1.1	0

3. Materials

The cement of 53 grade, river sand confirming to zone II as fine aggregate, crushed granite of 20 mm nominal size as coarse aggregate is used to make the panels with M35 grade concrete. Mix proportions are calculated as per IS 10262 2009 [12] and the weight of ingredients are mentioned in Table 3 per one cubic meter of the concrete.

Table 3. Mix proportions of concrete(by weight)

w/c ratio	wt. of cement	wt. of water	wt. of FA	wt. of CA
0.5	440.1 kg	220.4 lit.	764.4 kg	1008.4 kg

Macro hooked end steel fibres and Recron 3S micro polypropylene fibres are used in this experimental study. Physical properties of the fibres are mentioned the Table 4 and 5, respectively.

Table 4. Physical properties of hooked end steel fibres

Sl. No.	Property	Details
1	Fibre Length	35mm
2	Fiber diameter	0.60mm
3	Young's modulus	210000Mpa
4	Tensile strength	1100Mpa
5	Relative density	7.8 g/cm^3

Table 5. Physical properties of Recron 3S microfibres

Sl. No.	Property	Details
1	Specification	Microfiber
2	Colour	White
3	Cut length	12mm
4	Tensile Strength	4000kg/cm^2
5	Relative density	0.9 g/cm^3
6	Modulus of Elasticity	17500 MPa

4. Casting and testing of specimens

Panels were cast under controlled laboratory conditions using standard equipment. For each batch of concrete, the quantities of required materials were kept ready in the required proportion and were mixed well using concrete mixer. The inside surface of the steel mould was coated with shutter-release oil before pouring concrete into the mould. Along with every panel, three cubes and three cylinders were cast for determining the compressive strength of concrete. All the panel specimens were tested after a normal curing period of 28 days. The panel specimens were tested under single point load in a loading frame. The load was applied through hydraulic jack, and the applied loads were read using a 1000 kN capacity digital proving ring. Steel bearing plates of 120 mm × 100 mm × 15mm size are provided at the loading and supporting point as shown in figure 3. The load was applied at the centre for a width of 120 mm only. Loads were applied at small increments to capture the specimen behaviour as closely as possible. A hand-held optical microscope with a least count of 0.1mm was used to measure the crack widths with a least count of 0.1 mm. To observe the crack propagation, the panels were whitewashed and dried before testing. A 100 mm × 100 mm grid was drawn on the panel surface to mark the location of cracks more precisely.

Figure 3. Test set-up of the panel specimen

Table 6. Summary of experimental results

Sl. No.	% SF	% PPF	f_{ck} (MPa)	f_c' (MPa)	Load at First visible crack (kN)	Peak load (kN)	Experimental efficiency factors	% improvement
1	0	0.0	40.1	32.4	249*	249	0.64	--
2	0.7	0.0	43.4	35.2	190	365	0.86	34.9
3	0.9	0.0	43.8	35.7	200	400	0.93	45.8
4	0.9	0.2	44.0	36.1	240	450	1.04	62.2
5	0.9	0.4	44.1	36.7	250	470	1.07	66.6
6	1.0	0.0	40.5	33.1	161	351	0.88	38.0
7	1.1	0.0	44.4	35.7	146	345	0.81	25.7

5. Results and Discussion

The load-carrying capacity of bottle-shaped strut depends on the resistance to the transverse tension developed in the strut. In this study, strut efficiency factors are evaluated for different amounts of fibres in steel fibre reinforced concrete and hybrid fibre reinforced concrete by equating the theoretical strut capacity to the actual load resisted by testing the panels as shown in figure 3. For each fibre volume fraction, 3 panels are tested and the average value is reported for first cracking load, peak load, efficiency and improvement in Table 6.

5.1 Effect of steel fibres on efficiency factors

Based on the peak load, the experimental efficiency factor is calculated as per equation (2) and plotted the variation for a different amount of fibres proportion. Percentage change in the efficiency factor is calculated compared to the controlled specimen which has no fibres in the concrete.

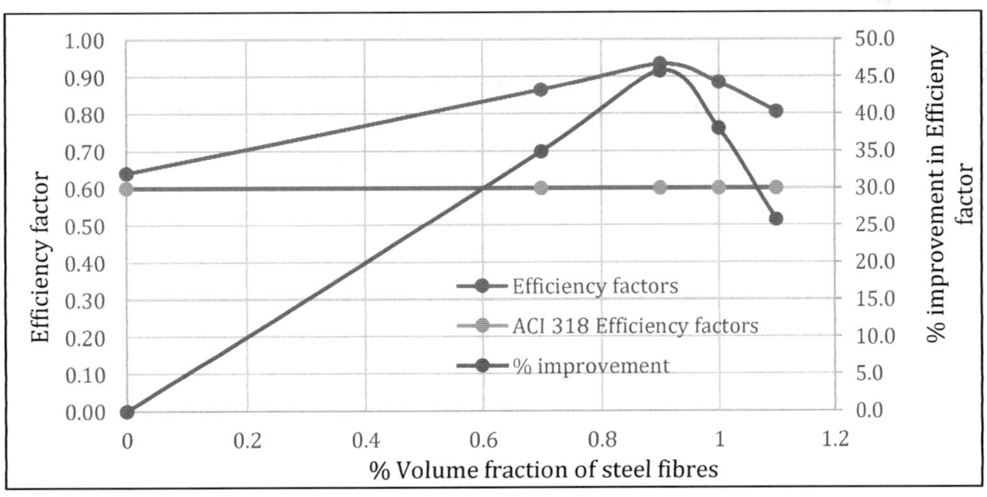

Figure 4. Graph showing the effect of steel fibres on efficiency factors

As shown in figure 3, the efficiency factor increased with the increase in the steel fibres content in the concrete and it has a peak value at 0.9% of the volume fraction. Around 45% improvement in efficiency factor is observed at 0.9% volume of steel fibres. From results shown in figure 4, it is observed that 0.9% volume fraction of steel fibres have shown good efficiency and improvement when compared with different amounts of steel fibres panels. So, the optimum percentage of steel fibres is 0.9% volume fraction.

5.2 Effect of microfibres on efficiency factors

Effect of hybrid fibres has been studied by taking steel fibres with 0.9% and varying polypropylene fibres as 0.1% and 0.2% volume fraction.

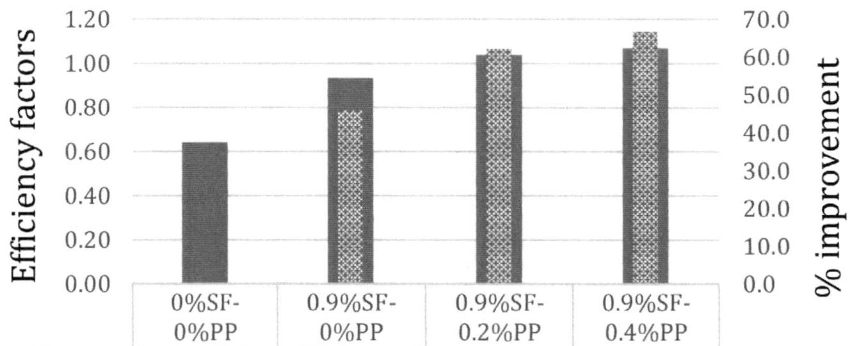

	0%SF-0%PP	0.9%SF-0%PP	0.9%SF-0.2%PP	0.9%SF-0.4%PP
Efficiency factors	0.64	0.93	1.04	1.07
% improvement	0.0	45.8	62.2	66.6

Figure 5. Effect of microfiber on the efficiency factor

As shown in figure 5, the efficiency of hybrid fibre reinforced concrete has been improved further when compared to mono steel fibre reinforced concrete.

5.3 Effect of steel fibres on crack propagation

Figure 6 represents the first crack load, peak load and the difference between them. Specimens with no fibres fail suddenly with a single splitting crack. Therefore, the same load value is reported for first visible crack and at failure. It has been observed that there is an increase in the ultimate load-carrying capacity of specimens with steel fibres up to 0.9% volume fraction, and from there, the reduction is observed. Though there is a decrease in load carrying capacity beyond 0.9% volume of steel fibres we can observe that the difference in the first crack load and the peak load increased with the percentage of steel fibres which indicates that steel fibres start bridging action once the concrete cracks which delay the ultimate failure after the initial crack and therefore, the ductility is increased.

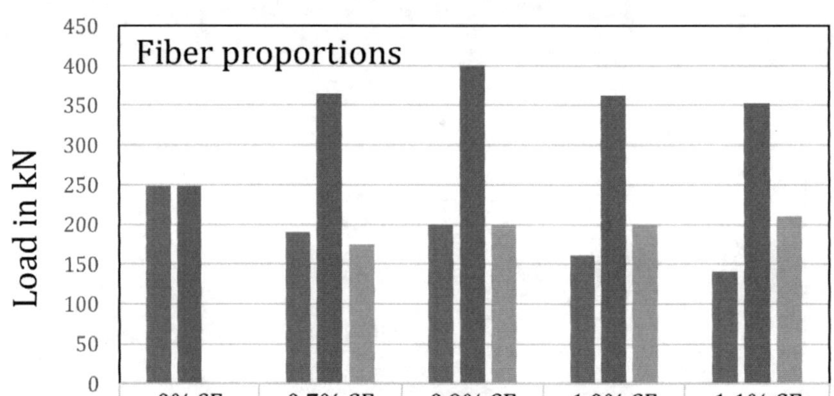

	0% SF	0.7% SF	0.9% SF	1.0% SF	1.1% SF
Load at first visible crack	249	190	200	161	141
Peak load	249	365	400	362	352
Difference	0	175	200	201	211

Figure 6. Effect of steel fibres on crack propagation

5.4 Effect of micro polypropylene fibres on crack propagation

Figure 7 represents the cracking behaviour at 0.9% of steel fibres with varying polypropylene microfibres at 0.1% and 0.2% proportions.

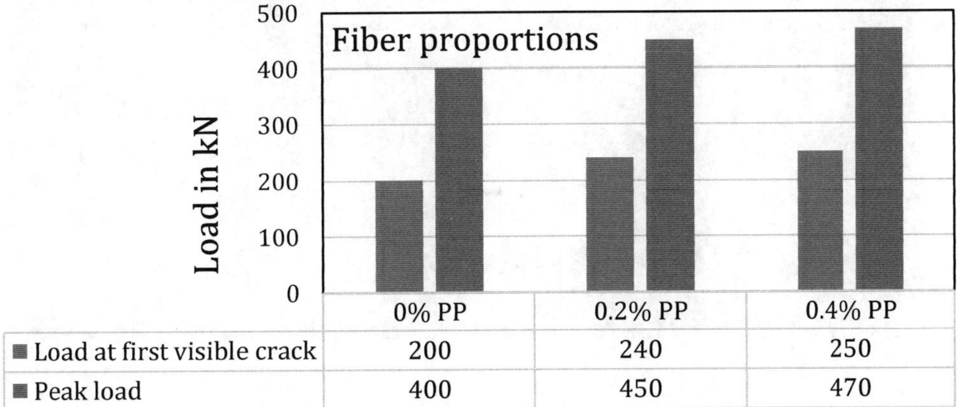

Figure 7. Graph showing the effect of polypropylene fibres on cracking behaviour at 0.9% SF with varying PP fibres

It is observed in figure 7 that, with an increase in the proportion of polypropylene fibres, load carrying capacity has been increased. Due to the addition of micro polypropylene fibres to the macro fibres, load at first visible crack increased compared to the specimens with no microfibres. This shows that the inclusion of microfibres arrests the formation of initial cracks and improves the ductility.

5.5 Crack widths and failure modes

Details of all specimens with crack widths and failure modes are listed below in Table 7. From Table 7, it can be observed that the addition of fibres to the unreinforced bottle-shaped strut changed the failure mode of the bottle-shaped strut and made it more ductile. Specimen with no fibres failed suddenly with a single splitting crack at the centre of the strut and it broken into two pieces as shown in figure 8 to 13. Further, an increase in the fibre percentage in concrete, the width of cracks reduced at service and ultimate load also. Addition of microfibres will further reduce the crack width. Pictures of all the specimens after failure are shown in figure 8 to 13.

Table 7. Crack widths and failure modes

% Steel Fibres	% of PP fibres	Maximum crack width at service load (mm)	Maximum crack width at peak load (mm)	Type of failure
0%	0%	—	--	Splitting failure (brittle failure with single crack)
0.7%	0%	0.4	0.5	Splitting failure with multiple cracks
0.9%	0%	0.4	0.5	Splitting failure with multiple cracks
0.9%	0.2%	0.2	0.3	Splitting failure with multiple cracks
0.9%	0.4%	0.2	0.2	Splitting failure with multiple cracks
1%	0%	0.3	0.5	Splitting failure with multiple cracks
1.1%	0%	0.3	0.5	Splitting failure with multiple cracks

IOP Conf. Series: Materials Science and Engineering 1025 (2021) 012023 doi:10.1088/1757-899X/1025/1/012023

Figure 8. Failure of the panel without fibres

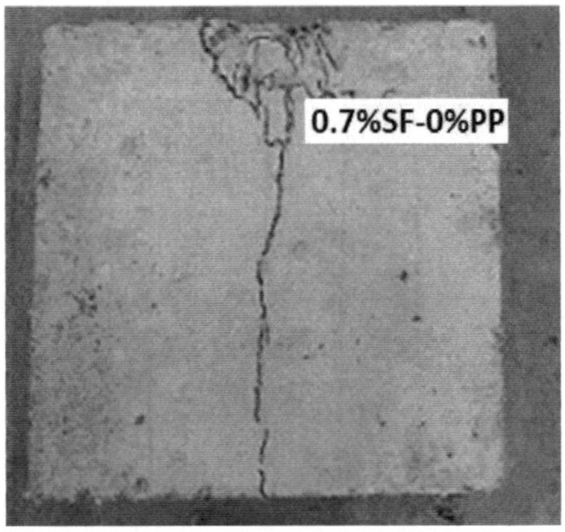

Figure 9. Failure of the panel with 0.7% macro steel fibres only

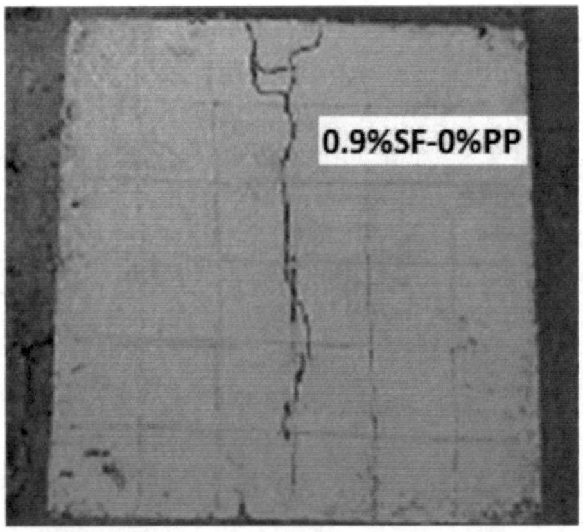

Figure 10. Failure of the panel with 0.9% macro steel fibres only

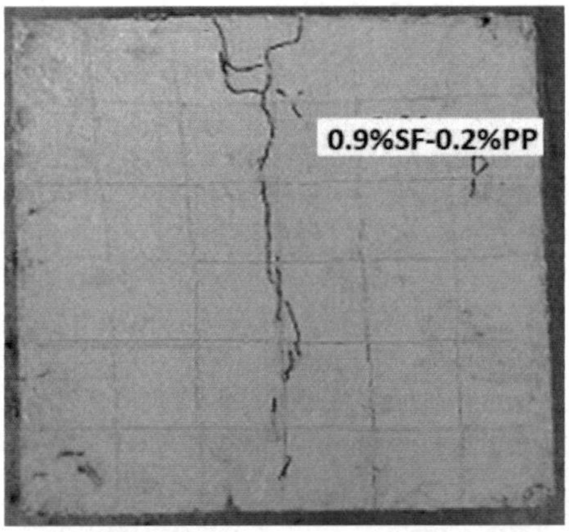

Figure 11. Failure of the panel with 0.9% macro steel fibres and 0.2% micro polypropylene fibres

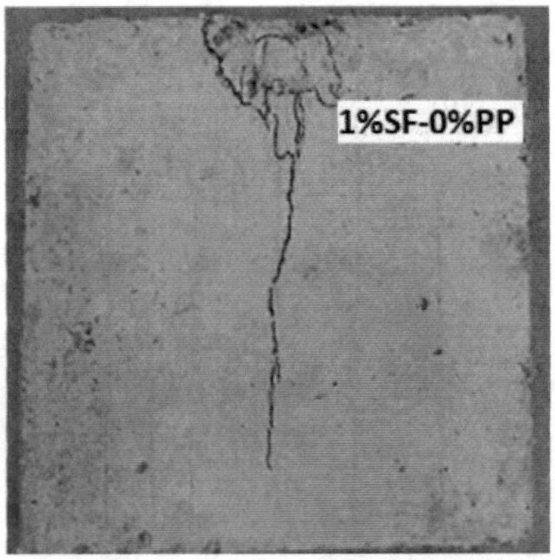

Figure 12. Failure of the panel with 1% macro steel fibres only

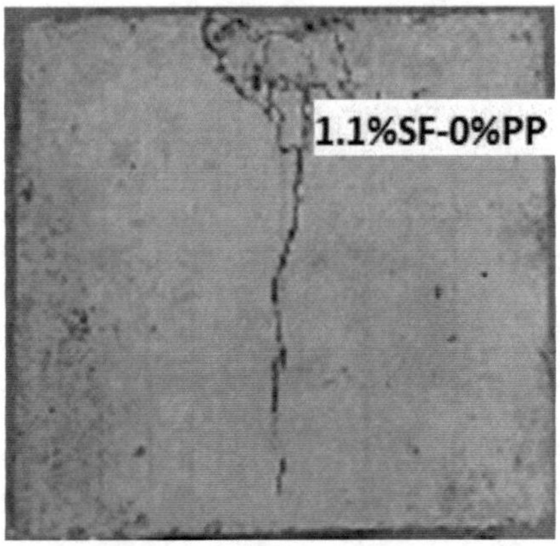

Figure 13. Failure of the panel with 1.1% macro steel fibres only

6. Summary and Conclusions

1. Discrete steel fibres in concrete significantly improved the Efficiency factors of bottle-shaped struts. Around 45% increase in the efficiency factor is observed at 0.9% volume fraction. Macro steel fibres also delayed the ultimate failure of the bottle-shaped strut after the first crack by developing the bridging action after the crack initiation.
2. Plain concrete bottle-shaped strut failed suddenly with single splitting crack. Addition of macro steel fibres develops bridging action after the crack occurs and change the mode of failure to ductile nature.
3. Microfibres in the concrete was found effective in controlling the initial cracks and also the crack width at service and ultimate loads.
4. The good synergy between macro steel fibres and micro polypropylene fibres is observed with the improved performance of bottle-shaped strut in terms of load-carrying capacity and serviceability.
5. Addition of fibres can be considered as the alternative the transverse reinforcement in the bottle-shaped strut to meet the serviceability requirements.

References

[1] ACI 318-11 2011 Building code requirements for structural concrete and commentary ACI Committee and International Organization for Standardization *American Concrete Institute* Farmington Hills, MI, USA
[2] Sahoo, Dipak Kumar 2010 Effect of detailing of web reinforcement on the behaviour of bottle-shaped struts *Journal of Advanced Concrete Technology* **8 (3)** p 303-314.
[3] Li, Zhong-Xian 2017 Experimental investigation on mechanical properties of Hybrid Fibre Reinforced Concrete *Construction and Building Materials* **157** p 930-942.
[4] Thomas, Job, Ananth Ramaswamy 200 Mechanical properties of steel fibre-reinforced concrete *Journal of Materials in Civil Engineering* **19 (5)** p 385-392
[5] Brown M D, Sankovich C L, Bayrak O, Jirsa J O 2006 Behaviour and efficiency of bottle-shaped struts *ACI Structural Journal* **103 (3)** p 348–355
[6] Ghanei A H, Aghayari R 2016 Experimental investigation on effect of aspect ratio on behaviour of isolated reinforced concrete struts *ACI Structural Journal* **113 (4)** p 643–654
[7] Laughery L, Pujol S 2015 Compressive strength of unreinforced struts *ACI Structural Journal* **112 (5)** p 617–623
[8] Pujol S Rautenberg J M, Sozen M A 2011 Compressive strength of concrete in non-prismatic elements *Concrete International* **33 (9)** p 42–49
[9] Sahoo D K, Gautam R K, Singh B, Bhargava P 2008 Strength and deformation characteristics of bottle-shaped struts *Magazine of Concrete Research* **60 (2)** p 137–144
[10] Sahoo D K, Singh B, Bhargava P 2009 An appraisal of the ACI strut efficiency factors *Magazine of Concrete Research* **61 (6)** p 445–456
[11] Su R K L, Looi D T W 2016 Revisiting unreinforced strut efficiency factor *ACI Structural Journal* **113 (2)** p 301–312
[12] IS 10262 2009 Indian Standard Concrete Mix Proportioning – Guidelines *Bureau of Indian Standards* New Delhi

IOP Conf. Series: Materials Science and Engineering 1025 (2021) 012024 doi:10.1088/1757-899X/1025/1/012024

Effect of ferrochrome slag and fly ash on the mechanical properties of concrete

B P R V S Priyatham[1], D V S K Chaitanya[2], G Prasanna Kumar[3]

[1]Assiatant Professor, Department of Civil Engineering, GMR Institute of Technology, Rajam, Andhra Pradesh, India
[2]Assistant Professor, Department of Civil Engineering, Acharya Nagarjuna University, Guntur, India
[3]Research Scholar, Department of Civil Engineering, Andhra University, Visakhapatnam, India
E-mail: priyatham.phani@gmail.com

Abstract: Now a day's construction industry is focusing on various wastes from other industries to use them as substitutes for construction materials which will reduce their initial cost and impact on the environment. Use of industrial waste materials in concrete is a substitute for the shortage of natural resources. Waste disposal from steel manufacturing industries and power plants are increasing day by day and creating a lot of environmental pollution due to lack of utilizing them. There is a lot of wastes from industries like fly ash, GGBS, and slag-based compounds which can be used as replacement materials for aggregates. The present experimental investigation examines the possibility of replacing coarse aggregate with ferrochrome slag waste obtained from steel manufacturing industries and cement with fly ash obtained from power plants in concrete mix design for many construction practices. The experiments were conducted on the concrete mix in which natural coarse aggregate is replaced with ferrochrome slag to achieve optimal replacement of ferrochrome slag and the effect of fly ash along with the optimal ferrochrome slag on the mechanical properties of concrete.

1. Introduction

Increased infrastructure projects due to rapid industrialization, population growth and continuous dependence on conventional materials of concrete making results in the scarcity of the construction materials which increases the construction costs. Natural resources were used to produce various construction materials from traditional methods. Due to the heavy development in the industrial and urban management systems, the generation of the wastes was more. These wastes were disposed of in the open places which lead to pollution of the entire environment resulting in the hazardous effects [1]. Hence as an alternative, those industrials wastes can be recycled and reused in concrete. Industrial wastes like fly ash, GGBS, steel slag [2], copper slag [3], bagasse ash and ferrochrome slag [4] etc. are generated from various industries like thermal power plants, steel plants, sugar industries etc. By utilizing these industrial wastes as alternative building materials in concrete may reduce some impact on the environment.

Concrete is an eco-friendly material in the environment throughout its life span in all the stages considering from the starting stage of production of raw materials to the ending stage of demolition thus results in converting it into a natural choice for the construction purposes to make it sustainable. Concrete primarily constitutes of cement, coarse aggregate and fine aggregate. Usually, cement is used as a binder, natural river sand and conventional stone are used as fine aggregate and coarse aggregate. Usage of cement in concrete leads to the emission of CO_2 into the environment which further leads to the greenhouse effect and global warming. Hence, we can choose alternative materials for concrete constituents to reduce the scarcity of natural resources and as well as adverse effects on the environment. The main purpose of this work is to study the efficient usage of different industrial solid wastes for aiming sustainable concrete.

FeCr alloy is a primary constituent in stainless steel manufacturing. During this process, the major by-product obtained is ferrochrome slag. About 1.1 –1.6 tons of slag is generated for a ton of ferrochrome production [5]. Majorly Ferrochrome slag constitutes silica, alumina, magnesium and chromium oxides which give strength to the ferrochrome slag. We can use this slag material as coarse aggregate in concrete and its applications [6]. Ferrochrome slag is obtained as a liquid at 1700 Degree Celsius, which comprises of chrome, ferrous/ferric oxides and CaO. Ferrochrome slag is used for recycling, reuse and in addition to this, the slag plays a vital role in attaining unique properties like strength and stability. Ferrochrome slag exhibits more advantages when compared to other materials based on physical properties. This slag is porous and hence have a high-water absorption rate. It has good adhesion and good abrasion resistance.

Fly ash is a substance which is consisting of finely divided particles. It is waste material that is formed during the pulverized coal-burning process in power plants for generation of electricity. It constitutes silica, alumina, calcium oxide contents and some metal oxides. Fly ash is a pozzolana which contains aluminous materials as well as siliceous materials that result in the formation of cement in the water presence. Hence, fly ash is considered as replaceable material for cement as it has some binding properties. It is of two types (class C, class F), based on the calcium oxide content and the type of coal. Generally, in India Class F fly ash will be available and it is having minimum CaO content.

2. Literature Review

Naga Rajesh et al 2019 researched on sustainable micro-concrete. Devi and Gnanavel studied the properties of Concrete Manufactured Using Steel Slag. Wei et al. determined the optimum content of copper slag as a fine aggregate in high strength concrete. Panda et al. assessed the feasibility of ferrochrome slag as a concrete aggregate material. Kumar et al. 2014 studied the effect of industrial waste ferrochrome slag in conventional and low cement castables. Rajashekar, et al. (2015) had attempted replacing the natural coarse aggregate in concrete with ferrochrome slag and tested for compressive strength, which has shown the increased strengths of concrete. Ponnada 2014 experimented on the combined effect of flaky and elongated aggregates on strength and workability of concrete.

3. Materials and Methodology

3.1. Cement

Cement is a binding material which binds the fine and coarse aggregate particles together by setting as well as hardening firmly when water is added. In this experimental work, OPC 53 grade confining to IS: 12269 - 2013 is used. Tests on the cement were performed according to IS: 4031 - 1988 and the properties of cement are listed in table 1.

Table 1. Properties of Cement.

Test	Test Specifications	Result	Recommended Values
Specific Gravity	IS 4031 (XI)	3.12	3.0 - 3.2
% Fineness	IS 4031 (I)	2%	< 10%
Normal Consistency	IS 4031 (IV)	32%	27 - 33%
Initial Setting Time	IS 4031 (IV)	46 min	> 30 min
Final Setting Time	IS 4031 (V)	330 min	< 10 hrs
Soundness	IS 4031 (III)	2 mm	< 10 mm

3.2. Fine Aggregates

The fine aggregate used in the study is river sand collected from Nagavali river region, the sample is sieved through 4.75mm IS sieve to remove deleterious and oversized particles and tested for its properties as per IS 383 2016. Tests on the fine aggregate were performed according to IS 2386 1963 and fine aggregate properties are shown in table 2

Table 2. Properties of Fine Aggregate.

Test	Test Specifications	Result	Recommended Values
Specific Gravity	IS 2386 (III)	2.66	2.50 - 3
Sieve analysis	IS 2386 (I)	Zone - III	N. A
Fineness modulus	IS 2386 (I)	2.4	2.2 - 2.6
Water absorption	IS 2386 (III)	1%	0.1 - 2%

3.3. Coarse Aggregate

The coarse aggregate in the present study is the crushed granite stone collected from a quarry at Rapaka village. It acts as a crucial constituent in developing strength. Its properties are tested as per IS 383 2016. 20mm and 10mm sized aggregates are used in the study. Coarse aggregates were tested according to IS 2386 1963 and coarse aggregate properties are tabulated in table 3.

Table 3. Properties of Coarse Aggregate.

Test	Test Specifications	Result	Recommended Values
Specific Gravity	IS 2386 (III)	2.87	2.50 - 3.0
Impact Value	IS 2386 (IV)	15.9%	< 35%
Flakiness Index	IS 2386 (I)	8.42%	< 15%
Elongation Index	IS 2386 (II)	14.28%	< 15%
Water Absorption	IS 2386 (III)	0.4%	0 – 2%

3.4. Ferrochrome Slag

Ferrochrome slag (FC) is the byproduct obtained from the manufacturing of stainless steel in which chromium is the important constituent. It is collected from the dumping yard of Facor Alloys Limited, Garividi. Ferrochrome Slag is of different sizes and is crushed manually to obtain 20mm and 10mm sized particles as coarse aggregates. The specific gravity of slag is 2.94, impact value is 13.7% and water absorption is 0.5%

Preliminary tests on the materials are carried out and the mix design is done as per IS: 10262 – 2019. The replacement of ferrochrome slag for coarse aggregate is optimized and for that mix, the replacement is done for cement with fly ash and properties of concrete were studied.

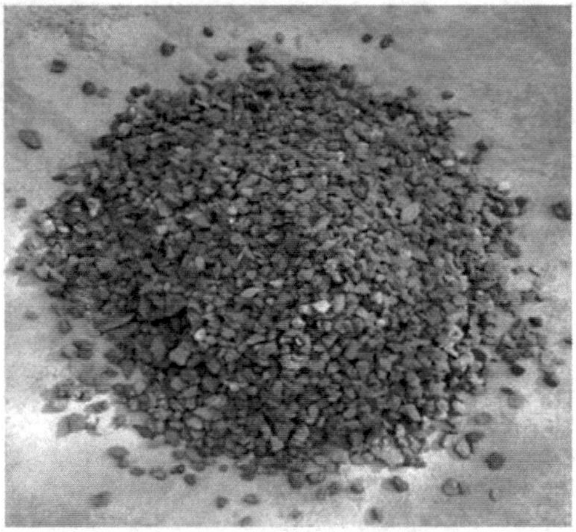

Figure 1. Ferrochrome Slag

4. Mix Proportions

Mix design is performed as per IS 10262 2019 and the grade of concrete adopted is M25. Coarse aggregate is replaced by 25%, 50%, 75% and 100% of ferrochrome slag to get the optimum mix. For this optimum mix, cement is replaced by 10%, 20% and 30% of fly ash to study the concrete properties.

To maintain the workability, superplasticizer is added. The details of quantities for mix design are detailed in table 4

Table 4. Quantities for Mixes.

Ingredients (kg/m³)	**CC**	**FC25**	**FC50**	**FC75**	**FC100**	**FC75 + FA10**	**FC75 + FA20**	**FC75 + FA30**
Cement	375	375	375	375	375	337.5	300	262.5
Fly ash (%)	0	0	0	0	0	10	20	30
Ferrochrome Slag (%)	0	25	50	75	100	0	0	0
Fly ash	0	0	0	0	0	37.5	75	112.5
Ferrochrome Slag	0	322.5	645	967.5	1290	967.5	967.5	967.5
Fine Aggregate	661	661	661	661	661	661	661	661
Coarse Aggregate	1290	967.5	645	322.5	0	322.5	322.5	322.5
Water	180	180	180	180	180	180	180	180

5. Preparation of samples

Strength characteristics of the samples replaced with ferrochrome slag and fly ash for coarse aggregate and cement respectively are studied by casting the prisms, cubes and cylinders. As per the quantities calculated based on mix design from table 4, the quantities of different ingredients are weighted and concrete is prepared using the mixer and the samples are cast in the moulds and compacted in three different layers using the tamping rod to remove the air voids and the top surface is finished with the trowel. For each mix 6 cubes (150x150 mm), 6 prisms (500x100x100 mm) were cast to find out the compressive strength and flexural strength respectively and cylinders (150 mm diameters x 300 mm height) were cast for fly ash mixes to know the split tensile strength. After 24 hours the casted samples are demoulded and placed in the tank for water curing. The samples are taken out and are dried before testing for 7- and 28-days strengths.

6. Experimental Results and Discussion

6.1. Compressive Strength

Compressive strength of the specimens are determined using ACTM which is having a capacity of 2000kN and the rate of loading is applied at 140kg/cm²/min as per IS: 516 – 2004. The test result for the strength of concrete in compression at 7 and 28 days of concrete is shown in table 5 and the strength variation is plotted in figure 2.

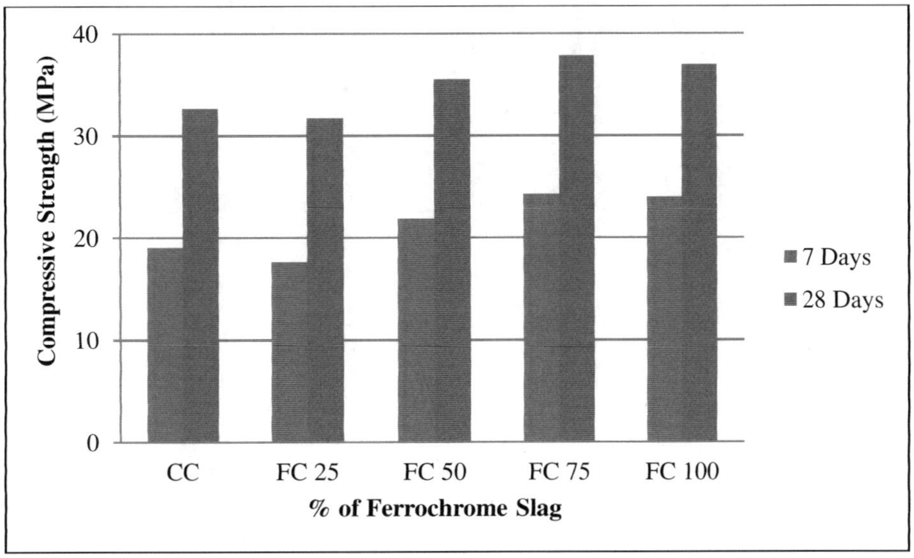

Figure 2. Variation of Compressive strength of cubes with ferrochrome slag

Table 5. Compressive Strength for Ferrochrome Slag replaced mixes.

Mix	Compressive Strength (MPa)	
	7 days	28 days
CC	19	32.65
FC25	17.59	31.72
FC50	21.83	35.46
FC75	24.26	37.86
FC100	24.01	36.94

6.2. Flexural Strength

Flexural strength of concrete is determined by using UTM which checks the resistance of concrete towards bending. If the beam is stressed uniformly along with one-third of its mid surface then beam fails in its weakest point of that one-third parts, so three-point loading is used here. The load is applied at the rate 1.8kN/m and the test are performed in as per IS: 516 – 2004. Table 6 represents the strength of concrete in flexure at 7 and 28 days of curing.

Table 6. Flexural Strength for Ferrochrome Slag replaced mixes.

Mix	Flexural Strength (MPa)	
	7 days	28 days
CC	3.78	4.95
FC25	3.71	5.03
FC50	3.89	5.27
FC75	4.27	5.4
FC100	3.91	5.31

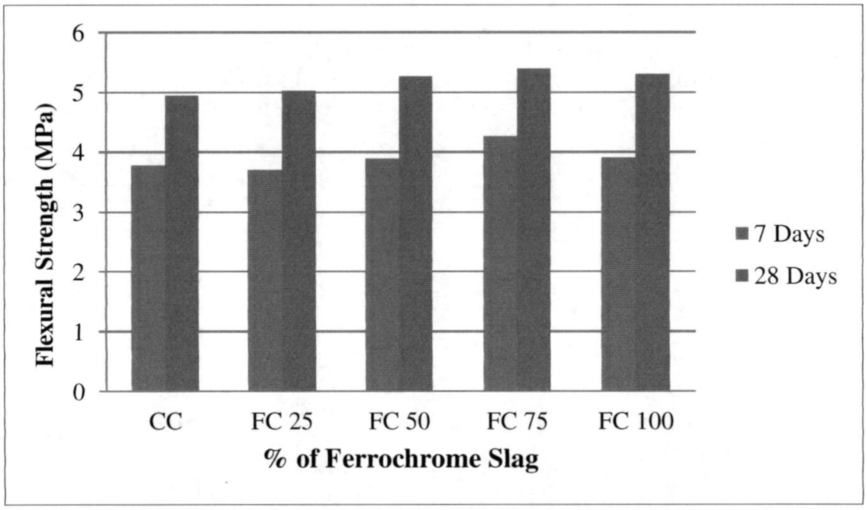

Figure 3. Variation of flexural strength with ferrochrome slag

From figure 2 and 3 corresponding to the test results at 7 and 28 days, the optimal replacement of ferrochrome slag is found to be 75% both in compression as well as flexural. The compressive strength is increased by 8.6%, 15.95% and 13.93% for FC50, FC75 and FC100 respectively and decreased by 2.8% for FC25 when compared with the conventional mix. The flexural strength is increased by 1.6%, 6.46%, 9.09% and 7.27% for FC25, FC50, FC75 and FC100 respectively when compared with conventional mix. Now for the optimum ferrochrome slag mix, cement is replaced with fly ash at proportions of 10%, 20% and 30%. Table7 represents the strength characteristics of concrete.

Table 7. Strength characteristics of concrete for optimum ferrochrome slag and fly ash replaced mixes.

Mix	Compressive Strength (MPa)		Flexural Strength (MPa)		Split Tensile Strength (MPa)	
	7 days	28 days	7 days	28 days	7 days	28 days
CC	19	32.65	3.78	4.95	2.29	2.99
FC75	24.26	37.86	4.27	5.4	2.49	3.58
FC75 + FA10	19.51	29.18	3.77	4.82	2.28	3.25
FC75 + FA20	15.6	24.82	3.26	4.35	2.19	2.7
FC75 + FA30	12.19	20.75	2.95	4.16	2.01	2.28

Figure 4. Test setup for flexure and compression

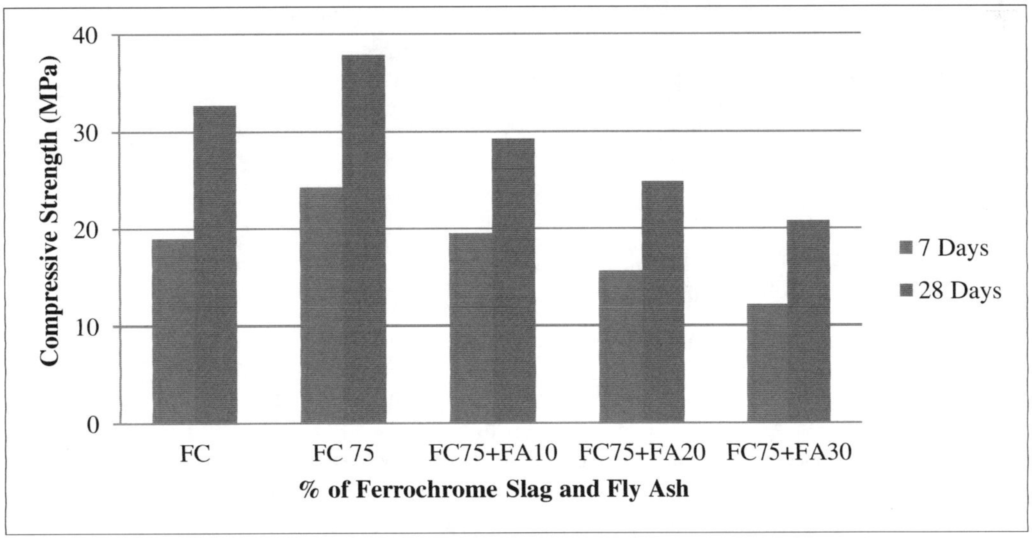

Figure 5. Variation of compressive strength with optimum ferrochrome slag and fly ash

From figure 5, 6 and 7 it is noticed that the strengths in compression, flexural and split tension decreases for the optimum ferrochrome slag mixes with the replacement of cement by fly ash. From figure 5 it was inferred that the compressive strength is decreased by 10.6%, 23.9% and 36.4% for FC75+FA10, FC75+FA20 and FC75+FA30 respectively when compared with the conventional mix. From figure 6, the strength of concrete in flexure is decreased by 2.62%, 12.12% and 15.95% for FC75+FA10, FC75+FA20 and FC75+FA30 respectively when compared to conventional mix. From figure 7, the split tensile strength is increased by 19.7%, 8.7% for FC75, FC75 + FA10 respectively and decreased by 9.7%, 23.7% for FC75+FA 20, FC75+FA 30 respectively when compared with conventional mix.

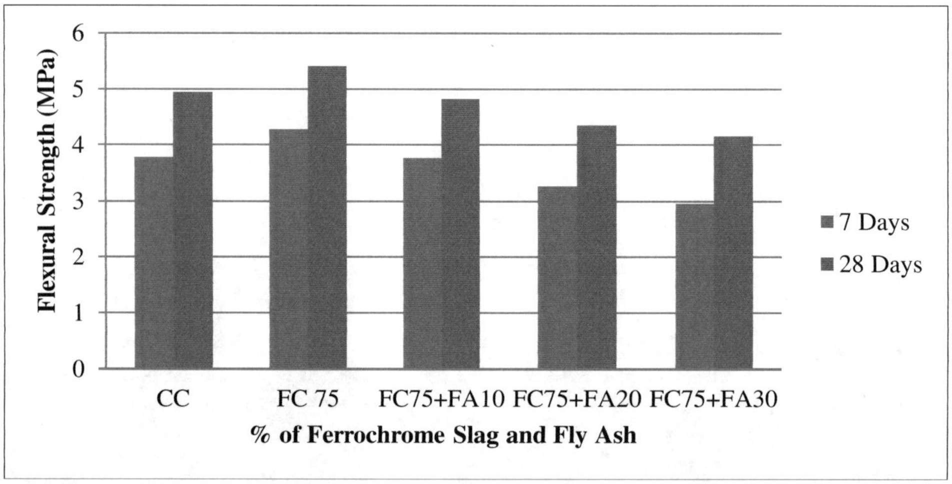

Figure 6. Variation of flexural strength with optimum ferrochrome slag and fly ash

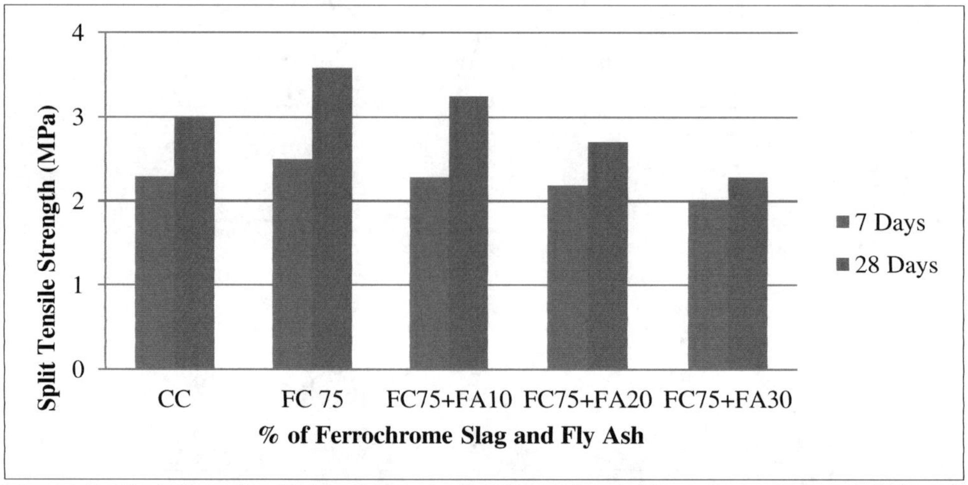

Figure 7. Variation of split tensile strength with optimum ferrochrome slag and fly ash

Concrete which is obtained from the optimum mix of ferrochrome slag has exhibited higher compressive strength values when compared to conventional mix, as FeCr is having more specific gravity than the coarse aggregate. This shows that it is suitable for replacing coarse aggregate to achieve sustainable concrete. As the surface of ferrochrome slag is rough, the cement paste has a tendency to fill the rough surfaces which leads to an increase in the strength. On the other side when the optimum mix with ferrochrome slag in replaced by fly ash, the strength decreases gradually.

7. Conclusions

Nowadays in most of the civil engineering practices, many industrial waste materials are used as alternatives to produce sustainable concrete. The strengths of concrete in compression and flexure increases significantly when ferrochrome slag is replaced with the coarse aggregate of up to 75%. The concrete losses its strength as the cement replacement with fly ash increases. The strength in flexure and tension also decreases with the increase of fly ash to the ferrochrome slag mix. Hence ferrochrome slag is recommended as a substitute material for natural coarse aggregate replacement as it yields better strength properties.

References

[1] Naga Rajesh K, Rath M K, Markandeya Raju P 2019 A research on sustainable micro-concrete *International Journal of Recent Technology and Engineering* **8 (2 Special Issue 3)** p 1137-9

[2] Devi S V, Gnanavel 2014 B K Properties of Concrete Manufactured Using Steel Slag *Procedia Engineering* **97** p 95-104

[3] Wei W, Zhang W, Guowei M 2010 Optimum content of copper slag as a fine aggregate in high strength concrete *Materials and Design* **31 (6)** 2878-83

[4] Panda C R, Mishra K K, Panda K C Nayak B D and Nayak B B 2013 Environmental and technical assessment of ferrochrome slag as concrete aggregate material *Construction and Building Materials* **49** p 262-71

[5] Kumar P H, Srivastava A, Kumar V, Singh V K 2014 Implementation of industrial waste ferrochrome slag in conventional and low cement castables: Effect of calcined alumina *J. Asian Ceram. Soc.* **2 (4)** p 371-9

[6] Rajasekar K, Satyanarayana Reddy C N V 2015 An experimental study on use of ferrochrome slag aggregate in concrete making *ICI Journal* **15 (4)** p 25-9

[7] Ponnada M R 2014 Combined effect of flaky and elongated aggregates on strength and workability of concrete *International Journal of Structural Engineering* **5 (4)** p 314-25

[8] IS 10262 2019 Concrete mix proportioning – Guidelines *Bureau of Indian Standards* New Delhi India

[9] IS 516 1959 Indian standard Methods of test for strength of concrete *Bureau of Indian Standards* New Delhi India

[10] IS 4031 1988 Methods of physical tests for hydraulic *Bureau of Indian Standards* New Delhi India

[11] IS 2386 1963 Methods of test for aggregates for concrete *Bureau of Indian Standards* New Delhi India

[12] IS 383 2016, Coarse and fine aggregate specifications for concrete specifications *Bureau of Indian Standards* New Delhi India

IOP Conf. Series: Materials Science and Engineering 1025 (2021) 012025 doi:10.1088/1757-899X/1025/1/012025

Study on cement paste blended with calcined granite powder

Rajendra Prasad Singh[1,2], K Rajasekhar[3], S Adiseshu[3]

[1] Research Scholar, Department of Civil Engineering, College of Engineering (Autonomous), Andhra University Visakhapatnam, India
[2]Assistant Professor, Department of Civil Engineering, MVGR College of Engineering (Autonomous), Vizianagaram, Andhra Pradesh 535005 India
[3]Professor, Department of Civil Engineering, College of Engineering (Autonomous), Andhra University Visakhapatnam, India
Email: rp.singh130@gmail.com

Abstract. Granite in India is famous for flooring and also used as concrete aggregate. Granite mineral is processed in the industry and large sheets are made which are used as the flooring of dwelling units. During the processing of granite, the mineral is cut into suitable pieces and polished which produces very fine powder. The fine granite powder is a by-product of the granite industry and which is often used for landfill. On the other hand, the granite crushers industries break the granite to suitable sizes to make concrete aggregates also have this fine granite powder as its by-product. Granite is a hard mineral-rich in silica. The silica-rich mineral can be used in concrete as a mineral additive. The hydration compounds of cement in concrete have free lime and mineral admixtures rich in silica are useful in secondary reacting with the excess lime to form C-S-H gel. In the present study, experimental results are discussed for the suitability of granite powder in concrete manufacturing. In the present work, the cement by partial replacement with granite powder in normal state and calcined state up to 50% and the compressive strength of cement paste is found at 28 days. The results indicate that the partial replacement of cement with granite powder results in only a 10% decrease in strength when cement is replaced up to 30% with granite powder.

1. Introduction

With increasing environmental awareness, it is essential to explore alternatives to waste disposal. Similarly, the production of cement requires the processing of various raw materials for which natural minerals are being exploited. On the other hand, the by-products that have been produced from various industries such as Fly ash, granite powder, ground granulated blast furnace, rice husk ash, rock dust, quarry waste, marble powder etc. are used as landfilling materials. The by-products consist of various mineral and chemical compositions that are not advised to be disposed of openly as they may cause pollution. Disposal of such industrial by-products is a challenge.

Many studies are being carried out to find the suitability of various industrial by-products into the cement to minimize the consumption of minerals required for the production of cement and safe disposal of industrial by-products. The chemical reaction when cement is mixed with water is already understood and the formation of secondary products such as free lime is undesired for obvious reasons. Because of this, the by-products from industry are utilized to make secondary reactions of minerals added with the existing free lime to form a desirable product in the concrete.

1.1 Granite Powder

Granite is an igneous rock formed from cooled magma. The major chemical composition of the granite is silica and alumina with traces of sodium and potassium. Granite powder is a grey fine powder. The granite powder is formed due to cutting or polishing of granite slabs used for flooring. The granite powder is initially wet and is available in the form of sludge. This sludge upon drying is formed into a fine powder. The product being fine, it is difficult to handle and dispose of.

1.2 Calcined Granite Powder

The free lime formed due to the reaction of cement and water. The excess free lime is an undesirable product and over some time will leach out of the concrete and make very huge pores in concrete causing reduction in strength. Reactive silica added to concrete will react with this free lime and forms C-S-H gel which improves not only strength but also prevents leaching in concrete. Granite powder consists of silica which is inert in nature. Upon burning the granite powder to 800°C, it is observed that the colour changes to light pink. Figure 1 shows Unburnt and burnt granite powder.

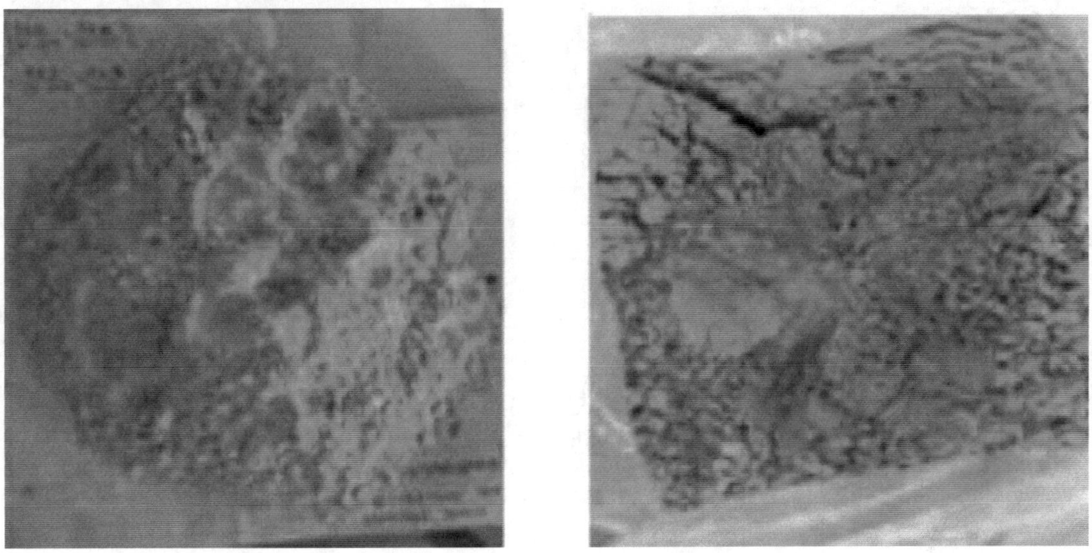

(a) (b)

Figure 1. (a) Un-Burnt Granite powder and (b) Burnt Granite powder

1.3 Inert and reactive fillers in concrete.

The concrete is a matrix consisting of particles of various shapes and sizes. Larger particles being coarse aggregate and packing of coarse aggregate cause voids, the voids caused by coarse aggregates are filled by fine aggregates and further, the voids caused by fine aggregates are occupied by cement. Cement acts as binding materials and binds aggregates and makes stiff and rigid matrices. The voids further caused by the cement can be occupied by the smaller filler material which can be either inert or reactive in nature.

2. Literature Review

Many researchers have worked on partial replacement of cement and fine aggregate with granite dust. A few works have been presented here. Zhang and Liu [1] indicated in his study that the effect of substituting the cement with granite powder by 5% to 30% at every 5% interval and has observed that the compressive strength reduces beyond 10%. The microstructure study indicates that the pore diameter and throat formation are small for a lower percentage of cement replacement. Nithya Nairet et al. [2] reported the rheological qualities of cementitious materials made with Limestone - Calcined Clay Cement, in correlation with conventional Portland concrete and Portland - Fly Ash concrete. It is seen that the calcined clay has increased the plasticizers dosage. Sing et al. [3] reported the study on strength of M30 grade concrete by substituting cement with fly ash and fine aggregate with granite powder. It is observed that the strength of concrete has improved by 15% when there is the replacement of cement up to 20% with fly ash along with fine aggregate being replaced up to 20%. Manpreet et al. [4] observed

IOP Conf. Series: Materials Science and Engineering 1025 (2021) 012025 doi:10.1088/1757-899X/1025/1/012025

that the strength of concrete on partial replacement of cement with dried waste marble powder slurry. Four distinctive substitution rates (0, 10, 15, and 20 and 25%) and three water-binder proportions (0.35, 0.40 and 0.45) have been tried. The tests suggest that the replacement of 15% has increased the strength and durability of concrete. Yaswanth Kumar et al. [5] explains that the Granite Slurry (GS) utilized as a partial substitute in percentages from 5% to 20% by weight to cement in concrete and found its compressive, tensile and flexural strength. It was seen that replacement of 10% of concrete with GS in concrete brought about an increase in compressive strength to 48 N/mm^2 contrasted with 35 N/mm^2 of ordinary cement. To accomplish a similar quality of conventional cement, a 20% replacement with GS is effective.

3. Objectives

The objectives of the present study are as follows

1. To understand the suitability of granite powder to be an inert filler by partial substitution in cement.
2. To determine the effectiveness of using calcined granite powder as partial substitution in cement.
3. To study the variations in the compressive strength of cement mortar by partial substitution of cement with granite powder.

4. Scope

The scope of the work is as follows

1. To burn the part of granite powder to 800°C and air cool the material to get calcined granite powder.
2. To perform SEM and EDAX tests on normal granite powder and calcined granite powder to study the microstructure.
3. To prepare a mix by partial replacement of cement at 0%, 30% and 50% with normal granite powder and cast 7.06cm cube specimens with 1:3 proportions (Cement: Sand) for determining its compressive strength at 7days, 14days and 28days.
4. Similarly, a mix is prepared by partial replacement of cement at 0%, 30% and 50% with calcined granite powder and cast 7.06cm cube specimens with 1:3 proportions (Cement: Sand) for determining its compressive strength at 7 days, 14 days and 28 days.

5. Methodology

The granite powder (GP) and the calcined granite powder (CGP) is sieved over a 90 microns sieve and is used a blending material in cement for the proportions mentioned below. SEM and EDX tests are performed on granite powder and calcined granite powder to understand its properties.

The proportions for which the cement compressive strength is tested to know the suitability of inert filler in concrete is done by casting and testing the samples at 7, 14 and 28days. The mix proportions are presented in table 1.

Table 1. Nomenclature of mixes

Nomenclature	Cement (%)	Granite powder (%)	Burnt Granite powder (%)
100C	100	0	0
70C-30CGP	70	0	30
70C-30GP	70	30	0
50C-50CGP	50	0	50
50C-50GP	50	50	0

6. Results

With the above-mentioned mixes, the 7.06cm cube specimens are cast with 1: 3 (Cement blend: Sand) cured and tested at 7, 14 and 28 days.

6.1 Physical properties of cement, granite powder and calcined granite powder

The physical properties of the cement, granite powder and calcined granite powder are determined in the laboratory and the results are mentioned in table 2.

Table 2. Physical properties of cement, granite powder and calcined granite powder

Property	Cement	Granite powder	Calcined Granite powder
Fineness (Dry Sieving) (in % Passing on 90 microns sieve)	96 %	76%	80%
Specific gravity	2.95	2.21	2.56
Normal Consistency	32%	-	-
Initial setting time	36 min.	-	-
Final setting time	565 min.		

6.2 Average compressive strength of cement blends

The samples cast are cured for the desired days and the compressive strength is determined as per IS 4031 (Part 6) 1988 [6]. The results obtained are as presented in table 3 and figure 2.

Table 3. Average compressive strength results

Sample	Compressive strength (MPa)		
	7 day	14 day	28 day
100C	41.73	45.13	61.93
70C-30CGP	36.90	45.73	52.03
70C-30GP	25.57	38.43	45.70
50C-50CGP	16.13	22.77	29.10
50C-50GP	15.77	19.50	20.23

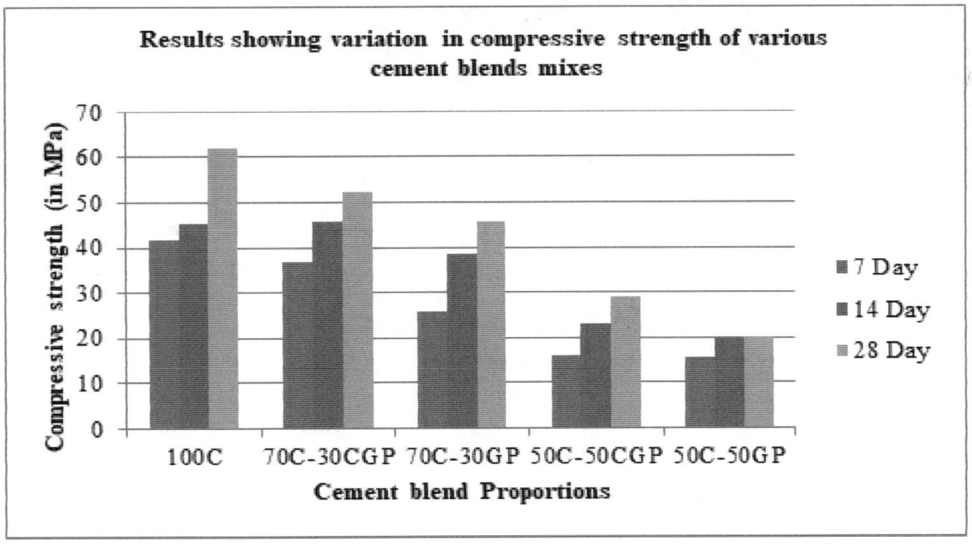

Figure 2. Variation of compressive strength of cement blends at 7, 14 and 28 days

From results presented in figure 2, it can be observed that the Cement blended with 30% of calcined granite powder the 28days average compressive strength is reduced up to only 16%. For the same percentage replacement of normal granite powder, the average compressive strength is reduced up to 27% when compared with no replacement in cement.

6.3 Comparison of compressive strength results obtained for 30% substitution of cement with granite powder and calcined granite powder

The results obtained for 30% replacement of cement with granite powder and calcined granite powder is discussed below

Table 4. Average compressive strengths for 30% replacement of cement with granite powder and calcined granite powder

Age of testing	Average compressive strength of 30% replacement of CGP in cement (in MPa)	Average compressive strength of 30% replacement of GP in cement (in MPa)	Average compressive strength of cement alone (in MPa)
7 days	36.90	25.57	41.73
14 days	45.73	38.43	45.13
28 days	52.03	45.70	61.93

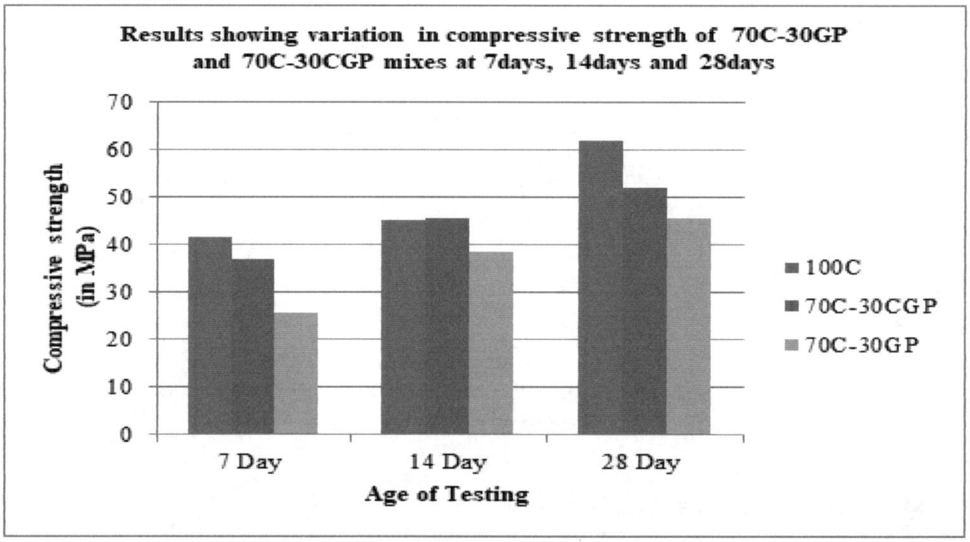

Figure 3. Variation of compressive strength of partial replacement of 30% cement with granite powder and calcined granite powder at 7, 14 and 28days

From the above results, it can be understood that the reduction of strength happens upon partial replacement of cement with granite powder and calcined granite powder. However, calcination of granite powder is observed to have a positive influence on strength when compared with the usage of normal granite powder. The strength of the calcined granite powder sample is 12% more than normal granite powder replacement of up to 30% with cement

6.4 Comparison of compressive strength results obtained for 50% replacement of cement with granite powder and calcined granite powder

The results obtained for 50% substitute for cement with granite powder and calcined granite powder are presented in table 5 and figure 4.

Table 5. Average compressive strengths for 50% replacement of cement with granite powder and calcined granite powder

Age of testing	Average Compressive strength (in MPa)		
	50% replacement of CGP in cement	50% replacement of GP in cement	Cement alone
7 days	16.13	15.77	41.73
14 days	22.77	19.50	45.13
28 days	29.10	20.23	61.93

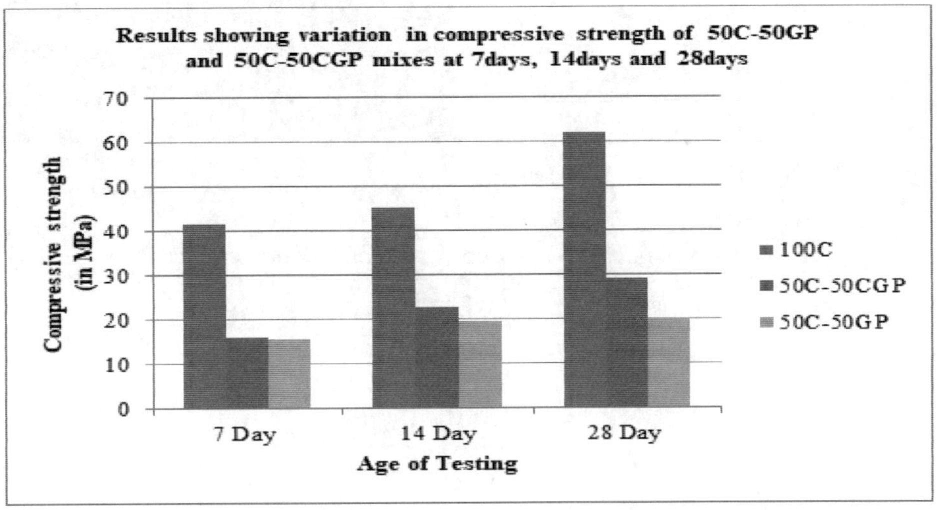

Figure 4. Variation of compressive strength of partial replacement of 50% cement with granite powder and calcined granite powder at 7, 14 and 28days

Similarly, the replacement of cement up to 50% with granite powder has drastically reduced the strength of 67%. Eventually, the reduction of strength is only 53% for the same replacement with calcined granite powder. Further, it is also observed that the calcined granite powder replacement has 30% more strength normal granite powder.

6.5 Microstructure study of granite powder and calcined granite powder:

The Microstructure analysis of granite powder and calcined granite powder presented in table 6, table 7, figure 5, figure 6, figure 7 and figure 8.

Table 6. Chemical composition of Granite powder

Element	Wt. %	At. %
O	28.70	46.63
Na	00.18	00.20
Mg	08.65	09.25
Al	01.34	01.30
Si	30.10	27.86
P	00.29	00.24
K	00.34	00.23
Ca	00.72	00.47
Ti	00.21	00.11
Mn	00.93	00.44
Fe	28.54	13.28

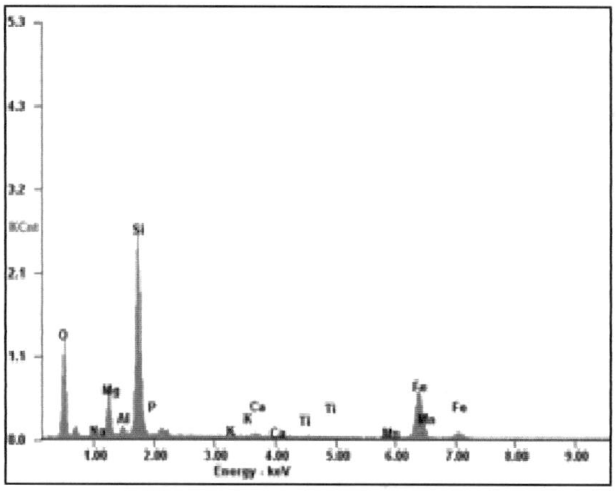

Figure 5. EDX Results of granite powder

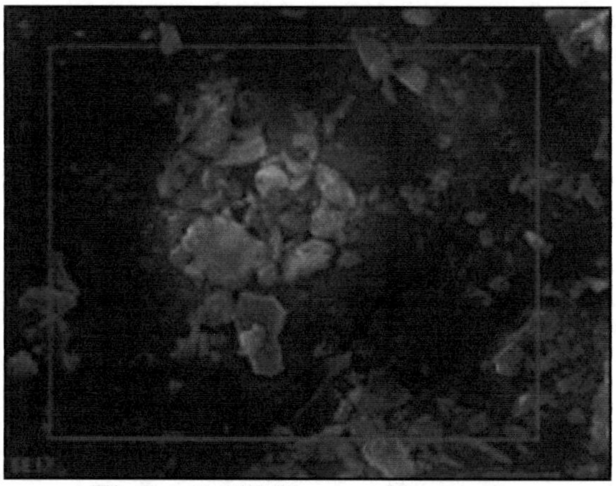

Figure 6. SEM image showing the microstructure of granite powder

Table 7. Chemical composition of Calcined granite powder

Element	Wt. %	At. %
O	37.56	51.60
Na	00.18	00.17
Mg	00.13	00.12
Al	00.65	00.53
Si	59.78	46.78
P	00.23	00.16
K	00.31	00.17
Ca	00.00	00.00
Ti	00.09	00.04
Mn	00.50	00.20
Fe	00.57	00.22

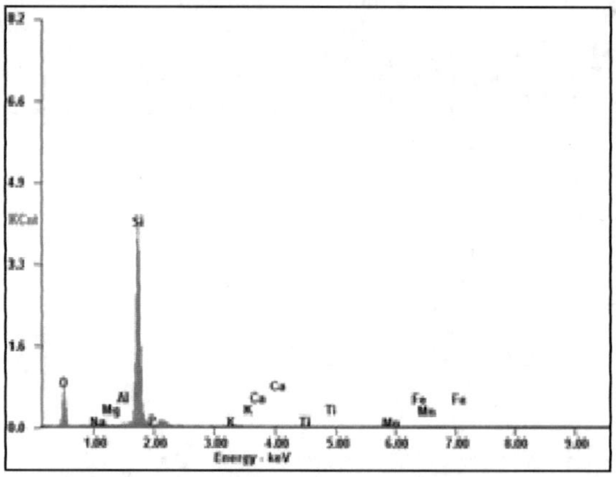

Figure 7. EDX Results of calcined granite powder

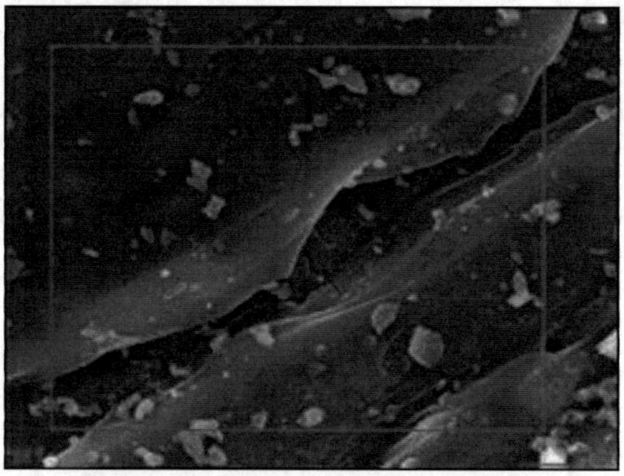

Figure 8. SEM image showing the microstructure of calcined granite powder

From the chemical composition obtained for the granite powder and calcined granite powder, it can be understood that the silica composition is about 30% in normal granite powder and the same granite powder upon burning up to 800 degrees, the silica content is observed to be about 60%. This can be the reason for improved compressive strength. The microstructure of the calcined granite powder has smaller particles compared to normal granite powder, hence can have a much higher void filling ability.

7. Conclusion

The cement when fractional substitution with inert filler material like granite powder shows appreciable suitability.

1. Compressive strength of 7.06 cm cube of cement mortar tested at 07 days is 41.73Mpa, 14 days is 45.13 MPa and 28 days is 61.93 MPa.
2. When the cement is fractional substituted up to 30% with granite powder shows that there is a decrease of 25% of strength. For the similar replacement of calcined granite, powder shows only a 10% decrease in strength at 28 days.
3. Later, when the cement is fractional substituted up to 50% with granite powder shows that there is a decrease of 65% of strength. But when the same granite powder is replaced in the calcined state shows only a 50% reduction in strength by 28 days.
4. From the experiment, it is understood that, upon burning the silica-rich granite powder, it becomes reactive and reacts with cement compounds to contribute to its strength. The increase in strength could be due to secondary reactions in cement with the free lime and the reactive silica in calcined granite powder.
5. From the microstructure of the calcined granite powder, it can be understood that the silica reactivity increases and more fine particles are observed when compared with normal granite powder.

Hence, studies on partial replacement of cement with calcined and un-calcined granite powder have a great deal of promise in making cement more sustainable (Naga Rajesh et al. 2019 [7]).

References

[1] Zhang, Kunqiang, Fusheng Liu, Qiang Yue, Jianguo Feng 2020 Effect of granite powder on properties of concrete *Asia-Pacific Journal of Chemical Engineering* e2468

[2] Nair Nithya, K Mohammed Haneefa, Manu Santhanam, Ravindra Gettu 2020 A study on fresh properties of limestone calcined clay blended cementitious systems *Construction and Building Materials* 254 p 119326.

[3] Singh R P, Rajasekhar K, Adiseshu S 2019 Experimental Study on Performance of M30 Grade Concrete by Partial Replacement with Fly Ash and Granite Powder *In: Das B., Neithalath N. (eds) Sustainable Construction and Building Materials. Lecture Notes in Civil Engineering* Springer, Singapore 25 p 627-634

[4] Singh Manipreet, Anshuman Srivastava, Dipendu Bhunia 2019 Long term strength and durability parameters of hardened concrete on partially replacing cement by dried waste marble powder slurry *Construction and Building Materials* 198 p 553-569.

[5] Kumar Y Yashwanth, Vivek Vardhan C M, Anitha A 2015 Use of granite waste as partial substitute to cement in concrete *International Journal of Engineering Research and Application* **5 (4)** p 25-31

[6] IS 4031 (Part 6) 1988 Indian Standard - Methods of physical tests for hydraulic cement Bureau of *Indian Standard* New Delhi

[7] Naga Rajesh K, Rath M K, Markandeya Raju P 2019 A research on sustainable micro-concrete *International Journal of Recent Technology and Engineering* ***8 (2 Special Issue 3)*** p 1137-9

Seasonal effect on compressive strength of ambient cured, nominal mix proportioned alkali-activated slag concrete

Markandeya Raju P[1], Lakshmi V [2], Santosh Kumar K [3, 4]

[1]Professor and Head, Department of Civil Engineering, MVGR College of Engineering (Autonomous), Vizianagaram Andhra Pradesh, India
[2]Professor, Department of Civil Engineering, University College of Engineering (Autonomous), JNTU Kakinada Andhra Pradesh, India
[3]Research Scholar (Civil Engineering), JNTUK College of Engineering, JNTU Kakinada Andhra Pradesh, India
[4]Assistant Professor, Department of Civil Engineering, MVGR College of Engineering (Autonomous), Vizianagaram Andhra Pradesh, India
Email: [1]markandeyaraju@mvgrce.edu.in

Abstract. Strength of the concrete is a very important factor for the safe design of concrete and safe construction. Nowadays high strength concrete has an important role in the construction of different structures. It is difficult to enhance the strength of the concrete and enhancing strength requires a lot of changes in the mix design. The present study is to develop Geopolymer concrete to increase the compressive strength of low nominal mix proportions using alkali actuators like NaOH and Na_2SiO_3 solutions. Geopolymer concrete is one of the best alternative concrete for conventional and other cement concretes for many reasons. The Geopolymer concrete is prepared with ground granulated blast furnace slag (GGBS) and silica fume (SF) as a binder. NaOH and Na_2SiO_3 solutions are adopted as alkali activators. Quartz sand is used as fine aggregate along with normally used coarse aggregate. Cubes are prepared with different combinations of ingredients and after 28 days of ambient curing conditions, the cubes are tested for compressive strength. And this study is conducted in summer, monsoon and winter seasons to study the effect of seasons on ambient curing of the Geopolymer concrete.

1. Introduction

Concrete plays a significant role in the development of infrastructure that is taking place all over the world. In past decades so many varieties of concretes were developed based on the requirements in construction. High strength concrete is one of the most needed concrete for structures which are affected to heavy loads and the need for high strength Cement increasing day by day. Strength is the most important property of concrete. It is difficult to enhance the strength concrete with low conventional concrete mix proportions. Generally, the characteristic compressive strength up to 20 N/mm^2 the concrete is called ordinary concrete, 25 to 55 N/mm^2 is called standard concrete and 60 to 80 N/mm^2 is called high strength concrete. Generally, with conventional mix and ingredients, it is difficult to increase the concrete strength. Another important problem with the conventional ingredient is pollution. The process of producing cement not only consumes a significant amount of natural resources but is also highly internal energy-intensive. About 3 billion tons of raw materials are needed for cement manufacture every year. It is also responsible for large emission of CO_2. On average, approximately one Ton of cement is being produced each year by every human being in the world. Hence to protect the environment, the main concern of minimizing CO_2 emission can be realized by reducing the percentage of cement used in making concrete. To mitigate these environmental problems, Civil Engineers have strived to produce many alternatives and more eco-friendly concretes. In this concrete, the methodology is to replace the ingredients with industrial by-product materials and still possesses excellent mechanical properties. Every year millions of tons of industrial waste are generated of which is no correct process to dispose of them environmentally friendly. Many efforts are being made to reduce the use of Portland cement in concrete by replacing these industrial wastes [2].

Normal water curing is required for any structure which is constructed with concrete. The requirement of water is so high that the concrete industry uses over one trillion gallons of water each year globally excluding wash water. This becomes a problem in that region where the freshwater is not easily available. So, if we use concrete which can be cured in ambient air, the problem in many regions can be solved. So, there is a need to replace concrete with such a material which is environmentally friendly and can be cured in ambient air condition.

1.1 Geopolymer concrete

Producing Geopolymer by a polymerization (it involves a chemical reaction of aluminosilicate minerals under the alkaline condition that results in a three-dimensional polymeric chain [5]) reaction of alkaline liquid with the source of industrial by-product materials rich in silica and alumina. It exhibits many excellent properties such as low creep, low shrinkage, good acid resistance and high compressive strength [3, 4]. The theoretical basis of Geopolymerization as a major reaction mechanism of cementless concrete was established for the first time by the French researcher JOSEPH DAVIDOVITS in 1978. In Geopolymer concrete, alkali activator solution plays a major role in the dissolution of Si and Al oxides. Hence, Geopolymers are synthesized by mixing the most common alkaline activators that are strong alkaline solutions such as Sodium Hydroxide (NaOH), Potassium Hydroxide (KOH), Potassium Silicate or Sodium Silicate with aluminosilicate reactive materials [8]. The compressive strength of the Geopolymer is significantly affected by the concentration of the activator that increases the strength of Geopolymer.

1.2 Materials for Geopolymer concrete

Supplementary Cementitious materials are used as a partial or full replacement of ordinary Portland cement. These are industrial by-products and are environmentally friendly. Some of the supplementary Cementitious materials are Fly ash, GGBFS, Silica Fume, Rice husk ash, Metakaolin etc., In the preparation of geopolymer concrete adopted in this study, GGBFS and Silica Fume are used as a replacement of cement. The main ingredients with which Geopolymer mortar is made up of in this study are Binder (GGBS, silica fume), Alkali activators (Sodium hydroxide and Sodium silicate), Coarse aggregate, Fine aggregate (quartz sand) and Admixture (Gypsum).

2. Methodology

2.1 Principle steps involved in the Step-by-step methodology

The following methodology was adopted to conduct the experimental program.

1. Procuring the required materials for Geopolymer concrete and study their physical and chemical properties to establish their suitability in making Geopolymer concrete.
2. The mix proportion and alkali binder ratios are determined by conducting trail mixes to obtain an ideal mix proportion of maximum strength.
3. Three types of nominal mix proportions were considered that is 1:1:2, 1:1.5:3 and 1:2:4 with different combinations of NaOH and to Na_2SiO_3. Three sets of concrete cubes were prepared in summer, monsoon and winter to check the effect of weather in ambient air curing.
4. Alkali solutions are prepared one hour before their use in Geopolymer concrete. From the ideal mix proportion and alkali binder ratios, cubes of size 150mm^3 are cast in moulds and demolded after 24 hours by keeping them at room temperature. Compressive strength at 28 days is determined for ambient air-cured concrete cubes using the compressive testing machine.

2.2 Modal mix calculation.

The following details are modal mix proportion for a Nominal mix of 1:2:4 of Geopolymer concrete mix for one cube of size (150 mm ×150 mm × 150 mm) with A/B ratio of 0.8

IOP Conf. Series: Materials Science and Engineering 1025 (2021) 012026 doi:10.1088/1757-899X/1025/1/012026

Mix proportion = 1: 2: 4 (Binder: Fine aggregate: Coarse aggregate)
Weight of binder = (8.438/7) × (1) = 1.2053 kg
GGBS (80% of binder) = 0.8 × 1.2053 = 0.9642 kg
Silica fume (20% of binder) = 0.2 × 1.2053 = 0.2410 kg
Gypsum (5% of binder) = 0.05 × 1.2053 = 0.01026 kg
Alkali solution = 0.8 × 1.2053 = 0.9642 kg
The ratio of NaOH to Na_2SiO_3 = 2: 5
Mass of NaOH solution = (2/7) × (0.9642) = 0.2754 kg
Mass of Na_2SiO_3 solution = (5/7) × (0.9642) = 0.6887 kg

2.3 Quantities of different mix proportions

For the different mixes considered in this study, the are calculated and presented in table 1

Table 1. Quantities of different mix proportions for one cubic meter of Geopolymer concrete (kg)

Mix	Mix Code	GGBS	Silica fume	Gypsum	Quartz sand	Coarse Aggregate 20 mm	Coarse Aggregate 10 mm	Mass of NaOH W_{water}	Mass of NaOH W_{SHf}	Mass of Na_2SiO_3 W_{water}	Mass of Na_2SiO_3 W_{ss}
1:1:2	Z1	500	125	31	625	750	500	102	40	285	71
	Z2	500	125	31	625	750	500	102	40	232	125
	Z3	500	125	31	625	750	500	102	40	178	178
	Z4	500	125	31	625	750	500	85	57	285	71
	Z5	500	125	31	625	750	500	85	57	232	125
	Z6	500	125	31	625	750	500	85	57	178	178
	Z7	500	125	31	625	750	500	72	70	285	71
	Z8	500	125	31	625	750	500	72	70	232	125
	Z9	500	125	31	625	750	500	72	70	178	178
1:1.5:3	Y1	363	90	22	681	818	545	75	29	207	52
	Y2	363	90	22	681	818	545	75	29	168	91
	Y3	363	90	22	681	818	545	75	29	130	130
	Y4	363	90	22	681	818	545	62	41	207	52
	Y5	363	90	22	681	818	545	62	41	168	91
	Y6	363	90	22	681	818	545	62	41	130	130
	Y7	363	90	22	681	818	545	52	51	207	52
	Y8	363	90	22	681	818	545	52	51	168	91
	Y9	363	90	22	681	818	545	52	51	130	130
1:2:4	X1	285	71	17	714	857	571	58	23	163	41
	X2	285	71	17	714	857	571	58	23	132	71
	X3	285	71	17	714	857	571	58	23	102	102
	X4	285	71	17	714	857	571	49	32	163	41
	X5	285	71	17	714	857	571	49	32	132	71
	X6	285	71	17	714	857	571	49	32	102	102
	X7	285	71	17	714	857	571	41	40	163	41
	X8	285	71	17	714	857	571	41	40	132	71
	X9	285	71	17	714	857	571	41	40	102	102

2.4. Preparation of concrete specimens

1. The alkali solutions made of required Sodium hydroxide and required Sodium silicate solution, are prepared one hour before for making Geopolymer concrete.

2. The binder materials are mixed in sequence i.e. GGBS, silica fume and gypsum respectively. Next, the quartz sand and coarse aggregate are added to the mix and all the materials are combined by electrical mixer for 2-3 minutes.
3. After proper mixing of materials, alkali activators like sodium hydroxide and sodium silicate are added in required proportions.
4. After proper mixing of Geopolymer concrete, the concrete mix is filled into moulds and each layer is compacted by 25 blows using tamping rod having specifications of diameter 16mm & length 610 mm length. After filling the moulds and proper compaction, place the moulds on a vibration table for better compaction.
5. After completion of casting placed the specimens are demolded after 24 hours from cast time after keeping the specimens in ambient curing for 28 days.

3. Results

As per IS 516 (1959), the compressive strength of concrete was performed on the compressive testing machine having a capacity of 3000 kN. For testing, an average of three cube specimens of each size 150 mm^3 that were cast and tested under compressive load is considered. 28 days compressive strength results are presented in Table 2 for ambient cured cube specimens in summer, monsoon and winter.

Table 2. Compressive Strengths of different mix proportions in different seasons

Mix	Code	NaOH	Na_2SiO_3	Compressive strength (MPa) in		
				Summer	Monsoon	Winter
1:1:2	Z1	9M	20%	21.6	23.8	24.9
	Z2		35%	29.8	29.4	30.7
	Z3		50%	25.5	31.8	33.6
	Z4	14M	20%	25.9	22.4	23.7
	Z5		35%	33.3	30.1	30.5
	Z6		50%	33.1	33.3	36.8
	Z7	19M	20%	25.1	20.8	26.2
	Z8		35%	42.2	32.8	32.8
	Z9		50%	44.3	38.6	39.4
1:1.5:3	Y1	9M	20%	20.9	26.2	27.3
	Y2		35%	22.6	24.7	25.8
	Y3		50%	25.6	33.8	35.4
	Y4	14M	20%	35.5	24.3	25.6
	Y5		35%	40.5	33.7	36.1
	Y6		50%	38.7	39.0	41.0
	Y7	19M	20%	25.6	25.3	26.5
	Y8		35%	39.8	29.1	30.9
	Y9		50%	44.2	40.0	38.1
1:2:4	X1	9M	20%	17.5	22.5	23.5
	X2		35%	28.9	27.8	30.9
	X3		50%	30.9	35.7	36.3
	X4	14M	20%	25.6	28.7	28.6
	X5		35%	43.9	36.5	38.3
	X6		50%	39.5	36.5	36.5
	X7	19M	20%	26.9	27.5	28.7
	X8		35%	43.2	43.0	43.2
	X9		50%	46.9	45.8	45.6

4. Discussion of results: Effect of alkali activators and seasons on compressive strength

For conventional concrete, the mix proportion 1:1:2 represents M25 concrete. The Geopolymer concrete developed with mix proportion of 1:1:2 with different combinations of NaOH and Na_2SiO_3 produced higher compressive strength values than conventional concrete. The effect of alkali solution and seasons on compressive strength is represented in figure 1. Highest compressive strength value 44.3 MPa has obtained for19M NaOH solution and 50% Na_2Sio_3 concentration in the summer season. Out of three seasons summer season exhibit good compressive strength values at higher concentration of alkali solution. Monsoon and winter seasons exhibit approximately equal strengths for all concentrations of alkali solution.

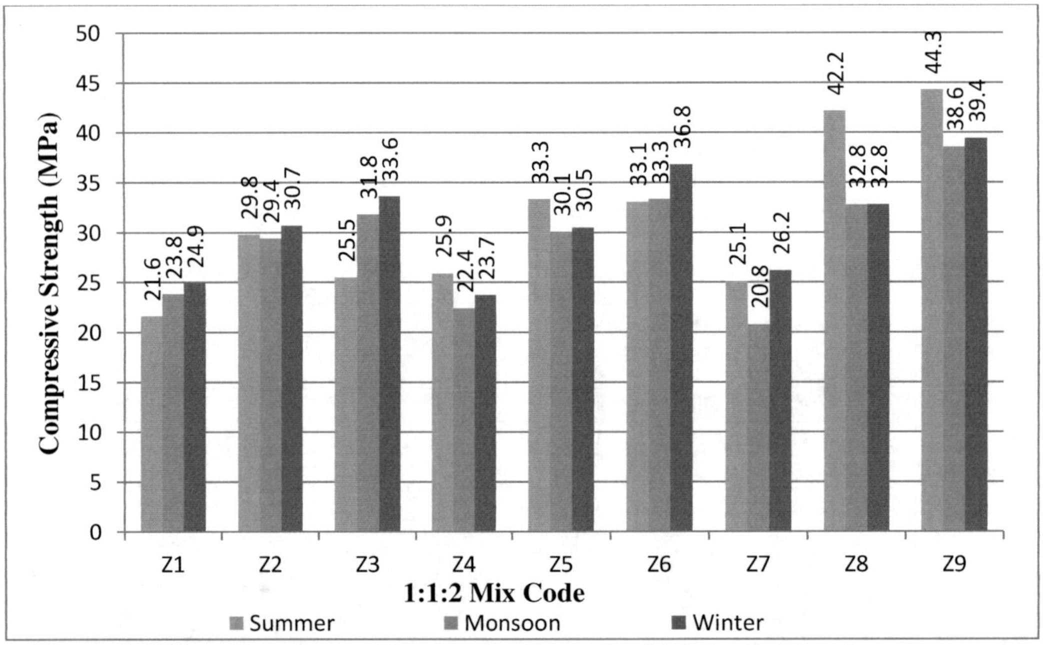

Figure 1. Effect of seasons on Compressive Strength of 1:1:2 mix Geopolymer concrete

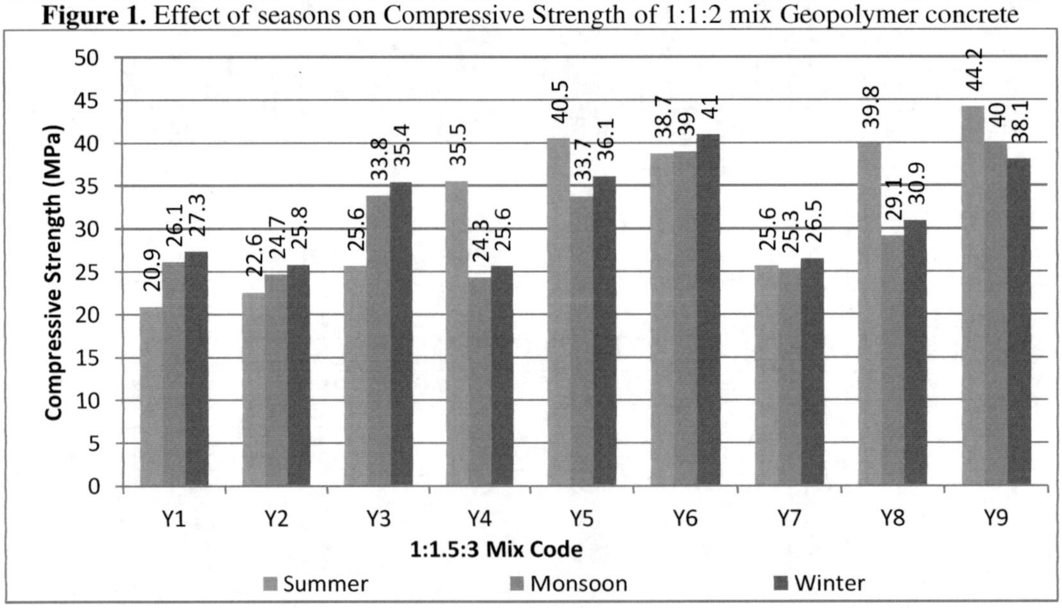

Figure 2. Effect of seasons on Compressive Strength of 1:1.5:3 mix Geopolymer concrete

For conventional concrete, the mix proportion 1:1.5:3 represents M20 concrete. The Geopolymer concrete developed with mix proportion of 1:1.5:3 with different combinations of NaOH and Na_2SiO_3 produced higher compressive strength values than conventional concrete. The effect of alkali solution and seasons on compressive strength is represented in figure 2. Highest compressive strength value 44.2 MPa has obtained for19M NaOH solution and 50% Na_2Sio_3 concentration in the summer season. Out of three seasons summer season exhibit good compressive strength values at higher concentration of alkali solution. Monsoon and winter seasons exhibit approximately equal strengths for all concentrations of alkali solution.

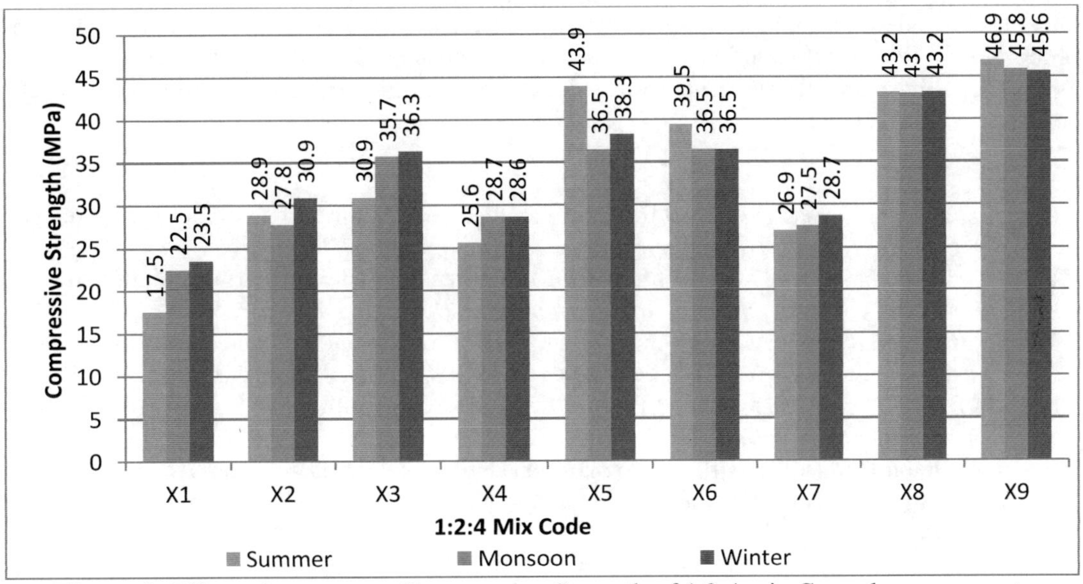

Figure 3. Effect of seasons on Compressive Strength of 1:2:4 mix Geopolymer concrete

For conventional concrete, the mix proportion 1:2:4 represents M15 concrete. The Geopolymer concrete developed with mix proportion of 1:2:4 with different combinations of NaOH and Na_2SiO_3 produced higher compressive strength values than conventional concrete. The effect of alkali solution and seasons on compressive strength is represented in figure 3. Highest compressive strength value 46.9 MPa has obtained for19M NaOH solution and 50% Na_2Sio_3 concentration in the summer season. Out of three seasons summer season exhibit good compressive strength values at higher concentration of alkali solution. Monsoon and winter seasons exhibit approximately equal strengths for all concentrations of alkali solution.

5. Conclusions

From the experimental results, the following conclusions are acquired

1. The experimental study is conducted for different mix proportions, they are 1:1:2, 1:2:4 and 1:1.5:3 and all these mix proportions produces good compressive strength results out of these 1:2:4 mix produces higher results.
2. Compressive strength of Geopolymer concrete increases with the increasing of sodium silicate concentration and maximum results obtain at 50 % concentration.
3. Compressive strength of geopolymer concrete also increases with the increasing of molarities of sodium hydroxide and maximum results obtain at 19M solution.
4. 9M and 14M sodium hydroxide mixes produce the same results in all mix proportions. But 19M varies very highly in every mix proportion.
5. Compressive strength of Geopolymer concrete will not depend on the mix proportion. It depends on the Alkali activators only

6. Summer, monsoon and winter seasons also affect the Ambient curing of the Geopolymer concrete which leads to changes in compressive strength.
7. Summer season exhibits good compressive strength values at higher concentration of alkali solution. Monsoon and winter seasons exhibit approximately equal strengths for all concentrations of alkali solution.
8. Maximum compressive strength value 44.3 MPa is obtained for 1:1:2 mix proportion of 19M NaOH solution and 50% Na_2Sio_3 concentration in the summer season
9. Maximum compressive strength value 44.2 MPa is obtained for 1:1.5:3 mix proportion of 19M NaOH solution and 50% Na_2Sio_3 concentration in the summer season
10. Out of all condition's highest compressive strength value, 46.9 MPa is obtained for 1:2:4 mix proportion of 19M NaOH solution and 50% Na_2Sio_3 concentration in the summer season
11. With 1:2:4 low nominal mix proportions higher compressive strength Geopolymer concrete is achieved.

References

[1] Daniel L Y Kong, Sanjayan G 2010 Effect of elevated temperatures on Geopolymer paste, mortar and concrete *Elsevier Cement and Concrete Research* **40** p 334-339
[2] Vijai K, Kumutha R and Vishnuram B G 2010 Effect of types of curing on the strength of geopolymer concrete *International Journal of the Physical Sciences* **5** p 1419-23
[3] Temuujin J, Van Riessen A, MacKenzie K J D 2010 Preparation and characterisation of fly ash-based geopolymer mortar *Elsevier Construction and Building Materials* **24** p 1906-10
[4] Prakash R Vora, Urmil V Dave 2013 Parametric Studies on Compressive Strength of Geopolymer Concrete *Elsevier Procedia Engineering* **51** p 210-9
[5] Pradip N, Prabir Kumar 2014 Effect of GGBFS on setting, workability and early strength properties of fly ash geopolymer concrete cured in ambient condition *Elsevier Construction and Building Materials* **66** p 163-71
[6] Platias S, Vatalisa K, Charalampidesa G 2014 Suitability of quartz sands for different industrial applications *Elsevier Procedia Economics and Finance* **14** p 491-498.
[7] Pradip N, Prabir Kumar S 2014 Effect of GGBS on setting, workability and early strength properties of fly ash geopolymer concrete cured in ambient condition *Construction and Building materials* Elsevier **66** p 163-171
[8] Gorhan G, Kurklu G 2014 The influence of the NaOH solution on the properties of the fly ash based Geopolymer mortar cured at different temperatures *Elsevier Composites* **58** p 371-7
[9] Santosh Kumar K, Markandeya Raju P and Rama Rao G 2015 Strength and durability studies on GGBS based concrete *International Journal of Civil Engineering* **2** p 10
[10] Amer M, Farzadnia N, Abdullah A, Ramazan D 2015 Development of high strength alkali activated binder using palm oil fuel ash GGBS at ambient temperature *Construction and Building Materials Elsevier* **93** p 289-300
[11] Shabarish V, Veeresh B, Manojkumar C 2018 Granulated Blast-Furnace slag (GGBS) based Geopolymer Concrete *International Journal of Advanced Science and Engineering* **5** p 879-85
[12] IS 456 2000 Indian standard code of practice for Plain and Reinforced Concrete (Fourth Revision) *Bureau of Indian Standards* New Delhi, India.
[13] IS 516 1959 Indian standard Methods of Tests for Strength of concrete *Bureau of Indian Standards* New Delhi India.
[14] IS 12089 1987 Indian standard Specifications for GGBS (Ground Granulated Blast Furnace Slag) natural sources for Geopolymer concrete *Bureau of Indian Standards* New Delhi, India.
[15] IS 15388 2003 Indian standard Specifications for Silica Fume from natural sources for concrete *Bureau of Indian Standards* New Delhi, India.
[16] IS 383 1970 Specifications for coarse and fine aggregate from natural sources for concrete *Bureau of Indian Standards* New Delhi, India

The potency of IS 875 (Part 3) 2015 code provisions on A-type trusses in the cyclonic regions of India

Santhosh Kumar Baswa[1], BALAJI KVGD[2], Chandan Kumar Patnaikuni[3]

Department of Civil Engineering, GITAM, Visakhapatnam, 532001, Andhra Pradesh, INDIA
Email. santhoshamie@gmail.com

Abstract: SP 38:1987 Handbook for the design of trusses have been widely used by the structural engineers in India. The designs were recommended 1987 version wind load provisions. The revised version of IS 875 was released in 2015 with certain additional recommendations for computation of design wind pressures in static load consideration. In this paper comparative analysis was made on A-Type trusses for 9m and 24 m spans with STAAD Pro (8i) software to determine the extent of increase of truss forces. It was inferred that the trusses forces for smaller spans are more increased when compared to larger spans.

1. Introduction.

The typified designs illustrated for various spans with different slopes in the handbook SP (S&T) 38 1987 have the economical configuration for A-Type roof steel trusses. Because of the consideration of basic wind speed in the east coast, the region shall not be more than 50 m /s, the IS 875 2015 code has very clearly stated that the wind speed for the cyclonic region is to be modified by the cyclonic importance factor (k_4 factor) for industrial structures and post cyclonic importance structure category. The off Shore wind speeds, which are generally extended up to 60 km inland after striking the coast, are already reflected in the code as a safety measure in 1987 and 2015 wind code versions.

2. Literature Review

Prem Krishna (2006); Subash and Tamurai (2007) as cited in Suresh Kumar, Cini et al. (2012) [1] pinioned for further investigation of the applicability of basic wind speed map of the east coast of India because of the occurrence of high-speed wind cyclonic storms on the east coast of India. N Laxman, S. Gomathinarayana et al. (2009) [2] suggested for revision of the wind map in India with the significant conclusion that the maximum wind speed in coastal areas of the code of practice of IS 875 (Part 3) 1987 [3] is conservative. Kumar et al. (2018, 2017, 2017) [4], [5], [6] studied the effect of wind on transmission towers, monopole towers and steel chimneys. Markandeya Raju et al. (2017) [7] compared communication towers with different bracings subjected to wind.

Even though, the revised IS 875 (Part 3) 2015 [12] defines the same basic wind speed of the IS 875 (Part 3) 1987, the design wind speed in the cyclonic region is modified by the wind speed multiplication factor (k4) - importance factor. This k4 factor has numerical value 1.15 for industrial structures, and maximum value 1.30 for structures of post-cyclone importance structures. IS 875 (Part 3) 2011 Proposed draft and commentary [9] explained the factor 1.15 is referred for economic consideration of industrial structures and 1.30 is referred for emergency service purpose building categories such as cyclone shelters, hospitals, schools, communication towers in the cyclonic region considering to minimize the human loss during these cyclones. These cyclonic importance factors were already reflected in IS 15498 2004 [10].

3. Methodology

The height of the building less than 20m is defined as a low-rise building by the IS 875 (Part 3) 2015. By this definition, all spans with the roof slopes of A-type roof trusses illustrated from the SP 38 (S&T) 1987 handbook are related to low rise building category. These truss configurations for various spans and different roof slopes published by the Bureau of Indian standard through SP 38 (S&T) 1987 [11] have the minimum weight over different truss configurations. This standard considers the compound combination of N-type truss and fink/fan truss for economic weight consideration. In this paper, 9 m and 24 m span with roof slopes 1 in 3 and 1 in 5 of A-type roof angular type steel trusses are illustrated for computing the impact of cyclone factor. The analysis is made with STAAD Pro 8i software. The general methodology for load combinations is adopted from IS 800-2007 [12]. Wind load on individual members (Roofs, walls, etc.,) is computed from the equation (1)

$$F = (Cpe - Cpi) \times A \times Pd. \tag{1}$$

Where
F = wind load acting in a direction normal to the individual structural element.
Cpe = external Pressure Coefficients, (max at 45° wind angle)
CPI = internal pressure coefficients.
A = roof area under consideration,
Pd = design wind pressure.

The design wind pressure (Pd) is computed with equation (2).

$$Pd = Kd \times Ka \times Kc \times Pz \tag{2}$$

Where
Kd = wind directionality factor which has a maximum numerical value of 1.00 in the cyclonic region.
Ka = area averaging factor, which depends on the slope areas of the truss under the wind force consideration.
Kc = combination factor. The code specified the combination factor has a value 0.90 for the frames of clad buildings since the suction /pressures inside and outside of structures are not fully correlated.
Pz = The design wind pressure at any height above the mean ground level is computed by the relationship between wind pressure and wind speed as shown in equation (3).

$$Pz = 0.6 \times Vz^2 \tag{3}$$

Where
Vz = designed wind speed in m/s at height z.

The basic wind speed (Vb) for important cities is obtained from figure 1 of IS 875 (Part 3) 2015 [12] and shall be modified to include the following effects to get design wind speed, Vz at any height, Z for the chosen structure: (a) risk level, (b) terrain roughness and height of structures, (c) local topography, and (d) importance factor for the cyclonic region.

It can be mathematically expressed with the equation (4)

$$V_z = V_b \times k_1 \times k_2 \times k_3 \times k_4 \tag{4}$$

Where
Vz = design wind speed at any height z in m/s;

k1 = probability factor (risk coefficient) (clause 5.3.1 of IS 875 (Part 3) 2015 [13] for general and industrial structures with 50 years of return period is 1.00 and for the most cyclonic Importance structures with the return period of 100 years is 1.08;
k2 = terrain roughness (Category 2) and height factor (clause 5.3.2 of wind code 2015) varying according to the height of the structure.;
k3 = (1.00) topography factor (clause 5.3.3 of wind code 2015); and
k4 = (1.15 for Industrial structures and 1.30 for post cyclonic importance structures) cyclonic importance factor for the cyclonic region (clause 5.3.4 of IS 875 (Part 3) 2015 [13]);
Vb is the basic wind speed in the east coast region is specified as 50m/s.

With the above information and equations (1) to (4) are utilized in computing the wind force on the truss. Then the combination of dead load and wind loads are computed with the general methodology. In this context, the live load cannot be taken into consideration, because in the cyclones period there is no possibility of expecting live load on the roof trusses.

The A-type roof steel truss for 9.00 m span with roof slopes of 1 in 3 and 1 in 5 and 24.00 m span with roof slopes 1 in 3 and 1 in 5 for low permeability, medium permeability and large permeability conditions are considered for this study. up to 5 % building opening (Cpi ± 0.20) is known as low permeability, for building openings from 5 to 20% (Cpi ±0.50) is referred as medium permeability and building openings more than 20% (Cpi ±0.70) are known as large permeability condition. The spacing of truss is 6 m and a column height of 9 metres are common for all spans and slopes. The truss diagrams are shown in figures A and B for 9m span and 24 m span. With the above data, all the models are simulated in STAAD Pro 8i software. The dead load and wind loads acting at the nodal points on the trusses are represented in figures 1 and 2. The percentage variation of truss forces for industrial structures category and post cyclonic category are produced in figures 3 and 4 respectively.

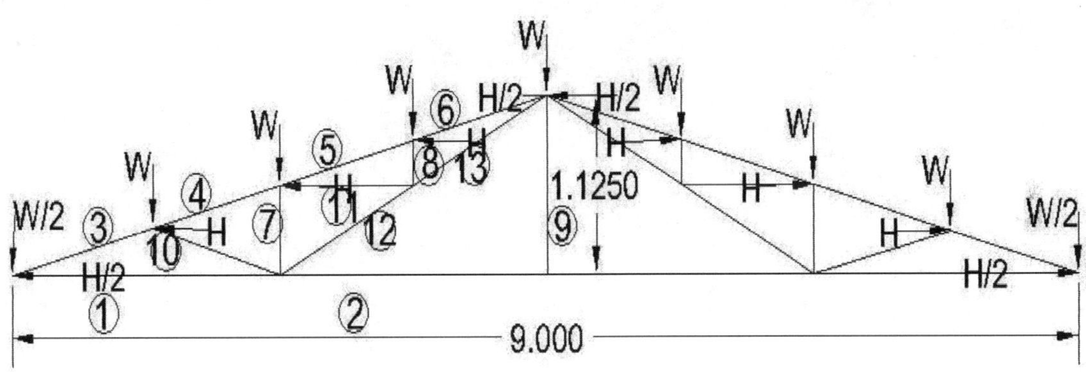

Figure 1. 9m span with 1 in 3 roof slope

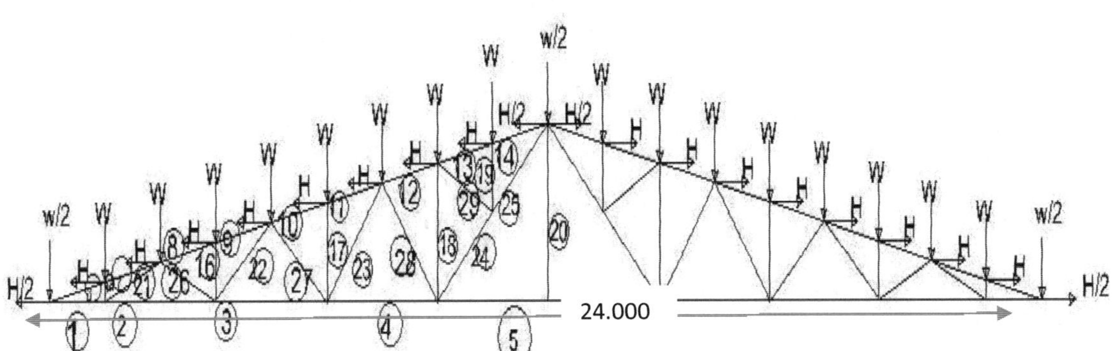

Figure 2. 24m span with 1 in 3 roof slope

IOP Conf. Series: Materials Science and Engineering 1025 (2021) 012027 doi:10.1088/1757-899X/1025/1/012027

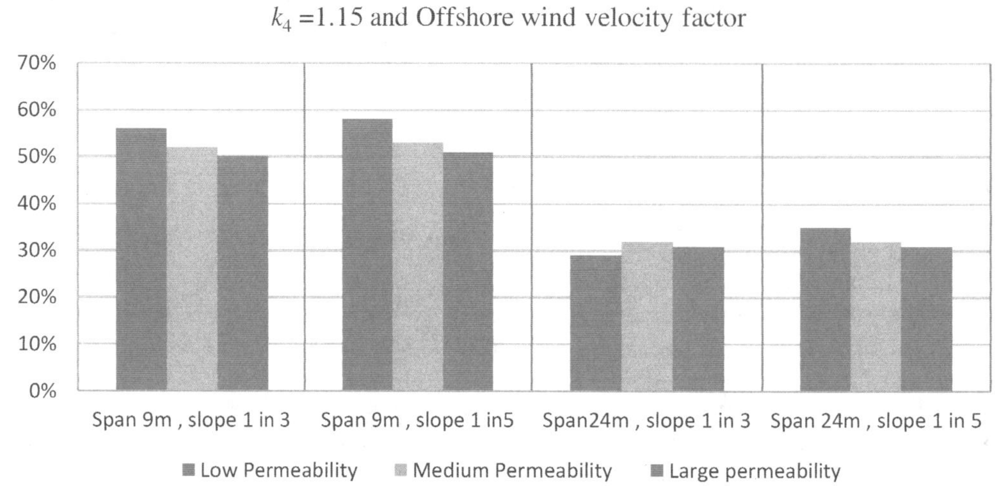

Figure 3. Variation of Truss forces for Industrial trusses

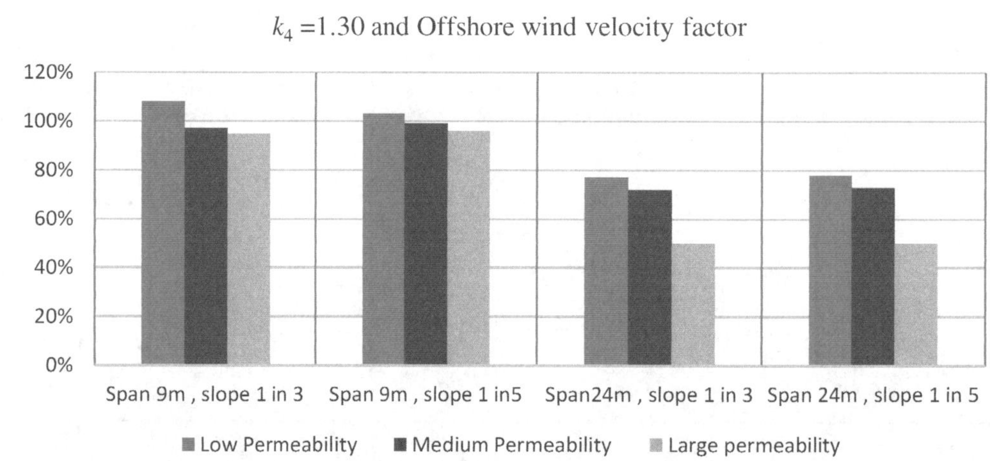

Figure 4. Variation of Truss forces for Post cyclonic Importance factor consideration

4. Discussions and Results

A type roof truss spans 9m and 15 m with roof slopes 1 in 3 and 1 in 5 have been simulated for total truss forces with IS 875 (Part 3) 1987 provisions. However, IS 875 (Part 3) 2015 revised version specified the k_4 factor for industrial and post cyclonic importance structures. Hence, the above-mentioned models with the k_4 factor have been stimulated for truss forces. The results obtained for both the models, i.e., 9 m and 24 m stimulated as per the provisions of IS 875 1987 and 2015 versions have been compared and presented in the figures 3 and 4.

For Industrial category of structures

The variations of truss forces are in the range of 51% to 56% for 9 m span and 29% to 32% for 24 m spans.

For post cyclonic importance structures

The variations of truss forces are in the range of 96% to108 % for 9 m span and 50% to 78% for 24 m spans.

5. Conclusions

With the above discussions, the following conclusions are drawn.

1. The highest variation of truss forces for Industrial condition occurs for 9 m spans with slope 1 in 5 for low permeability condition. The recent review and observation of Tecle et al (9) showed that Low internal permeability has a larger variation of truss forces.
2. The highest variation of truss forces for post cyclonic importance condition has occurred for 9 m span with slope 1 in 3 for low permeability condition.
3. The larger spans show the lesser variation of truss forces and there is a little variation of truss forces for change of slopes from 1 in 3 to 1 in 5.

References

[1] Suresh Kumar K, C Cini, Valerie Sift 2012 Assessment of design wind speeds for metro cities of India *The seventh International Colloquium on Bluff Body Aerodynamics and Applications (BBAA7)* Shanghai China September 2-6 2012

[2] Lakshmanan N, Gomathinaygam S, P Hari Krishna, A Abraham, Chitra Ganapathi S 2009 The Basic wind speed map of India with long- term, hourly wind data *Current science* **96 (7)** p 911-22

[3] IS 875 (Part 3) 1987 (Reaffirmed 1997*)* Code of practice for design loads (other than Earthquake) for Buildings and Structures *Bureau of Indian Standards* New Delhi India

[4] Kumar M P, Vishalakshi D, Raju P M, Rambabu D 2018 Progressive collapse study of 220 KV transmission line tower with different bracing patterns *Disaster Advances* **11 (10)** p 16-25

[5] Pavan Kumar M, Markandeya Raju P, Navya, M, Naidu G T 2017 Effect of wind speed on structural behaviour of Monopole and self-support telecommunication towers *Asian Journal of Civil Engineering* **18 (6)** p 911-27

[6] Kumar M P, Raju P M, Babu N V, Roopesh K 2017 A parametric study on lateral load resistance of steel chimneys *International Journal of Civil Engineering and Technology* **8 (7)** p 858-75

[7] Markandeya Raju P, Pavan Kumar M, Vishalakshi, D, Manoharini K 2017 Parametric comparison of communication towers with different bracings *International Journal of Civil Engineering and Technology* **8 (10)** p 235-54

[8] IS 875 (Part 3) 2015 Design Loads (other than Earthquake) for Buildings and Structures-code of Practice *Bureau of Indian Standards* New Delhi India

[9] Prem Krishna, Krishen Kumar, Bhandari N M *IS 875 (Part 3) 2011 Wind loads on Buildings and structures Proposed draft & Commentary*

[10] IS 15498 2004 Guidelines for improving the cyclonic resistance of low-rise Houses and other buildings/structures *Bureau of Indian Standards* New Delhi India

[11] SP 38 (S&T) 1987 Hand Book of Typified designs for structures with steel roof Trusses (with & without crane) based on IS Codes *Bureau of Indian Standards* New Delhi India

[12] IS 800 2007 General Construction in steel Code of Practice *Bureau of Indian Standards* New Delhi India.

[13] Tecle Amaneul S, Girma T Bitsuamalk, Mousaad Aly 2013 Internal pressure in a low–rise building with existing envelope openings and sudden breaching *Wind and Structures* **16 (1)** p 25-46

Verification of irregularities in a reinforced concrete tall structure

Penumatcha Krishnam Raju[1], Vipparthy Ravindra[2]

[1]Professor, Department of Civil Engineering, Gayatri Vidya Parishad College of Engineering (A), Visakhapatnam - 530048, INDIA
[2]Professor, Department of Civil Engineering, Jawaharlal Nehru Technological University Kakinada, Kakinada - 533003, INDIA
E-mail: drpkraju1962@gmail.com

Abstract. Building plans with simple, regular and compact arrangements are preferred to perform well during the occurrence of earthquakes. In practice, planning a regular layout may not be possible in all occasions due to irregular shape of building sites in the plan as well as in elevation due to uneven ground conditions. It is a challenge both for the architect and structural designer to plan and design such buildings. Further, the planning is to be made in compliance with the rules and regulations of the local approving authorities, which sometimes may lead to irregular shapes both in plan and elevation. Most of the times, the owner's and / or builder's requirement is to be fulfilled such that the project is viable to them. This is a challenge for the practising professionals designing the structures that are built in earthquake zones of III, IV and V. The earthquake resistant design code i.e., IS 1893 (Part 1) was revised in the year 2016 and adopting the provisions of building irregularities with more restrictions on their allowable limits. This paper addresses the identification or verification of the existence of building irregularities in a 17 storey multipurpose tall structure with two parking, two commercial and 13 residential floors. From the results, it is opined that there is a need for a review of the procedure and some of the limits set on irregular building configurations particularly Stiffness, Excessive openings, Out-of-plane offsets, Strength, In-plane discontinuity and Irregular modes of oscillation.
Keywords: Irregular building configurations; Stiffness; Excessive openings; Out-of-plane offsets; Strength; In-plane discontinuity; Irregular modes of oscillation.

1. Introduction

According to IS 1893 (Part 1): 2016 (Code) [1] building irregularities are broadly classified into two types i.e., (a) Plan irregularities and (b) Vertical irregularities. The Code recommends that the buildings should be made with simple, regular geometry and uniformly distributed mass and stiffness both in plan and elevation. Limits on irregularities are also specified for irregular configurations indicating some special requirements that are to be followed for the buildings constructed in seismic zones III, IV and V. Specific reference is also made to certain irregularities that are to be avoided in seismic zone II in the Code.

Tso and Myslimaj [2] Myslimaj and Tso [3] and Mario and Barbara [4] studied the effects of interdependence between strength and stiffness irregularities. Magliulo and Ramasco [5] studied the seismic response of the design of RC frames with strength discontinuities in elevation. They have obtained irregularities by assigning over strengths either to the beams or to the columns of a regular frame. The reinforcements in beams or in columns at different floors were modified. The storey strengths are computed by two different methods i.e., (i) considering flexural resistance of columns

only (ii) considering both flexural resistance of columns and beams. The methods of evaluation of strength are elaborated as given below.
1) storey strength is equal to the sum of the yielding bending moments at the ends of the columns belonging to the storey divided by the storey height;
2) evaluation of storey strength is same as above but, the yielding bending moment at each end of the storey columns is multiplied by the ratio between the sum of the yielding bending moments of the beams and the sum of the yielding bending moments of the columns converging into the same joint if such ratio is lower than one. Shreyasvi and Shivakumaraswamy [6] Pathan and Dhamge [7] and Livian et al. [8] investigated the irregularities due to "Re-entrant corners", "mass" and "Non-parallel systems" respectively. Ismail et al. [9] studied the stress and vertical geometric irregularities.

2. Objective and scope of the study

This study is made to identify various irregularities in a seventeen storied multipurpose reinforced concrete tall structure with two parking, two commercial and 13 residential floors. The architectural plan of the residential floor is shown in figure 1.

Figure 1. Typical floor plan of residential floor

A schematic cross-section showing the floor heights of the building structure is presented in figure 2. The height of each residential floor, commercial floor and the lower parking floor is kept at 3.05m. However, the floor height of the upper parking floor is maintained to be 3.96m to facilitate entry of parking vehicles into basement floors. The beam layout for parking and commercial floors is shown in Figure 3 and of the typical residential floor in Figure 4. After several trials of analysis, the column and beam dimensions are finalized and adopted as shown in Table 1, Figure 3 and Figure 4.

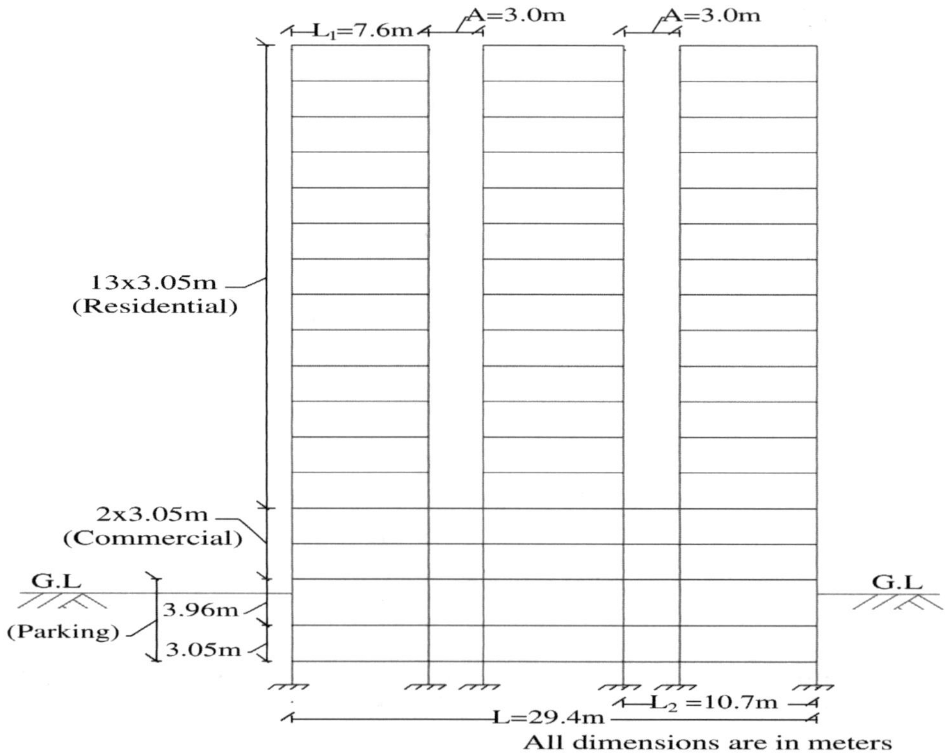

Figure 2. Schematic cross-section

Table 1. Dimensions of columns

Sl. No	Storey No.	Grids A1 to A6 (mm×mm)	Grids B1 to B6 (mm×mm)	Grids C1 to C6 (mm×mm)	Grids D1 to D6 (mm×mm)	Grids E1 to E6 (mm×mm)	Grids F1 to F6 (mm×mm)
1	17, 16, 15, 14	760×380	760×380	760×380	760×380	760×380	760×380
2	13, 12	760×380	B1, B6 = 760×380 B2 to B5 = 910×380	C1 to C5 = 910×380 C6 = 760×380	D1 to D5 = 910×380 D6 = 760×380	E1, E6 = 760×380 E2 to E5 = 910×380	760×380
3	11, 10, 9	910×380	B1, B6 = 910×380 B2 to B5 = 910×460	C1 to C5 = 910×460 C6 = 910×380	910×460	E1, E6 = 910×380 E2 to E5 = 910×460	910×380
4	8, 7	910×380	910×460	C1 to C5 = 910×530 C6 = 910×460	910×530	910×460	910×380
5	6, 5	910×460	910×460	910×530	910×530	910×460	910×460
6	4, 3	910×460	910×530	910×530	910×530	910×530	910×460
7	2, 1	910×460	910×530	910×610	910×610	910×530	910×460

IOP Conf. Series: Materials Science and Engineering 1025 (2021) 012028 doi:10.1088/1757-899X/1025/1/012028

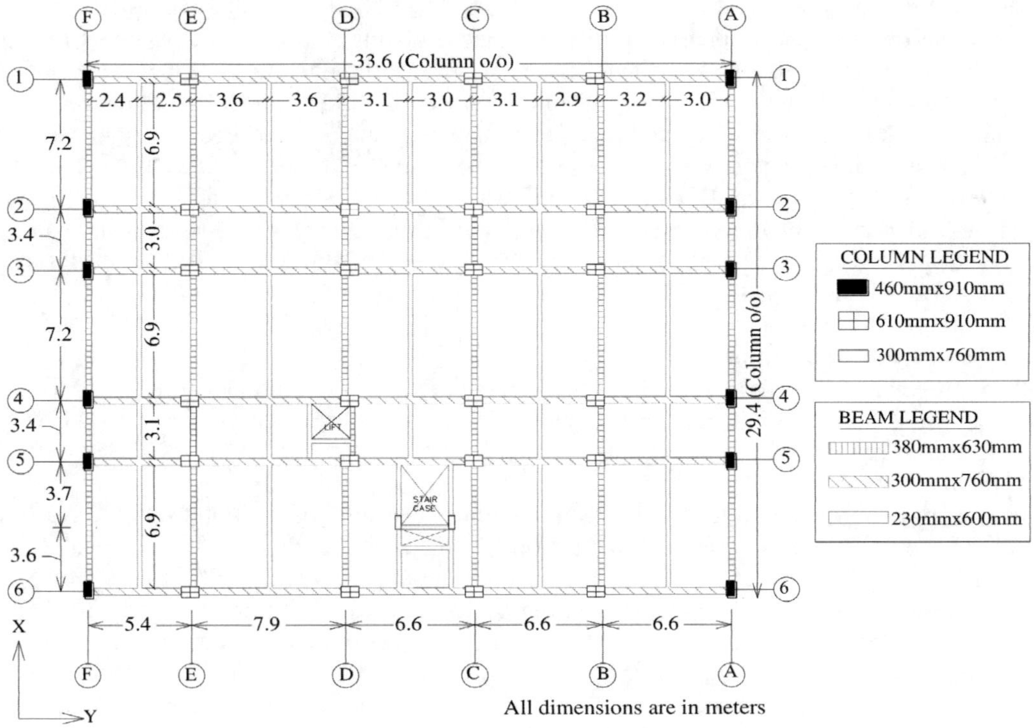

Figure 3. Beam layout of commercial/parking floor

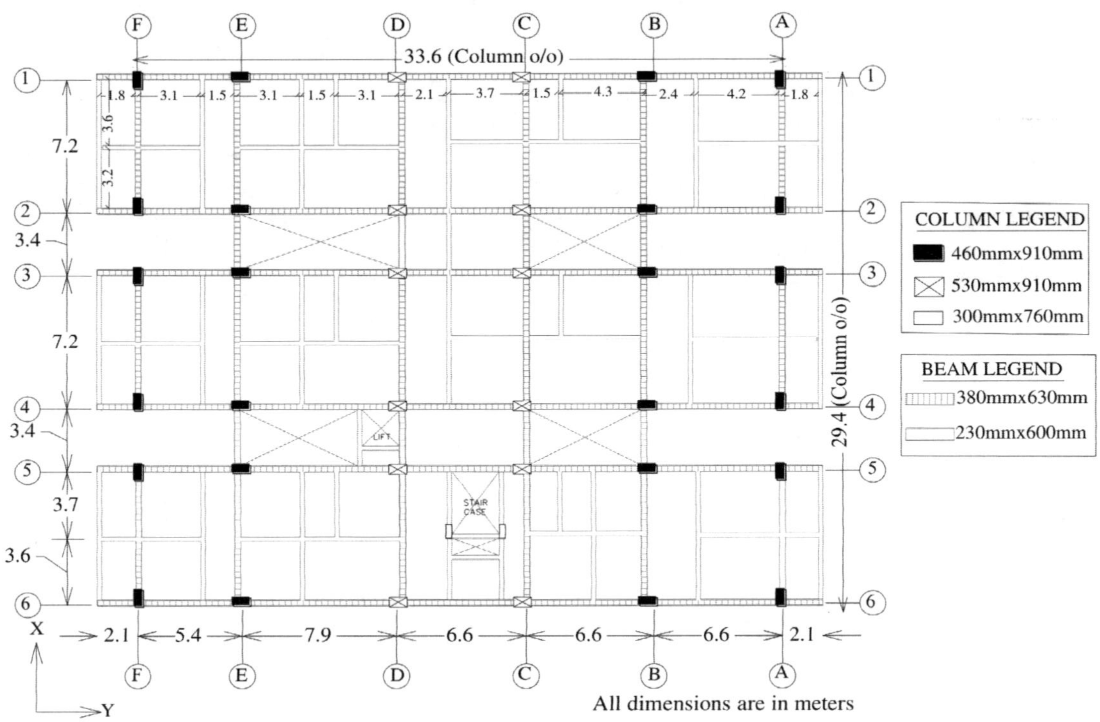

Figure 4. Beam layout of residential floor

The ETABS Evaluation Version is used for analysis and design of the building structure considered for the study. The study also covers all the seismic zones of India i.e., Zone II to Zone V.

The dead loads are considered as per IS 875 (Part 1): 1987 [10]. The live loads on parking, commercial and residential floors are taken as 4, 5 and 2 kN/m^2 respectively as per IS 875 (Part 2): 1987 [11]. The structure is analyzed for all the possible load combinations that include both Limit state of collapse and Limit state of serviceability as per IS 456: 2000 [12]. However, the load combinations with wind are not included in the analysis as the study is specific to earthquake-resistant design. The results of the earthquake analysis are so arrived duly considering the strong column-weak beam effect as per cl. 7.2 of IS 13920: 2016 [13]. However, the presentation of results is limited to building irregularities only in this paper. The Importance factor and Response reduction factor are taken as 1.2 and 5 respectively. The structure adopted is assumed to be founded on medium soil as classified in Table 2 of IS 1893 (Part 1): 2016 [1].

3. Study on Plan irregularities

The Plan irregularities that cover (i) Torsional irregularity, (ii) Re-entrant corners, (iii) Floor slab having excessive cut-outs or openings, (iv) Out-of-plane offsets in vertical elements and (v) Non-parallel lateral force system are discussed as follows.

3.1. Torsional irregularity

This irregularity can only be identified after obtaining the results of post-analysis for the building considered. The building Rules permit constructions for the accommodation of commercial activity in some of the lower floors and with the remaining upper floors for residential purpose. However, the parking is normally allowed in basement floors in a multipurpose tall building structure. Depending upon the configuration, the building is to be analyzed several times by modifying the positions of vertical elements to bring the torsional irregularity within the specified limit. The plan aspect ratio is arrived less than 3 as per the requirement of the code. Two specifications are given defining the Torsional irregularity in Table 5 of IS 1893 (Part 1): 2016 [1].

a) If $\Delta_{max} > 1.5\ \Delta_{min}$

Then, Torsional irregularity is present in the building.

Where,

Δ_{max} = Maximum horizontal displacement at the one end

Δ_{min} = Minimum horizontal displacement at the far end

The Δ_{max} and Δ_{min} values in both X and Y directions of the building for all seismic zones are tabulated in Table 2. It can be seen from this table that the ratios of Δ_{max} and Δ_{min} are less than 1.5 in both the directions and hence Torsional irregularity is not present. Figure 5 shows the checks for Torsional irregularity in X and Y axis respectively for Zone II.

b) If TR > TTX and TR > TTY

Then, Torsional irregularity exists in the building.

Where,

TR = Natural time period corresponding to the fundamental torsional mode

TTX = Natural time period corresponding to translational mode in X-direction

TTY = Natural time period corresponding to the translational mode in Y-direction

Table 2. Check for Torsional irregularity

	Lateral Displacement (mm)						**Time period (Sec.)**		
Zone	**X-direction (mm)**		$\left(\frac{\Delta_{max}}{\Delta_{min}}\right)$	**Y-direction (mm)**		$\left(\frac{\Delta_{max}}{\Delta_{min}}\right)$	**Translational**		**Rotational**
	(Δ_{min})	(Δ_{max})		(Δ_{min})	(Δ_{max})		**TTX**	**TTY**	**TR**
Zone - II	18.94	19.43	1.03	17.70	18.58	1.05	2.33	2.29	2.10
Zone - III	30.12	30.52	1.01	27.50	28.55	1.04	2.33	2.29	2.10
Zone - IV	45.02	45.30	1.00	40.50	41.85	1.03	2.33	2.29	2.10
Zone - V	67.37	67.48	1.00	60.03	61.79	1.03	2.33	2.29	2.10

It can be seen from the Table 2, that the values of TR are less than the values of both TTX and TTY in all seismic zones and hence no torsional irregularity is present in any of the floors of the building.

In addition to the above provisions, if the ratio of $\left(\frac{\Delta_{max}}{\Delta_{min}}\right)$ is in the range of (1.5 to 2.0), the building configuration shall be revised to ensure that the natural time period of the fundamental torsional mode of oscillation shall be smaller than those of the first two translational modes. Even though the code recommended 3D dynamic analysis for the above case, it is not possible to assess the torsional irregularity in the building without 3D dynamic analysis. The building configuration shall be revised also in case of $\left(\frac{\Delta_{max}}{\Delta_{min}}\right)$ more than 2.0 to ensure that the natural time period of the fundamental torsional mode of oscillation is less than those of the first two translational modes along with each principle plan directions. In the present study, the above situation does not arise.

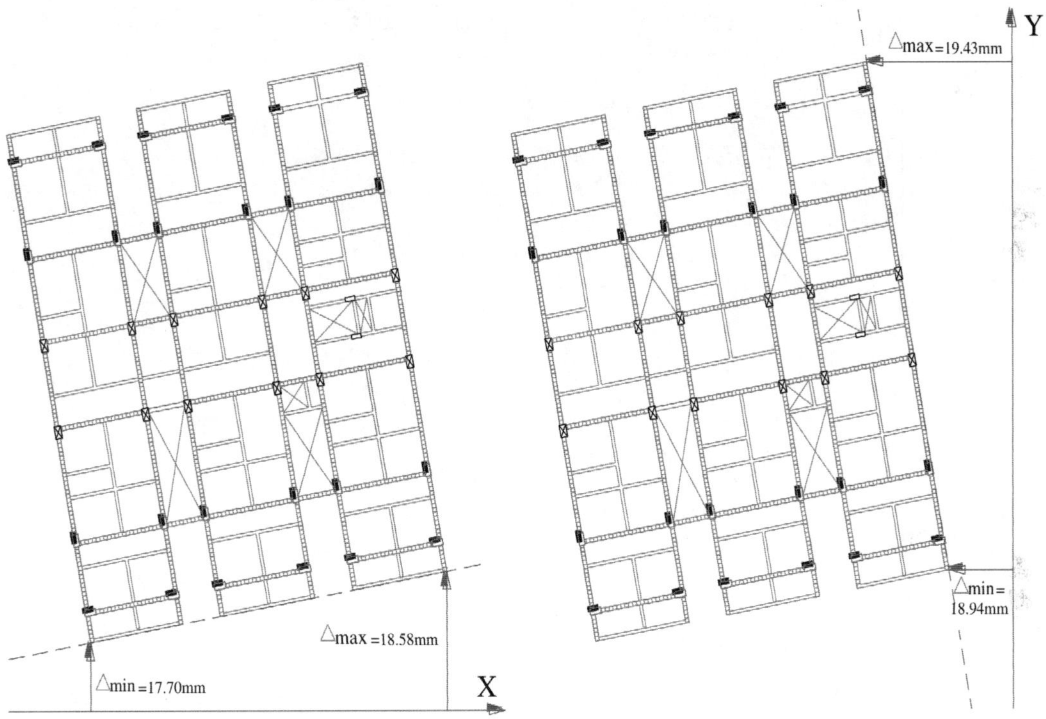

(a) With respect to X-axis (b) With respect to Y-axis

Figure 5. Check for Torsional irregularity (top storey)

3.2. Re-entrant corners

It is a pre-determined irregularity based on the planning of the structure. If irregularity due to Re-entrant corners exists, IS 1893 (Part 1): 2016 [1], recommends the adoption of 3D dynamic analysis. If the ratio of $\frac{A}{L} > 0.15$ in any plan direction, then it is said to be irregularity due to Re-entrant corner.

Where A = length of Plan projection in the direction considered

L = Overall plan dimension in the same direction

In the present study, Re-entrant corners are identified as shown in Fig. 6 that cannot be eliminated in this plan configuration particularly along Y-axis. However, along X-axis, no such irregularity is present as the $\left(\frac{A_1}{L_1}\right)$the ratio is less than 0.15.

Along X-axis

Ratio of $\left(\frac{A_1}{L_1}\right) = \frac{3.0}{29.4} = 0.102 < 0.15$, Irregularity due to Re-entrant corner does not exist

Along Y-axis

Ratio of $\left(\frac{A_2}{L_2}\right) = \frac{5.20}{33.6} = 0.155 > 0.15$, Irregularity due to Re-entrant corner marginally exists

Ratio of $\left(\frac{A_3}{L_2}\right) = \frac{7.20}{33.6} = 0.214 > 0.15$, Irregularity due to Re-entrant corner exists

Where, A_1, A_2 and A_3 are plan projections and L_1 and L_2 are overall plan dimensions in X and Y directions as shown in Figure 6.

As residential flats are planned with at-least three sides ventilation and also acceptable as per building bye-laws such irregularities do exists in many of the structures. However, code recommends 3D dynamic analysis for such buildings and the same is carried out for the present structure.

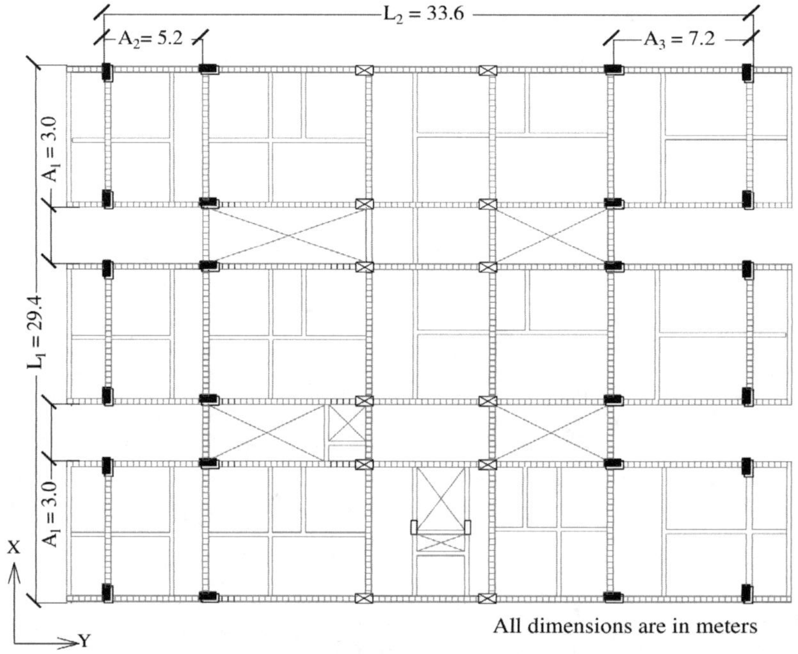

Figure 6. Check for Re-entrant corners

3.3. Floor slabs having excessive cut-outs or openings

It is a plan irregularity that can be determined based on the planning of the structure. This irregularity can be verified based on plan dimensions before actual analysis is carried out. In the present case, hardly any openings or cut-outs both in commercial and parking floors except lift and staircases present. However, openings or cut-outs are present in the residential floors of the building as shown in Figure 7. According to IS 1893 (Part 1): 2016 [1] the building is said to have a discontinuity in their in-plane stiffness and is considered as "Flexible diaphragm" behaviour if the area of cut-outs or openings is more than 50% of the total floor area of the slab. In respect of openings located along any edge of the slab, the above provision is restricted to 10% only to consider the "Rigid diaphragm" effect.

This provision is verified for the study under consideration.

Area of Total openings (m^2), Ao = (a_1+ a_2+ a_3+ a_4+ a_5+ a_6+ a_7+ a_8+ a_9+ a_{10}+a_{11})

=(15.6+24.3+17.4+21.6+15.6+17.7+3.8+17.4+21.6+8.16+2.16)

Where, a_1, a_2, a_3 etc., are areas of openings as marked in Figure 7.

Total floor area

A_t = Floor area surrounded by grids A1, A6, F6, F1 and A1 = 29.4 × 33.6 = 987.84m^2

Ratio = $\left(\frac{A_O}{A_t}\right) = \frac{165.32}{987.84} = 0.167 < 0.5$ (Hence, rigid diaphragm effect can be considered)

Therefore, the irregularity due to "Excessive openings or cut-outs" is not present in this structure and the floor acts as a rigid diaphragm.
In this building, there are no openings in individual floor slab panels close to the edge of the slab and hence, the check for the ratio of $\left(\frac{A_o}{A_t}\right) > 0.1$ does not arise. It is not clear in the code, whether the portions of openings extended beyond column line and up to edge of balconies are to be taken into consideration while calculating the "opening area". According to cl. 5.6.2.2 of IS 16700: 2017 [14], the maximum area of openings in any floor diaphragm is limited to 30% of the plan area of the diaphragm, probably to achieve Rigid diaphragm behaviour. Hence, this provision of IS 1893 (Part 1): 2016 [1] needs further review.

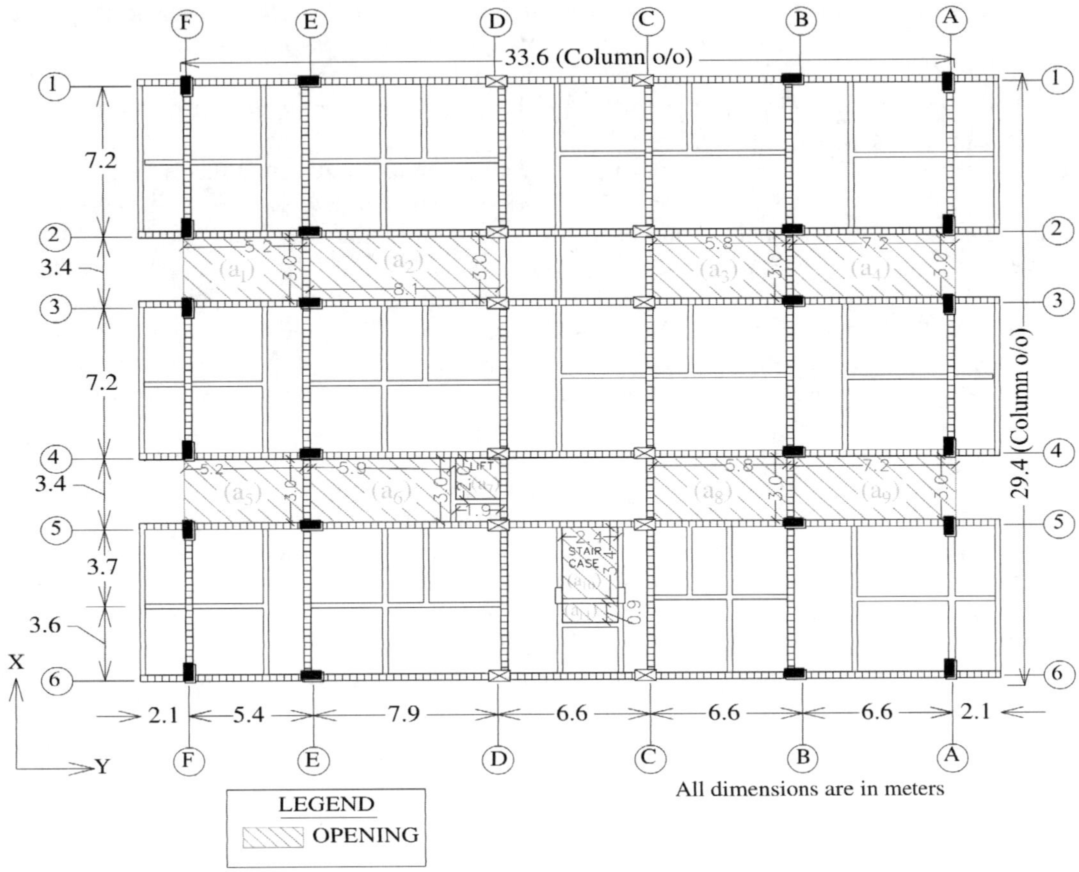

Figure 7. Check for irregularity due to Floor slabs having excessive cut-outs or openings

3.4. Out-of-plane offsets in vertical elements

According to the code, a building is considered to be out of plane offset in the vertical direction, when the structural walls or frames are moved out of the plane in any storey along the height. The amount of out-of-plane offset is not quantified. Instead, it was advised to refer specialist literature for all seismic Zones of II to V. However, for the buildings located in Zone III, IV and V, it is recommended that the lateral drift should be less than 0.2% in the storey having the offset and in storeys below them. Whereas, no such provision is included for Zone II. Post analysis verification is only possible to identify this irregularity.

In the present study, the above provision is verified with the columns in residential floors and Parking and Commercial floors. Even though the sizes of columns in the upper floors are reduced with respect to the lower floors, the centroid of the column is maintained throughout the height of the building. Hence, the above irregularity is not present in this case. In case, where the irregularity is present, specialist literature is to be referred as recommended in IS 1893 (Part 1): 2016 [1]. The code should also refer to such specialist literature to follow uniformly by the practising structural designers.

3.5. Non-parallel lateral force system

If vertically oriented structural systems resisting lateral forces are not oriented along the two principal orthogonal axes in the plan, the code recommends to carry out analysis of such systems for load combinations as mentioned in cl. 6.3.2.2 or 6.3.4.1. However, Non-parallel lateral force system does not exist in the present building under consideration.

4. Study on vertical irregularities

The vertical irregularities such as Stiffness irregularity (Soft storey), Mass irregularity, Vertical geometric irregularity, In-Plane discontinuity in vertical elements resisting lateral force, Strength irregularity (Weak storey), Floating or Stub columns and Irregular modes of oscillation in two principal plan directions are discussed below with respect to the structure under verification.

4.1. Stiffness irregularity (soft storey)

According to the code, if the lateral stiffness of the storey considered is less than that of the storey above, then stiffness irregularity exists and is to be eliminated. The lateral stiffness of the storey can be assessed only through 3D analysis and hence this type of irregularity can only be established after obtaining analysis results. One of the controlling factors in stiffness is storey height. Considering a uniform horizontal diaphragm in all the floors with varying height even in any lower or intermediate floor causes the stiffness irregularity. But in practice, the basement floors or lower floors of the building are catered for parking or commercial purposes and the upper floors are catered for residential or other purposes. Higher floor height is needed for basement and commercial floors compared to residential floors particularly in multipurpose buildings.

The other aspect is the consideration of diagonal strut action due to Un-Reinforced Masonry (URM) infills for internal and external walls. As per the clause 7.9 of IS 1893 (Part 1): 2016 [1], the advantage due to the infills for thickness 230mm, 200mm, 150mm and 115mm (to match with standard brick size) is limited compared to the regular spacing of columns that are maintained for residential buildings. The same is demonstrated below for most commonly adopted unreinforced masonry infill with a sample calculation.

4.1.1. Check for the advantage of URM infill. The following two limits are set for calculation of thickness of "equivalent diagonal strut" and both these limits should satisfy to consider the infill effect.

$$\left(\frac{h}{t}\right) < 12 \text{ and } \left(\frac{l}{t}\right) < 12$$

Where,

h = Clear height of the URM infill wall between the top beam and bottom floor slab

l = Clear length of the URM infill wall between the columns/walls

t = Thickness of unreinforced masonry infill

a) For 230mm thick wall (maximum infill thickness that is being considered in practice)

(i) The spacing of columns, $l < 12t$

$$l < 12\times230 = 2760\text{mm} = 2.76\text{m}$$

Hence, this restricts maximum spacing of columns to 2.76m, and same is not practically being followed.

(ii) The clear height of the URM in-fill wall, $h < 12t$

$$h < 12\times230 = 2760\text{mm} = 2.76\text{m}$$

This limit may be satisfied for residential buildings with regular beam depths. Even if one of the above two limits are not satisfied, the contribution of URM infill wall for improvement of storey stiffness is not achieved.

4.1.2. Check for soft storey effect.

From the results of the study, the lateral stiffness of different storeys as obtained in the 3D dynamic analysis are tabulated for both X and Y axes in Tables 3. The soft storey effect is observed in second (upper parking) and third (lower commercial) floors in X - direction as the storey stiffness of these floors is less than that of its upper floors. But, as per cl. 7.1 (Table 5) of IS 1893 (Part 1): 2002 [15], the above two storeys do not fall under soft storey effect as explained below.

Lateral stiffness (LS) of sixth storey = 1171069 kN/m
Lateral stiffness (LS) of fifth storey =1306403 kN/m
Lateral stiffness of fourth storey = 1460919 kN/m
Lateral stiffness of third storey = 1438802 kN/m
Lateral stiffness of second storey = 1004628 kN/m

4.1.2.1. Verification of soft storey effect for second storey (as per IS 1893 (Part 1): 2002) [15].

a) As 1004628> 0.6 ×1438802 > 863281 kN/m and

b) As 1004628>0.7 × $\left(\frac{1438802 + 1460919 + 1306403}{3}\right)$> 981429 kN/m,

it is to be considered that the Stiffness irregularity is not present in the second storey as the building is planned with stilts.

4.1.2.2. Verification of soft storey effect for third storey (as per IS 1893 (Part 1): 2002) [15].

a) As 1438802> 0.6 ×1460919 > 876551kN/m and

b) As 1438802>0.7 × $\left(\frac{1460919 + 1306403+ 1171069}{3}\right)$> 918958 kN/m,

it is to be considered that the Stiffness irregularity is not present in the third storey also as the building is planned with stilts.

Similarly, the soft storey effect found exists from second to the sixth storey in Y- direction as per IS 1893 (Part 1): 2016 [1] as shown in Table 3. But such an effect is not found as per IS 1893 (Part 1): 2002 [15]. The above situation calls for further review on Stiffness irregularity provisions.

Table 3. Check for Stiffness irregularity in X and Y directions

Type	Storey	Storey height	X-direction		Y-direction		Soft Storey IS 1893 (Part-1): 2016		Soft Storey IS 1893 (Part-1): 2002	
			Storey Stiffness (kN/m)	Diff. (%)	Storey Stiffness (kN/m)	Diff. (%)	X	Y	X	Y
Residential	17	3.05	324068	NA	535310	NA	No	No	No	No
	16	3.05	537012	165.71	837017	156.36	No	No	No	No
	15	3.05	631327	117.56	955676	114.18	No	No	No	No
	14	3.05	697501	110.48	1042352	109.07	No	No	No	No
	13	3.05	771670	110.63	1134823	108.87	No	No	No	No
	12	3.05	844698	109.46	1179321	103.92	No	No	No	No
	11	3.05	913283	108.12	1224948	103.87	No	No	No	No
	10	3.05	977674	107.05	1269840	103.66	No	No	No	No
	9	3.05	1002363	102.53	1301649	102.50	No	No	No	No
	8	3.05	1039985	103.75	1325843	101.86	No	No	No	No
	7	3.05	1107218	106.46	1347341	101.62	No	No	No	No
	6	3.05	1171069	105.77	1345628	99.87	No	**Yes**	No	No
	5	3.05	1306403	111.56	1244157	92.46	No	**Yes**	No	No
Commercial	4	3.05	1460919	111.83	1174904	94.43	No	**Yes**	No	No
	3	3.05	1438802	98.49	1132984	96.43	**Yes**	**Yes**	No	No
Parking	2	**3.96**	1004628	69.82	842794	74.39	**Yes**	**Yes**	No	No
	1	3.05	1616927	160.95	1498875	177.85	No	No	No	No

4.2. Mass irregularity

The structure considered at present has three different types of occupation i.e., Residential, Commercial and Parking purposes. Different live loads are considered for these occupations i.e., 2 kN/m^2, 5 kN/m^2 and 4 kN/m^2 respectively. The code considers that the mass irregularity exists when the seismic weight of any floor is more than 150 % of that of the floors below. The code further recommended that the earthquake effects are to be estimated by adopting 3D dynamic analysis for the buildings located in Zone III, IV and V. In the present building, the seismic weights are estimated for Residential floor, Commercial floor and Parking floors and Mass irregularity is calculated and shown in Table 4. It is found that there is no Mass irregularity present in the building.

Table 4. Check for Mass irregularity

Mass Irregularity calculation			
Storey	**Seismic weight (kg×1000)**	**Difference (%)**	**Limit (%)**
Residential	1247	91.3	<150
Residential-Commercial	1365	121.7	<150
Commercial -Parking	1122	104.3	<150
Parking	1077		

4.3. Vertical geometric irregularity

The Vertical geometric irregularity of the building under consideration is verified as follows.

The parking and commercial floors occupy the total area of the building whereas the residential towers are projected above fourth slab and up to 17th slab with a gap (A) of about 3.0m maintained between towers as shown in figure 2.

The length of the building (L) in X-axis is 29.4m. Accordingly, the ratio of $\left(\frac{A}{L}\right)$ is $\left(\frac{3.0}{29.4}\right) = 0.102$, and the same is less than 0.125 and hence no vertical geometric irregularity exists in the building. The code recommends estimation of earthquake effects by adopting dynamic analysis for the buildings located in Zone III, IV and V in case Vertical geometric irregularity exists.

4.4. In-Plane discontinuity in vertical elements resisting lateral force

The code does not permit the buildings with In-plane discontinuity particularly in seismic Zones III, IV and V. However, the buildings with in-plane discontinuity are permitted in zone II with a limitation of the lateral drift of the building under-designed lateral force to be less than 0.2% of the building height. It is considered that the in-plane discontinuity in vertical elements resisting lateral force exists when the offsets are greater than 20% of the plan dimension of those elements. In the present case, the column sizes are reduced in upper floors by maintaining centroid throughout the height of the building. Therefore, the above irregularity is not present in this case. In practice, the size of the peripheral column in upper floors are normally reduced in tall structures concerning the outer side and thereby the above irregularity may sometimes exist. Further research in this area is necessary.

4.5. Strength irregularity (Weak Storey)

The strength irregularity is said to exist when the lateral strength of the storey considered is less than that of the storey above. The code does not specify the method of estimation of lateral strength. However, two methods are referred in 'Magliulo and Ramasco' [5], for estimating the lateral strength of the storey. In the first method, flexural resistance of columns is only considered and in the second method, both flexural resistance of columns and beams are considered. A typical column at grid C2 (refer Figure 3 or 4) is considered for estimation of column strength in 3rd and 4th storeys and the details are presented in Figure 8 and Appendix.

It can be seen from the Appendix that the column strength (C2) in 3rd storey is more than the column strength in 4th storey according to both the methods. Hence, Strength irregularity is not present in this case.

The storey strengths of 3rd and 4th are also estimated adopting the Method 1 procedure and the summary is tabulated in Table 5. As can be seen from the table that the lateral strength of 3rd storey is more than the lateral strength of 4th storey in both X and Y directions. Hence, no Strength irregularity (weak storey) is present in the structure. However, the procedure is to be extended to other floors also to ascertain the storey strengths. The manual calculations for estimation of storey strength are very tedious and hence the software should directly provide the results of strength like storey stiffness.

Table 5. Check for Storey strength

Method 1 (with Column moments)		**Storey strength (kN-m/m)**	
Storey No.	**Storey Height (m)**	**X-Direction**	**Y-Direction**
Storey 4	3.05	6966	7091
Storey 3	3.96	7409	7166

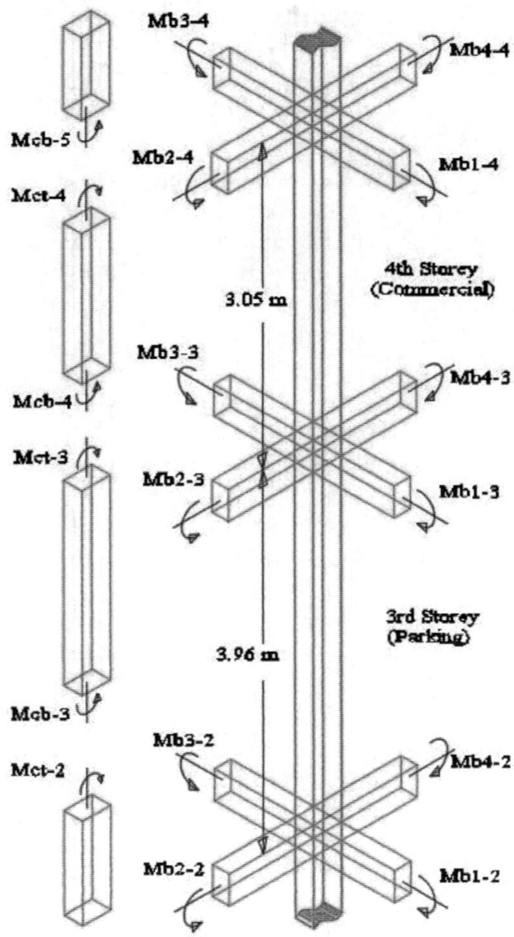

Figure 8. Estimation of storey strength

4.6. Floating or Stub columns

Floating columns as a part of supporting the primary lateral load resisting system shall be avoided as recommended by the code. In practice, Floating columns or Stub columns are being adopted widely particularly in duplex houses and also where the columns of upper floors obstruct the parking in residential buildings.

The code does not specify any limit in height or any other parameter to be followed for adoption of Floating columns. In the present study, no Floating columns were provided in the structure.

4.7. Irregular modes of oscillation in two principal plan directions

The above irregularity is related to mass participation of the structure and limitation of difference in Time period in mode 1, 2 and 3. The mass participation factor and time periods both in X and Y axes are obtained from the analysis and the same are reported in Table 6.

Table 6. Irregular modes of oscillation in two principal plan directions

Zone-II and Zone-III					Zone-IV and Zone-V			
Mass participation (%)			Check	Irregularity	Time Period T (sec)	Diff with Max T (Sec)	Difference w.r.t Max. T (%)	Irregularity
Mode	X	Y						
1	80.10	-	> 65 %	Not Present	X-2.332	0.047	2 < 10	Present
2	80.11	84.37	> 65 %	Not Present	Y-2.285	-	-	-
3	80.11	84.75	> 65 %	Not Present	-	-	-	-

As the mass participation factor for all the three modes is greater than 65% for all seismic zones, it is considered that there is no irregularity present in the structure.

For Zones IV and V, in addition to the mass participation factor, the difference of time periods along X and Y axes with respect to the maximum time period either X or Y directions should be more than 10% and same is not satisfied in the present case. This limit may need a review by experts as the buildings with square shape may have the same time period.

5. Conclusions

An attempt was made to identify and verify the existence of various irregularities present in a 17 storey tall structure. Based on the results and discussion, the following conclusions have arrived.

5.1. Plan irregularities

1. There is no "Torsional Irregularity" present in 50m high Tall structure in both principal plan directions and all seismic Zones, even-though the Building system is planned only with "Special moment resisting frames".
2. Building Irregularity due to Re-entrant corners is found to be in existence in this structure along Y-axis only. However, 3D dynamic analysis is adopted as recommended in IS 1893 (Part 1): 2016 to take care of such irregularity.
3. In respect of Excessive Cut-outs or Openings, there is no such Irregularity present in the building. However, it is recommended to review the limits set in the code to this irregularity to consider rigid diaphragm effect compared to the provision given in cl. 5.6.2.2 of IS 16700: 2017.
4. There are no "Out-of-plane offsets" present in any of the Columns of the structure considered as the centroid of all upper columns are coinciding with the lower columns and hence such Irregularity does not exist. Code is silent on quantification of Out-of-plane offsets and hence there is a scope for further research in this area.
5. As the structure considered for the study is rectangular with certain openings, no "Non-Parallel lateral forces" arise on the system.

5.2. Vertical irregularities

6. The advantage of equivalent diagonal strut action due to un-reinforced masonry infill of even for 230mm thick is very less as the limits of (l/t) and (h/t) (cl.7.9 of IS 1893 (Part 1): 2016) is restricting the maximum spacing between columns not more than 3m. The same is not desirable in most of the residential buildings.
7. No Strength irregularity (weak storey) is present in the structure as the lateral strength of 3rd storey is more than that of lateral strength of 4th storey in both X and Y directions. However, the procedure is to be extended for other floors also to ascertain all storey strengths. The structural designers are facing a lot of difficulty in the estimation of storey strength to verify the "Strength irregularity" (Weak storey) as IS 1893 (Part 1): 2016 is silent on the method of estimation of Storey strength. The

applicability of the methods as indicated in the Appendix is also to be confirmed by more rigorous studies.

8. The two methods that are available in the literature for estimation of lateral Story strength involves a lot of calculations and the regular structural analysis and design software is not directly providing the "Storey strength" unlike "Storey stiffness". There is a scope for software developers to consider this aspect.

9. No Mass irregularity and Vertical geometric irregularity exists in the present structure.

10. There is no Irregularity present in the structure due to In-plane discontinuities in vertical elements. However, further research is needed in case of column offsets that are unavoidable in certain cases.

11. There is no Irregularity as the floating columns are absent in the structure. The code completely restricts the presence of floating columns that may not be avoidable in practice in certain cases.

12. Irregular modes of oscillation in two principal plan directions need a review by experts as the provision contradicts for the square-shaped building in respect of percentage of difference in time periods.

Future scope of Research

A similar study can be conducted on Irregular building configurations studied by Markandeya Raju et. al. (2018), [16], Siva Bhanu et. al. (2016) [17] and Markandeya Raju and Reddi Poornima (2020) [18].

Appendix

Methods of estimation of storey strength

Method I

a) Column location = C2, 4th storey

Storey height, H = 3.05 m

Column moments:

4th storey column moment at the top, M_{ct-4} = 217 kN-m

4th storey column moment at the bottom, M_{cb-4} = 400 kN-m

Strength of column = $\frac{(M_{ct-4}+M_{cb-4})}{H} = \frac{(217+400)}{3.05} = \mathbf{202}$ kN-m/m

b) Column location = C2, 3rd storey

Storey height, H = 3.96 m

Column moments:

3rd storey column moment at the top, M_{ct-3} = 316 kN-m

3rd storey column moment at bottom, M_{cb-3} = 593 kN-m

Strength of the column = $\frac{(M_{ct-3}+M_{cb-3})}{H} = \frac{(316+593)}{3.96} = \mathbf{241}$ kN-m/m

Method II

a) Column location = C2, 4th storey

Storey height, H = 3.05 m

Column moments at the top joint:

4th storey column moment at the top, M_{ct-4} = 217 kN-m

5th storey column moment at bottom, M_{cb-5} = 416 kN-m

Column moments at bottom joint:

4th storey column moment at bottom, M_{cb-4} = 400 kN-m

3rd storey column moment at the top, M_{ct-3} = 361 kN-m

Beam moments at the top joint of 4th storey:

4th storey beam, M_{b1-4} = 467 kN-m, M_{b2-4} = 407 kN-m, M_{b3-4} = 288 kN-m and M_{b4-4} = 353 kN-m

Beam moments at bottom joint of 3rd storey:

3rd storey beam, M_{b1-3} = 487 kN-m, M_{b2-3} = 456 kN-m, M_{b3-3} = 433 kN-m and M_{b4-3} = 408 kN-m

The ratio of the sum of beam moments to the column moments at the top joint is

$$\frac{\sum M_b}{\sum M_c} = \frac{(467+407+288+353)}{(217+416)} = 2.39 > 1,$$

Consider the ratio 2.39 since it is more than 1 as outlined in the method.

Strength of the column at the **top** joint $= \frac{M_{ct\text{-}4}}{H} \times \frac{\sum M_b}{\sum M_c} = \frac{217}{3.05} \times 2.39 =$ **170** kN-m/m

The ratio of the sum of beam moments to the column moments at the bottom joint is

$\frac{\sum M_b}{\sum M_c} = \frac{(487+433+456+408)}{(400+361)} = 2.34 > 1$, Consider the ratio 2.34 since it is more than 1 as outlined in the method.

Strength of the column at the **bottom** joint $= \frac{M_{cb\text{-}4}}{H} \times \frac{\sum M_b}{\sum M_c} = \frac{400}{3.05} \times 2.34 =$ **307** kN-m/m

Total column strength = (170+307) = 477 kN-m/m

Check for Column strength (C2) in 3rd and 4th Storey

4th storey					3rd storey				
Column No. (C2)					Column No. (C2)				
Floor height (H) = 3.05m					Floor height (H) = 3.96m				
	At top of the column		At bottom of the column			At top of the column		At bottom of the column	
Beam moments (kN-m)	$M_{b1\text{-}4}$	467	$M_{b1\text{-}3}$	487	Beam moments (kN-m)	$M_{b1\text{-}3}$	487	$M_{b1\text{-}2}$	445
	$M_{b2\text{-}4}$	407	$M_{b2\text{-}3}$	433		$M_{b2\text{-}3}$	433	$M_{b2\text{-}2}$	425
	$M_{b3\text{-}4}$	288	$M_{b3\text{-}3}$	456		$M_{b3\text{-}3}$	456	$M_{b3\text{-}2}$	419
	$M_{b4\text{-}4}$	353	$M_{b4\text{-}3}$	408		$M_{b4\text{-}3}$	408	$M_{b4\text{-}2}$	439
Column moments (kN-m)	$M_{cb\text{-}5}$	416	$M_{cb\text{-}4}$	400	Column moments (kN-m)	$M_{cb\text{-}4}$	400	$M_{cb\text{-}3}$	593
	$M_{ct\text{-}4}$	217	$M_{ct\text{-}3}$	361		$M_{ct\text{-}3}$	361	$M_{ct\text{-}2}$	223
Ratio	$\frac{\sum M_{bt}}{\sum M_c}$	2.39	$\frac{\sum M_{bb}}{\sum M_c}$	2.34	Ratio	$\frac{\sum M_{bt}}{\sum M_c}$	2.34	$\frac{\sum M_{bb}}{\sum M_c}$	2.11
Strength (kN-m/m)		170		307	Strength (kN-m/m)		213		317
Total Strength as method I = 202 (kN-m/m)					Total Strength as method I = 241 (kN-m/m)				
Total Strength as method II = 477 (kN-m/m)					Total Strength as method II = 530 (kN-m/m)				

b) Column location = C2, 3rd storey

Storey height, H = 3.96 m

Column moments at the top joint:

3rd storey column moment at the top, $M_{ct\text{-}3}$ = 361 kN-m

4th storey column moment at bottom, $M_{cb\text{-}4}$ = 400 kN-m

Column moments at bottom joint:

3rd storey column moment at bottom, $M_{cb\text{-}3}$ = 593 kN-m

2nd storey column moment at the top, $M_{ct\text{-}2}$ = 223 kN-m

Beam moments at the top joint of 3rd storey:

3rd storey beam, $M_{b1\text{-}3}$ = 487 kN-m, $M_{b2\text{-}3}$ = 456 kN-m, $M_{b3\text{-}3}$ = 433 kN-m and $M_{b4\text{-}3}$ = 408 kN-m

Beam moments at bottom joint of 2nd storey:

2nd storey beam, $M_{b1\text{-}2}$ = 445 kN-m, $M_{b2\text{-}2}$ = 425 kN-m, $M_{b3\text{-}2}$ = 419 kN-m and $M_{b4\text{-}2}$ = 439 kN-m

The ratio of the sum of beam moments to the column moments at the top joint is

$\frac{\sum M_b}{\sum M_c} = \frac{(487+433+456+408)}{(361+400)} = 2.39 > 1$, Consider ratio 2.39 as it is more than 1 as outlined in the method.

Strength of the column at the **top** joint $= \frac{M_{ct\text{-}3}}{H} \times \frac{\sum M_b}{\sum M_c} = \frac{361}{3.96} \times 2.34 =$ **213** kN-m/m

The ratio of the sum of beam moments to the column moments at the bottom joint is

$\frac{\sum M_b}{\sum M_c} = \frac{(445+425+419+439)}{(593+223)} = 2.12 > 1$, Consider ratio 2.12 as it is more than 1 as outlined in method.

Strength of the column at the **bottom** joint = $\frac{M_{cb\text{-}3}}{H} \times \frac{\sum M_b}{\sum M_c} = \frac{593}{3.96} \times 2.12 = \mathbf{317}$ kN-m/m

Total column strength = (213+317) = **530** kN-m/m

References

[1] IS 1893 (Part 1) 2016 Indian Standard Criteria for Earthquake Resistant Design of Structures *Bureau of Indian Standards* New Delhi

[2] Tso W K, Myslimaj B 2003 A yield displacement distribution-based approach for strength assignment to lateral force-resisting elements having strength dependent stiffness *Earthq. Eng. Struct. Dyn.* **32** p 2319–51

[3] Myslimaj B, Tso WK 2005 A design-oriented approach to strength distribution in single story asymmetric Systems with elements having strength-dependent stiffness *Earthq Spectra* **21** p 197–212

[4] Stefano De M, Pintucchi B 2007 A review of research on seismic behaviour of irregular building structures since 2002 *Bull. Earthquake Eng.* (2008) **6** p 285–308

[5] Magliulo G, Ramasco R 2008 seismic performance of r/c frames with regular and irregular strength vertical distributions *14th World Conference on Earthquake Engineering* October 12-17 2008 Beijing China

[6] Shreyasvi C, Shivakumaraswamy B 2015 Seismic response of buildings with re-entrant corners in different seismic zones *IJRET* **4 (Special Issue: 04)**

[7] Khan I P, Dhamge R N 2016 Seismic analysis of multistoried RCC building due to mass irregularity *IJEDR* **4 (3)**

[8] Teddy L, Hardiman G, Nuroji, Sri Tudjono 2017 The effect of earthquake on architecture geometry with nonparallel system irregularity configuration. *IOP Conf. Series: Earth and Environmental Science* **99** 012004

[9] Ismail R, Mahmud A N, Ishak S I 2018 Seismic performance for vertical geometric irregularity frame structures *IOP Conf. Series: Earth and Environmental Science* **140** 012101

[10] IS 875 (Part 1) 1987 Indian Standard Code of Practice for Design Loads (Other than Earthquake) for Buildings and Structures – Dead Loads – Unit weight of building materials and stored materials *Bureau of Indian Standards* New Delhi

[11] IS 875 (Part 2) 1987 Indian Standard Code of Practice for Design Loads (Other than Earthquake) for Buildings and Structures – Imposed loads *Bureau of Indian Standards* New Delhi

[12] IS 456 2000 Indian Standard Plain and Reinforced Concrete- Code of Practice *Bureau of Indian Standards* New Delhi

[13] IS 13920 2016 Indian Standard Ductile Design and Detailing of Reinforced Concrete Structures Subjected to Seismic Forces- Code of Practice *Bureau of Indian Standards* New Delhi

[14] IS 16700 2017 Indian standard criteria for structural safety of tall concrete buildings *Bureau of Indian Standards* New Delhi

[15] IS 1893 (Part 1) 2002 Indian Standard Criteria for Earthquake Resistant Design of Structures *Bureau of Indian Standards* New Delhi

[16] Markandeya Raju P, Kumar M P, Sahithichandr B 2018 Parametric study on effect of wind on geometrical configurations of RC space frames *Disaster Advances* **11 (4)** p 28-45

[17] Siva Bhanu Sai Kumar S, Rama Rao G V, Markandeya Raju P 2016 Seismic fragility analysis of regular and setback RCC frames - A few hypothetical case studies *Asian Journal of Civil Engineering* **17 (5)** p 551-69

[18] Ponnada Markandeya Raju, Reddi Poornima 2020 Linear static analysis of multi storey building with horizontally asymmetric architectures. *Journal of Building Pathology and Rehabilitation* **5 (25)** p 1-10 s

IOP Conf. Series: Materials Science and Engineering 1025 (2021) 012029 doi:10.1088/1757-899X/1025/1/012029

Behaviour of buildings with coupled Reinforced Concrete columns

P Krishnam Raju[1], V Ravindra[2] and M Gopi[3]

[1]Department of Civil Engineering, Gayatri Vidya Parishad College of Engineering (A), Visakhapatnam - 530048, INDIA
[2]Department of Civil Engineering, Jawaharlal Nehru Technological University Kakinada, Kakinada - 533003, INDIA
[3]Department of Civil Engineering, Gayatri Vidya Parishad College of Engineering (A), Visakhapatnam - 530048, INDIA
E-mail: drpkraju1962@gmail.com

Abstract. Columns play a very crucial role in transferring forces to the foundation. This study aims to compare the influence of Coupled columns in place of Regular (conventional) columns in Reinforced concrete framed structures. A Coupled column is one with two individual parallel members connected at their ends with deep beams so that it behaves as a single member. The parallel members are also some times connected at intermediate levels. A six-storey parking structure is considered for the study with three alternatives of different column configurations. They are analysed for gravity loads, seismic (Zone-II) and wind (basic speed of 50m/s) loads. The configurations are Regular Column with Regular Beam (RCRB), Coupled Column with End Connecting Beam (CCEB) and Coupled Column with End and Intermediate Connecting Beams (CCIB). The cross-sectional area of the Coupled column is maintained equal to that of conventional Regular column. Linear Dynamic analysis of the structure is performed for load combinations as per IS 456: 2000. It is observed from the results that the lateral sway is reduced in CCIB by 7%, while quantities of concrete and steel reinforcement are 7% and 13% less respectively compared to RCRB. Considering the advantages of strength and serviceability, apart from the economy, it was concluded that the structure with Coupled columns performs better than conventional column structure.

1. Introduction

The configuration of any building structure is generally decided based on functional aspects. For high-rise buildings, the effect of wind and earthquake forces are predominant on the lateral load resisting system of a structure. As the columns are primary lateral load resisting elements, the size and spacing of columns depend upon functional requirements duly following the prescriptive guidelines laid down in respective codes of practice. Alternatively, shear walls and bracing systems need to be introduced to satisfy the lateral load capacity of buildings which may have an impact on the economy of the building. Aditya and Prakash [1] made a comparison between the buildings with specially shaped RC columns and rectangular-shaped RC columns of 6 storeys, 9 storeys, 12 storeys and 15 storeys. The authors reported that seismic performance of the building with specially shaped columns behaves better compared to the building with rectangular shaped columns. The authors reported that the stiffness of the frame with a rectangular column of the same area is inferior to the frame of a specially shaped column. Atickur et al. [2] studied the effect of the special-shaped column on lateral load resistance capacity of Reinforced Concrete building. The authors have considered four different buildings (6 storeys, 10 storeys, 15 storeys and 20 storeys) which were analyzed with conventional rectangular columns and specially shaped columns of L, Tee and Cross. They reported that the maximum storey displacements in building with rectangular shaped columns increased by 11.78%, 8.83%, 34.84 and 5.51% respectively for 6, 10, 15 and 20 storey buildings. Yang et al. [3] carried out a comparative study of seismic behaviour between specially shaped column frame structure and rectangular column frame structure as per Chinese code and technical specifications. Yang and Tang [4] investigated seismic behaviour of specially shaped columns with a cross-section of L, T, + and compared with rectangular columns. The authors reported that the cross-section shape of a column, axial compression ratio and reinforcement are the main parameters sensitive to seismic behaviour of these specially shaped columns. Zhou et al. [5] investigated the behaviour of a specially shaped column composed of concrete-filled steel tubes (SCFST column) experimentally subjected to a constant axial load and a cyclically varying flexural load. They concluded that mono columns of SCFST worked together well and behaves effectively when subjected to lateral loads.

2. Aim, Objective and Scope

2.1. Aim

To study the influence of Coupled Columns vis-a-vis Regular columns in a normal Reinforced Concrete framed structure.

2.2. Objective

1) To assess the variations in the Stiffness and Member forces in columns.
2) To study Lateral sway, Inter-Storey Drift behaviour.
3) To compare the concrete and reinforcement requirements.
4) To report the influence of Coupled Column.

2.3. Scope

The study conducted on a six-storey Parking structure accommodating 80 cars in each floor is considered in seismic Zone-II (Visakhapatnam). Accordingly, the basic wind speed of 50 m/s is considered. ETABS Evaluation version software was adopted for the analysis and design.

The Equivalent Static method and Linear dynamic analysis (Response Spectrum Method) of the structure are performed for seismic design. The pressure coefficient method is adopted for wind load analysis.

3. Details of building considered for the study

A six-storey parking structure is planned as per the National Building Code of India 2016 (Volume-1) [6] and considered for the study. The size of the building is 72m × 35m excluding ramps and the building is assumed to be located in Visakhapatnam. The ramps are provided on either end of the building to access the structure for parking and also climbing to different storey levels. The ends of the parking structure block are not integrated with the ramps provided on either end to study the behaviour of parking structure independently. The height of the structure is about 20 m with a storey height of 3.3 m each. Plan of the parking structure is shown in figure 1. The parking structure beam layouts with Rectangular and Coupled columns are shown in figure 2. Figure 3 shows Rendered view of the parking structure.

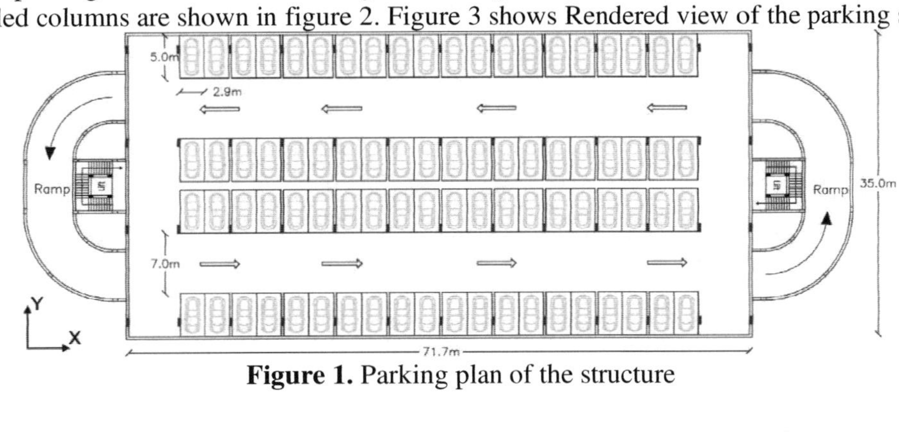

Figure 1. Parking plan of the structure

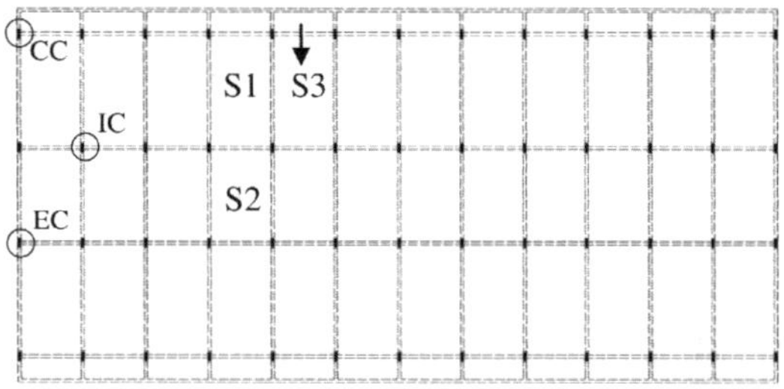

a) Beam layout with Regular columns

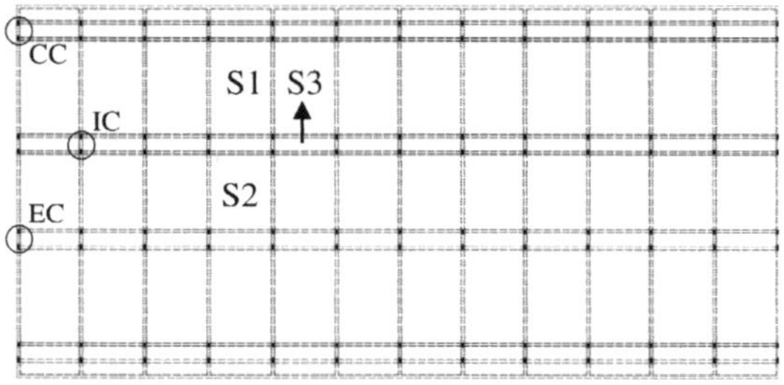

b) Beam layout with Coupled columns

IC = Intermediate Column; EC = Edge Column; CC= Corner Column; Slab-1; Slab-2; Slab-3

Figure 2. Beam layouts of the Parking structure

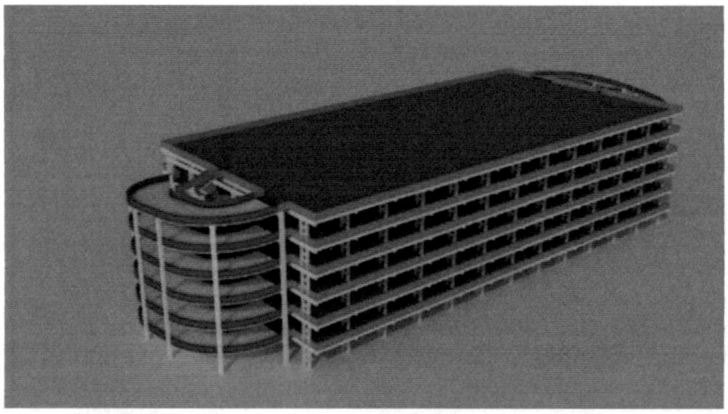

Figure 3. Render view of the Parking structure

3.1. Aim

In this study, coupled columns are introduced as a replacement to Regular rectangular columns to behave structurally better and providing large column-free space, particularly in parking structures. Since it is a new concept that is being introduced, the definition of the Coupled column is explained as follows.

A Coupled column is one with two individual parallel members connected at their ends with deep beams so that it behaves as a single member. The parallel members are also connected at intermediate levels if necessary. The vertical configurations of the alternatives considered are presented in figure 4. The cross-sections of both rectangular and coupled columns are shown in figure 5. The sizes of elements in a Coupled column is fixed in such a way that the area of cross-section is constant with that of a Regular column. However, the sizes of individual elements may not be equal and will be decided based on adjoining spans. Clear distance between the two elements of the Coupled column is computed based on the trial and error method to meet the Coupled behaviour (Deep beam).

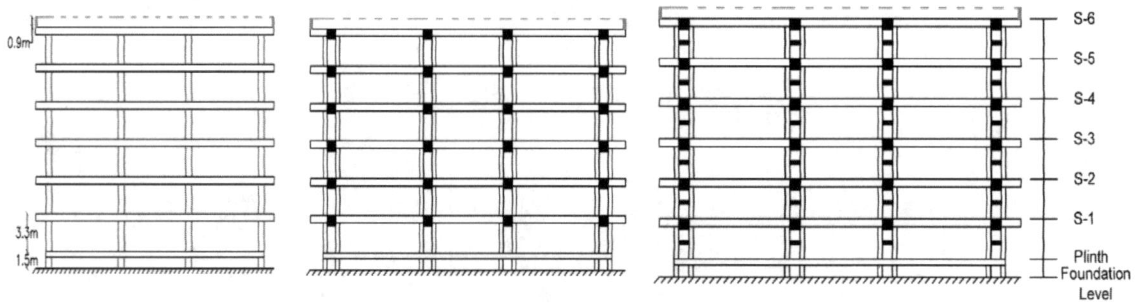

a) RCRB b) CCEB c) CCIB

Figure 4. The vertical configuration of three Alternatives

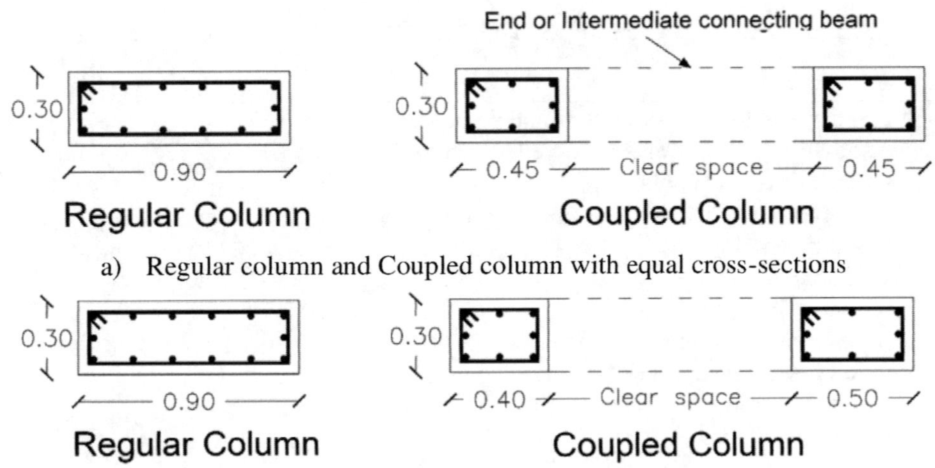

a) Regular column and Coupled column with equal cross-sections

b) Regular column and Coupled column with unequal cross-sections

Figure 5. Cross-section of Regular and Coupled columns

The member dimensions and material properties are shown in Table 1 and Table 2 respectively and load cases considered are shown in Table 3.

Table 1. Member Dimensions (mm)

Type		RCRB	CCEB		CCIB	
			C1	C2	C1	C2
Columns	1-3 storey	300 × 800	300 × 300	300 × 500	300 × 300	300 × 500
	4-6 storey	300 × 600	300 × 250	300 × 350	300 × 250	300 × 350
Beam - X direction		300 × 450	230 × 350		230 × 350	
Beam – Y direction		300 × 650	300 × 650		300 × 650	
Plinth Beam		300 × 500	300 × 500		300 × 500	
End Connection		-NA-	300 × 900		300 × 900	
Intermediate Connection		-NA-	-NA-		300 × 300	
Slab	S1	180	160		160	
	S2	180	160		160	
	S3	130	130		130	

Table 2. Material Properties

Parameter	Detail
Grade of concrete for all RCC structural elements	M30
Grade of concrete for Parapet wall	M25
Grade of steel	Fe415

Table 3. Design load parameters

Type	Description	Reference
Dead	Self-weight of the structure to be considered by the software (Density of Reinforced Concrete = 25 kN/m^3) Floor finish = 1.5 kN/m^2	Table-1 of IS 875 (Part 1): 1987 [7]
Live load	Live load = 4 kN/m^2 (Fixed based on the gross weight of the vehicle)	IS 875 (Part 2): 1987 [8]
Earthquake load	Seismic load parameters: Zone = II Importance factor = 1 Soil type = Medium (II) Response reduction factor = 3	IS 1893 (Part-1): 2016 [9]
Wind load	Wind load parameters: Basic wind speed,V_b = 50 m/s Risk factor, K_1 = 1 Terrain Category = 3 (K_2 varies with the height of the building) Topography factor, K_3 = 1 Importance factor for Cyclone region K_4 = 1	IS 875 (Part-3): 2015 [10]

*3.1.1. End connections.*Effective span, l = 1 m; Overall depth of beam, = D = 0.9 m

$$\frac{l}{D} = \frac{1}{0.9} = 1.11 < 2.5$$

Hence, as per Clause 29.1 of IS 456: 2000 [11], the End Connection is designed as a deep beam.

4. Analysis and design of the structure

The three alternatives, viz., Regular columns and Regular beams (RCRB), Coupled column structure with End connecting beam (CCEB), Coupled column structure with End and Intermediate connecting beams (CCIB) were analyzed for gravity and lateral loads as indicated in Table 3. The load combinations are taken as per IS 456: 2000 [11].

The analysis was carried out using ETABS Evolution version software. The wind analysis of the alternatives is performed using Equivalent Static pressure coefficient method. The seismic design of the alternatives is carried out based on results of Linear Dynamic analysis (Response Spectrum Method).

5. Results and discussion

The results of the analysis and design for all the three alternatives considered are presented. The parameters such as stiffness, lateral sway, storey drift and quantity of materials (concrete and steel reinforcement) are compared among the alternatives for the whole building.

The flexural behaviour of three typical columns i.e., Intermediate column (IC), Edge column (EC), Corner column (CC) as marked in figure 2 is also presented.

5.1. Lateral sway

According to Cl. 20.5 of IS 456: 2000 [11] the lateral sway at the top of the structure should not exceed (H/500), where H is the total height of the building. Accordingly, the permissible lateral sway is 40 mm for a 20 m total building height. The lateral sway along the principal direction (Y-axis) of the structure for all the three alternatives is shown in figure 6. The alternative with CCIB has less lateral sway than RCRB followed by CCEB. When compared with RCRB, the sway in CCIB is less by 7%. However, it is 15% more in the case of CCEB. The sway in CCEB and CCIB followed a similar pattern whereas a regular trend is observed in RCRB. However, the lateral sway is within the permissible limits in all the three alternatives.

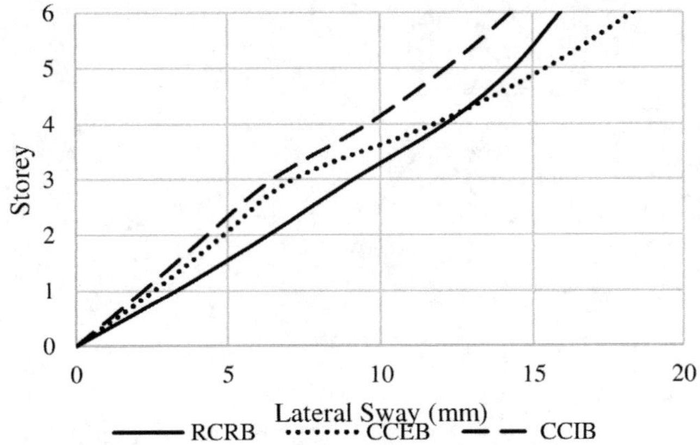

Figure 6. Variation of Lateral sway in principle direction (Y-axis)

5.2. Inter-storey drift

According to Clause 7.11.1 of IS 1893 (Part 1): 2016 [9], the Inter-storey drift shall not exceed 0.004 times the storey height. Accordingly, permissible storey drift is 13.2 mm for a storey height of 3.3 m.

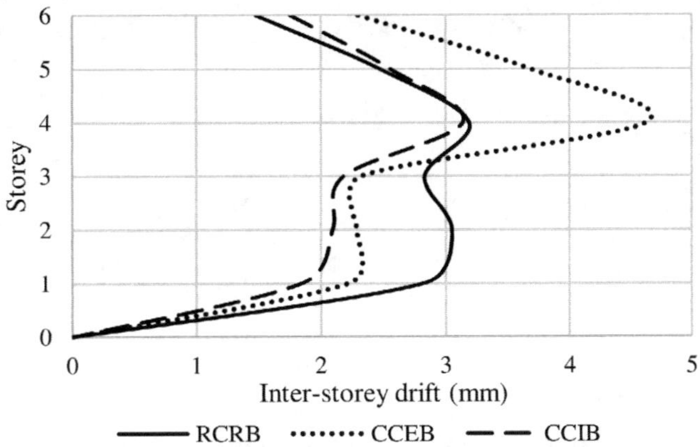

Figure 7. Variation of Storey Drift in principle direction (Y-axis)

The storey drifts along the principal direction (Y-axis) of the structure for all the three alternatives are shown in figure 7. Storey drift in RCRB and CCIB is very close at 4th storey level, whereas drift in CCEB is more by about 45% at the same level. However, the storey drift in all the three alternatives is less than the permissible limits.

5.3. Storey stiffness

Variation in storey stiffness along the principal direction (Y-axis) of the structure for all the three alternatives is shown in figure 8. It is observed that the storey stiffness is higher in CCIB almost throughout the height of the building compared to RCRB and CCEB. The stiffness of RCRB and CCEB is almost similar up to 3rd storey. Thereafter the stiffness of CCEB is gradually reduced compared to RCRB.

The Axial force in the Intermediate column for the three alternatives are shown in figure 9. Reduction of an Axial force is noticed in CCEB and CCIB in the order of about 13% compared to RCRB. This may be due to reduction in self-weight of the structure.

The bending moment for the three alternatives in the Intermediate column, Edge column and Corner column for the total height of the structure is shown in figure 10. When compared with RCRB, the bending moment in the Intermediate column is less in CCIB and followed by CCEB. A maximum reduction of 70% in CCIB and CCEB compared to RCRB at plinth level is observed in the load combination (1.5 Dead loads +1.5 R.S.M).

IOP Conf. Series: Materials Science and Engineering 1025 (2021) 012029 doi:10.1088/1757-899X/1025/1/012029

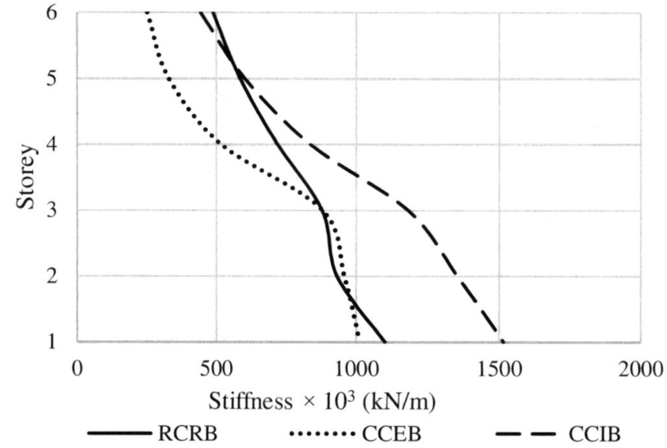

Figure 8. Variation of Storey Stiffness in principle direction (Y-axis)Member forces

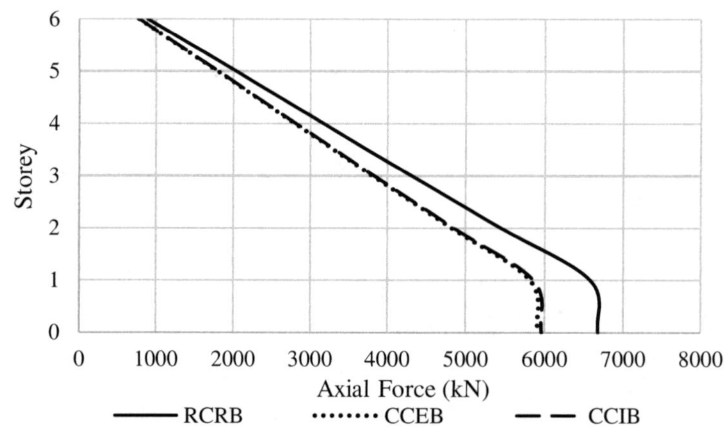

Figure 9. Axial Force in Intermediate column

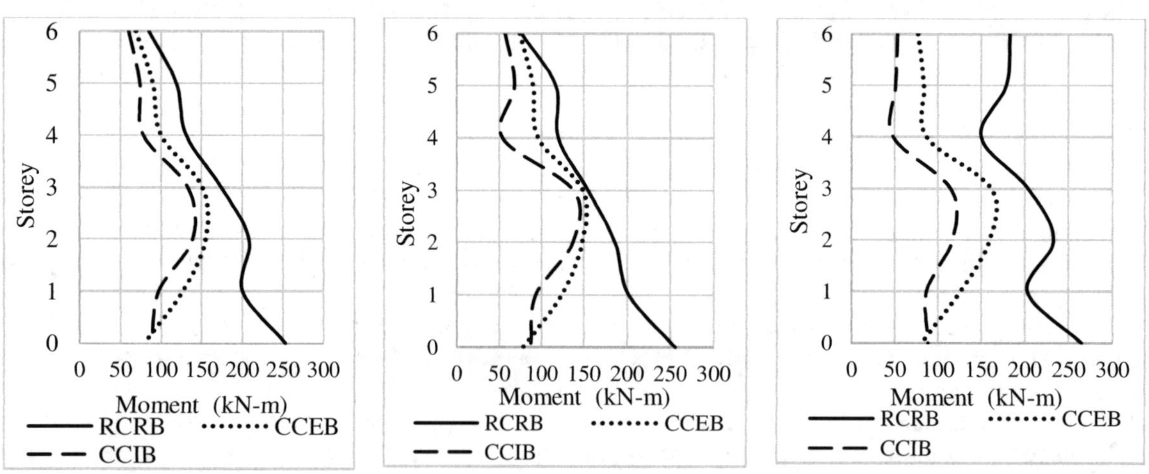

a) Intermediate column b) Edge column c) Corner column

Figure 10. Bending Moments along with IC, EC and CC

5.4. Longitudinal rebar percentage (LRP)

Longitudinal rebar percentage in an Intermediate column, Edge column and Corner column is shown in figure 11 for the three alternatives. The LRP in CCEB and CCIB was reduced to 21% compared to RCRB at plinth level in respect of Intermediate column.

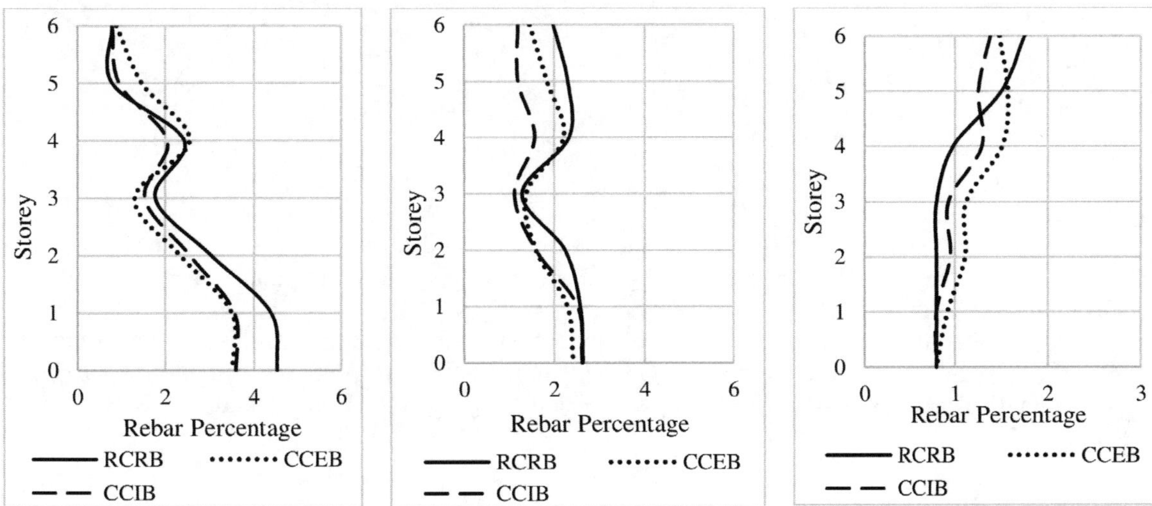

a) Intermediate column b) Edge column c) Corner column

Figure 11. Longitudinal Rebar Percentage along with IC, EC and CC

In the case of the Edge column, the LRP is also less for CCIB and CCEB compared to RCRB. The maximum reduction is found to be 48% and 20% for CCIB and CCEB respectively at 5th storey level. LRP is less in RCRB compared to CCEB and CCIB up to 4th storey in case of Corner column. At 4th storey level, LRP reduction is 24%.

The longitudinal reinforcement in a typical beam along Y-axis is shown in figure 12. The LRP at Top (support) of the beam is less for CCEB and CCIB compared to RCRB and the reduction was up to 22% at supports. Similarly, the LRP at the Bottom (mid-span) was up to 35%.

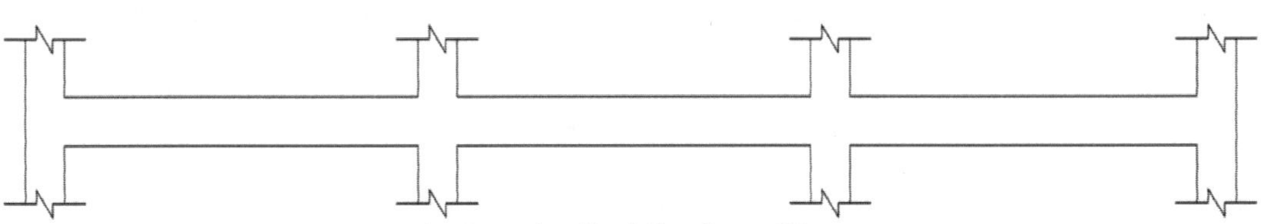

a) Longitudinal Section of Beam

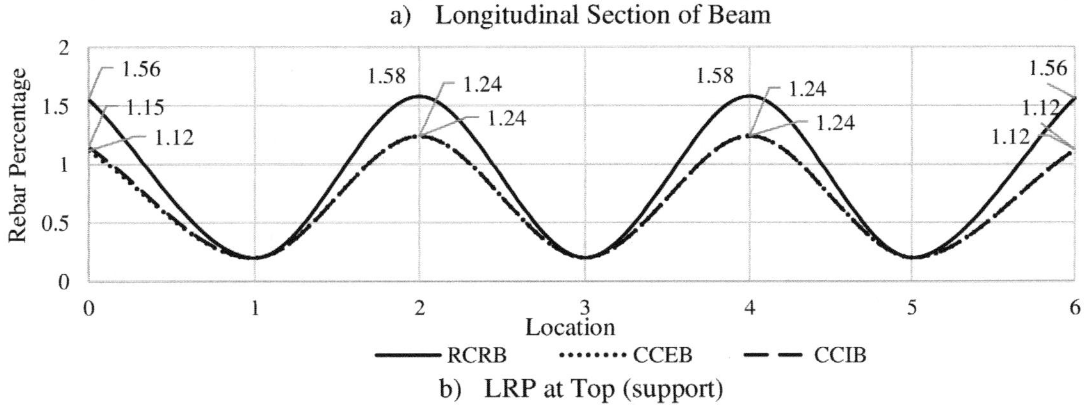

b) LRP at Top (support)

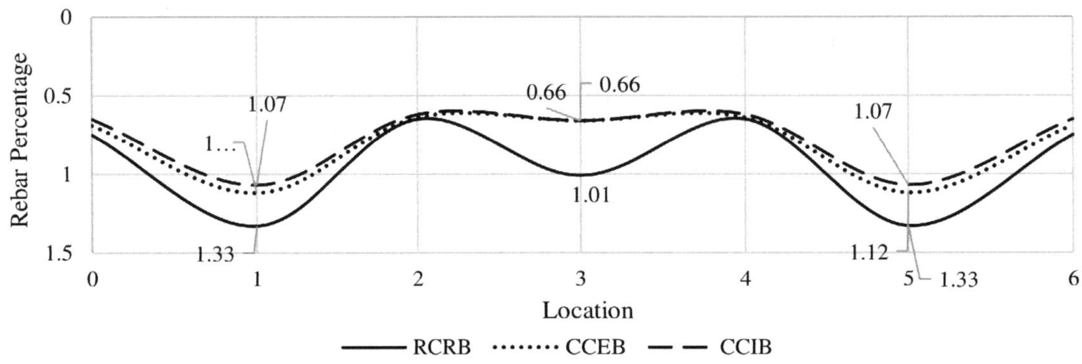

c) LRP at Bottom (mid-span)

Figure 12. Variation of Longitudinal rebar percentage in Beam along Y-axis

5.5. Quantity of materials

The total material quantity for all the three alternatives is presented in figure 13. When compared to RCRB, the concrete quantity is less by 8% and 7% in CCEB and CCIB respectively. Similarly, total steel quantity in CCEB and CCIB is less by 16% and 13% respectively compared to RCRB.

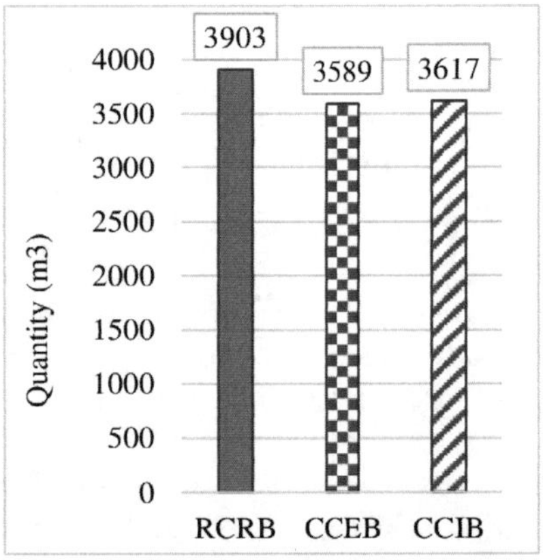

a) Total concrete quantity

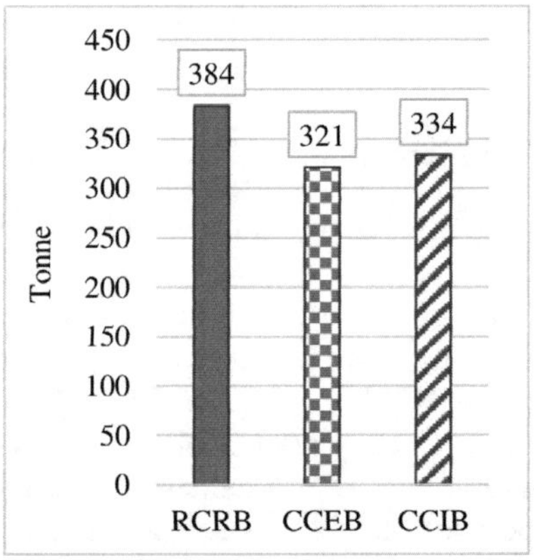

b) Total steel quantity

Figure 13. Variation of Material quantity (Excluding Foundation)

6. Conclusions and recommendations

After a comparative study, as per engineering knowledge and best practices, it has been established that the building with Coupled columns with End and Intermediate connecting beams (CCIB) performs better under seismic and wind load conditions than building with Regular rectangular columns (RCRB) under same loadings. The following conclusions are derived based on results and discussion.

6.1. Conclusions based on serviceability aspects

1. The structure with CCIB has less lateral sway (7%) compared to RCRB. However, it is more in CCEB (15%) when compared with RCRB.
2. The maximum inter-storey drift is more (45%) in CCEB compared to RCRB. However, in the case of CCIB, it is marginally less (2%). In all the three cases maximum storey drift occurs at the fourth storey. However, the storey drift in all three cases is within the maximum permissible limit.
3. The structure with CCIB provides more storey stiffness than CCEB and RCRB.

IOP Conf. Series: Materials Science and Engineering 1025 (2021) 012029 doi:10.1088/1757-899X/1025/1/012029

6.2. Conclusions based on collapse behaviour

1. The axial force in intermediate columns is less by about 13% in both CCEB and CCIB compared to RCRB.
2. The bending moments in Intermediate, Edge and Corner columns in CCEB and CCIB structures are less compared to RCRB structure throughout their height.
3. The span and support bending moments in beams along Y-axis are less in CCIB and CCEB structures compared to RCRB.
4. Coupled column structure with “End connecting beam” (CCEB) and with “End and Intermediate connecting beams” (CCIB) require less quantity of concrete (by 8% and 7% respectively) than the structure with “Regular columns” (RCRB). In respect of steel reinforcement also, CCEB and CCIB need less quantity than RCRB by 16% and 13% respectively.

6.3. Recommendations

To bring Coupled column behaviour, Deep beams are required to be introduced at the column ends as Connecting beams instead of Regular beams.

1. An advantage of using Coupled columns is demonstrated through a case study on a Parking structure. It is recommended to introduce such columns to improve the economy in construction and safety in serviceability parameters.
2. Coupled column arrangement would facilitate the running of various common services such as pipelines, fire lines, and plumbing, etc., in multi-storied buildings.

References

[1] Aditya N G, Prakash S P 2015 Comparison between Specially shaped Columns and Rectangular Columns in R.C. Structure *IJAERD* **2 (5) p** 679-84

[2] Rahaman A, Anik M A, Serker N H M K 2018 Effect of Special Shaped Column on Lateral Load Resistance Capacity of Reinforced Concrete (RC) Building *AJCE***6(5)** 147-153

[3] Yang P, Liu H, Haung Z 2008 A comparison of seismic behaviour between specially shaped column frame structure and rectangular column frame structures, The *14th World Conference on Earthquake Engineering* October 12-17

[4] Yang P, Tang J 2010 A Study on Parameters Affect Seismic Behaviour of Reinforced Concrete Specially Shaped Columns AMR 163-167 1300-1306

[5] Zhou T, Chen Z, Liu H 2012 Seismic behaviour of special shaped column composed of concrete filled steel tubes*JCSR***75**131-141

[6] National Building Code of India 2016 (Volume 1) *Bureau of Indian Standards* New Delhi.

[7] IS 875 (Part 1) 1987 Indian Standard Code of Practice for Design Loads (Other than Earthquake) for Buildings and Structures – Dead loads – Unit weight of building materials and stored materials *Bureau of Indian Standards* New Delhi

[8] IS 875 (Part 2): 1987 Indian Standard Code of Practice for Design Loads (Other than Earthquake) for Buildings and Structures – Imposed loads *Bureau of Indian Standards* New Delhi

[9] IS 1893 (Part 1) 2016 Indian Standard Criteria for Earthquake Resistant Design of Structures *Bureau of Indian Standards* New Delhi

[10] IS 875 (Part 3) 2015 Indian Standard Criteria for Design Loads (Other than Earthquake) for Buildings and Structures – Wind loads *Bureau of Indian Standards* New Delhi

[11] IS 456 2000 Indian Standard Criteria for Plain and Reinforced Concrete Code of Practice *Bureau of Indian Standards* New Delhi

Influence of wind speed on a Reinforced Concrete tall building

P Krishnam Raju[1], V Ravindra[2], V Susmitha[3]

[1]Department of Civil Engineering, Gayatri Vidya Parishad College of Engineering (A), Visakhapatnam - 530048, INDIA,
[2]Department of Civil Engineering, Jawaharlal Nehru Technological University Kakinada, Kakinada - 533003, INDIA,
[3]Department of Civil Engineering, Gayatri Vidya Parishad College of Engineering (A), Visakhapatnam - 530048, INDIA,
E-mail: drpkraju1962@gmail.com

Abstract. This study is conducted to assess the influence of wind on a 17-storey multipurpose Reinforced Concrete Tall Building according to the revised wind code of IS 875 (Part 3): 2015 compared to its previous version i.e., IS 875 (Part 3): 1987. The study covers all basic wind speeds of 33, 39, 44, 47, 50 and 55m/s as specified in the Indian standards. Dynamic effects due to 'Along wind' and 'across wind' are considered in the analysis using ETABS evaluation software. The Lateral load, Lateral sway and Longitudinal Rebar Percentage (LRP) are found to be increased as per the revised version. The LRP in the Middle, Edge and Corner columns are also reported. An increase of total Rebar quantity by about 3.7% (beams and columns) in the entire structure for a basic wind speed of 50m/s is observed.

1. Introduction

Tall buildings are being constructed in major cities for the optimum utilization of limited land area and also to accommodate commercial, residential and corporate space in a single building.

Tall building construction nowadays is an indication of prosperity, socio-economic status, technological and engineering advancement. Therefore, the effect of wind load must be considered to obtain the strength and stability of the structure to resist wind in addition to gravity loads.

Gordan et al. [1] studied about wind interaction on the Tall building and concluded that 'across wind response' is greater than 'along wind response' because of vortex shedding due to which the lateral displacement in 'across wind direction' is more than 'along wind direction'.

Halder and Dutta [2] suggested that the buildings with the number of storeys more than 25 and aspect ratio < 0.5 in Terrain category 3 should be analysed by the gust factor method to predict the response of the building. Likewise, they suggested a method to be preferred in other terrain categories also.

Kumar and Rajan [3] observed that the Gust Response Factor values on the windward side are almost comparable at all levels of the building considered while on the leeward side, fewer values are obtained at 8th level than other. This may be due to the terrain roughness factor and vortex shedding phenomenon.

Mendis et al. [4] discussed the advanced levels of wind design, in the context of the Australian Wind Code and different flow situations that arise due to the interaction of wind with the structure and concluded that significant dynamic response can result from both buffeting and across wind loading excitation mechanisms.

Zhou and Kareem [5] derived the advantage of Gust loading factor method while assigning the wind loads to a tall structure based on Base bending moment over the traditional Gust loading factor which is based on displacement approach.

An earlier revision of IS 875 (Part 3): 1987 [6] has only accommodated the provisions for along wind load in gust factor method in which dynamic parameters like peak factor, roughness factor, background factor, size reduction factor, turbulence length scale and measure of available energy are derived from graphs and across wind load procedure is not mentioned.

Importance factor, wind directionality factor, area averaging factor and combination factor are now introduced in IS 875 (Part 3): 2015 [7] in addition to across wind load in gust factor method. Dynamic parameters in along wind response are now obtained using formulae and procedures as given in codes.

The study aims to compare the impact of wind on a Tall Building as per the revised provisions of the IS 875 (Part 3): 2015 [7] code and its previous version IS 875 (Part 3): 1987 [6]. The objective of the study is to compare the Code provisions of IS 875 (Part 3): 1987 [6] and IS 875 (Part 3): 2015 [7] that influence the Structural Design. Further, Lateral sway and its pattern are evaluated as per the above Codes for all the basic wind speed values of 33, 39, 44, 47, 50 and 55m/s.

The Longitudinal Rebar Percentage (LRP) is assessed for Middle column, Edge column and Corner column. The total longitudinal rebar quantity is estimated for all the columns and beams in the entire building for a basic speed of 50m/s using the above-mentioned codes.

The design of the structure was carried out for combinations of wind in addition to dead and live loads as per IS 875 (Part 1) [8] and IS 875 (Part 2) [9] under a limit state of Collapse and Serviceability as per IS 456: 2000 [10].

Pressure coefficient method and Gust factor method were adopted for the analysis of the structure to assess the Lateral load, Lateral sway and Longitudinal Rebar Percentage and Quantity in the superstructure. ETABS Evaluation Version software is used for analysis and design.

2. Building configuration

A multipurpose 17 storey building consisting of two parking, two commercial and 13 residential floors with M30 grade concrete and Fe500 steel is considered for the study.

The total height of the building is about 53m from the plinth level including its basements. The overall plan dimensions of the building are 29.0m × 37.5m. The Typical floor plan of the building is shown in figure 1.

The total height of the building =53m > 50m

$$\frac{\text{Height of the building}}{\text{Width of the building}} = \frac{53}{29} = 1.82 < 6$$

According to the above criteria, the building is categorised as a tall structure as per Cl.4.13 of IS 875 (Part 3): 2015 [7].

Dimensions of the members were fixed to cater for all basic wind speeds satisfying the reinforcement and serviceability limitations. The member sizes adopted in the structure are given in Table 1.

Table 1. Cross-sectional dimensions of structural elements

Description	Size
Column Size (mm × mm)	530 × 910, 450 × 910, 380 × 910, 450 × 830, 380 × 830
Beam Size (mm × mm)	380 × 760, 300 × 760, 300 × 610, 300 × 530 and 300 ×450 230 × 580, 300 × 450, 380 × 630 and 300 × 760
Slab Thickness (mm)	125

Three typical columns viz., Middle column (MC), Edge column (EC), and Corner column (CC) in the building are considered for the study as shown in figure 1.

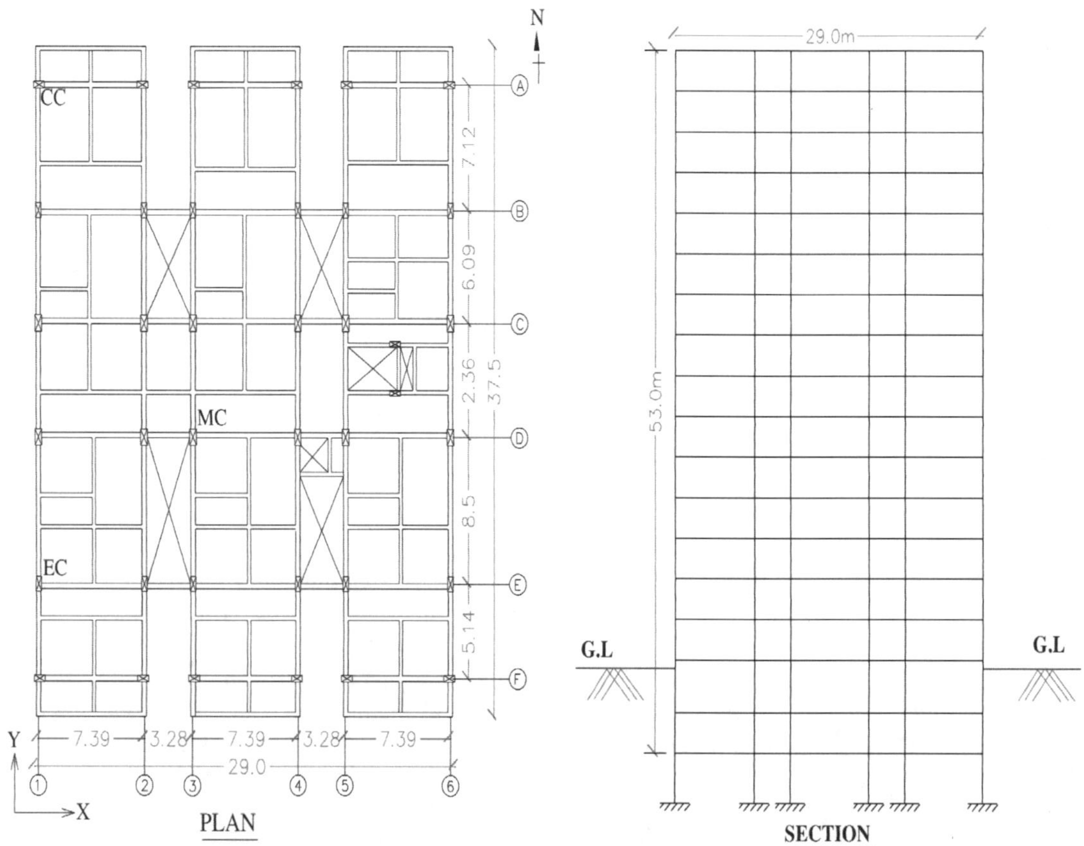

Figure 1. Typical floor plan with Column positions and Beam layout (All dimensions are in m)
MC: Middle Column; EC: Edge Column; CC: Corner Column

3. Results and discussion

The results are presented in terms of Lateral load, Lateral Sway, Longitudinal Rebar Percentage and Quantity of reinforcement. Comparison is made with the results obtained from the analysis as per IS 875 (Part 3): 1987 [6] and IS 875 (Part 3): 2015 [7] and design as per IS 456: 2000 [10].

3.1. Lateral load due to wind

The lateral load due to wind is estimated taking a basic wind speed of 50 m/s as per IS 875 (Part 3): 1987 [6] and IS 875 (Part 3): 2015 [7] for both 'along' and 'across' winds in X and Y directions are shown in figure 2.

The wind load 'Along wind' is gradually increased above ground level in both X and Y directions and the maximum is observed at 16th storey.

The maximum increase is found to be 12% and 13% in X and Y directions as per IS 875 (Part 3): 2015 [7] compared to its earlier version. However, as per IS 875 (Part 3): 1987 [6] the lateral load is found shot up at the roof level of the ground floor and same is due to the changes as adopted in hourly mean wind speed factor.

The wind load due to 'Across wind' is gradually increased and reached to about 13kN and 10kN in X and Y directions respectively at the roof level which is an additional load as per IS 875 (Part 3): 2015 [7]. A similar trend is observed for the remaining wind zones also.

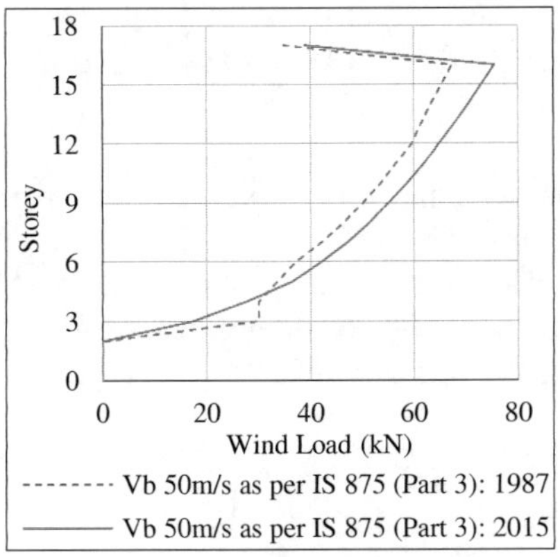

(a) Along wind in X- direction

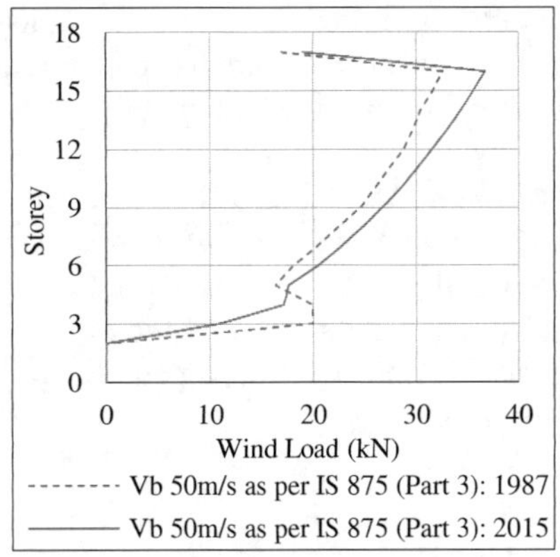

(b) Along wind in Y- direction

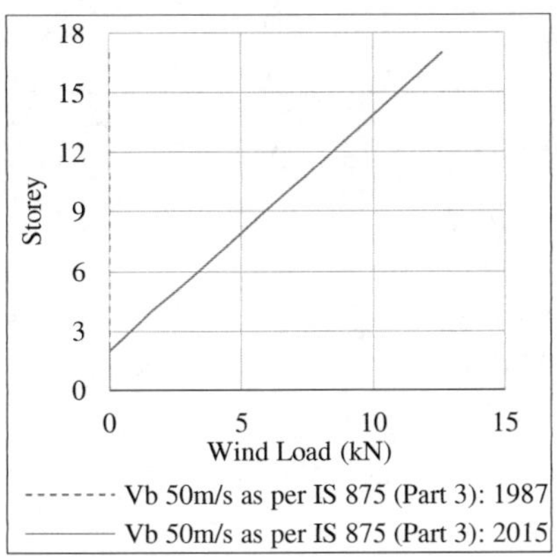

(c) Across wind in X- direction

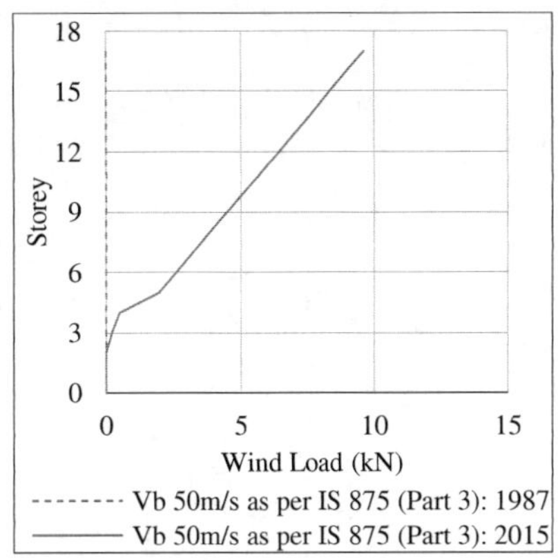

(d) Across wind in Y- direction

Figure 2. Comparison of lateral load – "Along wind" and "Across wind"

3.2. Lateral sway

Lateral sway results obtained are absolute maximum from load combinations of Limit state of serviceability and the same is reported in Table 2. These values are below the permissible limit of 106mm as per Clause 20.5 of IS 456: 2000 [10]

Table 2. Maximum Lateral sway (mm) for various basic wind speeds

IS 875(Part 3)	V_b as 33 (m/s)	V_b as 39 (m/s)	V_b as 44 (m/s)	V_b as 47 (m/s)	V_b as 50 (m/s)	V_b as 55 (m/s)
1987 rev.	21.2	30.5	40.2	46.2	53.3	66.4
2015 rev.	25.0	35.7	45.4	54.8	59.5	73.0

There is an increase in lateral sway obtained as per IS 875 (Part 3): 2015 [7] compared to its previous version. This increase is found to be 18, 17, 13, 19, 12 and 10% respectively for V_b as 33, 39, 44, 47, 50 and 55 m/s. However, the maximum lateral sway obtained at the top storey for all the basic wind speeds considered is within the permissible limits.

3.3. Longitudinal rebar percentage (LRP) in columns

The change in wind parameters from the IS 875 (Part 3): 1987 [6] to the IS 875 (Part 3): 2015 [7] also resulted in changes to the reinforcement in beams and columns.

 doi:10.1088/1757-899X/1025/1/012030

The Longitudinal Rebar Percentage in three typical columns viz., Middle Column (MC), Edge Column (EC) and Corner Column (CC) for basic wind speeds of 33, 39, 44, 47, 50 and 55m/s are shown in figure 3, figure 4 and figure 5 respectively.

3.3.1 Middle Column (MC). From figure 3, it is observed that the Longitudinal rebar percentage with V_b as 33m/s is the same up to the7[th] storey and also from 15th to 17th storey. A maximum difference of 7.7% is observed in the 13th storey. For V_b as 39m/s, LRP is same up to the 5th storey. The variation is noticed from 6th to the 16th storey in which a maximum of 9% is observed in the 14th storey. For V_b as 44, 47, 50 and 55m/s, the maximum percentage increase of LRP are found to be 8%, 15%, 13% and 11% respectively in 15th storey.

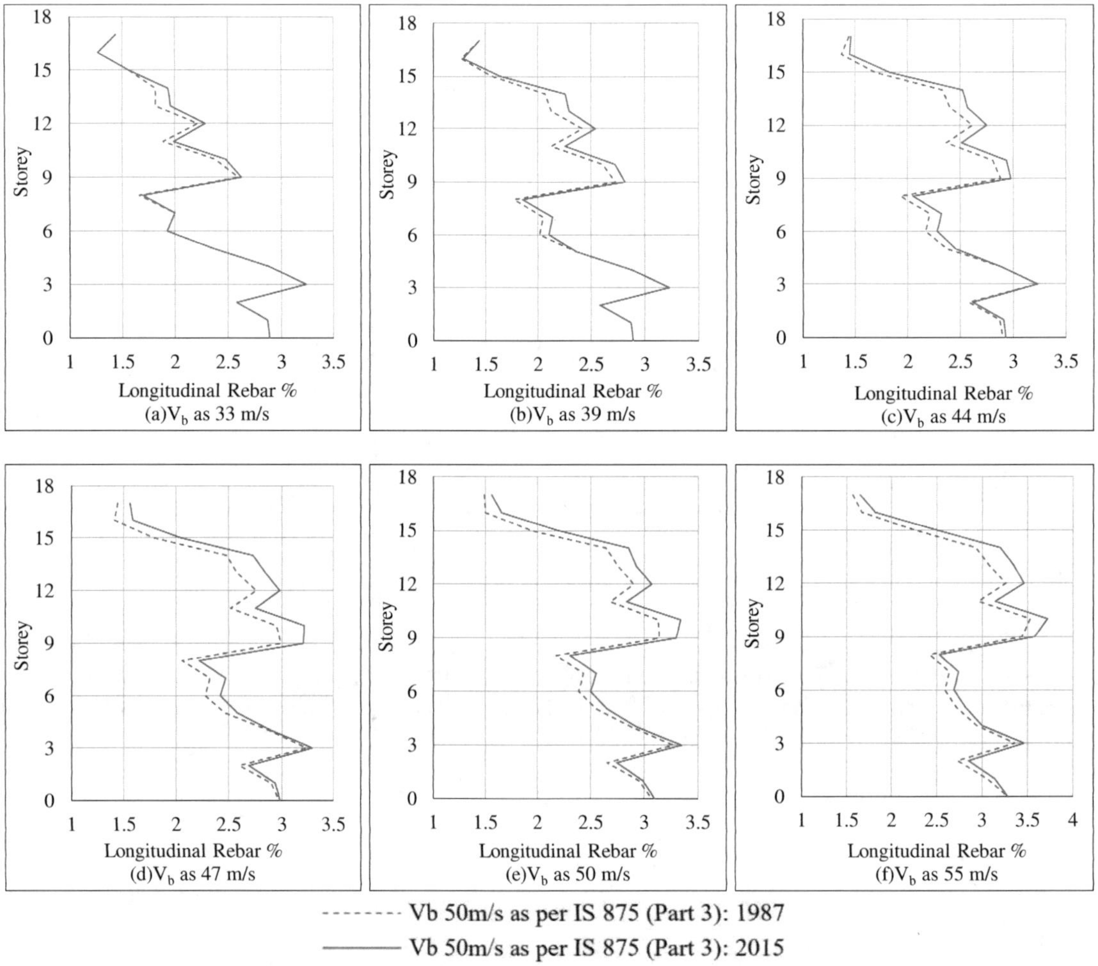

Figure 3. Longitudinal Rebar Percentages in Middle Column (MC)

3.3.2 Edge Column (EC). From figure 4, an increase of 2.2% is observed in storey 6, whereas rebar percentage in all other storeys remained the same for both the codes with a basic wind speed of 33m/s. LRP due to previous code has increased up to the maximum of 7.5% in 12th storey and it remains same above 13th storey for V_b as 39m/s. For basic wind speed of 44m/s, 47m/s and 50m/s; the maximum percentage difference is 9.5%, 10.4% and 9% respectively at 14th storey. However, it remains the same up to the top of the building. For V_b as 55m/s, the maximum variation in LRP is 9% in 14th storey while no variation is observed in 16th and 17th storey.

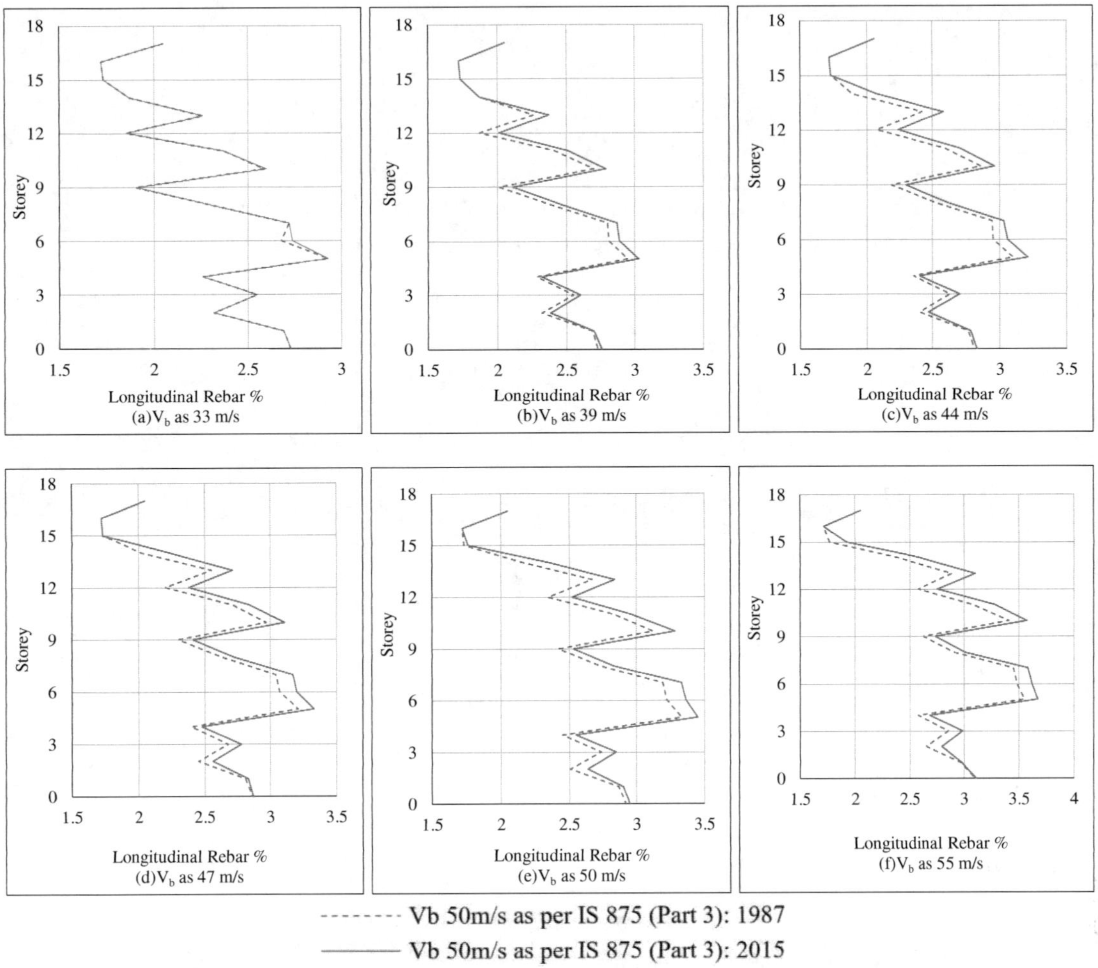

Figure 4. Longitudinal Rebar Percentages in Edge Column (EC)

3.3.3 Corner Column (CC). A zigzag variation is seen in Corner Column throughout the building height for all the basic wind speeds as shown in figure 5.

An increase in the percentage of longitudinal rebar of 6% is observed in 6th storey. Then after, no difference in LRP is observed for a basic wind speed of 33m/s.

For V_b as 39m/s, a maximum increase of 8% is obtained in both 6th and 13th storey. Similarly, there is an increase of LRP by14% as seen in 14th storey for V_b of 44 m/s whereas the maximum difference of 13% and 11% are obtained respectively for a basic wind speed of 47 and 50 m/s in both 12th and 14th storey.

For V_b of 55m/s, an increased by 17% is observed in the 15th storey. Around 10% increase is observed until 14thstorey.

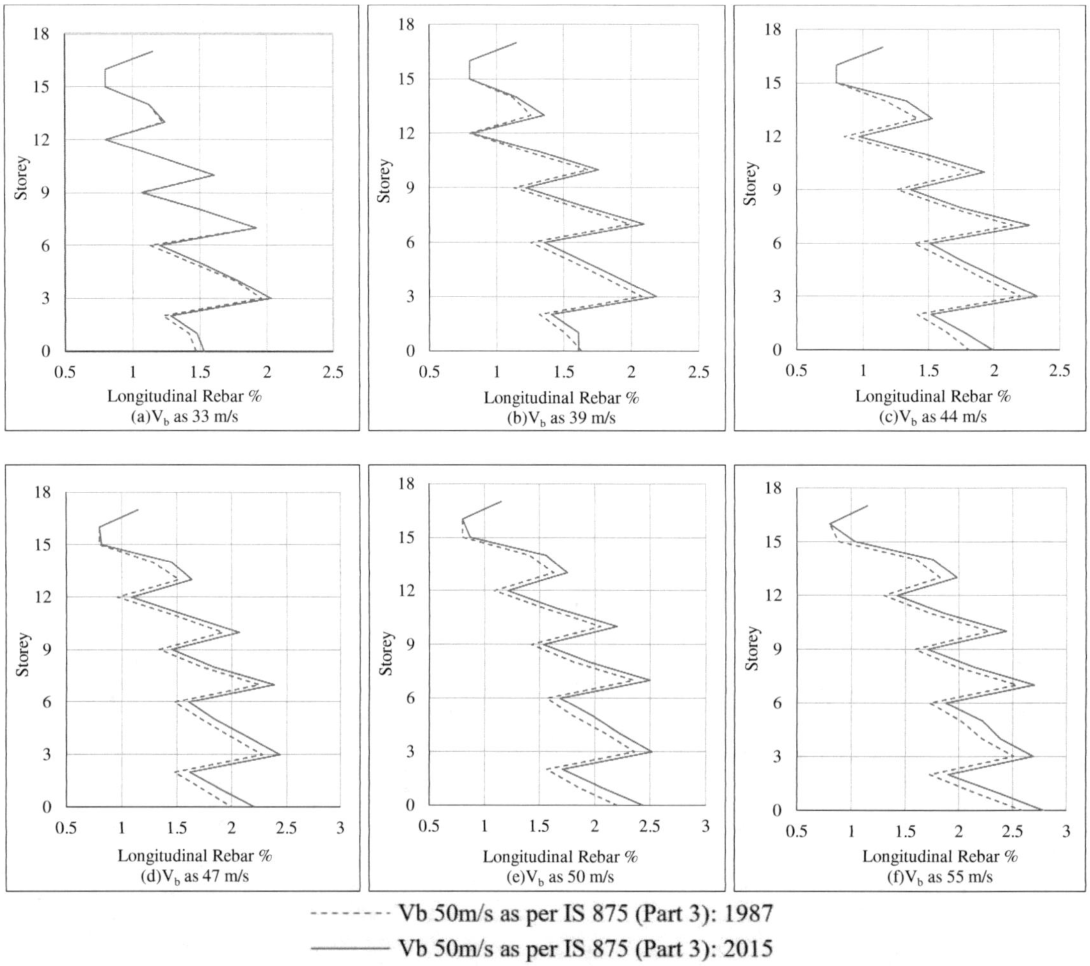

Figure 5. Longitudinal Rebar Percentages in Corner Column (CC)

3.3.4 Quantity of longitudinal rebar in columns. Variation in the Quantity of longitudinal rebar in columns in terms of mass is estimated in the entire building and presented in Table 3 for the basic wind speeds of 33m/s, 39m/s, 44m/s, 47m/s, 50m/s and 55m/s and found that there is an increase of 1.0, 2.5, 3.5, 5.0, 4.7 and 5.1% respectively.

Table 3. Total Rebar quantity in Columns for the entire building

Basic wind speed	IS 875(Part 3): 1987	IS 875(Part 3): 2015	Percentage increase
33m/s	84722 kg	85578 kg	1.0
39m/s	86969 kg	89185 kg	2.5
44m/s	90633 kg	93847 kg	3.5
47m/s	93620 kg	98275 kg	5.0
50m/s	97304 kg	101851 kg	4.7
55m/s	105086 kg	110475 kg	5.1

3.3.5 Quantity of longitudinal rebar in beams. The total rebar quantity in beams for the entire structure is estimated for a basic wind speed of 50m/s (Visakhapatnam). The total rebar quantity for all the beams in the entire building is presented in Table 4.

IOP Conf. Series: Materials Science and Engineering 1025 (2021) 012030 doi:10.1088/1757-899X/1025/1/012030

It can be observed from Table 4 that the total rebar quantity in beams for the entire building for a basic wind speed of 50m/s increases by 2.7%.

Table 4. Quantity of Longitudinal rebar (kg) in beams

Storey	IS 875 (Part 3): 1987	IS 875 (Part 3): 2015	Percentage increase
17	3922	3920	-0.1
16	5195	5248	1.0
15	5253	5480	4.3
14	5195	5412	4.2
13	5383	5535	2.8
12	5580	5700	2.2
11	5788	5862	1.3
10	5788	5962	3.0
9	5899	6032	2.3
8	6088	6153	1.1
7	6197	6351	2.5
6	6268	6410	2.3
5	6207	6460	4.1
4	5680	5850	3.0
3	4939	5112	3.5
2	4989	5160	3.4
1	4757	4970	4.5
Total	93127	95615	2.7

3.3.6 Total quantity of rebar. The total rebar quantity in both beams and columns excluding foundation and slabs is shown in Table 5. It is observed that the percentage increase in the total quantity of rebar (beams + columns) is 3.7% as per IS 875 (Part 3): 2015 [7] compared to IS 875 (Part 3): 1987 [6].

Table 5. Total Rebar quantity of beams and columns (V_b as 50m/s)

Structural element	IS 875(Part 3): 1987	IS 875(Part 3): 2015	% increase
Beams	93127 kg	95615 kg	2.7
Columns	97304 kg	101851 kg	4.7
Total quantity	190431 kg	197466 kg	3.7

4. Conclusions

The results are compared and conclusions on the following attributes are made.

4.1 Conclusions on lateral load and lateral sway.

1. The wind load "Along wind" is gradually increased above 6m onwards from ground level in both X and Y directions and maximum is at 16th storey in all basic wind speed cases whereas, the "Across wind" is an additional load on the structure as per IS 875 (Part 3): 2015 compared to its previous revision.
2. The lateral sway is increased by about 10% to 20% against basic wind speeds of 33 to 55m/s as per the current revision (IS 875 (Part 3): 2015) of the code and same is due to changes in the dynamic analysis procedure.

4.2 Conclusions on longitudinal rebar percentage.

The Longitudinal rebar percentage (LRP) and Quantity of steel reinforcement obtained in three typical columns viz., Middle column, Edge column and Corner column for various basic wind speeds of 33m/s, 39m/s, 44m/s, 47m/s, 50m/s and 55m/s are compared between IS 875 (Part 3): 1987 [6] and IS 875 (Part 3): 2015 [7].

3. The overall quantity of rebar in the Middle column and Edge columns is increased up to 4% when basic wind speeds are increased to 55m/s.

4. The overall quantity of rebar in corner column is increased by 2.2%, 5.5%, 7%, 7.6%, 8% and 8.2% for basic wind speeds of 33, 39, 44, 47, 50 and 55m/s respectively.

5. The total Rebar quantity for the entire building (excluding foundations and slabs) is increased by 3.7% as per IS 875 (Part 3): 2015 compared to IS 875 (Part 3): 1987.

5. Recommendation

As the overall increase in rebar quantity on a 17-storey multi-purpose building is only 3.7%, as per the current wind code involving rigorous calculations in Dynamic analysis, the possibility of simplified methods may be explored.

References

[1] Gordan M, Izadifar M, Haddadiasl A, Javanmardi A, Abadi R and Hosseini M H 2014 Interaction of Across-Wind and Along-Wind with Tall Buildings *Australian Journal of Basic and Applied Sciences* 8(19) Special 96-101

[2] Halder L and Dutta C S 2010 Wind effects on Multi-storied buildings: A critical review of Indian Codal provisions with special reference to American standard *Asian Journal of Civil Engineering (Building and Housing)* **11(3)** 345-370

[3] Kumar S H and Rajan S S 2017 Estimation of gust response factor for a tall building model with 1:1.5 plan ratios *Materials Science and Engineering* **225** 1-11

[4] Mendis P, Ngo T, Haritos N and Hira A 2007 Wind loading on tall buildings *EJSE* Special Issue Loading on Structures 41-54

[5] Zhou Y and Kareem A 2001 Gust Loading Factor: New Model *Journal of Structural Engineering* **127** 168-175

[6] Indian Standard Code of Practice for Design loads (Other than Earthquake) for Buildings and Structures – wind loads IS 875 (Part 3): 1987 *Bureau of Indian Standards* New Delhi

[7] Indian Standard Code of Practice for Design loads (Other than Earthquake) for Buildings and Structures – wind loads IS 875(Part 3): 2015 *Bureau of Indian Standards* New Delhi

[8] Indian Standard Code of Practice for Design Loads (Other than Earthquake) for Buildings and Structures – Dead Loads – Unit weight of building materials and stored materials IS 875 (Part 1): 1987 *Bureau of Indian Standards* New Delhi

[9] Indian Standard Code of Practice for Design Loads (Other than Earthquake) for Buildings and Structures – Imposed Loads IS 875 (Part 2): 1987 *Bureau of Indian Standards* New Delhi

[10] Indian Standard Plain and Reinforced Concrete- Code of Practice IS 456: 2000 *Bureau of Indian Standards* New Delhi

Seismic analysis of Reinforced concrete frames with stiffness irregularity

Sripathi Siva Bhanu Sai Kumar [1,2], G V Rama Rao[3]

[1]Research Scholar, Department of Civil Engineering, College of Engineering (Autonomous), Andhra University, Visakhapatnam, Andhra Pradesh, India
[2]Assistant Professor of Civil Engineering, MVGR College of Engineering (Autonomous), Vizianagaram, Andhra Pradesh, India
[3]Professor, Department of Civil Engineering, College of Engineering (Autonomous), Andhra University, Visakhapatnam, Andhra Pradesh, India
Email: sivabhanu.sripathi@gmail.com

Abstract. The earthquake occurs due to severe seismic in a particular region and causes devastation and havoc to human life. So, there is a need to assess vulnerability characteristics of the structures which are located in severe seismic regions. In this present work,seismic performance of stiffness irregular structure was studied by using the response spectrum method as per IS 1893 (Part 1) 2002. And the performance of structures with shear wall and bracings placed in the least lateral stiffness direction was studied. From the present study by comparing the seismic performance of the conventional structureand structure with shear wall/ bracing it can be concluded that performance of RC frames with stiffness irregularity is more vulnerable than that of RC frames with shear wall/ bracings in place. Also, RC frames with shear wall have reduced the seismic impact more than RC frames with bracings.
Keywords: R.C building; stiffness irregular frame, bracings, shear wall.

1. Introduction

Earthquake is a natural phenomenon which cannot be avoided and predicted but the damaged caused by an earthquake can be minimised by improving lateral force resistance of the structure through design and construction practices different from conventional methods. The failure of the structure begins at points of weakness in the structure. These weaker spots are may be due to irregularity/discontinuity in geometry, stiffness and mass of the structure. These irregular structures hold a large proportion of the urban infrastructure.Considerable difference in stiffness and mass from Storey to Storey makes the dynamic attributes of these structures not quite the same as the regular structure. According to IS 1893 (Part 1) 2002 [1] for regular structures of constrained height Linear static analysis can be used, as in this procedure, lateral forces are determined according to the code-based central timespan of the structure. The linear dynamic analysis gives more reliable results than that of direct static analysis, as dynamic analysis creates the higher modes vibrational impact and force distribution in a superior way. The present investigation focuses on seismic performance assessment of different RCframes with and without stiffness irregularity situated on hard soil and in Zone II utilizing Response Spectrum Method.

2. Literature Review

Rakshit Patil et al [2] concluded that in high rise structures by placing the shear wall at the centre core of the structure has reduced in storey drifts and displacements effectively. Cimellaro et al [3] conducted bidirectional push over analysis (BPA) technique to evaluate the seismic reaction of irregular structures exposed to bidirectional ground motion. And it is concluded that BPA technique given better outcomes when contrasted with NTHA strategy and torsion issue can be overcome in irregular structures by exposing to bidirectional ground motion. Chatpan Chintanapakdee et al [4] have analysed the seismic requirements for vertically sporadic and conventional frames assessed by non-linear time history method. By utilizing strong column and soft Storey model structures with stiffness and strength irregularity were designed. Drift demands in upper stories are considerably more sensitive to irregularities in the lower stories than that of lower stories is influenced by irregularities in upper stories. Krishnan et al [5]studied different parameters such as Storey drift, Storey shear, deflection, in column reinforcement etc., of a structure subjected to lateral loading by changing the position of shear walls. Both linear and nonlinear analyses were carried out and the effect of shear wall position on different parameters was studied by carrying out both linear and non-linear analysis. In medium ascent structures arrangement of shear walls is seen as viable in improving the seismic capacity of the structure. Ravi Kumar et al [6] considered two kinds of irregularity (with geometry and diaphragm discontinuityand three different lateral load patterns on the performance of various irregular buildings in pushover analysis. And concluded in pushover analysis the vertical distribution of lateral force as per IS code was found more detrimental in low rise models. Dhinam Basu et al [7] studied the application of super position-based analysis procedure to implement code-specified torsional specifications for flexible diaphragm floored structures. Thiswas proven to be optimal for unsymmetrical buildings withand without orthogonality. Jack P Mohele et al [8] conducted an experimental and analytical investigation of the reaction to strong base movements of reinforced concrete structures having vertical irregularity. It was concluded inelastic static analyses were capable of accurately reproducing the measured envelope relations between top displacement and base shear. The investigation showed that that frame-wall interaction can be influenced altogether by inelastic part behaviour. The inelastic static and dynamic strategies were better in deciphering impacts of the structural discontinuities. Bhanu Sai Kumar et al [9] performed the Seismic fragility analysis of regular and setback RCC frames. Ponnada and Reddi [10] performed the Linear static analysis of multi-storey building with horizontally asymmetric architectures,

3. Objectives

The objectives of this study are

1. Seismic analysis of the stiffness irregular structures.
2. Evaluation of Storeydisplacement and inter Storey drift of the RC framesby conductingdynamic analysisusing SAP 2000.
3. Performance assessment of RC frames after adopting different lateral force resisting systems.

4. Methodology

Step 1: Modelling using SAP 2000.
Step 2: Applying different load combinations.
Step 3: Analysing the frames by changing the position ofsoft Storey.
Step 4: Assessing Storey displacements and inter Storey drifts for all the cases and identifying the most vulnerable RC frame.
Step 5: Adopting different lateral load resisting systems to the most vulnerable frame and calculating Inter Storey drifts and Storey displacements.
Step 6: Comparison of the results and suggesting the effectivelateral load resisting systems with minimum displacements and drifts.

5. Modelling

5.1 Details of the Building considered

For the present study 'G+5' (Ground+5 Storeys) building located in the seismic zone II and soil zone III (hard strata) are considered. The elevation and plan of the building considered is shown in figure 1.

Height of the Building	= 24.7 m
Length of the Building	= 16 m
Width of the Building	= 16 m
Number of bays in X and Y directions	= 4
Number of Storeys in Z-direction	= G+5
Spacing of bays in X and Y directions	= 4
Length of each bay	= 4 m

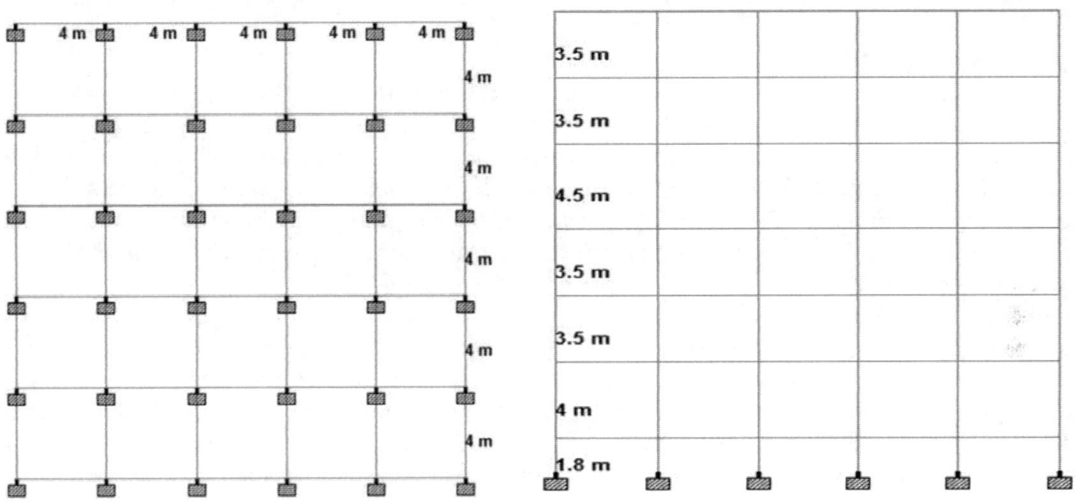

Figure 1. Plan and Elevation of the Building

5.2 Details of Structural Elements considered

Dimensions of the beams and columns of the building frame considered as follows

Table 1. Dimensions of Beams and Columns of the building frame

Structural Member	Section	Dimensions	Material
Beam	Rectangle	300 × 230	Concrete
Column	Rectangle	450 × 300	Concrete
Bracing	I-Section	ISMB 250	Steel

5.3 Details of the Slab and Walls Considered as follows

Thickness of the Slab	= 0.180m
Thickness of the external wall	= 0.300m
Thickness of the internal wall	= 0.150m
Thickness of the parapet wall	= 0.100m
Height of the parapet wall	= 1.0m

5.4 Loads Considered

Unit weight of the Brick masonry		= 20 kN/m^3
Unit weight of the reinforced cement concrete		= 25 kN/m^3
Live load -	Soft Storey	= 4.5 kN/m
	Normal Storey	= 3.5 kN/m
Super Dead –	Wall load (external)(Normal Storey)	= 25.6 kN/m
	Wall load (internal)(Normal Storey)	= 12.8 kN/m

Dead load – Slab load(Normal Storey) = 28.8 kN/m
Slab load(Soft Storey) = 57.6 kN/m
Parking load(stilt) = 4.5 kN/m
Parapet wall load = 4.8 kN/m

5.5 Storey Stiffness and Stiffness Irregularity

Stiffness of normal storey (storey with 3.5m storey height) (K_N)= 2361.52 N/mm
Stiffness of soft storey (storey with 4.5m storey height) (K_S)= 1111.11 N/mm
From this it is evident that $(K_S) < 0.7(K_N)$ and $(K_S) < 0.8[(K_N+ K_N+K_N)/3]$
So according toIS 1893-2002 Table No.5, the above structure is a stiffness irregular structure.

6. Results and Discussion

6.1 Seismic Analysis of Stiffness IrregularStructures with Soft Storey at Different Heights

In this seismic analysis, the analysis was conducted by placing soft storey at particular storey levels and then displacement& inert storey drifts are calculated at each storey level. The same procedure was repeated by placing the soft storey at different storey levels. Finally, the storey displacement and inter storey drifts are compared. The vulnerable position of the soft storey was identified.

6.2 Comparison of all displacements and drifts in all the frames to identify the frame with maximum displacements and drifts

From figure 2, it is observed that for a particular frame the storey displacements are increased from bottom storey to top storey. And as the soft storey moved from lower storey levels to higher storey level storey displacements have increased. In the frame with soft storey placed at 5th storey level maximum storey displacements are observed and they are, at 1st, 2nd, 3rd, 4th, 5th storey levels are 2.05mm, 15.93mm, 30.71mm, 45.04mm, 57.73mm, 67.86mm, 76.84mm respectively. The minimum storey displacements are observed when the soft story is placed at 1st storey level and they are, at 1st, 2nd, 3rd, 4th, 5th storey levels are 1.11mm, 8.66mm, 16.89mm, 25.19mm, 32.9mm, 39.32mm, 45.09mm.

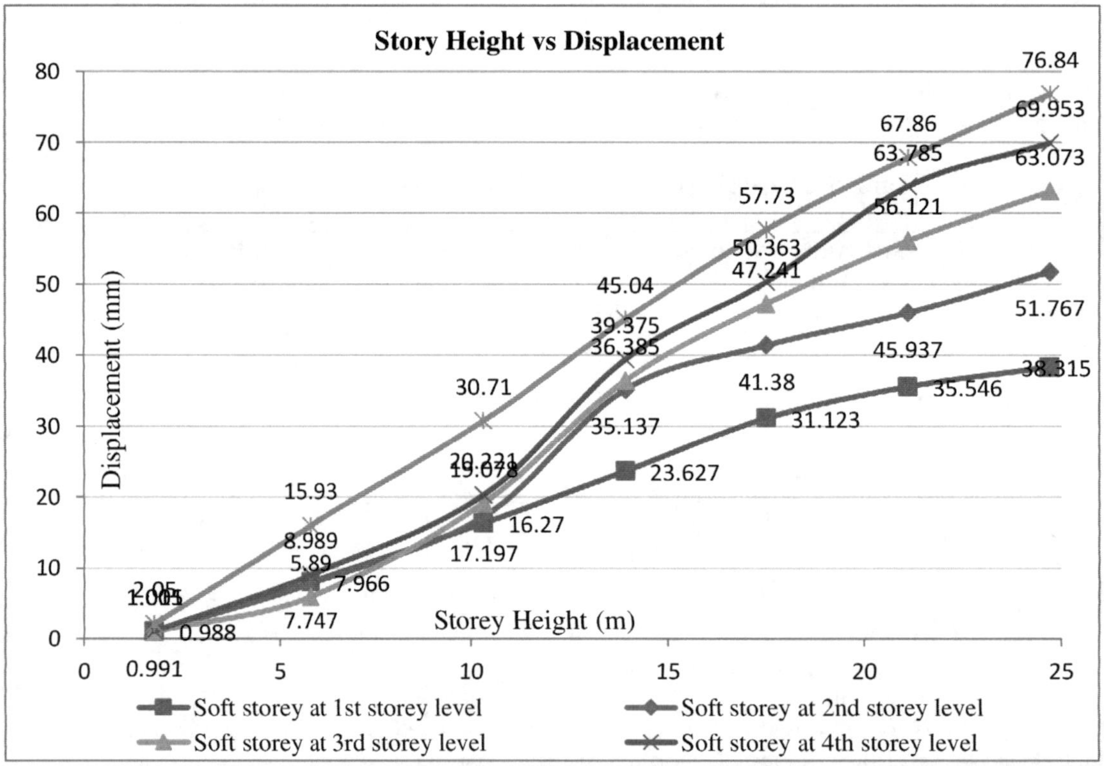

Figure 2. Displacement Curves of Different RC frames with Soft Storey at different levels

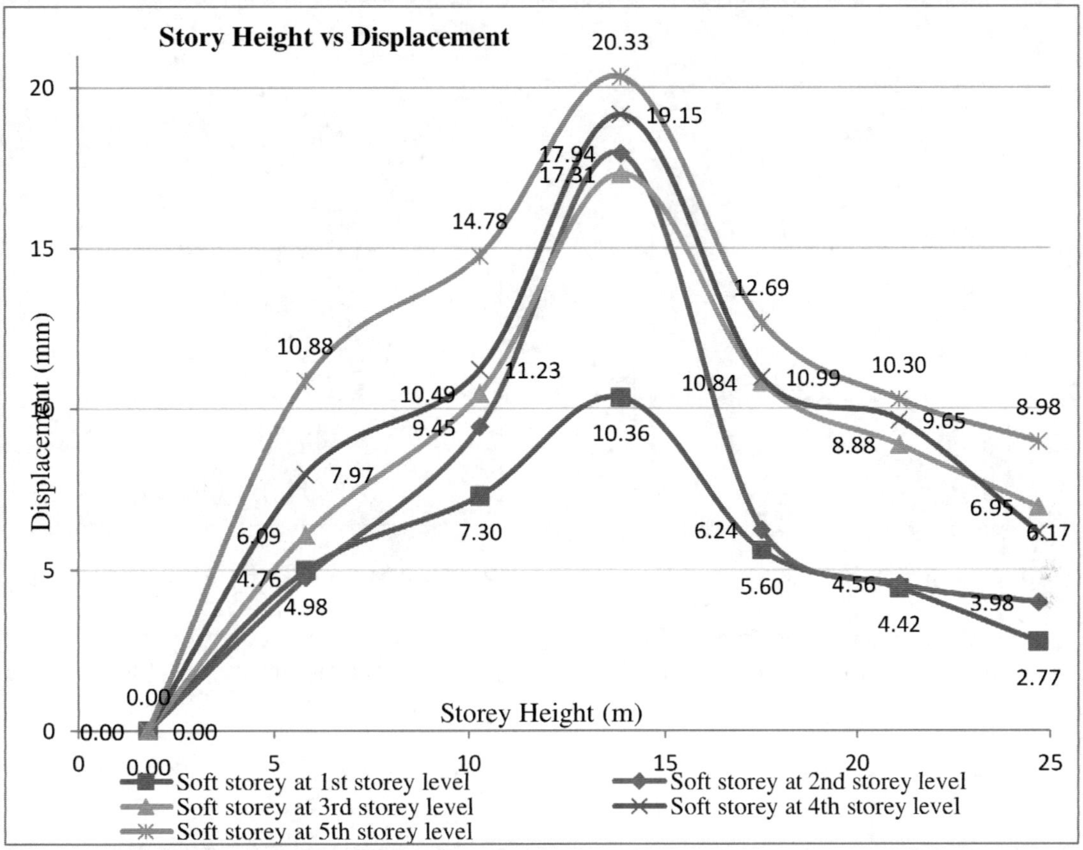

Figure 3. Drift Curves of different RC frames with Soft Storey at different levels

From figure 3, it is observed that inter storey drifts are maximum when soft storey placed at 5th storey level and the minimum inter storey drift is observed when the soft story is placed at 1st storey level. It is also observed that the soft storey moved from lower storey levels to higher storey level inter storey drifts have increased. Maximum inter storeydrift of 20.33mm is observed at 2nd and 3rd storey interface when soft storey placed at 5th storey level.

6.3 Provided lateral force resisting systems(Shear Walls and Bracings)

The plan and elevation of the frame with bracings placed on three external sides in the stilt is shown in figure 4. Table 2 shows the variation of storey displacement and storey drift with storey height.

Table 2. Storey Displacements and Drifts when Frame is provided with Bracings in the Stilt

Sl.No.	Storey Height (m)	Storey Displacement(mm)	Storey Drift (mm)
1	1.8	2.09	0.00
2	5.8	16.25	14.16
3	10.3	31.38	15.12
4	13.9	46.05	14.68
5	17.5	59.02	12.97
6	21.1	69.32	10.30
7	24.7	74.39	5.06

IOP Conf. Series: Materials Science and Engineering 1025 (2021) 012031 doi:10.1088/1757-899X/1025/1/012031

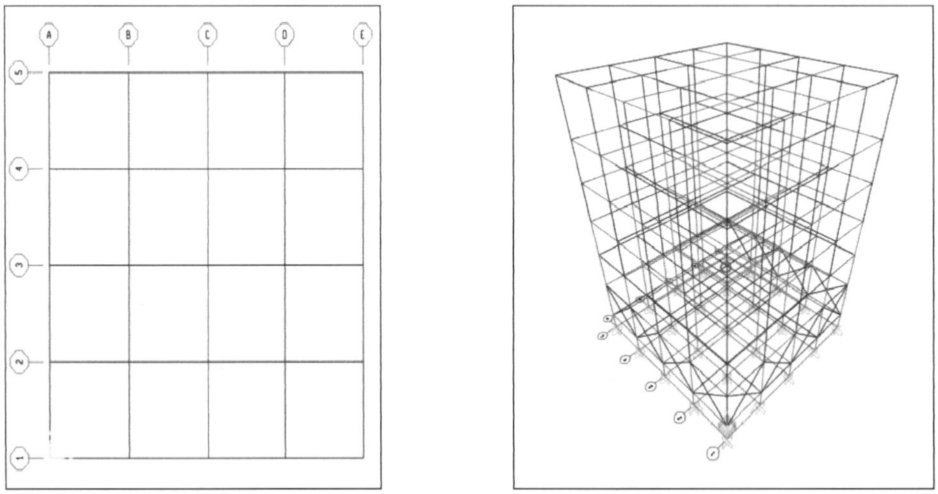

Figure 4. Plan and Elevation of Frameprovided with Bracings on three sides of the stilt

The plan and elevation of the frame with a shear wall provided on two opposite external sides is shown in figure 5. Table 3 shows the variation of storey displacement and storey drift with storey height.

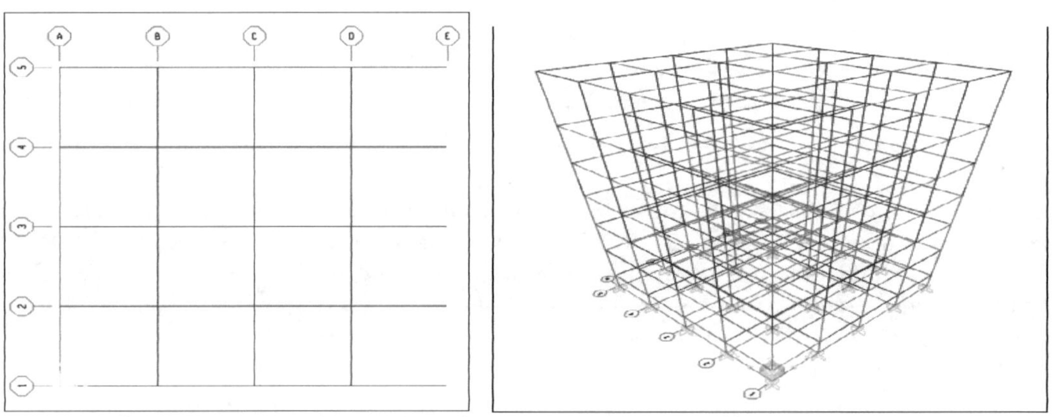

Figure 5. Plan and Elevation of Frame provided with Shear walls

Table 3. Storey Displacements and Drifts for Frame with Shear Walls on Opposite Sides

Sl. No.	Storey Height (m)	Storey Displacement (mm)	Storey Drift (mm)
1	1.8	2.70	0.00
2	5.8	18.13	15.42
3	10.3	35.24	17.11
4	13.9	51.03	15.79
5	17.5	62.09	11.07
6	21.1	68.43	6.34
7	24.7	73.72	5.29

The plan and elevation of the frame with Shear walls at the centre and bracings placed on three external sides in the stilt is shown in figure 6. Table 4 shows the variation of storey displacement and storey drift with storey height.

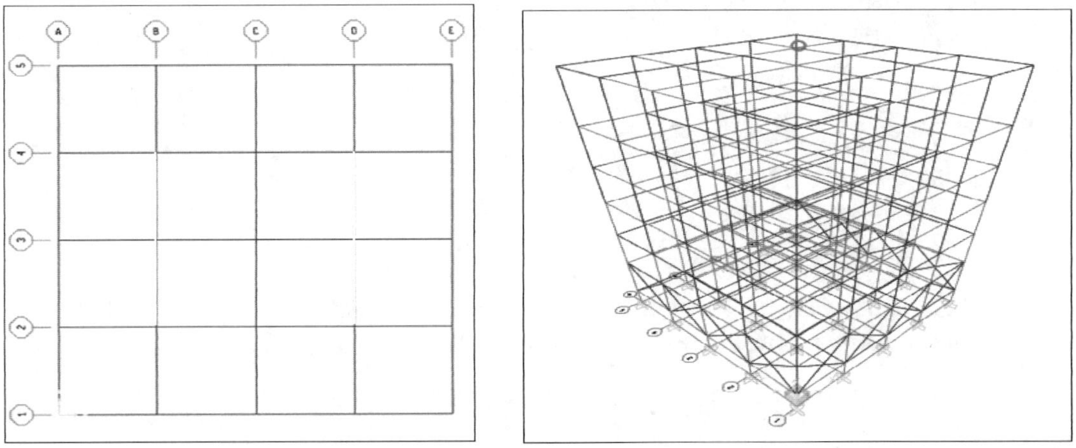

Figure 6. Plan and Elevation of Frame provided with Shear walls at the centre and external bracings

Table 4. Storey Displacements and Drifts when Frame is provided with Centre Shear walls and External Bracings at Stilt

S.No.	Storey Height(m)	Storey Displacement(mm)	Storey Drift(mm)
1	1.8	1.11	0.00
2	5.8	8.66	7.56
3	10.3	16.89	8.23
4	13.9	25.19	8.30
5	17.5	32.90	7.71
6	21.1	39.32	6.42
7	24.7	45.09	5.77

The plan and elevation of the frame with Shear walls on two external opposite Sides and Bracings at Centre is shown in figure 7. Table 5 shows the variation of storey displacement and storey drift with storey height.

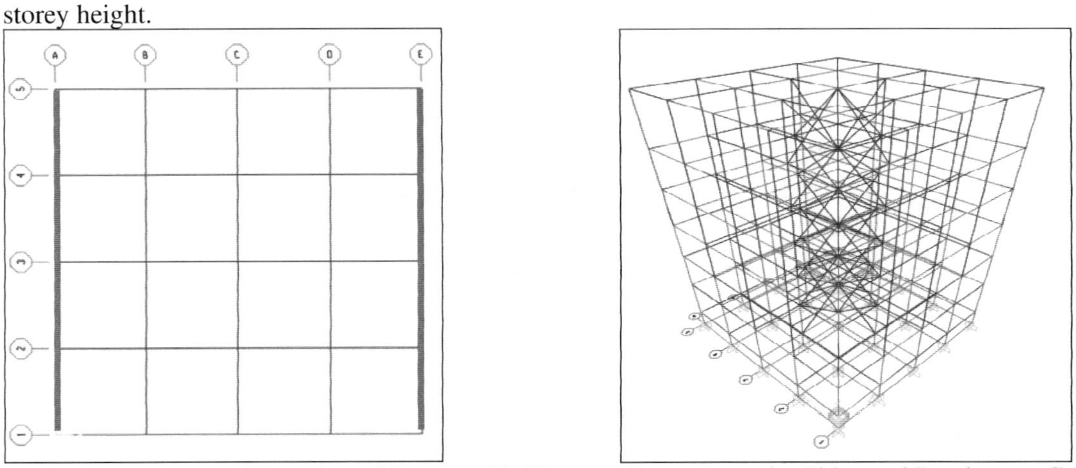

Figure 7. Plan and Elevation of Frame with Shear walls on Opposite Sides and Bracings at Centre

Table 5. Storey Displacements and Drifts when Frame is provided with Shear walls on Opposite Sides and Bracings at Centre

Sl. No.	Storey Height (m)	Storey Displacement (mm)	Storey Drift (mm)
1	1.8	4.38	0.00
2	5.8	12.48	8.11
3	10.3	25.06	12.58
4	13.9	39.46	14.39
5	17.5	56.31	16.86
6	21.1	65.81	9.50
7	24.7	72.67	6.86

The plan and elevation of the frame with bracings at the centre is shown in figure 8. Table 6 shows the variation of storey displacement and storey drift with storey height.

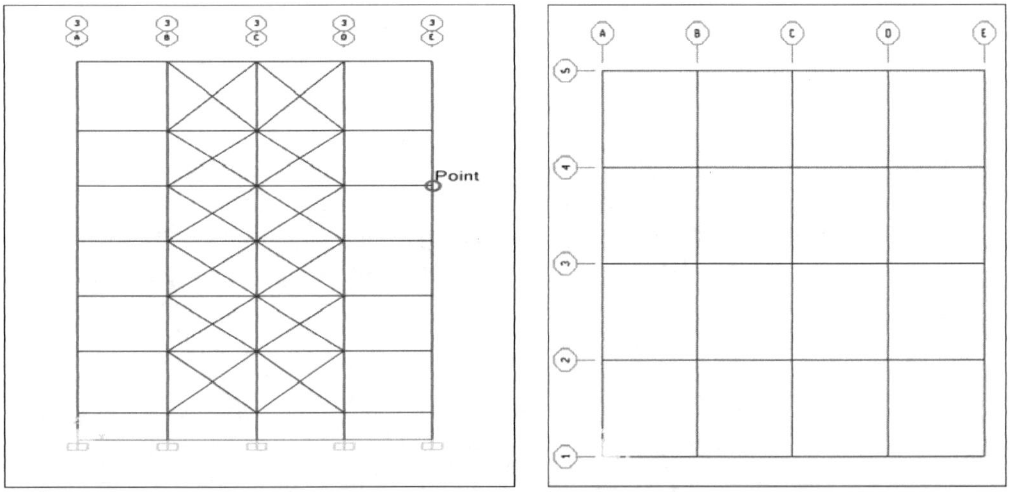

Figure 8. Plan and Elevation of Frame provided with Bracings at Centre

Table 6 Storey Displacements and Drifts when Frame is provided with Bracings at Centre

Sl. No.	Storey Height (m)	Storey Displacement (mm)	Storey Drift (mm)
1	1.8	11.89	0.00
2	5.8	13.97	2.08
3	10.3	23.05	9.08
4	13.9	33.77	10.73
5	17.5	43.38	9.61
6	21.1	51.37	7.98
7	24.7	58.86	7.50

The plan and elevation of the frame with shear walls at the centre is shown in figure 9. Table 7 shows the variation of storey displacement and storey drift with storey height.

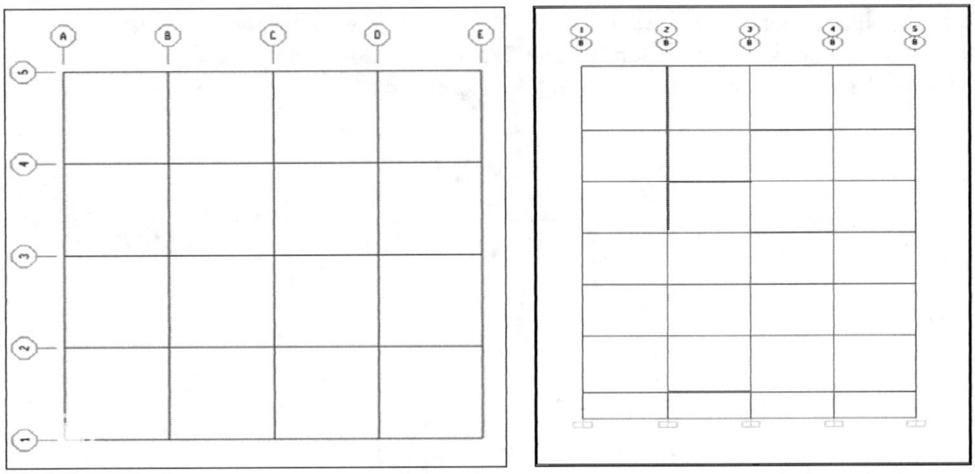

Figure 9. Plan and Elevation of Frame provided with Shear walls at Centre

Table 7. Storey Displacements and Drifts when Frame is provided with Shear walls at Centre

Sl. No.	Storey Height (m)	Storey Displacement (mm)	Storey Drift (mm)
1	1.8	11.89	0.00
2	5.8	13.97	2.08
3	10.3	23.05	9.08
4	13.9	33.77	10.73
5	17.5	43.38	9.61
6	21.1	51.37	7.98
7	24.7	58.86	7.50

From the figure 10, it is observed that for the vulnerable frame (frame with soft storey placed at 5th storey level) with shear walls are placed at centre and bracings at external sides in the stilt of the structure is causing minimum storey displacements when compared to that of the frame with any other lateral load resisting systems. And the storey displacements at 1st, 2nd, 3rd, 4th, 5th storey levels are reduced from 2.05mm, 15.93mm, 30.71mm, 45.04mm, 57.73mm, 67.86mm, 76.84mm respectively to 1.11mm, 8.66mm,16.89mm, 25.19mm, 32.90mm, 39.32mm, 45.09mm respectively. So, of all lateral loads resisting systems shear walls are placed at centre and bracings at external sides in the stilt is effective in reducing the storey displacements.

From the figure 11, it is observed that for the vulnerable frame (frame with soft storey placed at 5th storey level) with shear walls are placed at centre and bracings at external sides in the stilt of the structure is causing minimum inter storey drift is observed when compared to that of the frame with any other lateral load resisting systems. And the maximum inter storey drift in the frame without any lateral load resisting system is 20.33mm and this is reduced to 9.08mm by placing shear walls at centre and bracings at external sides in the stilt. So, this lateral load resisting system was effective in reducing inter storey drifts.

From figure 11, it is observed that there is reduction in storey displacements in the frames soft storey placed at 5th storey level without any lateral load resisting system and with shear wall placed at the centre and external bracings in stilt. The storey displacements at 1st, 2nd, 3rd, 4th, 5th storey levels are reduced by 46.1%, 45.65%, 44.99%, 44.06%, 43% and 41.32% respectively.

From figure 13, the reduction in inter storey drifts in the frames soft storey placed at 5th storey level without any lateral load resisting system and with shear placed at the centre and external bracings in stilt is observed. The maximum inter storey drift is reduced by 59.17%.

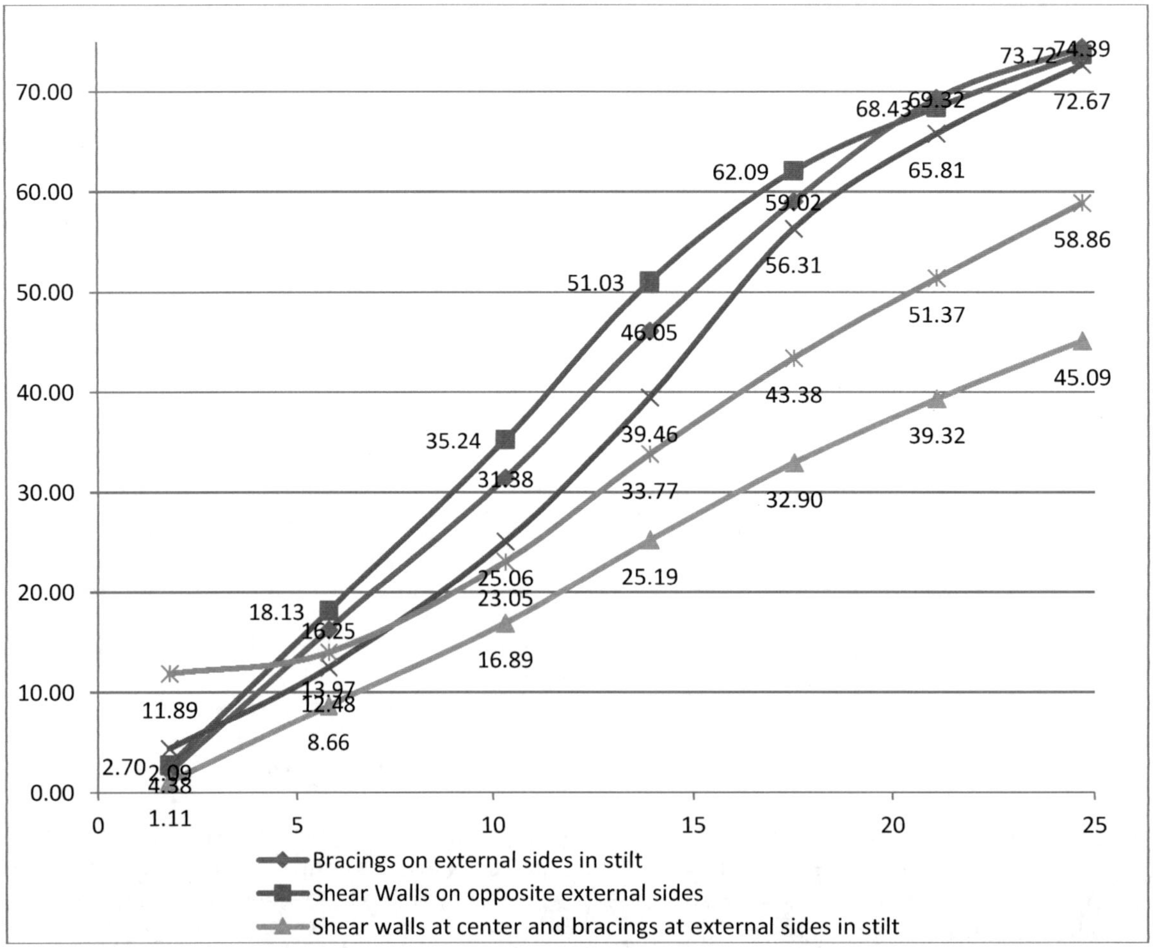

Figure 10. Displacements Curves of the Frames with Different Remedial Measures with Maximum Displacement

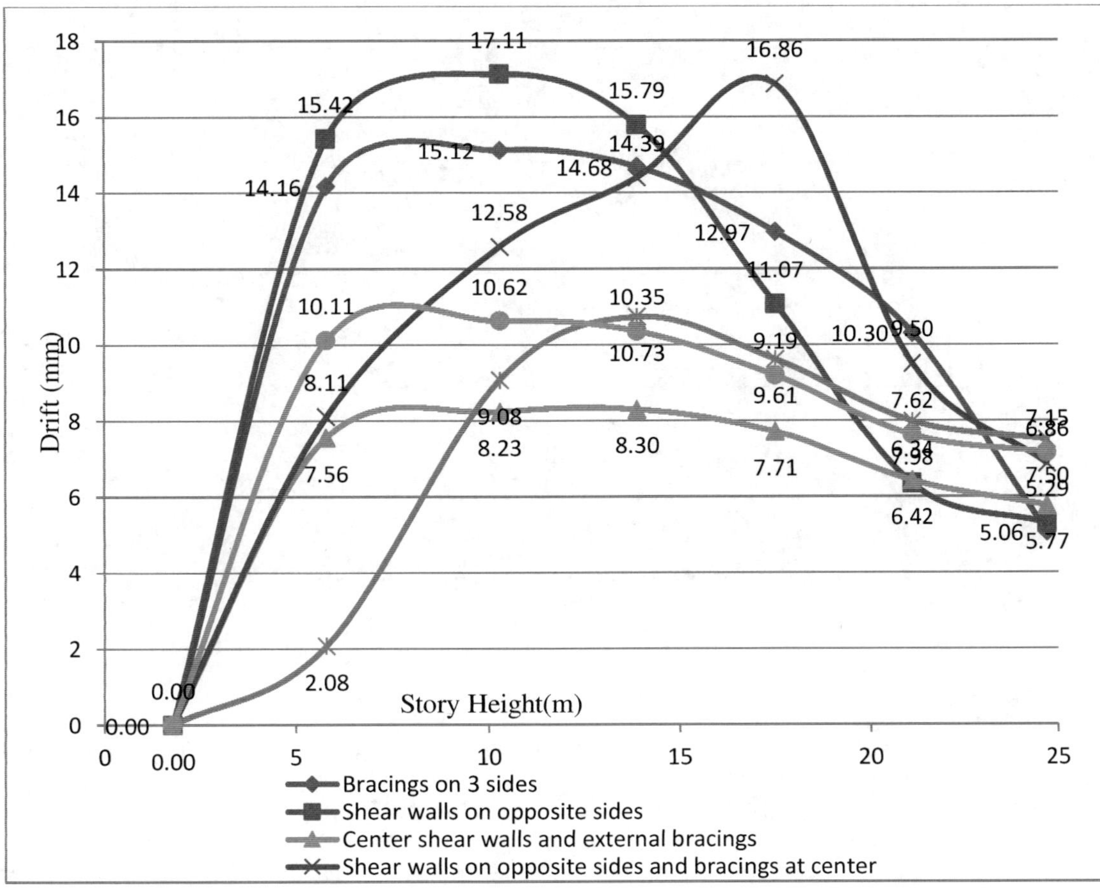

2

Figure 11. Drift Curves of the Frames with Different Remedial Measures with Maximum Displacement

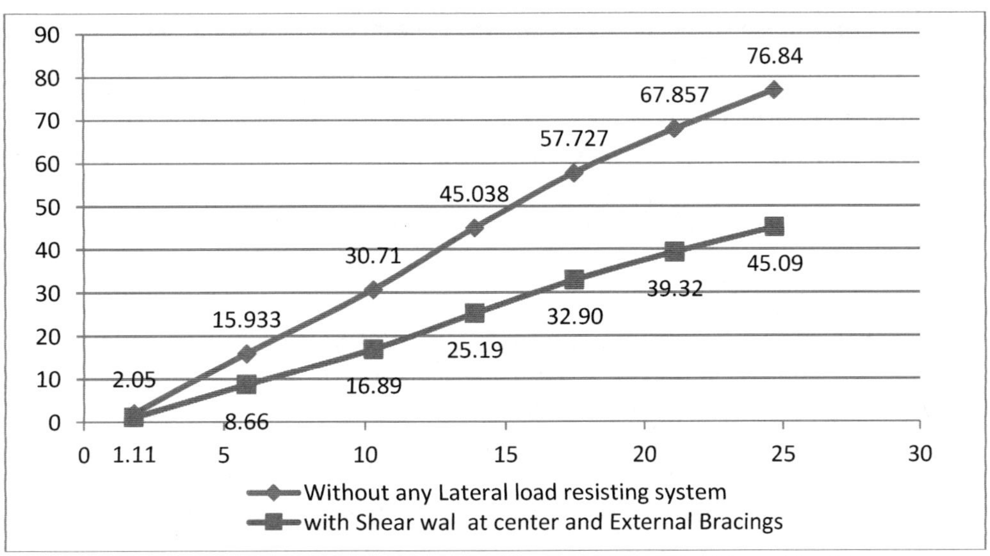

Figure 12. Displacement Curves of frames with and without Revision.

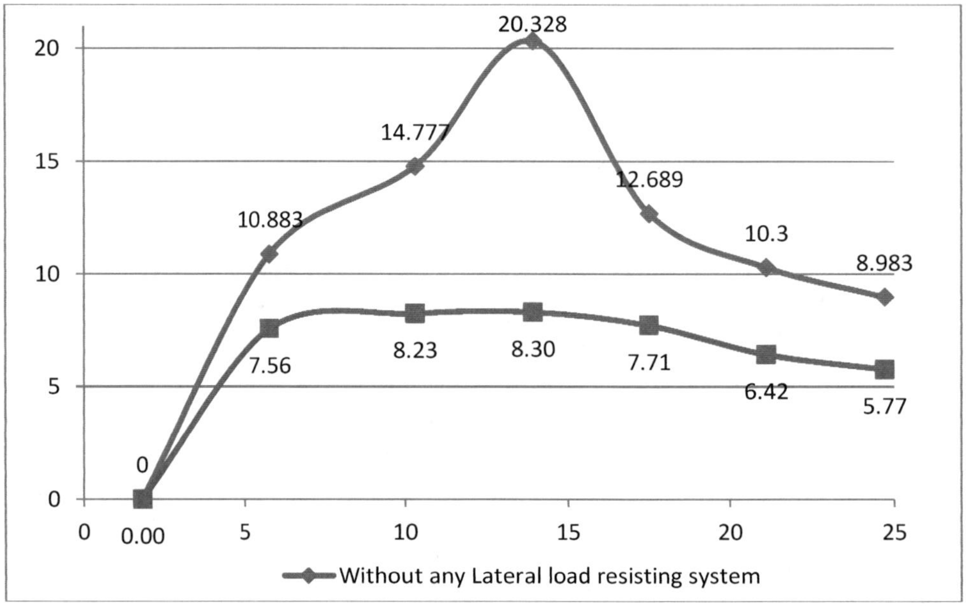

Figure 13. Displacement Curves of frames with and without Revision.

7. Conclusions

Seismic vulnerability assessment for stiffness irregular buildings with and without lateral load resisting systems was carried out and the following can be concluded

1. As the soft storey shifted from lower storey levels to higher storey level storey displacements and inter storey drifts have increased. In the frame with soft storey placed at 5th storey level maximum storey displacements are observed and they are, at 1st, 2nd, 3rd, 4th, 5th storey levels are 2.05mm, 15.93mm, 30.71mm, 45.04mm, 57.73mm, 67.86mm, 76.84mm respectively. And maximum inter storey drift of 20.8mm is obtained between 2nd& 3rd storey interface for the same frame.So, frame with soft storey placed at 5th storey level is the vulnerable frame.
2. Frames without any lateral force resisting systems have obtained more displacements when compared to that of frames with lateral load resisting systems.
3. Out of all lateral loads resisting systems when shear walls placed at the centre and external bracings in stilt of vulnerable frame minimum storey displacements and inter storey drifts are observed and they are, at 1st, 2nd, 3rd, 4th, 5th storey levels are 1.11mm, 8.66mm,16.89mm, 25.19mm, 32.90mm, 39.32mm, 45.09mm respectively.
4. By placing the shear wall at centre and external bracings in stiltfor a vulnerable frame the storey displacements at 1st, 2nd, 3rd, 4th, 5th storey levels are reduced by 46.1%, 45.65%, 44.99%, 44.06%, 43% and 41.32% respectively and maximum inter storey drift is reduced by 59.17%.

References

[1] Indian Standard 1893 (Part 1) 2002 Criteria for Earthquake Resistant Design of Structures General Provisions and Buildings (Fifth Revision) *Bureau of Indian Standards* New Delhi
[2] Rakshit Patil, Avinash S Deshpande, Shrishail Sambanni 2016 Optimal location of the Shear wall in high rise building subjected to Seismic Loading *International Journal for Technological Research in Engineering* **3(10)** p 2678-82
[3] G P Cimellaro 2014 Bidirectional push-over analysis of Irregular Structures *Journal of Structural Engineering* **140(9)** (04014059) p 07-13
[4] Chatpan Chintanapakdee Anil k Chopra 2004 Seismic Response of Vertically Irregular

Frames: Response History and Modal Pushover Analyses *Journal of Structural Engineering* **130 (8)** p 1177-84

[5] P A Krishnan, Anjaly Francis, V N Pradeep 2019 Effect of Shear Wall location in Buildings subjected to Seismic Loads *International Journal of Engineering Research and Management* **6 (7)** p 34-37

[6] Ravikumar CM, Babu Narayan K, Sujith B, Venkat Reddy V 2012 Effect of Irregular configuration on seismic vulnerability of RC Buildings *Architecture Research* **2 (3)** p 20-6

[7] Dhinam Basu 2004 Seismic Analysis of Asymmetric Buildings with Flexible Diaphragms *Journal of Structural Engineering* **130 (8)** p 1169-76

[8] Jack P Moehle, Luis F Alarcon 1986 Seismic analysis methods for irregular buildings *Journal of Structural Engineering* **112 (1)** p 129-36

[9] Siva Bhanu Sai Kumar S, Rama Rao G V, Markandeya Raju P 2016 Seismic fragility analysis of regular and setback RCC frames—a few hypothetical case studies *Asian Journal of Civil Engineering* **17 (5)** p 551–69

[10] Ponnada M R, Reddi P 2020 Linear static analysis of multi-storey building with horizontally asymmetric architectures *Journal of Building Pathology and Rehabilitation* **5 (25)** p 1-10

IOP Conf. Series: Materials Science and Engineering 1025 (2021) 012032 doi:10.1088/1757-899X/1025/1/012032

Enhancing Seismic Resistance of G+10 MSB by Introducing FVD System

T S D Phanindranath[1,2], K V G D Balaji[3], P Markandeya Raju[4]

[1] Research Scholar, Department of Civil Engineering, GITAM (Deemed to be University), Visakhapatnam, Andhra Pradesh, India
[2]Assistant Professor (C), University College of Engineering Vizianagaram (JNTU-K UCEV) Vizianagaram, Andhra Pradesh, India
[3]Professor, Department of Civil Engineering, GITAM (Deemed to be University), Visakhapatnam, Andhra Pradesh, India
[4]Professor and Head, Department of Civil Engineering, MVGR College of Engineering (Autonomous), Vizianagaram, Andhra Pradesh, India
Email: phani.civil485@gmail.com, balaji@gitam.edu, markandeyaraju@mvgrce.edu.in

Abstract
Lateral load resisting systems (LLRS) play a key role in reducing or eliminating the damage of the structure due to the earthquake. The damping system is one of the LLRS which controls the vibrations when an earthquake occurs. Fluid Viscous damping decreases tension and deflection because due to flexing the columns, the energy from the damping is totally out of phase with stresses. This is valid only for viscous fluid damping, where the damping force varies with the velocity of the stroke. Other types of damping do not vary their performance with velocities, such as yielding components, friction devices, plastic hinges, and visco-elastic elastomers; hence, they can increase column stress while reducing deflection, and typically do so. While coming to this work, under the seismic coefficient method for seismic zone-5, G+10 multistoried buildings were modelled and analyzed in ETABS along with limit state of collapse and limit state of serviceability load combinations. The response of the building was examined by introducing viscous dampers to the structure by varying the position of FVD in the building (S-1, S-2, and S-3), which are regular and irregular in plan, and by altering the width of the bay, by inserting the dampers at the four corners of the building in the first stage, by inserting the dampers at the four edges of the building in the second stage (S-2), and by inserting the dampers at the centre of the building in the third stage (S-3). Dampers are introduced from top to bottom storey of the building. Placement of FVD device system plays a key role against earthquake forces which are acting on the structure during the event of an earthquake. After the comparison in between above-defined stages and various parameters, when dampers are inserted at the edges of the building (S-3), the entire building is displaced identically with lesser values of displacement and inter-storey drifts, which are significant parameters in the event of an earthquake.
Key Words: Seismic Coefficient Method, Fluid Viscous Damper, Displacements, Inter-Storey Drift, Base Shear, limit state of collapse, limit state of serviceability.

1. Introduction

Earthquakes are natural hazards that are primarily caused by colliding tectonic plates and also for many reasons. Disruption or collapse of buildings and other man-made structures may happen due to such disasters. Past Experiences has shown that the establishment of earthquake-resistant guidelines and their enforcement is the essential protection against earthquake-induced damage for new constructions. As regards current systems, they need to be reviewed and strengthened based on pre-earthquake assessment requirements [1]. Damage of buildings due to earthquakes depends on many parameters, including ground motion strength, time period, frequency, geological and soil condition, construction quality, etc. The design of the building must ensure that the building has sufficient strength, high ductility and remains as one entity, even when it is subjected to very high deformation. Ground shaking which induces damage to the structure. Buildings on the ground surface experiences to the vibration to varying degrees as the earth vibrates. Therefore, in the design of earthquake-resistant structures, the impact of ground shaking on buildings is a key area of concern. According to the random nature of earthquake movements, seismic architecture loads are extremely hard to assess. Experiences from previous strong earthquakes have shown, however, that reasonable and careful practices during an earthquake can keep a building secure. Lateral load resisting systems play a key role in reducing or eliminating the damage of the structure due to the earthquake.[2]

1.1 Seismic Design: Seismic output describes the ability of a system to preserve its primary functions after a specific earthquake exposure, such as its stability and serviceability. If it does not endanger the lives and well-being of those in or around it by partially or completely collapsing, a structure is usually considered secure. A system may be called functional if it is capable of performing the operational functions for which it was planned. Basic concepts of the earthquake engineering, implemented in the major building codes, assume that a building should survive a rare, very severe earthquake by sustaining significant damage but without globally collapsing. On the other hand, it should remain operational for more frequent, but less severe seismic events. They are a lot of studies to determine or assess the behaviour of structure both theoretical and experimental studies. Experimental assessments are costly studies usually conducted by putting a (scaled) model of the structure on a shake-table that simulates the shaking of the earth and observes its behaviour. Seismic design is based on approved engineering techniques, principles and requirements for the design or retrofitting of earthquake-exposed structures [3]. These standards are only compatible with the current state of understanding of structures in earthquake engineering. Therefore, a building design that precisely meets the regulations of the seismic code does not assure protection against collapse or damage of the structure.

The cost of inadequate seismic design can be high. Nevertheless, whether based on physical laws or empirical knowledge of the structural performance of various shapes and materials, the seismic design has always been a trial and error process. Seismic design largely uses the same relatively small number of simple structural elements (to say nothing of vibration control devices) to construct complex structural systems as any non-seismic design project. Normally, buildings are built to resist the largest earthquake of a certain probability that is likely to occur at their site, according to construction codes. This suggests that the loss of life can be minimized by preventing the building from collapsing. Seismic design is carried out by understanding the possible failure modes of a structure and providing the structure with appropriate strength, stiffness, ductility, and configuration to ensure those modes cannot occur. The specifications for seismic design depend on the type of structure, the location of the project and the authorities that set down relevant seismic design codes and guidelines.

1.2 Damping: It is the process of dissipation of vibratory energy over time. Similar to sound absorption in air, damping takes place if there is some kind of friction that decreases motion and disperses energy. The damping capacity of each material is referred to as its loss factor, and this represents the ratio between the energy dissipated and the energy remaining during each step in the device. Damping is important in construction to limit vibrations and ensure protection and comfort in buildings and infrastructure. Damping is a way to minimize vibrations and is important for the device in which it works to be covered. This is what happens with door or drawer springs, in which, when opened/closed, damping prevents blows, maintains them and protects the mechanism. Structures take on a significant amount of earthquake-generated energy. The energy can cause significant deformations, even a structure fracture. A structure, either by permanent deformation or by dampers installed in the structure, must have the capacity to dissipate the energy. One dependent on viscous fluids, or hydraulic dampers, is the most commonly applied damper.

A frictional force created by rubbing one fluid layer against the other when they are in motion is viscosity or internal friction. The fluid is moved from one into the other chamber by a piston in such a system. Viscous dampers are devices that allow displacement due to changes in temperature, creep and shrinkage, without generating significant force, but dissipating large amounts of energy during the sudden occurrence of dynamic seismic energy entry [4]. The devices allow energy to be injected more uniformly into a system and prevent energy accumulation. Fluid viscous dampers have been commonly used by the various energy dissipation devices in the vibration regulation of different structural and mechanical systems. In the military and aerospace industries, these dampers have been commonly used for many years and have recently been modified for structural applications in civil engineering.

In a structure exposed to a transient, FVDs have the unusual ability to simultaneously minimize both deflection and stress. This is because FVD only varies its force with rate, which provides a response that is basically out of phase with stresses.

FVD consists of a piston rod in stainless steel with a bronze orifice head and a piston displacement accumulator that is self-contained. A compressible viscous fluid (silicone oil) fills the damper cylinder, which is usually non-toxic, non-flammable, thermally stable and environmentally safe. A standard fluid viscous longitudinal segment damper is shown in figure1.

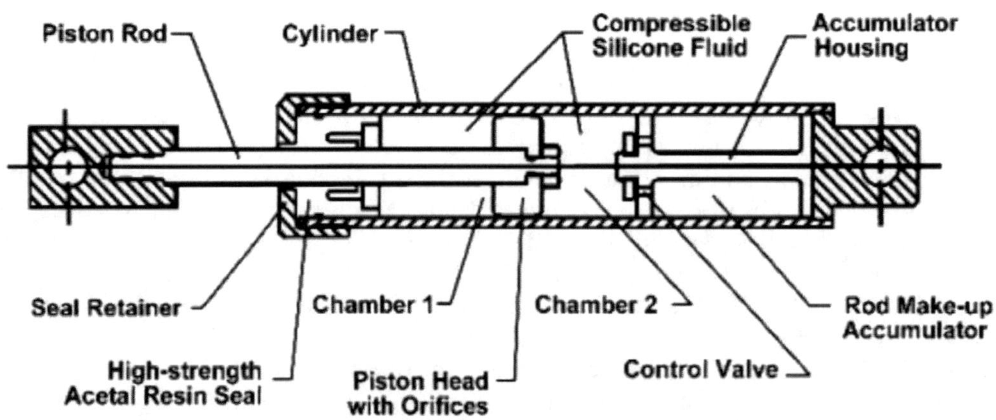

Figure 1. Skeleton of Fluid Viscous Damper
(Source: https://www.seminarstopics.com/seminar/6070/semi-active-control-system-seminar-ppt)

The typical shock absorber, such as those used in cars, is similar to these devices. FVDs consist of a piston made of stainless steel that passes through silicone oil-filled chambers. For very long periods, silicone oil is inert, non-toxic, non-flammable and stable. Due to the pressure difference between the two chambers, the silicone oil flows through an orifice in the piston head and the seismic energy is converted into heat which dissipates into the atmosphere. The piston rod with a piston will reciprocate movement in the cylinder when the FVD is subjected to external excitations to cause the damping medium to move back and forth between the two chambers separated by the piston. In this method, there was friction between the damping medium molecules, the medium and the shaft and piston, the damping medium and the cylinder, and the damping force created by the damping medium through the piston, all of which together constitute the damping force. FVD's function is to turn the mechanical energy generated by winds, earthquakes or other structural vibrations into the damping medium's inner energy. The dampers use the damping medium's rising temperature to temporarily store energy. By natural cooling, the heat is gradually absorbed. The fluid viscous dampers protect the structure from damage in this manner.

2. Literature Survey

Domenico et al. (2019) [5] opined that fluid viscous dampers (FVDs) are well-established supplementary energy dissipation devices commonly used to protect structures from earthquakes. In the last four decades, optimum design, positioning and sizing of FVDs have been thoroughly investigated. A summary of the most popular methodologies from the abundant literature in the field is provided in this review article. Key aspects and key features of the various methods to evaluate the optimal damping coefficients and the optimal location of FVDs are analyzed comparatively. By minimizing certain output parameters that are representative indicators of the system response, the optimal design problem is often solved by a numerical approach to a constrained optimization problem. These methods are based either on the desired (target) restriction of the damping ratio or on a fixed total cost, where the number of the damping coefficients of the added FVDs is roughly related. To interpret the seismic output in terms of the amount of energy dissipated by the FVDs, from the total input energy from the earthquake excitation, some energy-based perspectives are also given in this review paper. Another important factor concerns the cost estimation of the dampers, which has been loosely related to two simpler measurements in this paper, such as the sum of the damping coefficients and the sum of the damper strengths. In this respect, it would be beneficial to provide more effective cost functions that account not only for damper forces, but also for construction costs, prototype testing costs, or life-cycle costs, including initial, repair and failure costs. A challenging line of future research is the re-examination of the studied design methods under these more practical modelling assumptions. Michele and Stefano (2020) [6] investigated the dynamic behaviour of a multi-storey frame building, which the structure to be regulated, linked by horizontal fluid viscous dampers to an adjacent support structure. There are two key effects of the dampers attachment system. The relative contribution of the two effects is highly dependent on the coupling system's key properties. First, the minimal coupled dynamic system composed of two Single-Degree of Freedom systems connected by a viscous damper is analyzed to investigate this complex behaviour. Comparisons are made between one-storey and uniform multi-storey structures. The findings can be useful for the preliminary design and evaluation of the effectiveness of the dampers link system, based on the fundamental dynamic properties of the connected buildings. The patterns of peak roof displacement damping reduction factors are identical to those obtained for one-storey paired systems with the same main parameters. In the preliminary design process, the analytical formula proposed to determine the upper and lower bound values of the minimum values of the peak roof displacement damping reduction factor may be useful for a rapid assessment of the possible energy dissipation capabilities offered by the additional dampers connecting the two adjacent buildings. Sai et al. (2017) [7] opined that an earthquake affects the structure directly by increasing the amount of energy inside the structural framework. By implementing structural control systems, a large portion of this energy can be dissipated. In the earthquake resistant design of buildings, many structural control systems, such as passive, active and semi-active control systems, are now gaining importance. A type of passive energy dissipation system operating on the principle of fluid flow through orifices is a fluid viscous damper (FVD). For several years, this type of damper has seen many applications in the military and aerospace industries. Recently, high-capacity fluid viscous dampers have found commercial applications in the field of civil engineering on buildings and bridges subjected to seismic and or wind storm inputs. When the optimum damping coefficient and damping exponent values of the fluid viscous dampers have been obtained, the response quantities for the bare frame and the damped frame are obtained and the results are compared. Maximum storey displacements, maximum storey drifts, maximum storey shears, maximum axial loads in the columns due to earthquakes and maximum moments in the columns due to earthquakes in each storey stage are the response quantities considered in this analysis. At each storey step, the reduction in both of these responses is discovered and the percentage reduction is also calculated for both zone IV and zone V. Because the frame is symmetrical in both directions, both directions also have the same response amounts. Sinha and Sharad (2017) [8] explained about the need for structural response control to gain speed in application around the world has been needed by regular earthquakes around the globe and large numbers of structures susceptible to it. This paper explores the use and efficacy of one such unit, viscous fluid dampers, for structure response control and to reduce structural system damping demand. A non-linear time history study was carried out in this paper using a 3-directional synthetic accelerogram on a 3D model of a 12-storey RCC MRF building. Using ETABS, two separate cases of building models with and without supplementary damping were analyzed. The storey responses have been compared in terms of absolute maximum displacement and storey drift. The results of the non-linear modal time-history analysis carried out on a 12-storey RC frame structure with and without FVD, described using storey responses and time history plots for different parameters, show that the structure's storey response was significantly reduced by dampers. The displacement values are within the required restricted range, proving the efficacy of the dampers in reducing the structure's displacement response. For storeys 1, 5, 9 and 12, the time history plots of base shear and storey shears have a similar pattern to the time history plot of roof acceleration. In reducing shear forces in the structure, the dampers have been successful. Although the responses have been substantially reduced by the FVD's, the damping demand of the structure can be further reduced by optimal selection and installation of FVDs at different critical locations. Rakesh Goel (2005) [9] experimented on the seismic response of one-storey, one-way asymmetric linear and non-linear systems with non-linear viscous dampers is concerned. For a group of 20 ground movements produced for the SAC studies and the median values tested, the seismic responses are computed. The behaviour of single-degree of freedom systems for harmonic and earthquake loading has been checked. The presented harmonic loading outcomes are used to describe a few unusual patterns for earthquake loadings, such as decreasing deformation and rising damper force of short-period systems with increasing damper non-linearity. Subsequently, the seismic responses are contrasted with those having identical linear dampers for linear and non-linear asymmetric-plan structures with non-linear dampers. The action of SDOF systems with non-linear dampers has been checked. In particular, when subjected to earthquake loading, the device deformation decreases and damper force rises for short-period linear systems due to damper non-linearity. These trends cannot be clarified based on commonly used harmonic loading with a forcing frequency equal to the system's normal vibration frequency, suggesting the deformation and decrease of the force of the damper. The trends in earthquake loading have been clarified in this investigation based on an alternative method in which the response of systems over a range of vibration cycles (or frequencies) is computed for harmonic frequency loading. Furthermore, for non-linear systems, the rise in damper force due to damper non-linearity observed for short-period linear systems does not occur.

3. Objective

The main objectives of the work are to increase the seismic resistance of the structure by introducing the damping system into the building and thereby finding out the critical required location of the damping system by making comparisons in between with and without damping models which reports low inter storey drift and displacement values.

4. Scope of the work

By inserting dampers at three different locations called S-1, S-2, and S-3, different bay systems (change in the plinth region of the structure) and separate buildings were modelled for both regular and irregular in plan. Model S-1 indicates that the dampers are located at four corners of the buildings (figures 4 and 5), model S-2 indicates that the dampers are located at the edges of the buildings (figures 6 and 7), while model S-3 indicates that the dampers are located at in the center of the building (figures 8 and 9). Dampers are inserted from top storey to bottom storey for all models. These models are tested using the seismic coefficient method for the limit state of collapse and limit state of serviceability load combinations for seismic zone-5. The following plans will demonstrate the modelled buildings with and without a damping system.

4.1 Scope of the work is defined in the Following Table

Table 1. Models with each Case under the Limit State of Collapse

Limit State of Collapse (seismic zone-V)					
Stage-1		**Stage-2**		**Stage-3**	
Regular model	**Irregular model**	**Regular model**	**Irregular model**	**Regular model**	**Irregular model**
10×10	12×6	10×10	12×6	10×10	12×6

Table 2. Models with each Case under Limit State of serviceability

Limit State of Serviceability (seismic zone-V)					
Stage-1		**Stage -2**		**Stage -3**	
Regular Model	**Irregular Model**	**Regular Model**	**Irregular Model**	**Regular Model**	**Irregular Model**
10×10	12×6	10×10	12×6	10×10	12×6

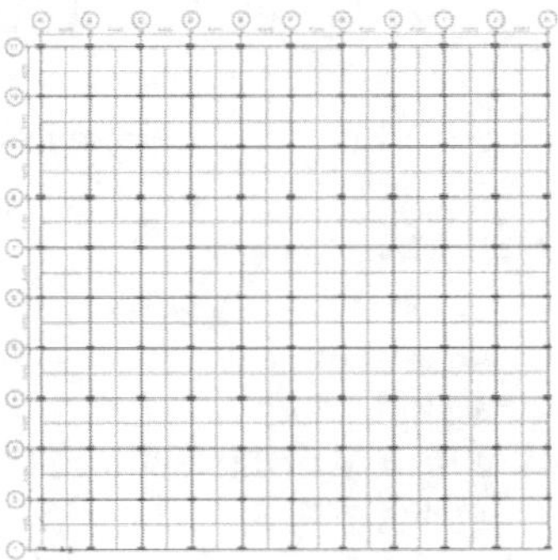

Figure 2. 10×10 Symmetric Bay System (without damping system)

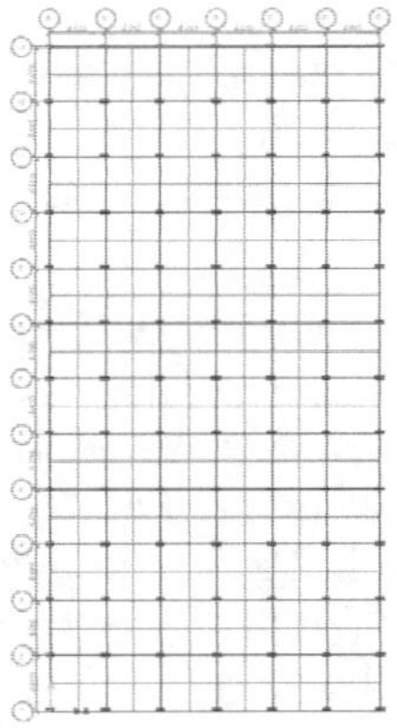

Figure 3. 12×6 Un-Symmetric Bay System (without damping system)

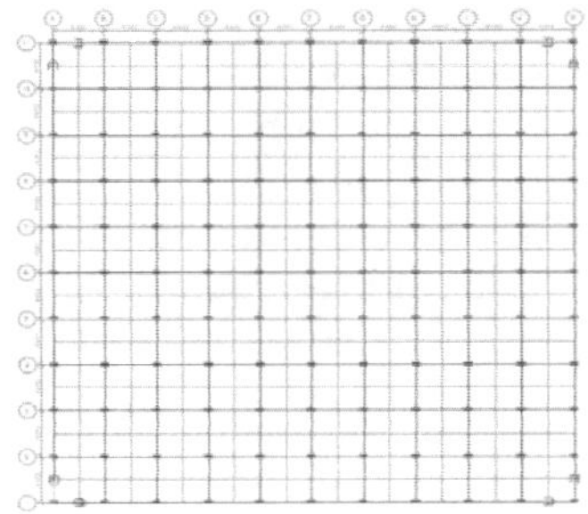

Figure 4. 10×10 Bay System (S-1)

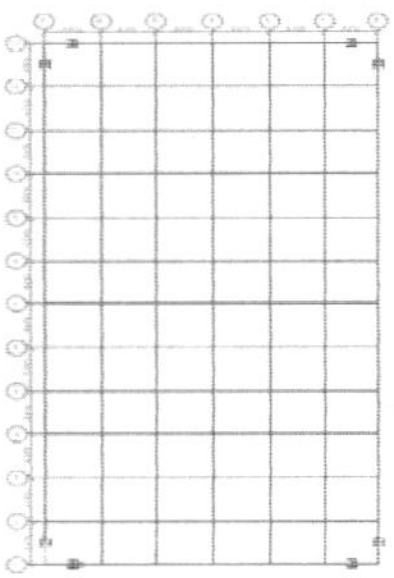

Figure 5. 12×6 Bay System (S-1)

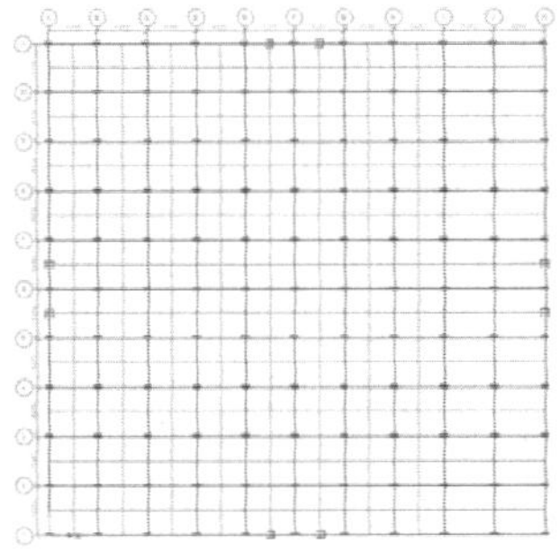

Figure 6. 10×10 Bay System (S-2 FVD at edges of the building)

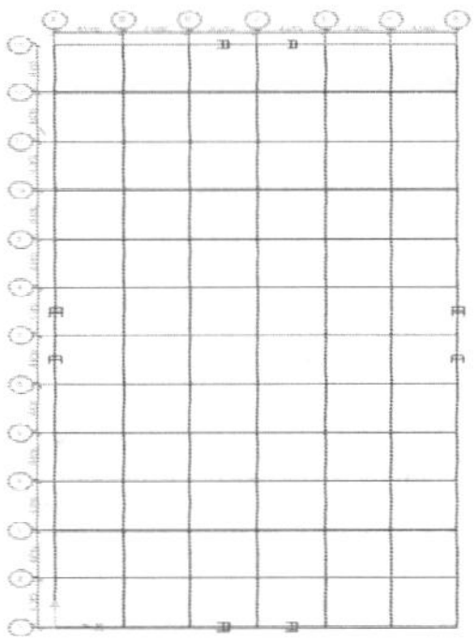

Figure 7. 12×6 Bay System (S-2)

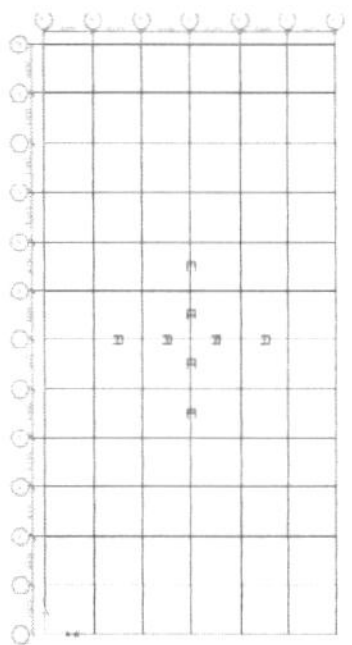

Figure 8. 10×10 Bay System (S-3)

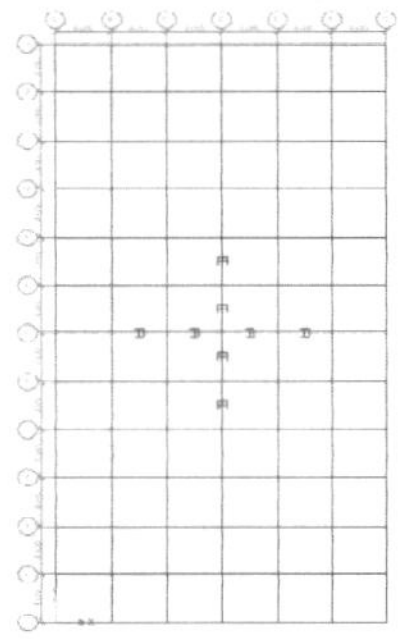

Figure 9. 12×6 Bay System (S-3)

5. Structure Modeling of G+10 RC Building in ETABS 2015

It is a G+10 RCC residential-type structure. The objective of the work is to determine the effects of building responses say storey displacements, inter-storey drifts and base shear due to insertion of FVD. The models were analyzed in many stages through the consideration of models as stages. Initially, the building was studied as a bare model, i.e. without a damping system, and dampers are then assembled at three separate positions in a building. In the first stage, the dampers were placed at the four corners of the model. In this case, the model mentioned was depicted as S-1. In the second stage, dampers were installed at the edges of the building (S-2). In stage-3, dampers were introduced at the centre of the building. Both regular and irregular plans were adopted for modelling the structure along with regular and irregular bay systems, say 10×10, 12×6. The seismic coefficient method according to IS 1893 (Part 1) 2002 [12] was evaluated for seismic zone-5 by the limit state of collapse and limit state of serviceability in ETABS-2015. Different properties of the elements and models are described in the following tables.

5.1.1 Material Property

Table 3. Material Properties of Model

Density of RCC	25 kN/ m^3
Density of steel	7850 kg/m^3
Yield strength of main reinforcement	415 N/mm^2
Yield strength of secondary reinforcement	415 N/mm^2
Compressive strength of concrete	30 N/mm^2
Grade of concrete	M30
Steel grade	Fe 415

5.1.2 Section Property

Table 4. Section Properties of Structural Elements

Beam size	230×450
Column size	300×600
Slab thickness	125mm

5.1.3 Description of Model

Table 5. Detailed Models Description

Number of stories	G+10
Storey height	3m
Building height	33
Bay width in x- direction	4m
Bay width in y-direction	4m
Regular bay system	10×10
Irregular bay system	12×6
Type of building	SMRF and OMRF (residential)
Fluid viscous damper	FVD 250
Placement of dampers	stage-1 (corners) stage-2 (edges) stage-3 (center)

5.1.4 Loading on Structure

Table 6. Load Intensity on Structure

Live load	3kN/m^2
Self-weight of slab	5.75 kN/m^2
Load on the external wall (9")	12.42 kN/m
Load on the internal wall (4.5")	6.21 kN/m
Live load on the terrace	1.5 kN/m^2
Parapet wall load	3kN/m

6. Load Combinations

The Loads are calculated according to IS 875 (Part 1 [13], and Part 2 [14]) for dead and live loads respectively. Loads are calculated in two ways

1. Limit state of collapse
2. Limit state of serviceability

Limit state of collapse	Limit state of serviceability	
1.5(DL+LL)	1.0(DL+LL)	0.9DL+1.5EQX
1.2(DL+LL+EQX)	1.0 DL+1.0 EQX	0.9DL+1.5EQZ
1.2(DL+LL-EQX)	1.0 DL-1.0 EQX	0.9DL-1.5EQZ
1.2(DL+LL+EQZ)	1.0DL+1.0EQZ	0.9DL-1.5EQX
1.2(DL+LL-EQZ)	1.0DL-1.0EQZ	
1.5(DL+EQX)	1.0DL+0.8LL+0.8EQX	
1.5(DL-EQX)	1.0DL+0.8LL-0.8EQX	
1.5(DL+EQZ)	1.0DL+0.8LL+0.8EQZ	
1.5(DL-EQZ)	1.0DL+0.8LL-0.8EQZ	

7. Results and Discussions

7.0 Limit State of Serviceability

7.1 Maximum Storey Displacement for Symmetric Frame (10x10) Located in Zone-5 with and without Damping System (UX)

Table 7. Storey Displacement of all Cases for Symmetric Frames in X-Direction

Frame type	Bay system 10×10
Bare model	102.36
S-1	78.63
S-2	63.74
S-3	81.62

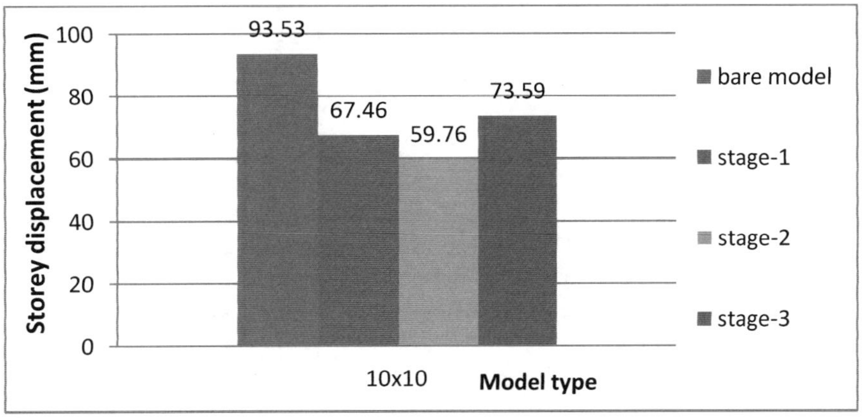

Figure 10. Maximum Storey Displacement of all cases for Symmetric Frames in X-direction

Decrease of maximum floor displacement in X-direction has been observed by 23.18 %, 37.72%, and 20.26% due to the insertion of the damping system at the edges (S-1), middle (S-2) and corner (S-3) for a building that has 10x10 bay systems compared to the bare model.

7.2 Maximum Storey Displacement for Symmetric Frame (10×10) Located in Zone-5 with and without Damping System (UY)

Table 8. Storey Displacement of all Cases for Symmetric Frames in Y-Direction

Frame type	Bay system 10×10
Bare model	**103.62**
S-1	84.53
S-2	71.26
S-3	93.24

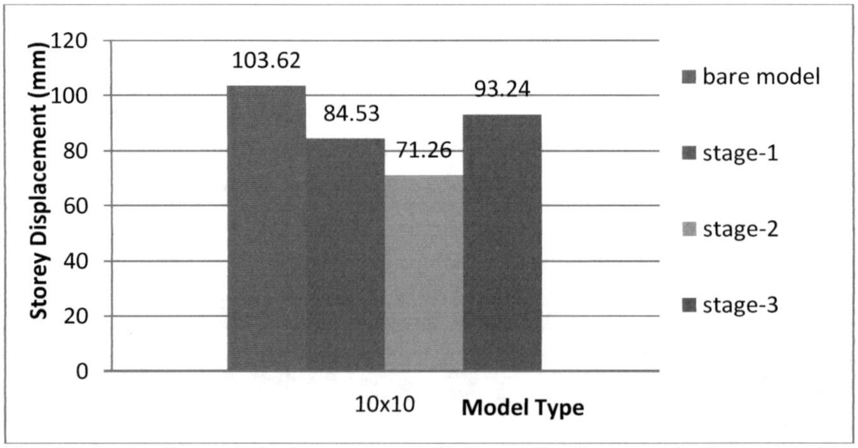

Figure 11. Maximum Storey Displacement of all Cases for Symmetric Frames in Y-direction

Decrease of maximum floor displacement in Y-direction has been observed by 18.42%, 30.88%, and 10.01% due to the insertion of damping system at edges (S-1), centre (S-2) and corner (S-3) for a building that has 10×10 bay systems as compared to the bare model.

7.3 Inter Storey Drift of all cases for 10x10 Bay System Located in Zone-5

Table 9. Inter-Storey Drifts of 10×10 Bay Systems for each case

Storey level	Bare model	S-1	S-2	S-3
Storey-13	2.94	5.14	4.41	2.14
Storey-12	3.41	5.86	4.84	3.54
Storey-11	4.96	5.47	5.19	4.19
Storey-10	5.51	6.45	5.45	5.35
Storey-9	6.72	6.72	5.92	5.74
Storey-8	7.14	6.89	5.86	6.76
Storey-7	7.54	6.74	5.64	7.14
Storey-6	8.36	6.41	5.21	7.87
Storey-5	8.51	5.14	4.74	7.96
Storey-4	8.72	4.86	4.87	7.21
Storey-3	8.45	3.47	3.64	6.41
Storey-2	7.34	2.12	2.98	6.16
Storey-1	3.14	0.00	1.14	3.41
Base	0.00	0.00	0.00	0.00

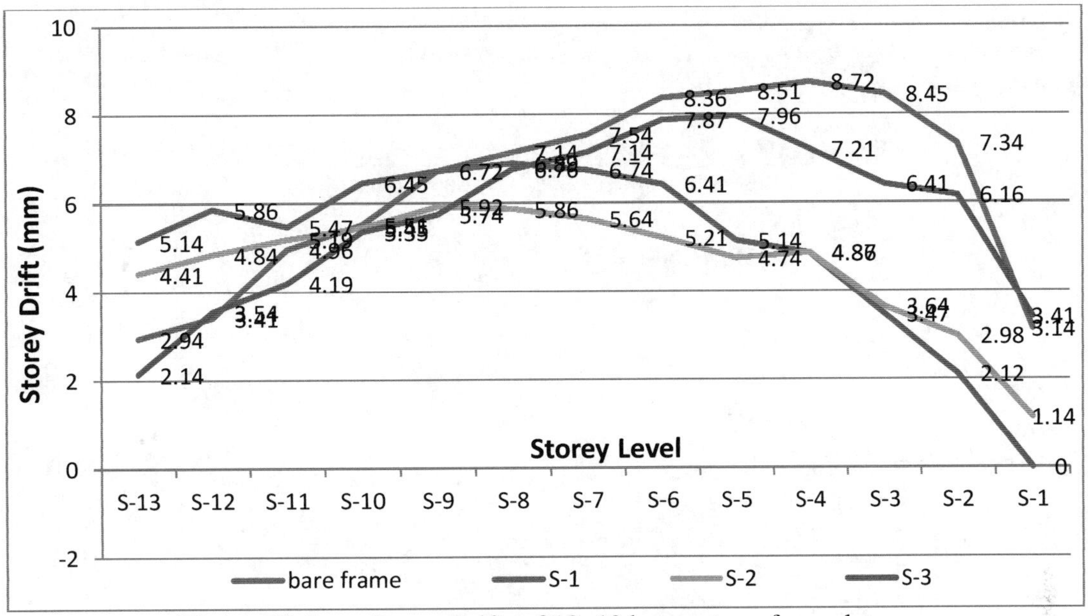

Figure 12. Inter-Storey Drifts of 10×10 bay systems for each case

Decrease of maximum inter-storey drift has been observed in between second storey and the first storey by 71.11%, 59.40%, 16.07%, due to the insertion of the damping system at the edges (S-1), middle (S-2) and corner (S-3) for a building that has 10x10 bay systems when compared to the bare model.

7.4.0 Limit state of Collapse

7.4.1 Maximum Base Shear for the symmetric frame (10x10) located in zone-5 with and without damping system (kN)

Table 10. Maximum Base Shear of Each Case for Symmetric Frames

Frame type	**Bay system 10×10**
Bare model	1462
S-1	3142
S-2	3964
S-3	3587

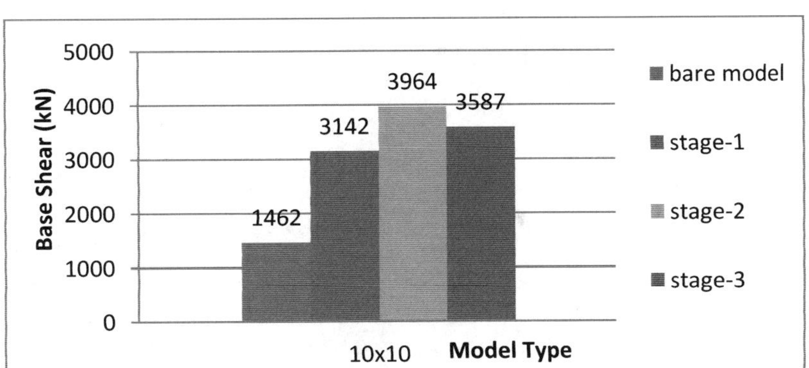

Figure 13. Base Shear of Symmetric Models for each case

Increase of base shear has been observed by 114.91% 171.13%, 145.34%, due to the insertion of the damping system at the edges (S-1), middle (S-2) and corner (S-3) for a building that has 10×10 bay systems in comparison with the bare model.

7.5. The Limit state of serviceability

7.5.1 Maximum Storey displacement for the un-symmetric frame (12x6) located in zone-5 with and without damping system (UX)

Table 11. Maximum Storey Displacement of each case in X-Direction

Frame type	Bay system 12×6
Bare model	102.32
S-1	76.56
S-2	68.68
S-3	79.32

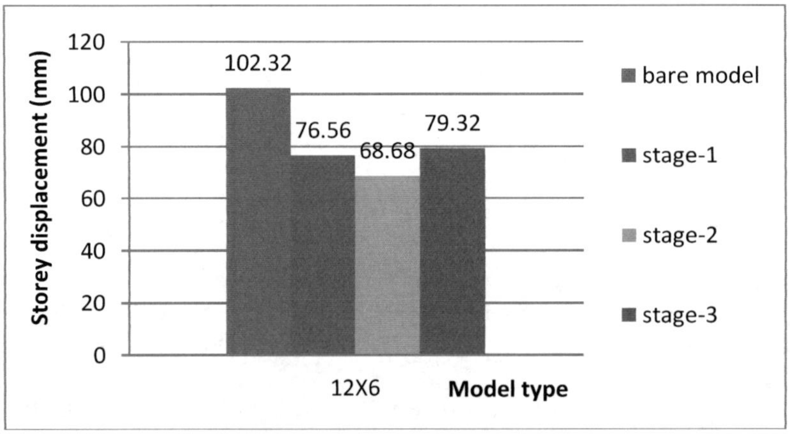

Figure 14. Maximum Storey Displacement of each case in X-Direction

Decrease of maximum floor displacement in X-direction has been observed by 25.17%, 32.87%, and 22.47% due to the insertion of the damping system at the edges (S-1), middle (S-2) and corner (S-3) for a building that has 12×6 bay systems as opposed to the bare model.

7.6.0 Maximum Storey displacement for the un-symmetric frame (12x6) located in zone-5 with and without damping system (UY)

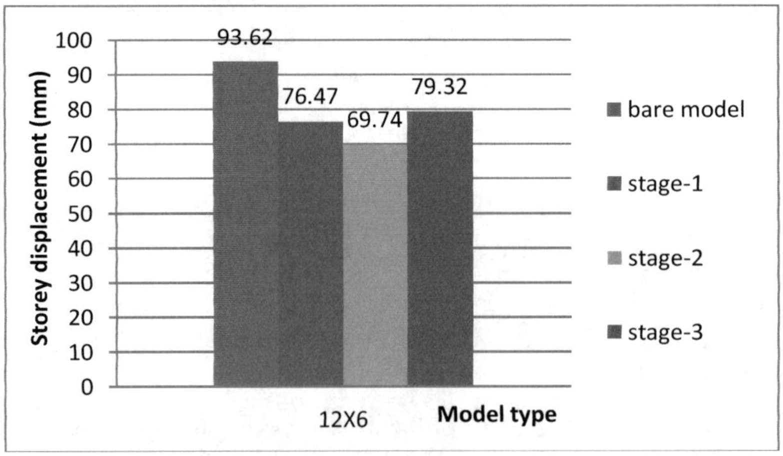

Figure 15. Maximum Storey Displacement of each case in Y-Direction

Table 12. Maximum Storey Displacement of each case in Y-Direction

Frame type	12×6
Bare model	93.62
S-1	76.47
S-2	69.74
S-3	79.32

Decrease of maximum floor displacement in Y-direction has been observed by 18.31%, 25.50%, 15.27%, due to the insertion of the damping system at the edges (S-1), middle (S-2) and corner (S-3) for a building that has 12×6 bay systems when compared to the bare model.

7.7 Inter Storey Drift of 12x6 Bay Systems for Each Case Located in Zone-5

Table 13. Inter Storey Drift of 12×6 Bay Systems for each case

Storey level	Bare model	S-1	S-2	S-3
Storey-13	2.92	6.08	4.53	9.93
Storey-12	3.98	6.36	4.74	9.91
Storey-11	5.24	6.48	5.35	6.54
Storey-10	6.46	6.83	5.83	6.96
Storey-9	7.39	6.92	6.34	6.91
Storey-8	6.58	6.98	6.59	7.53
Storey-7	7.86	6.84	6.87	7.61
Storey-6	8.99	6.63	6.91	7.34
Storey-5	9.17	5.38	6.51	6.90
Storey-4	10.69	4.92	5.96	6.35
Storey-3	8.93	3.87	4.92	5.63
Storey-2	7.87	1.92	3.63	4.51
Storey-1	3.98	0.00	1.54	1.98
Base	0.00	0.00	0.00	0.00

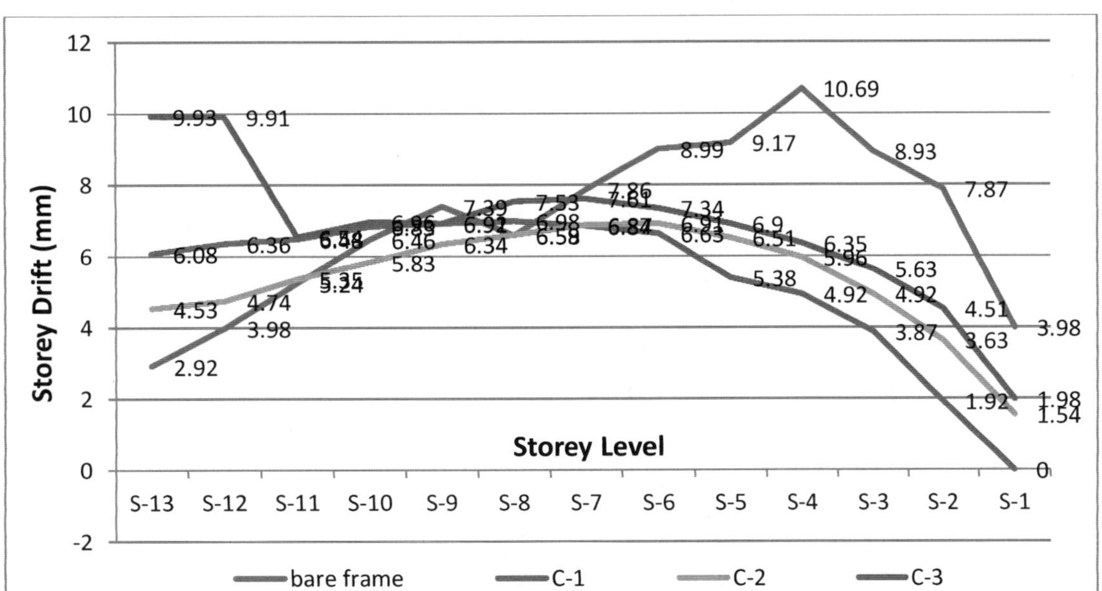

Figure 16. Inter Storey Drift of 12×6 Bay Systems for each case

Decrease of maximum inter-storey drift has been observed in between second storey and first storey by75.60%, 53.87%, 42.69 %, due to the insertion of the damping system at the edges (S-1), middle (S-2) and corner (S-3) for a building that has 12×6 bay systems in comparison to the bare model.

7.8 Limit state of Collapse

7.8.1Maximum Base Shear for the un-symmetric frame (12×6) located in zone-5 with and without damping system (kN)

Table 14. Maximum Base Shear of each case for un-symmetric frame

Frame type	Bay system 12×6
Bare model	703
S-1	1840
S-2	3621
S-3	963

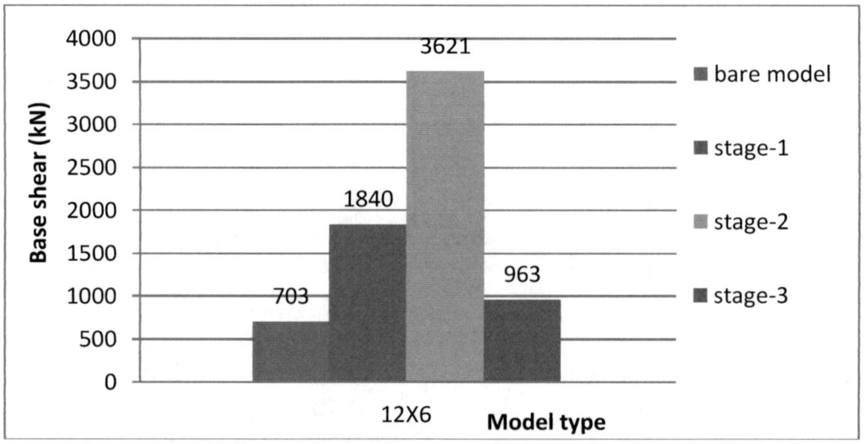

Figure 17. Maximum Base Shear of each case for un-symmetric frame

Increase of base shear has been observed by 161.73% 415.07%, 36.98%, due to the insertion of the damping system at the edges (S-1), middle (S-2) and corner (S-3) for a building that has 12×6 bay systems in comparison with the bare model.

8. Conclusions

1. Decrease of maximum floor displacement in X-direction has been observed by 23.18 %, 37.72%, and 20.26% due to the insertion of the damping system at the edges (S-1), middle (S-2) and corner (S-3) for a building that has 10x10 bay systems compared to the bare model.
2. Decrease of maximum floor displacement in Y-direction has been observed by 18.42%, 30.88%, and 10.01% due to the insertion of damping system at edges (S-1), centre (S-2) and corner (S-3) for a building that has 10x10 bay systems as compared to the bare model
3. Decrease of maximum inter-storey drift has been observed in between second storey and the first storey by 71.11%, 59.40%, 16.07%, due to the insertion of the damping system at the edges (S-1), middle (S-2) and corner (S-3) for a building that has 10x10 bay systems when compared to the bare model.
4. Increase of base shear has been observed by 114.91% 171.13%, 145.34%, due to the insertion of the damping system at the edges (S-1), middle (S-2) and corner (S-3) for a building that has 10x10 bay systems in comparison with the bare model.
5. Decrease of maximum floor displacement in X-direction has been observed by 25.17%, 32.87%, and 22.47% due to the insertion of the damping system at the edges (S-1), middle (S-2) and corner (S-3) for a building that has 12x6 bay systems as opposed to the bare model.

6. Decrease of maximum floor displacement in Y-direction has been observed by 18.31%, 25.50%, 15.27%, due to the insertion of the damping system at the edges (S-1), middle (S-2) and corner (S-3) for a building that has 12x6 bay systems when compared to the bare model.
7. Decrease of maximum inter-storey drift has been observed in between second storey and first storey by 75.60%, 53.87%, 42.69%, due to the insertion of the damping system at the edges (S-1), middle (S-2) and corner (S-3) for a building that has 12x6 bay systems in comparison to the bare model.
8. Increase of base shear has been observed by 161.73% 415.07%, 36.98%, due to the insertion of the damping system at the edges (S-1), middle (S-2) and corner (S-3) for a building that has 12x6 bay systems in comparison with the bare model.

8.1 Summary of conclusions

Placement of FVD device system plays a key role against earthquake forces which are acting on the structure during the event of an earthquake. When comparisons were made from the above three stages (varying the placement of FVD). After the comparison in between above-defined stages and various parameters, when dampers are inserted at the edges of the building (S-3), the entire building is displaced identically with lesser values of displacement and inter-storey drifts, which are significant parameters in the event of an earthquake.

References

[1] Devang B. Lad, Ashish K Sanghavi, Komal M Panchal, Prashant R Barbude 2017 Analysis of G+35 Storey RCC Building using Viscoelastic Damper on ETABS *International Journal of Engineering Technology Science and Research* **4 (5)** p 677-87
[2] M Rak, D. Mestrovic, A Nizic 2005 Damping and its importance to structure *WIT Transactions on The Built Environment* **81** p 1-7
[3] A Ras, N Boumechra 2016 Seismic energy dissipation study of linear fluid viscous dampers in steel Structure design *Alexandria Engineering Journal* **55 (3)** p 2821-32
[4] Wen-Hsiung Lin, Anil Chopra 2002 Earthquake response of elastic SDF systems with non-linear fluid viscous dampers *Earthquake Engineering and Structural Dynamics* **6 (7)** p 1623–42
[5] Domenico Ricciardi, Takewaki 2019 Design strategies of viscous dampers for seismic protection of building structures: A Review *Soil Dynamics and Earthquake Engineering* 118 **5 (4)** p 144–65
[6] Michele Palermo, Stefano Silvestri 2020 Damping reduction factors for adjacent buildings Connected by fluid-viscous dampers *Soil Dynamics and Earthquake Engineering* **138** p 1-12
[7] Sai Chethan K, Srinivas K, Ranjitha K P 2017 Seismic Performance Evaluation of Fluid Viscous Dampers *International Journal of Research in Engineering and Technology* **6 (6)** p 27-32
[8] A K Sinha, Sharad Singh 2017 Structural Response Control Of RCC Moment Resisting Frame Using Fluid Viscous Dampers *International Journal of Civil Engineering and Technology* (IJCIET) **8 1** p 900–10
[9] Rakesh K Goel 2005 Seismic response of linear & non-linear asymmetric systems with non-linear fluid viscous dampers *Earthquake Engineering and Structural Dynamics* **34 (7)** p 825-46
[10] Alireza Heysami 2015 Types of Dampers and their Seismic Performance During an Earthquake *Current World Environment* 10 (5) p 1002-15
[11] Robert J, Mc Namara Douglas, P Taylor 2003 Fluid Viscous Dampers For High-Rise Buildings *The Structural Design of Tall and Special Buildings* **12** p 145–54
[12] IS 1893 (Part-1) 2002 Indian Standard Criteria for Earthquake Resistant Design of Structures Part 1 General Provisions and Buildings *Bureau of Indian Standards* New Delhi
[13] IS 875 (Part 1) 1987 Indian Standard Code of Practice for Design Loads (Other than Earthquake) for Buildings and Structures Part 1 Dead Loads - Unit Weights of Building Materials and Stored materials *Bureau of Indian Standards* New Delhi
[14] IS 875 (Part 2) 1987 Indian Standard Code of Practice for Design Loads (Other than Earthquake) for Buildings and Structures Part 2 Imposed Loads *Bureau of Indian Standards* New Delhi

Defluorination of groundwater by low-cost adsorbents

K S B Prasad[1], P Jakeer Hussain[2], P Bharath Kumar[3]

[1]Assistant Professor, Department of Civil Engineering, GMR Institute of Technology, Rajam, Andhra Pradesh, India
[2]Graduate student, Department of Civil Engineering, GMR Institute of Technology, Rajam, Andhra Pradesh, India
[3]Graduate student, Department of Civil Engineering, GMR Institute of Technology, Rajam, Andhra Pradesh, India
E-mail: prasad.ksb@gmrit.edu.in

Abstract: India is among nation higher in percentage suffer s from fluorosis through consumption of groundwater. Fluoride content in natural water has become a major issue in major parts of the world and should be in 0.6 to 1.5 mg/l. Investigations in the removal of fluoride content have been carried in various techniques. This paper speaks on treating fluoride content and to utilize it as drinking water at low-cost techniques by natural adsorbents for a sustainable environment. In this process adsorption time effects, concentration of fluoride at varied adsorbent dosage with maintained temperature are studied. Adsorbents such as dry ginger powder, spinach, turmeric powder, orange peel powder were used.
Keywords: Defluorination, Adsorption, Fluorosis, Nanofiltration, coagulation

1. Introduction

Fluoride content in natural water has become a major issue in major parts of the world and should be in 0.6 to 1.5 mg/l [2] [18] [19] [20]. India is among nations high in percentage suffering from fluorosis through consumption of groundwater [1]. In the earth's crust, the most common element is fluorine where it occurs in amounts comparable to carbon and nitrogen [4] [13] [14] [15]. Fluorine at desirable amount will prevent dental issues [8] [16] [17]. Excess intake leads to dental, skeletal fluorosis and endemic diseases such as the risk of bone fractures, Immune deficiency, and cancer [5] [8].

Comparatively, rainwater will be having more purity, suits better for domestic and drinking purpose. Since it is expensive, it is difficult to store and maintain in large quantity through tanks, dams, canals, and rivers on earth surface [2] [18] [19] [20]. In Andhra Pradesh, Srikakulam and Prakasam districts are highly affected for high fluorine content in water. Due to water scarcity, people depend on groundwater in most of the villages of these districts where fluoride ranges between 0.4 to 20 ppm [2]. Various attempts showed little success in reducing fluorine in groundwater with alum precipitation technique [6].

Treatment of water and wastewater with fluoride ions content requires suitable, effective and economical methods. Methods such as membrane filtration, precipitation, Nanofiltration, ion exchange, electrocoagulation flotation and adsorption has been done [7] [10] [11] [12], where adsorption method is shown effectively at low concentration with less in cost. In such a way synthetic, biomass materials like fly ash, red mud, chitosan beads, and zeolite have been adopted. Using natural adsorbents with effective technique has been a new scope in reducing fluoride levels.

It is analysed that more than ten million people in India, over 16 states are affected with fluorosis, caused by continuous ingestion of water with the excess fluorine content, ranging from 1.5mg/l to 16 mg/l is caused mainly by deposition of fluoride in bones and cartilage including metabolic disorders, dental caries, and stiffened bones [3]. According to the University of Glasgow, a scientist found that there was a drastic decrease in white blood cells in people affected with fluorosis. It also becomes responsible for the malfunctioning of liver-kidney, respiratory system, digestive system. All these are the major effects of non-skeletal fluorosis [9].

In the last 5 years, 117 lakhs People were affected through Fluorosis in 19 states in India. In that the top 10 states in India who affected mostly are in Rajasthan around 40lakhs, Telangana around 20lakhs, Karnataka around 13lakhs, Andhra Pradesh around 10 lakhs, Maharashtra around 6lakhs, Bihar around 4 lakhs, Madhya Pradesh around 3lakhs, Kerala around 2lakhs, Uttar Pradesh around 1lakh and finally in Gujarat 50thousand [4][13][14][15].

Table 1. Effects of concentration of fluorine

CF (mg/l)	Effects
<1.0	Reduces dental caries
1.0 - 1.5	A marginal risk of fluorosis
1.5 - 3.0	Risk of dental fluorosis
3.0 - 10.0	Skeletal fluorosis with an adverse change in bone structure
>10.0	Crippling and skeletal fluorosis

2. Materials and Methods

Defluorination is a method which we used to remove the fluoride content by using low-cost natural adsorbents. There are many methods but this method is economically less, more efficient, naturally available, easy to operate and eco-friendly to nature.

Table 2. Adsorbents and sampling sites

Adsorbent	Sampling site
Dry ginger powder	Fields
Turmeric powder	Fields
Spinach powder	Fields
Orange peel powder	Fruit market

Equipment's and chemicals used

- Conical flasks of 250ml.
- NAF of 0.225grams
- Distilled water.
- Tisab solution of 50ml.
- Spectrometer which is used to detect the fluoride ions.
- Orbital shaker.
- Funnel.
- Filter paper.
- Volumetric flasks of 250ml and 50ml.
- Weighing and measuring jar
- Tissue paper

Adsorbents required (10 grams each and 90micro size)

- Ginger powder.
- Turmeric powder.
- Orange peel powder.
- Spinach powder

IOP Conf. Series: Materials Science and Engineering 1025 (2021) 012033 doi:10.1088/1757-899X/1025/1/012033

2.1 Preparation of stock solution

Four adsorbent samples of 100ml each weight of 0.5gram,1gram.1.5gram,2grams were added with 10ppm stock solution to measure the fluoride content and placed in an orbital shaker at room temperature for 24hours.

2.2 Procedure

In conical flasks, four varied samples of different weights were taken

Four adsorbent samples (dry ginger powder, turmeric powder, spinach powder, orange peel powder) of 100ml each weight of 0.5gram,1gram.1.5gram,2grams were added with 10ppm stock solutions (NAF solution) to measure the fluoride content and placed in an orbital shaker at room temperature for 24 hours and then adsorbents were filtered by using filter paper. 5ml of tisab solution (which separates the fluoride ions from NAF) is added to above 100ml adsorbent samples along with the mother sample (original sample). Now the samples were tested through spectrometer which is used to tell the fluoride content. Readings are noted in ppm.

Prepared adsorbent samples were kept in an orbital shaker for 24 hours to maintain room temperature as shown in figure 1. After removing from the orbital shaker as in figure 2, samples were made ready to test for fluoride range on fluoride electrode shown in figure 3.

Figure 1. Orbital shaker for temperature **Figure 2.** Samples for fluoride test

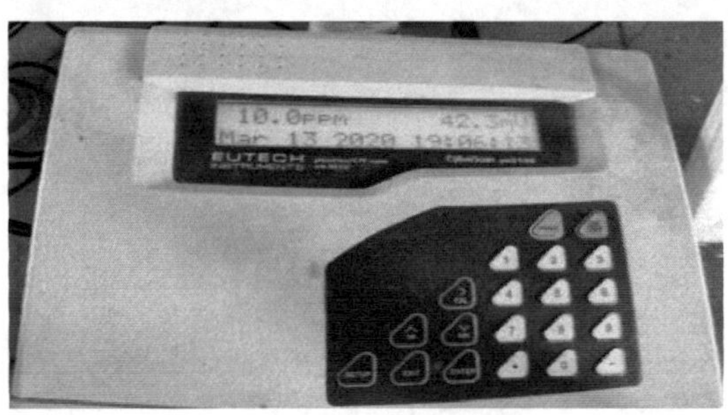

Figure 3. Fluoride electrode

2.3 Properties of natural adsorbents

Ginger belongs to the family zingiberaceae which is used in form of a paste or dry powder ginger has been used for cooking purposes due to its oil components and also for the treatment of various diseases like flu, stomach-ache and headache and also reduce the fluoride percentage in water [7].

The turmeric also belongs to Zingiberaceae and it is a ground russet. Due to its active components, turmeric can take as a household remedy for various diseases which is known as an antibiotic. So it can be used to treat fluoride [4] [13] [14] [15].

Spinach is taken as advantageous sources of various carotenoids and lipophilic active compounds. Taking spinach can have useful effects on cancer. Spinach leaves which contain a powerful water-soluble natural antioxidant mixture and reduce fluorine content [9].

Orange peel contains a wide of nutrients as well as phytonutrients which are extremely powerful to healing several skin problems. It can treat fluoride level of water to some extent [2] [18] [19] [20].

3. Results and discussion

3.1. Effect of Adsorbent

Quantity of adsorbent attempted is from 0.5 to 2 grams with a constant initial fluoride concentration of 10 mg/l and at time of agitation for 24 hours at room temperature. Crushed dry ginger powder after dried into the sunlight for 3-4 days it is sieved to 80micron to prepare as adsorbent. It is observed that as the dosage is increased then the removal of fluoride also increases which shown in table 3.

Table 3. Removal of fluoride with dry ginger powder

Sl. No.	Weight dosage of adsorbent in gm	Concentration of fluorine at initial mg/l	Concentration of fluorine in mg/l	Amount adsorbed in mg/ l	Percentage of removal
1	0.5	10	3.73	6.27	62.7
2	1.0	10	3.30	6.7	67
3	1.5	10	3.10	6.9	69
4	2.0	10	2.80	7.2	72

Crushed turmeric powder after dried into the sunlight for 3-4 days it is sieved to 80micron to prepare as adsorbent. It is found that percentage removal of fluoride is maximum at initial concentration only and then by increasing the weights the removal of fluoride is decreasing which shown in table 4.

Table 4. Removal of fluoride with turmeric powder

Sl. No.	Weight dosage of adsorbent in gm.	Concentration of fluorine at initial mg/l	Concentration of fluorine in mg/l	Amount adsorbed in mg/ l	Percentage of removal
1	0.5	10	5.56	4.44	44.4
2	1.0	10	5.62	4.38	43.8
3	1.5	10	5.71	4.29	42.9
4	2.0	10	5.65	4.35	43.5

Crushed spinach powder after dried into the sunlight for 3-4 days it is sieved to 80micron to prepare as adsorbent. The percentage removal of fluoride is maximum at initial concentration only which is like turmeric powder shown in table 5.

Table 5. Removal of fluoride with spinach powder

Sl. No.	Weight dosage of adsorbent in gm.	Concentration of fluorine at initial mg/l	Final concentration of fluorine in mg/l	Amount adsorbed in mg/ l	Percentage of removal
1	0.5	10	4.23	5.77	58
2	1.0	10	4.54	5.46	55
3	1.5	10	4.67	5.33	53.5
4	2.0	10	4.89	5.11	51

Crushed spinach powder after dried into the sunlight for 3-4 days it is sieved to 80micron to prepare as adsorbent. The percentage removal of fluoride is maximum at initial concentration only which is like turmeric and spinach powders shown in table 6.

Table 6. Removal of fluoride with Orange peel

Sl. No.	Weight dosage of adsorbent in gm	Concentration of fluorine at initial mg/l	Final concentration of fluorine in mg/l	Amount adsorbed in mg/ l	Percentage of removal
1	0.5	10	7.23	2.77	28
2	1.0	10	7.68	2.32	23
3	1.5	10	7.94	2.06	20
4	2.0	10	8.5	1.5	15

3.2. Effects of fluoride concentration

When the initial concentration of fluoride is changed with a constant amount of 0.5 grams, the amount adsorbents per mass was varied with the increase in weight dosage as mentioned in tables 3 to 4. Whereas an attempted has been made to increase the dosage up to 10 grams with four adsorbents as shown in figure 4. By the increase in dosage removal of fluoride content is increasing effectively. Ginger has shown more effective in a higher percentage.

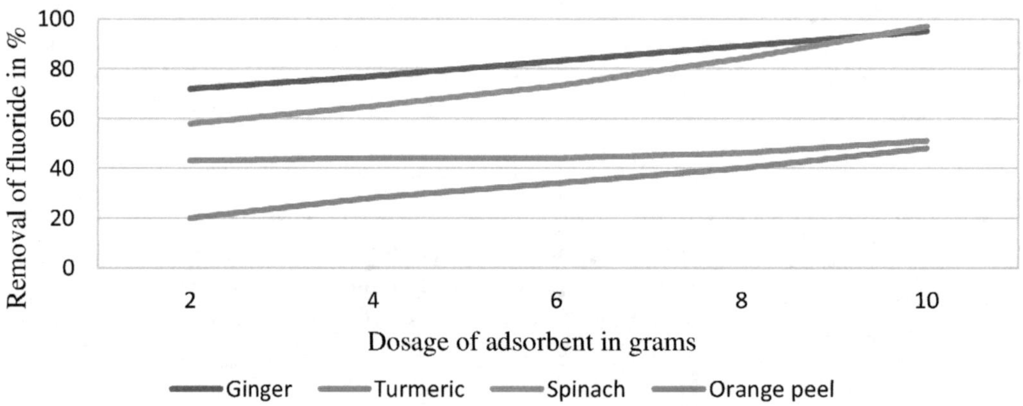

Figure 4. Variation of initial fluoride concentration

4. Conclusion

The objective of this paper is to develop a method for defluorination of drinking water at the rural level. Several methods have been developed for the removal of fluorine which is expensive. Among the adsorbents used for removal of fluoride from water, Ginger, Turmeric, Spinach are removing fluoride more effectively for about 40% to 75%. Orange peel remains the same not showing any kind of decreasing level of fluoride. Ginger powder identified as ideal adsorbent which acts as an excellent de fluoridating agent. Turmeric powder was found to be consistent for de fluoridation. Spinach powder when used as an adsorbent increased the concentration of fluoride. In comparison, dry ginger best suited with the greatest extent de fluoridating capacity where Orange peel powder was not within the permissible limits. However, all adsorbents can be used for the de fluoridation and methods are developed for the colour removal and turbidity. This study can reveal that many such adsorbents can be used for de fluoridation of water and it can be implemented at domestic and community levels within a low cost and also environmentally friendly.

References

[1] Maheshwari R, Meenakshi C 2006 Fluoride in drinking water and its removal *Journal of Hazardous Materials* **B137** p 456–63

[2] Gandhi N, Sirisha D 2012 Removal of fluorine from water and wastewater by using low cost adsorbents *International Journal of Chem Tech Research IJCRGG* **4 (4)** p 1646-53

[3] Raj Variramuthu, Alagumuthu Ganapathy 2015 Use of activated coconut nutshell carbon for the removal of fluoride contaminated drinking water *Chinese Journal of Chemical Engineering* **23** p 710-21

[4] Gayathri G, Kumar Raju B 2017 Defluoridation of groundwater using low-cost adsorbents **10 (5)** p 967-72

[5] Vivek C M, Karthikeyan 2011 Removal of fluorine by using low-cost Materials *International Water Technology Journal IWTJ* **I (2)** p 1-12

[6] Bhagyashree M Mamilwar, Bhole A G, Sudame A M 2012 Removal of Fluoride from Ground Water by Using Adsorbent *International Journal of Engineering Research and Applications* **2 (4)** p 334-8

[7] Sheetal Bandewar S J Mane, Tirkthakar S N Removal of fluorine from drinking water by using low-cost absorbent *International Journal of Research in Engineering and Technology IJRET* **4 (4)** p 349-51

[8] Jamode A V, Sapkal V S, Jamode V S 2004 Defluoridation of water using inexpensive adsorbents *Journal of Indian Institute of Science* **84 (5)** p 163–71

[9] Manoj Kumar M, Anantha Priya K 2018 Defluoridation of water using tamarind gel *International Journal of latest Technology and Engineering, Management and Applied Science IJLTEMAS* **7 (3)** p 209-10

[10] Chidambaram, Ramanathan S, Vasudevan S, 2001 Fluoride removal studies in water using natural material *Water, Air and Soil pollution* **29** p 378- 473.

[11] Saha B, Tai M.H Streat M 2001 Process safety and environmental protection 2001 **79** p 211-7.

[12] Jasmin Kaur, Haripada Bhunia, Anitha Rajor 2013 Removal of fluoride from drinking water by activated carbon, VSRD *International Journal of Technical and non-technical Research* **4 (8)** p 187-92

[13] Parlikar A R, Mokashi S S 2013 Defluoridation of Water by Moringa Oleifera - A Natural Adsorbent 2013 *International Journal of Engineering Science and Innovative Technology* IJESIT **2 (5)** p 245-51

[14] Killedar J, Bhargava D S 1988 An overview of Defluoridation methods Part-2, *Journal Institute Publication Health Engineering*, **2** p 37-42

[15] Koteswara Rao M, Mallikarjun 2014 Effective Low-Cost Adsorbents for Removal of Fluoride from Water *International Journal of Science and Research IJSR* **3 (6)** p 120-4

[16] Yadav K, Kaushik C P, Haritash A K, Kansal A, Neetu Rani, 2005 Defluoridation of Groundwater using Brick powder as an Adsorbent *Journal of Hazardous Materials* **128 (2-3)** p 289-93

[17] Bhargava D S, Killedar D J 1992 Fluoride adsorption on fishbone charcoal through a moving media absorber *Water Research* **26 (6)** p 781–8

[18] Ranjit N. Patil, Nagarnaik P B, Agrawal D K 2015 Removal of fluoride from water by using bio-adsorbents: A state of art *International journal of pure and applied research in Engineering and Technology* **3 (9)** p 272-79

[19] Patil A K, Shrivastava V S, Alternanthera bettzichiana Plant powder as Low-Cost Adsorbent for Removal of Congo red from Aqueous Solution *International Journal of Chem. Tech. Research* **2 (2)** p 842-50

[20] Mumtazuddin S, Azad A K 2012 Removal of Arsenic Using Mango, Java Plum and Neem Tree Barks *International Journal of Advances in Pharmacy, Biology and Chemistry IJAPBC* **1 (3)** p 372-6

Change detection analysis in LULC of the upstream Thandava reservoir using RS and GIS applications

B V Ramanamurthy[1], B Vijayasaradhi[2]

[1]Ph.D. Scholar, Department of Civil Engineering, Andhra University College of Engineering, Andhra University, Visakhapatnam, Andhra Pradesh, India.
[2]Professor, Department of Civil Engineering, Andhra University College of Engineering, Andhra University, Visakhapatnam, Andhra Pradesh, India.
Email: [1] vrmurthyb@gmail.com, [2] bvsaradhi70@gmail.com

Abstract. Change detection is a time series process to analyze Land use/land cover (LU/LC) changes within a specific period. This analysis will help to take the necessary decisions for policymakers. In this study for this purpose, we collect the data in the form of topographical maps from Survey of India (SOI), Satellite remote sensing data from USGS (United States Geological Survey) and software like Arc GIS 10.3 and ERDAS IMAGINE 2014 were used for analyzing the data between the periods of 1995-2008, 2008-2020 and 1995-2020. The study area was classified into eight land use and land cover categories based on the most likelihood classified approach and quantifying the changes in the above mentioned period time. It was observed that the forest area has vanished and barren land was increased during 1995-2020.
Keywords: Land use Land cover, Survey of India, remote sensing, USGS, Arc GIS, ERDAS IMAGINE

1. Introduction

The earth is covered by natural resources mainly forests, hilly areas, open lands, water bodies refers land cover whereas land use refers to how people are using these resources for development, conservation, or mixed uses. As the population of a given area increase, the demand for land cover such as built-up area also increases, while other land cover classes such as bare land, vegetation, decreases as a result of the increased demand for the built-up area [1]. The impacts of Land Use Land Cover (LULC) on the sustainability of the ecosystems are becoming increasingly important issues in global changes research [2]. In most cases, LULC Classification is the result of different anthropogenic activities, i.e., cutting down trees and conversion of forest land into agricultural land or human settlement which cause disturbance of biodiversity [3]. LULC and climate are two important aspects that govern the hydrological behaviour of any region. The slight change in LULC affects the resultant energy fluxes of the region thereby altering the climate [4]. The human and livestock population, different agricultural practices, urbanization, drought prevalence, and poor land-use planning are the main drivers of LULCC. However, different places have different driving factors and consequences [5]. Natural disasters like floods, hurricanes, volcanic eruptions, earthquakes, etc. are the factors that influence forest degradation, soil erosion, and damaging of manmade constructions. Hence, LULC classification is needed for planners, resource managers, scientists, and decision-makers to effectively manage the environment as well as living conditions. For analyzing change detection between past and present land use land cover, the preliminary data like surveying data, topographical data, necessary information collected from the surroundings people and analyzing existing conditions play a vital role everywhere. But getting the required time-related data is a tough job. In this regard, satellite data plays an important role throughout the world to fulfil the needs of decision-makers. Remote sensing has an important contribution to making and documenting the actual change in land use/land cover in regional and global scales [6]. So, the development of remote sensing plays a major role for risk analysis, early warning, monitoring, and mapping with the help of one of new software technology named as GIS (Geographical Information System). Land use and land cover classification can be done with the help of topographical maps at different scales, satellite images with different band combinations for accurate data acquisition, and analyzed with the help of field surveys and standard classification data in that region. From the output results of required time intervals, we can take preventive measures and mitigate the risk.

2. Literature review

Adel Shalaby and Ryutaro Tateishi [7] carried out supervised classification and post-classification change detection performed on the satellite images of 1987 and 2001. Based on six field visits and 1987 land cover map they had been done accuracy assessment and classified land cover classes accordingly. The major results showed vegetation degradation and waterlogging in the study area. Ashraf and Yasushi [8] focused on urban expansion in Greater Dhaka, Bangladesh and land use/cover changes between 1975 and 2003. The Supervised classification had been derived land cover results 85 to 90% using GIS techniques. The results showed that the growth of built-up areas resulted in a significant decrease in the area of water bodies, cultivated land, vegetation and wetlands. Aydoner and Maktav [9] analyzed the adverse effects of the Marmara earthquake on the land use/land cover (LULC) of Kocaeli province in Turkey using different satellite data. Damage in the urban texture and changes to the coastline had been studied through settlement suitability analysis and these results were interpreted by the information generated through satellite data before and after the earthquake. Canute et al. [10] used different software like Arc Map 10.1, ERDAS Imagine, SPSS and IDRISI Selva in their study to assess the factors influencing LULC change. The results showed in this study that the variations in annual rainfall, population density, and distance from road networks are the main factors for LULC change in this study. Christoph et al. [11] carried out the well-established Maximum Likelihood classifier and the random forest classifier were used to study LULC with a combination of x-and c-band microwave data. This study results showed a comparison between microwave and optical remote sensing images for LULC classifications. Chauhan and Nayak [12] studied land use/land cover changes and its implications on coastal systems using remote sensing and GIS. Erosion and deposition have been observed along with the declination of the forest area and agriculture area in this study. Daniel and Ayobami [13] have been assessed the changes in land use/land cover over a period of 16 years using the Maximum Likelihood classification method through GIS techniques in their study. From the results, they identified that the degraded forest has the most extensive type of land use/land cover in their study area. Khalid et al. [14] attempted analyses and monitor the LULC changes on the selected soil properties of cultivated land, fallow land and woodland. Texture, bulk density, organic matter, soil pH, electrical conductivity, sodium adsorption ratio, phosphorous, and potassium were analyzed from the collected soil samples for each of the LULC types by the laboratory tests. The results showed that Soil productivity has been declined through land degradation. Zope et al. [15] investigated the impact of land use–land cover (LULC) change and urbanization on flood during the years 1966 and 2009. The impact of LULC on flood hydrograph for different return periods was ascertained by using the HEC-GeoHMS and HEC-HMS models. The authors analyzed that the overall total flood hazard area is increased by 22.27% in this study.

3. Study Area

Thandava River (Bodderu) is the common boundary for Visakhapatnam and East Godavari districts. Thandava reservoir project was constructed across Thandava River during the years 1965-1975. The purpose of Thandava reservoir construction is to provide irrigation facilities for both Visakhapatnam and East Godavari districts. The project is completed and commenced for serving the needs of the ayacut since the year 1975 to provide irrigation facilities to an extent of 51,465 acres, out of which 32,689 acres are lying in Visakhapatnam district and 18,776 acres in East Godavari district. The ayacut is spread over in Kotauratla, Narsipatnam, and Nathavarammandals of Visakhapatnam district and Kotananduru,

Tuni and Routhulapudi mandals of East Godavari district. The reservoir is the most naturally formed reservoir due to its oval-shaped topographical configuration as it is surrounded by hills. Hence, the construction of an earth dam for a length of 201m facilitated for the formation of a reservoir of capacity 4960 Mcft. In this study, we selected upstream of Thanadava reservoir having 349 Square Kilometers study area consisting majorly, Pakalapadu, Cheedigummala, Yarakampeta, Yerravaram, Nallanki, Gundupala, KothaYellavaram, Gulugonda, Ammapeta, Jogumpeta, Sreerampuram, Narasinguballi, Pappusettipalem, Kothapalem, Hukkumpeta, Bangarammapeta, Gummalapalem, Teegalametta, Thotaluru, Chittimpadu, Patamallampeta, Lingampeta, krishnadevipeta, Konakasingi, Ratnampeta and Agency Laxmipuram. The map of the study area location is shown in figure 1.

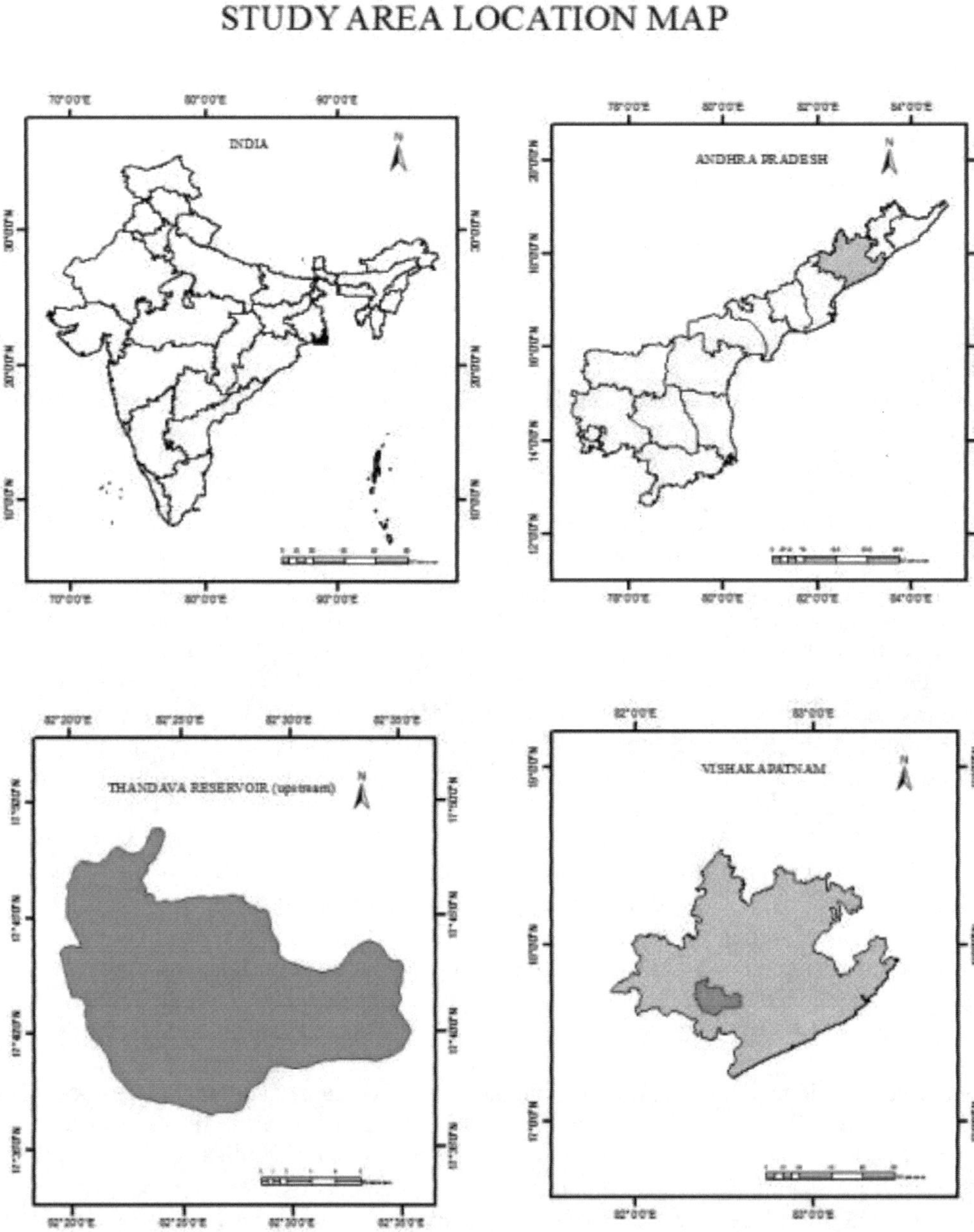

Figure 1. Study Area – Upstream Thandava Reservoir

IOP Conf. Series: Materials Science and Engineering 1025 (2021) 012034 doi:10.1088/1757-899X/1025/1/012034

4. Methodology

The following points have followed in the methodology

1. Toposheets of 65K5, 65K6, 65K9, and 65K10 (scale: 1:50000, First edition) were taken and geo rectification was done and mosaic all these top sheets.
2. Delineation of the study area from the mosaic image
3. Download satellite images from USGS
4. Delineation of the study area from geo-rectified satellite images
5. Ground truth
6. Image processing using ERDAS software
7. Supervised Classification and finalizing the Classes
8. Change detection Analysis
9. Results and Discussion
10. Conclusion

The study area was delineated from the mosaic image and ground truth was done for more accuracy. Based on ground truth accuracy, supervised classification was done individually in 1995, 2008, and 2020 geo-rectified satellite images and various land use/ land cover classes were defined using ERDAS IMAGINE 2014. For accuracy purpose, study area was converted into KMZ file and used this file in GOOGLE EARTH Pro. for identification of classes. Finally, the classified results were compared between the period time of 1995-2008, 2008-2020 and 1995-2020 and the work was concluded. The information about satellite data is presented in Table 1.

Table 1. Information about Satellite data

Sl. No.	Satellite	Sensor	Path /Row	Acquisition Date	Spatial Resolution	Spectral Bands
1	Landsat-8	OLI/TIRS	141/048	23-02-2020	15 meters	B2 to B8
2	Landsat 4-5	TM	141/048	06-02-2008	30 meters	B1 to B7
3	Landsat 4-5	TM	141/048	06-03-1995	30 meters	B1 to B7

5 Results and Discussion

5.1 Results

In the supervised classification with the help of ERDAS Imagine 2014, the land use land cover categories were defined and add area field for determination of the area of each class. After that accuracy assessment process was done with already defined class values that are replaced by the given reference values and then generates the reports. The magnitude change for each class was calculated by subtracting latest year values from the previous year values and percentage change is calculated as the magnitude change is divided by the base year and this value is multiplied by 100. For calculation of Annual rate of change, the magnitude change is divided by the number of corresponding study years.

Example:
B is the latest year and A is the previous year, study period between A and B is N, now
i) Magnitude change= B-A = C
ii) % change = (C/A) X 100
iii) Annual rate of change = (C/N)

The figures 2, 3 and 4 present the LULC of upstream thandava reservoir during 1995, 2008 and 2020.

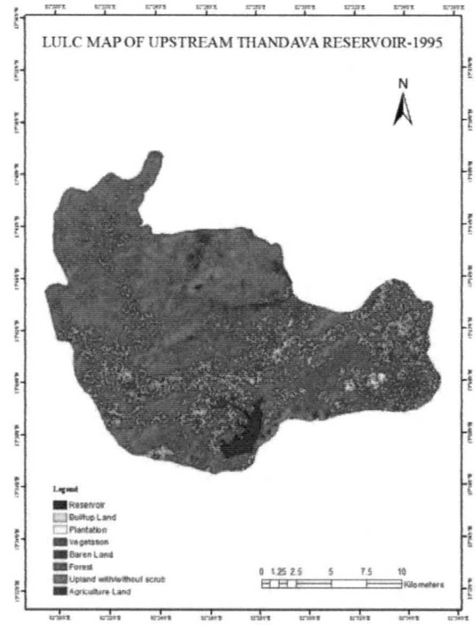

Figure 2. LULC map of upstream Thandava reservoir -1995

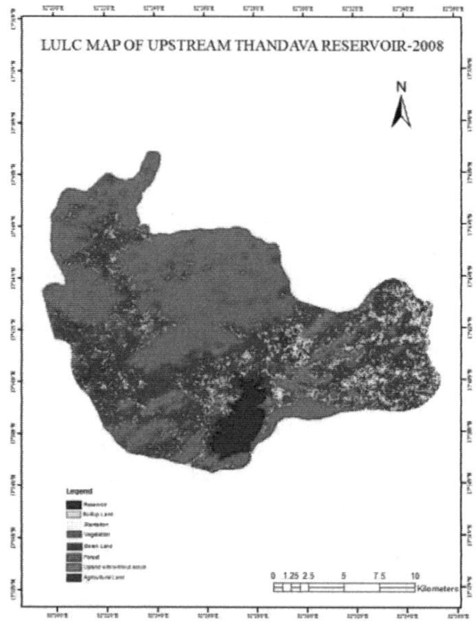

Figure 3. LULC map of upstream Thandava Reservoir -2008

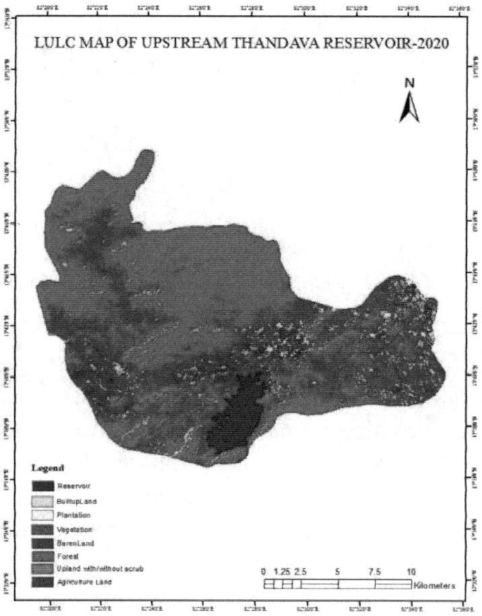

Figure 4. LULC map of upstream Thandava Reservoir-2020

5.2 Discussion
A total of 8 numbers of classes are commonly defined in the supervised classification of 1995, 2008 and 2020 images using ERDAS.

5.2.1 Land Use pattern of the study area in 1995, 2008 and 2020 (Table 2)
Forest class occupies 76.65% in the total study area. It is a major land occupancy when compared to the remaining classes. Barren land is 8.58% but later it was increased in the years 2008 and 2020. In 2008 the forest occupancy is decreased up to 42.4% but 9.46% of upland covers with scrub. At the same time, the vegetation vanished and most of the land up to 30.31% is converted into barren land. Agricultural land occupancy is almost the same in 1995 and 2008 that is 3% in total area. But development in the

plantation is almost double when compared to 1995. Unfortunately, agricultural activities are dropdown in the year 2020. At the same time protection of forestation is improved when compared to 2008. But plantation activities are dropdown when compared to 2008. There is no development in built-up land during the period 1995 to 2020. It is observed that nearly half of the built-up occupancy is faded in the years 1995, 2008 and 2020. Reservoir occupancy is the same in the years 2008 and 2020 and it is vast coverage when compared to 1995.

5.2.2 Relative change in Land use in the study area

Relative change in land use of study area was assessed from Tables 3, 4 and 5. The relative changes showed some irregular patterns in this study area from 1995-2020. Land-use changes from 1995-2008 showed positive changes in most of the categories. But a negative trend showed in 2008-2020 when compared to the 1995-2008 period. Barren land showed a positive change (+256.71%) in 1995-2008, whereas a negative change (-2.8%) showed in the same land use category in 2008-2020. Similarly, plantation, upland with/without scrub, reservoir classes showed a positive trend during 1995-2008 but a negative trend followed in the same classes during 2008-2020. The overall annual rate of change in the forest showed a negative trend (-342.1%) and barren land showed a positive trend (+292%) during 1995-2020. At the same time, it was observed that these two classes were highly varied and a positive annual rate of change observed in vegetation and reservoir during the same period time.

Table 2. Land use classification of the study area from 1995-2020

Sl. No.	Land Use Category	Land Use					
		1995		2008		2020	
		Area (Hectare)	% of Land	Area (Hectare)	% of Land	Area (Hectare)	% of Land
1	Plantation	1576.35	4.51	3195.24	9.14	1263.58	3.61
2	Upland with/ without scrub	927	2.65	3306.69	9.46	1350.62	3.86
3	Forest	26772.6	76.65	14808.1	42.4	18220.9	52.17
4	Barren Land	2998.62	8.58	10585.3	30.31	10291.6	29.47
5	Vegetation	410.94	1.17	210.795	0.6	1795.81	5.14
6	Built-up Land	848.16	2.42	485.505	1.39	301.737	0.86
7	Agricultural Land	988.02	2.82	1085.79	3.1	481.973	1.38
8	Reservoir	399.78	1.14	1244.05	3.56	1215.25	3.47
	Total Area	34921.47		34921.47		34921.47	

Table 3. Land-use change Assessment of study area on time frame data 1995-2008

Sl. No.	Land use category	Land-use change (1995-2008)		
		Area (Hectare)	% change	The annual rate of change
1	Plantation	(+)1618.89	(+)102.7	(+)124.53
2	Upland with/ without scrub	(+)2379.69	(+)256.71	(+)183.1
3	Forest	(-)11964.5	(-)44.7	(-)920.34
4	Barren Land	(+)7586.68	(+)253	(+)583.6
5	Vegetation	(-)200.145	(-)48.7	(-)15.4
6	Built up Land	(-)362.655	(-)42.8	(-)27.9
7	Agricultural Land	(+)97.77	(+)9.9	(+)7.52
8	Reservoir	(+)844.27	(+)211.2	(+)64.94

Table 4. Land-use change Assessment of study area on time frame data 2008-2020

Sl. No.	Land use category	Land-use change (2008-2020)		
		Area (Hectare)	% change	The annual rate of change
1	Plantation	(-)1931.66	(-)60.45	(-)161
2	Upland with/without scrub	(-)1956.07	(-)59.15	(-)163
3	Forest	(+)3412.8	(+)23.04	(+)284.4
4	Barren Land	(-)293.7	(-)2.8	(-)24.5
5	Vegetation	(+)1585.015	(+)752	(+)132.1
6	Built-up Land	(-)183.768	(-)38	(-)15.31
7	Agricultural Land	(-)603.817	(-)55.61	(-)50.31
8	Reservoir	(-)28.8	(-)2.31	(-)2.4

Table 5 Land use change Assessment of study area on time frame data 1995-2020

Sl. No.	Land use category	Land-use change (1995-2020)		
		Area (Hectare)	% change	The annual rate of change
1	Plantation	(-)312.77	(-)24.8	(-)13
2	Upland with/without scrub	(+)423.62	(+)31.4	(+)17
3	Forest	(-)8551.7	(-)46.93	(-)342.1
4	Barren Land	(+)7292.98	(+)71	(+)292
5	Vegetation	(+)1384.87	(+)77.11	(+)54
6	Built-up Land	(-)546.423	(-)181.1	(-)22
7	Agricultural Land	(-)506.047	(-)105	(-)20.24
8	Reservoir	(+)815.47	(+)67.1	(+)33

5.2.3 Overall accuracy and Kappa statistics (K^)

Kappa statistics is a measurement between users identified class data and reference data after the classification process is done. Kappa value is useful for accuracy check-in for each defined class. As per Cohen if this value between 0.20 to 1.00, he classified as 0.21–0.40 as fair, 0.41– 0.60 as moderate, 0.61–0.80 as substantial, and 0.81–1.00 as almost perfect agreement. The overall kappa statistics is 0.774 and overall classification accuracy is 80.24% obtained in supervised classification of the year 1995. It is substantial when compared to the 2008 and 2020 accuracy reports. Similarly, we obtained the overall kappa statistics and overall classification accuracy 0.84 and 86.72% in the 2008 classification and 0.88 and 89.84% in the 2020 classification.

6. Conclusions

The supervised classification had been performed and it showed satisfactory results. The authors faced difficulty in the identification of classes in the image classification of 1995 when compared to the 2008 and 2020 image classification. We had unable to understand the classes due to the lack of oral reports about the scenario of 1995. From the trend analysis of LULC in this study, majorly forestation, agricultural land declined by nearly 47% and 105% respectively from the year 1995 to 2020. we had noticed that the agricultural activities had been gradually declined and barren land increased in the overall study period. We suggest that using GPS instruments will give precise results along with RS and GIS techniques in a future study.

References

[1] K M Kafi, H Z M Shafri, A B M Shariff 2014 An analysis of LULC change detection using remotely sensed data; A Case study of Bauchi City *IOP Conf. Ser.: Earth Environ. Sci.* **20** 012056

[2] Kamrul Islama, Mohammed Jashimuddina, Biswajit Nathb, Tapan Kumar Natha 2018 Land use classification and change detection by using multi-temporal remotely sensed imagery: The case of Chunati wildlife sanctuary Bangladesh *The Egyptian Journal of Remote Sensing and Space Sciences* **21 (1)** p 37-47

[3] Masum Chowdhury, Mohammad Emran Hasanb, Abdullah-Al-Mamuna M M 2020 Land use/land cover change assessment of Halda watershed using remote sensing and GIS *The Egyptian Journal of Remote Sensing and Space Sciences* **23 (1)** p 63-75

[4] Vaibhav Garga, Bhaskar R. Nikam, Praveen K. Thakura, Shiv Prasad Aggarwal, Prasun Kumar Gupta, Sushil Kumar Srivastav 2019 Human-induced land use land cover change and its impact on hydrology *HydroResearch* **1** p 48–56

[5] Wakjira Takala Dibaba, Tamene Adugna Demissie, Konrad Miegel 2020 Drivers and Implications of Land Use/Land Cover Dynamics in Finchaa Catchment, Northwestern Ethiopia *Land* **9 (4)** p 1-20

[6] Olokeoguna O S, Iyiolab O F, Iyiolac K 2014 Application of Remote Sensing and GIS in Land use/Landcover Mapping and Change Detection in Shasha Forest reserve Nigeria *The International Archives of the Photogrammetry, Remote Sensing and Spatial Information Sciences* **XL-8** ISPRS Technical Commission VIII Symposium p 613-16

[7] Adel Shalaby, Ryutaro Tateishi 2007 Remote sensing and GIS for mapping and monitoring land cover and land-use changes in the Northwestern coastal zone of Egypt *Applied Geography* **27 (1)** p 28–41

[8] Ashraf M. Dewan, Yasushi Yamaguchi 2009 Land use and land cover change in Greater Dhaka, Bangladesh: Using remote sensing to promote sustainable urbanization *Applied Geography* **29 (3)** p 390–401

[9] Aydöner C 2009 The role of the integration of remote sensing and GIS in land use/land cover analysis after an earthquake, *International Journal of Remote Sensing* **30 (7)** p 1697-717

[10] Canute Hyandye, Christina Geoffrey Mandara, John Safari 2015 GIS and Logit Regression Model Applications in Land Use/Land Cover Change and Distribution in Usangu Catchment *American Journal of Remote Sensing* **3 (1)** p 6-16

[11] Christoph Hütt, Wolfgang Koppe D, Yuxin Miao, Georg Bareth 2016 Best Accuracy Land Use/Land Cover (LULC) Classification to Derive Crop Types Using Multitemporal, Multisensor, and Multi-Polarization SAR Satellite Images *remote sensing* **8 (8)** p 684

[12] Chauhan H B, Shailesh Nayak 2005 land use/land cover changes near Hazira region, Gujarat using remote sensing satellite data *Journal of the Indian Society of Remote Sensing* **33 (3)** p 413-20

[13] Daniel Ayalew Mengistu, Ayobami T. Salami 2007 Application of remote sensing and GIS in land use/land cover mapping and change detection in a part of southwestern Nigeria *African Journal of Environmental Science and Technology* **1 (5)** p 099-109

[14] Khalid Biro, Biswajeet Pradhan, Manfred Ferdinand Buchroithner and Franz Makeschin 2011 Land use/Land cover Change Analysis and its impact on soil properties in the northern part of Gadarif region Sudan *Land Degrad. & Develop.* **24 (1)** p 1-13

[15] Zope P E, Eldho T I, Vinayakam Jothiprakash 2016 Impacts of land use–land cover change and urbanization on flooding: A case study of Oshiwara River Basin in Mumbai India *Catena* **145** p 142–54

Land Use Land Cover (LULC) classification with wasteland demarcation using Remote sensing and GIS Techniques

B V Ramanamurthy[1], N Victorbabu[2]

[1]Assistant Professor, Department of Civil Engineering, Baba Institute of Technology and Sciences, PM palem, Visakhapatnam, Andhra Pradesh, India.
[2] Professor, Department of Civil Engineering, Baba Institute of Technology and Sciences, PM palem, Visakhapatnam, Andhra Pradesh, India.
Email: [1]vrmurthyb@gmail.com, [2]neelavictorbabu@gmail.com

Abstract
Land use and land cover changes are derived by natural process and anthropogenic interference in the ecosystem. Appropriate management of natural resources requires quick and up-to-date information for effective decision making. This study aims to generate land use land cover information using Remote sensing and Geographical Information System (GIS) technology to support the land use planning and strategy formulation of the newly emerging problem of the study area. Vizianagaram Mandal of the Vizianagaram district (Andhra Pradesh, India) covering 83°18'0"E to 83°29'0"E longitude and 18°2'0"N to 18°10'0"N latitude of total area 12437 hectares had been taken as a study area. The software Arc GIS and ERDAS used in this study for obtaining better outcomes. Two satellite imageries of Landsat TM, MSS 2000, and Landsat MSS 2010 were downloaded from USGS, and supervised classification carried out for the detection of Land use Land cover (LULC) classification. The ten-land use/land cover classes have been identified from the classification processes. The study focused to identify the wasteland demarcation and analyze the land cover changes between the years 2000 and 2010.
Keywords: Land use Land cover, Remote sensing, Arc GIS, ERDAS, supervised classification, USGS.

1. Introduction

LULC is a very complicated process, affected by natural and human dimensions [2] Due to population growth and their activities, most of the land is converted into urbanization, industrialization and some are turned up as wasteland that causes an impact on the environment. Land use practices generally develop over a long period under different environmental, political, demographic, and socio-economic conditions [10]. Technically, land use and land cover change (LULCC) mean quantitative changes (increase or decrease) in a particular Land Use Land Cover over some time [5]. Moreover, the mix of land use/cover in a place affects its sensitivity to climate change: e.g., dryland farming areas may be more sensitive than irrigated areas [7]. Land use data are needed in the analysis of environmental processes and problems that must be understood if living conditions and standards are to be improved or maintained at current levels [3]. For this reason, there is a need to understand the pattern and trends of LULC changes on the local, regional and global scales [6]. The questions like what changes occur where and when they occur, the rates at which they occur will give a better understanding of LULC change [1]. The timely, accurate and up-to-date information on land use/land cover can be obtained from various satellite data on a cost-effective basis at the shortest possible time [4]The minute changes on the earth's surface can be estimated fairly accurately [5] with advances in remote sensing science that made

IOP Conf. Series: Materials Science and Engineering 1025 (2021) 012035 doi:10.1088/1757-899X/1025/1/012035

it possible to obtain valuable spatiotemporal information on LULC [6]. Landsat Multispectral Scanner (MSS), Thematic Mapper (TM) data have been broadly employed in studies towards the determination of land cover since 1972[8]. By comparing two sets of images taken of the same area at different times the LULC change analysis can be carried out easily [9] For that, software like Arc GIS, ERDAS, and other software tools can be helpful for the display of spatial data and non-spatial attribute data to perform analysis and retrieval of the data.

2. Study Area

Vizianagaram District is situated within the geographical coordinates of 17-15' and 19-15' of the northern latitude and 83-00' and 83-45' of the eastern longitude. The district is bounded on the east by Srikakulam district on the west and south by Visakhapatnam district, on the south-east by the Bay of Bengal and North West by Odisha State. Vizianagaram Mandal located at 18.12°N, 83.42°E has been taken as a study area. This study area has a geographical area of 12437 hectares and has 23 revenue villages and 21 panchayats.

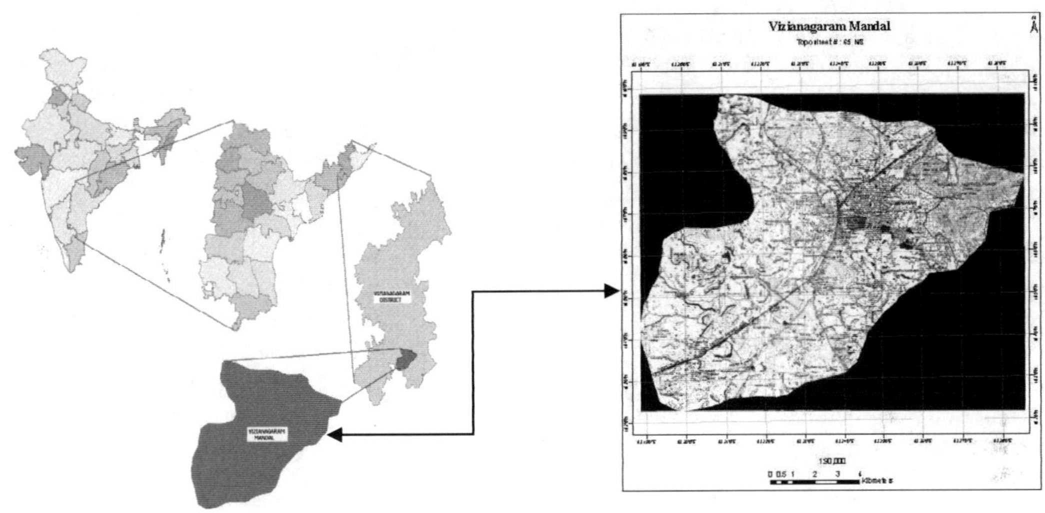

Figure 1. Location of the study area

3. Data source

The primary data collected from different sources include satellite imagery, statistical handbooks of Vizianagaram 2000 & 2010, information from local experts and surrounding community members. The secondary data such as a topographic map and some instruments like digital camera, compass and tape used to collect the field data. The software Arc GIS 9.0 and ERDAS 9.2 had been played a vital role in the demarcation of wastelands. Before the fieldwork, the necessary satellite imagery such as Landsat TM, MSS 2000 and TM 2010 was downloaded from USGS (United States Geological Survey). The following table depicts the source of satellite images that were used in this study.

Table 1. The data source of Satellite Images

Sl. No.	Sensor	Path /Row	Spatial Resolution
1	Landsat TM and MSS 2000	141/047	15 meters
2	Landsat TM 2010	141/047	13 meters

4. Methodology

The following steps had been followed in the methodology

1. INPUTS
   - -Toposheets
   - -Satellite Imageries (8-band Landsat-7) 2000 & 2010
2. PROCESS

   Geo-Rectification of Topo sheets
   - Un-Supervised Classification
   - Field visit (for Ground Truth)
   - Supervised Classification
   - Authentication with Statistical Handbook of the District
   - Finalizing the Classes
   - Delineation of Wastelands from both 2000 & 2010
   - Classifying the Wastelands
3. OUTPUT
   - Land use Land cover classes
   - Waste Land demarcation
   - Change Analysis

Initially, unsupervised classification was carried out to get some valuable information about the different land cover classes. After that, a field survey had been carried out and this data used in supervised classification for a better understanding of land covers. A total of 10 land use land cover classes were finalized and change analysis carried out between the years 2000 and 2010.

5. Results and Discussion

The results obtained from the land use land cover classification which is shown below (Figure 2 to 8). From the independent image classification results of the two-satellite imagery (Landsat TM, MSS2000 & TM 2010), ten different land use and land cover (LULC) classes had been identified. The classification results are shown in Table 2. In the 2000 image classification results, the built-up land covered 4.66% of the total study area. The cultivated land is 59.48% and the uncultivated land 16.46% covered in the total area. Forest area covered by 3.34% and scrubs occupied by 0.8%.

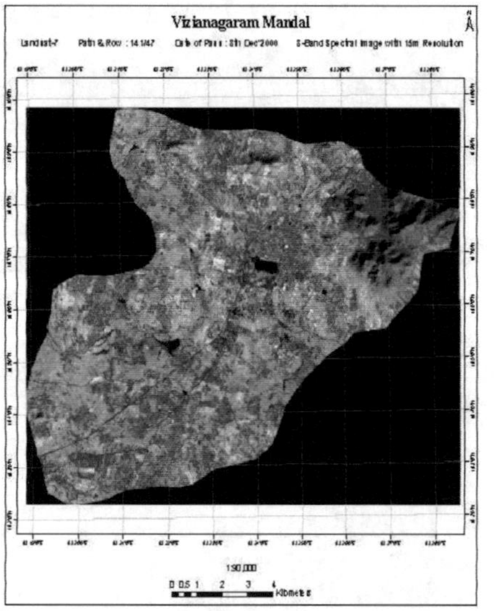

Figure 2. Landsat imagery of Vizianagaram Mandal (2000)

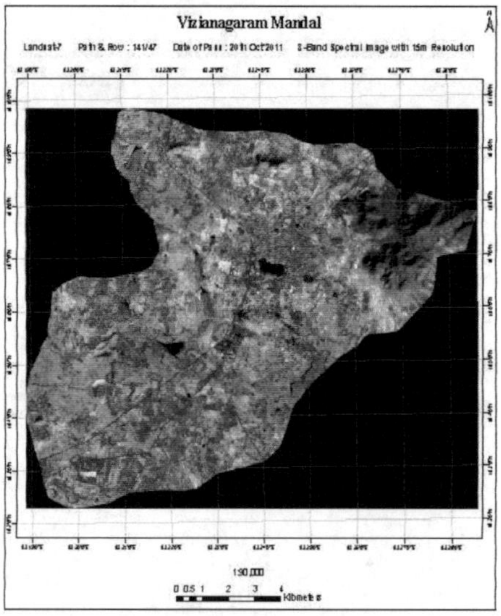

Figure 3. Landsat imagery of Vizianagaram Mandal (2010)

IOP Conf. Series: Materials Science and Engineering 1025 (2021) 012035 doi:10.1088/1757-899X/1025/1/012035

The image classification of 2010 showed that the cultivable land covered 29.46% in the total study area. The built-up land and the forest area covered 5.26% and 3.36% and the occupancy of scrubs and water bodies had been almost the same. At the same time, the wasteland is nearly the same in the years 2000 and 2010. It has been identified that, in change analysis, 0.94% (nearly 1%) of land was converted into wastelands. In change analysis, it was observed that built-up land and uncultivated land increased by 13% and 85% respectively.

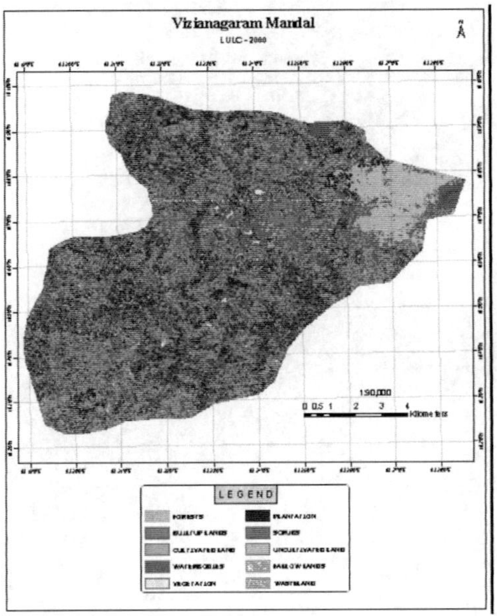

Figure 4. LULC map of Vizianagaram Mandal (2000)

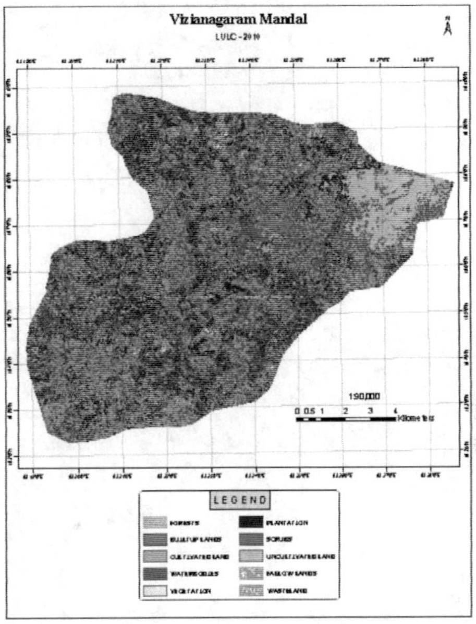

Figure 5. LULC map of Vizianagaram Mandal (2010)

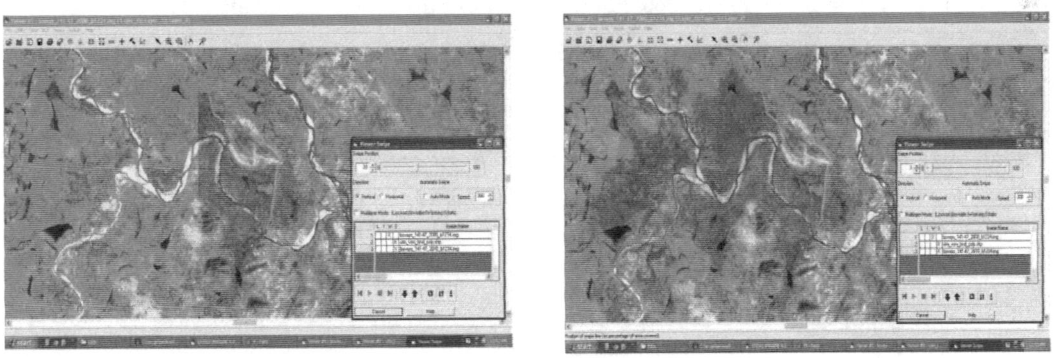

Figure 6. Change analysis of water body

Majorly the authors identified that the cultivated land is decreased by 30% in through 2010 classification. Due to inadequate rainfall on time, most of the farmers left their cultivation and some were migrated to other cities. If it continues most of the land will turn up into wasteland in the future.

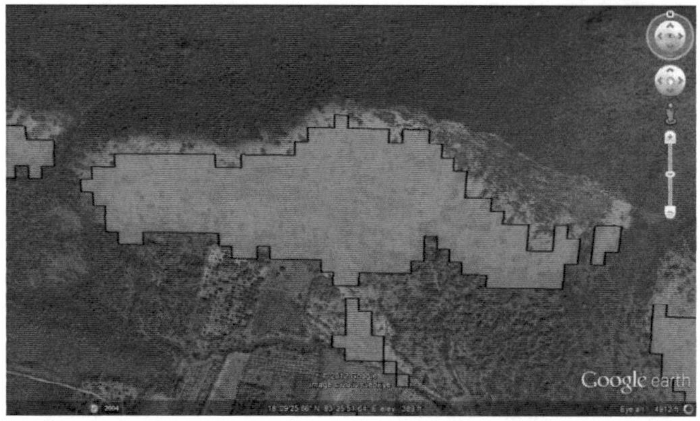

Figure 7. Wasteland demarcation using Google earth

Figure 8. Field photos (sample) of wastelands in the study area

Table 2 Land use classification of the study area in 2000 and 2010

Sl. No.	Land use category	In 2000		In 2010	
		Area (Hectare)	% of Total Area	Area (Hectare)	%of Total Area
1	Built-up Lands	580.22	4.66	654.28	5.26
2	Cultivated Lands scrub	7,397.55	59.48	3,664.44	29.46
3	Fallow Lands	501.48	4.03	2,626.90	21.12
4	Forests	416.00	3.34	418.00	3.36
5	Plantation	1,200.55	9.60	1,225.00	9.84
6	Scrubs	100.55	0.80	140.74	1.13
7	Un-Cultivated Land	2,047.54	16.46	3,781.68	30.40
8	Vegetation	642.00	5.16	652.25	5.24
9	Wasteland	264.02	2.12	267.09	2.14
10	Water bodies	153.22	1.23	167.99	1.35

IOP Conf. Series: Materials Science and Engineering 1025 (2021) 012035 doi:10.1088/1757-899X/1025/1/012035

6. Conclusions

The following conclusions are made in this study.

1) When compared to statistical reports of Vizianagaram Mandal of 2000 and 2010 the LULC classification had been done 80% with the help of Remote sensing and Arc GIS techniques in the present study.
2) The results obtained 98% in the demarcation of forest area and 85% in the wastelands had been successfully identified in this study.

6.1 Scope for further study

Using the latest surveying equipment and skilled teamwork, the Land use Land cover classification results will be achieved by 100%. Hence a detailed study with sophisticated equipment is suggested along with RS and GIS techniques.

References

[1] Daniel Ayalew Mengistu, Ayobami T. Salami 2007 Application of remote sensing and GIS in land use/land cover mapping and change detection in a part of southwestern Nigeria, *African Journal of Environmental Science and Technology* **1 (5)** p 099-109

[2] Hadeel A S, Mushtak T Jabbar, Xiaoling Chen 2009 Application of remote sensing and GIS to the study of land use/cover change and urbanization expansion in Basrah province, southern Iraq, *Geo-spatial Information Science* **12 (2)** p 135–41

[3] James R. Anderson, Ernest E. Hardy, John T. Roach, and Richard E. Witmer 1976 A Land Use and Land Cover Classification System for Use with Remote Sensor Data, *Geological Survey Professional Paper No. 964,* U.S. Government Printing Office, Washington, D.C. p. 28

[4] Jayakumar S, Arockiasamy D I 2003 Land use/Land Cover Mapping and Change Detection in part of Eastern Ghats of Tamil Nadu using Remote Sensing and GIS Journal of the Indian Society of Remote Sensing, **31 (4)** p 251-260

[5] Parthsarathi Roy, Arijit Roy 2010 Land Use and Land Cover Change: A Remote Sensing & GIS Perspective *Journal of the Indian Institute of Science* **90 (4)** p 489-502

[6] Ramita Manandhar, Inakwu O A Odeh, Tiho Ancev 2009 Improving the Accuracy of Land Use and Land Cover Classification of Landsat Data Using Post-Classification Enhancement, *Remote Sens.* **1 (3)** p 330-44

[7] Riebsame W E, Meyer WB, Turner B L II. 1994 Modeling Land-use and Cover as Part of Global Environmental Change *Climate Change* (**28**) p 45-64

[8] Seljuk Reis 2008 Analyzing Land Use/Land Cover Changes Using Remote Sensing and GIS in Rize, North-East Turkey *Sensors* **8 (10)** p 6188-202

[9] Singh A 1989 Digital Change Detection Techniques Using Remotely Sensed Data. *International Journal of Remote Sensing* **10 (6)**, p 989-1003.

[10] W Muttitanon, N K Tripathi 2005 Land use/land cover changes in the coastal zone of Ban Don Bay, Thailand using Landsat 5 TM data *International Journal of Remote Sensing* **26 (11)** p 2311–2323

AUTHOR INDEX